Nikolay I. Kolev

Multiphase Flow Dynamics 2

Nikolay I. Kolev

Multiphase Flow Dynamics 2
Thermal and Mechanical Interactions

2nd ed.

With 81 Figures

Springer

Dr.-Ing. habil. Nikolay I. Kolev
Framatome ANP GmbH
P.O. Box 3220
91050 Erlangen
Germany

ISBN 3-540-22107-7 **Springer Berlin Heidelberg New York**

Library of Congress Control Number: 2004111217

This work is subject to copyright. All rights are reserved, whether the whole or part of the material is concerned, specifically the rights of translation, reprinting, reuse of illustrations, recitation, broadcasting, reproduction on microfilm or in other ways, and storage in data banks. Duplication of this publication or parts thereof is permitted only under the provisions of the German Copyright Law of September 9, 1965, in its current version, and permission for use must always be obtained from Springer-Verlag. Violations are liable to prosecution under German Copyright Law.

Springer is a part of Springer Science+Business Media
springeronline.com

© Springer-Verlag Berlin Heidelberg 2002, 2005
Printed in Germany

The use of general descriptive names, registered names, trademarks, etc. in this publication does not imply, even in the absence of a specific statement, that such names are exempt from the relevant protective laws and regulations and therefore free for general use.

Typesetting: Digital data supplied by author.
Production: PTP-Berlin Protago-TeX-Production GmbH, Germany
Cover-Design: medionet AG, Berlin
Printed on acid-free paper   62/3020 Yu - 5 4 3 2 1 0

To Iva, Rali and Sonja with love!

Sun will be warmer, Feb. 2004, Nikolay Ivanov Kolev, 36× 48cm oil on linen

SUMMARY

This monograph contains theory, methods and practical applications for describing complex transient multi-phase processes in arbitrary geometrical configurations. It is intended to help applied scientists and practicing engineers to understand better natural and industrial processes containing a dynamic evolution of complex multi-phase flows. The book is intended also to be a useful source of information for students in advanced semesters and in PhD programs.

This monograph consists of two volumes:

Vol. 1 Fundamentals (14 Chapters and 2 Appendixes), 746 pages + CD-ROM
Vol. 2 Mechanical and thermal interactions (26 Chapters), 690 pages

In Volume 1 the concept of three-fluid modeling is presented in detail "from the origin to the applications". This includes derivation of local volume- and time-averaged equations and their working forms, development of methods for their numerical integration and finally finding a variety of solutions for different problems of practical interest.

Special attention is paid in Volume 1 to the link between the partial differential equations and the constitutive relations called closure laws without providing any information on the closure laws. Volume 2 is devoted to these important constitutive relations for mathematical description of the mechanical and thermal interactions. The structure of the volume is characterized by a the state-of-the-art review and selects of the best available approaches for describing interfacial transfer processes. In many cases the original contribution of the author is incorporated in the overall presentation. The most important aspects of the presentation are that it stems from the author's long years of experience developing computer codes. The emphasis is on the practical use of these relationships: either as stand-alone estimation methods or within a framework of computer codes.

In particular, Volume 2 contains information on how to describe the flow patterns and the specific mechanical and thermal interactions between the velocity fields in flight. In Chapter 1 the flow regime transition criteria in transient multi-phase flows are presented for the cases of pool flow, adiabatic flow, channel flow in vertical pipes, channel flow in inclined pipes, heated channels, porous media, particles in film boiling and rod bundles. The idea of flow pattern boundaries depending on the transient evolution of the particle number and particle size is presented. Chapter 2 collects information about modeling the drag forces on a single bubble, a family of particles in a continuum, droplets in a gas, solid particles in a gas in the presence of liquid, solid particles in a liquid in the presence of a gas, solid particles in a free particles regime, solid particles in bubbly flow, solid particles in a densely packed regime, annular flow, inverted annular flow, stratified flow in horizontal or inclined rectangular channels and stratified flow in horizontal or inclined pipes. Chapter 3 presents information about the friction pressure drop in single- and multi-phase flow. Historically algebraic correlations for describing the velocity difference preceded the development of interfacial interaction models. The large number of empirical correlations for the diffusion velocities for algebraic slip models for two- and three-phase flows is provided in Chapter 4. Chapters 5 and 6 are devoted to the entrainment and deposition in annular two-phase flow, respectively. Chapter 7 gives an introduction to the fragmentation and coalescence dynamics in multi-phase flows. Acceleration-induced droplet and bubble fragmentation is described in Chapter 8. Chapter 9 is devoted to the turbulence-induced particle fragmentation and coalescence and Chapter 10 to the liquid and gas jet disintegration. Chapter 11 presents the state of the art on the fragmentation of melt in coolant in a variety of its aspects. Chapter 12 presents nucleation in liquids. Chapter 13 presents different aspects of the bubble growth in superheated liquid and the connection to computational system models. Condensation of pure steam bubbles is considered in Chapter 14. New information with respect to the bubble departure diameter and nucleate boiling is presented in Chapters 15 and 16, respectively. An interesting result of the theory presented in these chapters is the prediction of a critical heat flux without empirical correlation for critical heat flux as a result of the mutual bubble interaction for increasing wall superheating. Applying the theory for the inverted problem of the flashing of superheated water in pipes as given in Chapter 17 surprisingly supported the validity of the new approach. The state of the art in boiling theory is presented in Chapters 18 to 21 where boiling in subcooled liquid, natural convection film boiling, critical heat flux, forced convection film boiling, and film boiling on vertical plates and spheres is presented. The emphasis is on the elaboration of all coupling terms between the fluids, and between the wall and fluid required for closure of the overall description. Chapter 22 provides information on all heat and mass transfer processes across a droplet interface starting with the nucleation theory, and going through the droplet growth, self-condensation stop, heat transfer across a droplet interface without mass transfer, direct contact condensation of pure steam on a subcooled droplet, spontaneous flushing of a superheated droplet, evaporation of a saturated droplet in superheated gas, and droplet evaporation in a gas mixture. A similar approach is applied to the description of the interface processes at a film-gas interface in Chapter 23 with a careful treatment of the influence of the turbulent pulsation on the interfacial heat and mass transfer. A set of empirical methods for prediction of condensation on cooled walls with and without a non-condensable is presented in Chapter 24. Chapter 25 provides information on the implementation of the discrete ordinate method for radiation transport in multi-phase computer codes. In this chapter the dimensions of the problem, the differences between micro and macro interactions and the radiation transport equation are discussed. Then the finite volume representation of the radiation transport is given and different aspects of the numerical integration are discussed. The computation of some material properties is discussed. Then three specific radiation transport cases of importance for the melt-water interaction are discussed in detail: a spherical cavity of gas inside a molten material; concentric spheres of water droplets, surrounded by vapor surrounded by molten material; clouds of spherical particles of radiating material surrounded by a layer of vapor surrounded by water. For the last case the useful *Lanzenberger* solution is presented and its importance is demonstrated in two practical cases.

The book ends with Chapter 26, which provides information on how to verify multi-phase flow models by comparing them with experimental data and analytical solutions. The complexity of the problems is gradually increased from very simple ones to problems with very complex melt-water interaction multi-fluid flows with dynamic fragmentation and coalescence and strong thermal and mechanical interactions. In particular the following cases are described and compared with the prediction using the basics presented in different chapters of this book: material relocation – gravitational waves (2D), U-tube benchmarks such us adiabatic oscillations, single-phase natural convection in a uniformly heated vertical part of a U-tube, single-phase natural convection in a uniformly heated inclined part of a U-tube, single-phase natural convection in a U-tube with an inclined part heated by steam condensation, steady state single-phase nozzle flow, pressure waves in single phase flow, 2D gas explosion in a space filled previously with gas, 2D gas explosion in space with internals previously filled with liquid, film entrainment in pipe flow, water flashing in nozzle flow, pipe blow-down with flashing, single pipe transients, complex pipe network transients, boiling in pipes and rod bundles, critical heat flux, post critical heat flux heat transfer, film boiling, behavior of clouds of cold and very hot spheres in water, experiments with dynamic fragmentation and coalescence like the FARO L14, 20, 24, 28, 31 experiments, PREMIX 13, 15, 17, 18 experiment, RIT and IKE experiment. In addition an application of a powerful method for investigation of the propagation of input and model uncertainties on the final results by using the Monte Carlo method and regression analysis is demonstrated for the prediction of non-explosive melt-water interactions. Benchmarks for testing the 3D capabilities of computer codes are provided for the rigid body steady rotation problem, pure radial symmetric flow, radial-azimuthal symmetric flow. Examples of very complex 3D flows are also given such as small break loss of coolant, asymmetric steam-water interaction in a vessel, and melt relocation in a pressure vessel.

Chapter 26 of this volume together with Chapter 14 of Volume 1 are presented on the compact disc (CD) attached to Volume 1. The reader can read these two documents with Acrobat Reader software, which is freely available on the internet. In addition many animated sequences (movies) have been presented on the CD. The reader can execute them directly from the already-opened document with Acrobat Reader by simply double clicking on the figures numbers marked in blue. HTML documents are then executed using any Web browser available on the local

X SUMMARY

computer of the reader. The reader can see also these movies with any other software able to open animated GIF files, for instance with PaintShopPro 7 software.

Table of Contents

1 Flow regime transition criteria 1
    1.1 Introduction 1
    1.2 Pool flow 3
    1.3 Adiabatic flows 6
        1.3.1 Channel flow – vertical pipes 6
        1.3.2 Channel flow – inclined pipes 10
    1.4. Heated channels 18
    1.5. Porous media 19
    1.6 Particles in film boiling 20
    1.7 Rod bundles 21
    Nomenclature 23
    References 25

2 Drag forces 27
    2.1 Introduction 27
    2.2 Drag coefficient for single bubble 28
    2.3 Family of particles in continuum 32
    2.4 Droplets-gas 36
    2.5. Solid particles-gas in presence of liquid. Solid particles-liquid in presence of a gas 38
        2.5.1 Solid particles: free particles regime 38
        2.5.2 Solid particles in bubbly flow 40
        2.5.3 Solid particles: dense packed regime 44
    2.6. Annular flow 49
    2.7. Inverted annular flow 55
    2.8. Stratified flow in horizontal or inclined rectangular channels 56
    2.9. Stratified flow in horizontal or inclined pipes 60
    Nomenclature 65
    References 68

3 Friction pressure drop 71
    3.1 Introduction 71
    3.2 Single-phase flow 71
    3.3 Two-phase flow 74
    3.4 Three-dimensional flow in a porous structure 79
    3.5 Heated channels 80

    3.6 Three-phase flow 82
    Nomenclature.. 84
    References 86

4 Diffusion velocities for algebraic slip models 89
    4.1 Introduction 89
    4.2 Drag as a function of the relative velocity 90
        4.2.1 Wall force not taken into account... 90
        4.2.2 Wall forces taken into account 94
    4.3 Two velocity fields 95
        4.3.1 Single bubble terminal velocity... 95
        4.3.2 Single particle terminal velocity.. 99
        4.3.3 Cross section averaged bubble rise velocity in pipes – drift flux
        models 100
        4.3.4 Cross section averaged particle sink velocity in pipes – drift flux
        models 119
    4.4 Slip models 122
    4.5 Three velocity fields – annular dispersed flow 124
    4.6 Three-phase flow 125
    Nomenclature.. 128
    References 130

5 Entrainment in annular two-phase flow.. 133
    5.1 Introduction 133
    5.2 Some basics 134
    5.3 Correlations 135
    5.4 Entrainment increase in boiling channels 143
    5.5 Size of the entrained droplets.. 144
    Nomenclature.. 146
    References 149

6 Deposition in annular two-phase flow 153
    6.1 Introduction 153
    6.2 Analogy between heat and mass transfer.. 153
    6.3 Fluctuation mechanism in the boundary layer 155
    6.4 *Zaichik*'s theory... 156
    6.5 Deposition correlations... 157
    Nomenclature.. 161
    References 164

7 Introduction to fragmentation and coalescence...................................... 167
    7.1 Introduction 167
    7.2 General remarks about fragmentation... 170
    7.3 General remarks about coalescence 171
        7.3.1 Converging disperse field.. 171
        7.3.2 Analogy to the molecular kinetic theory 172

7.4 Superposition of different droplet coalescence mechanisms 178
7.5 Superposition of different bubble coalescence mechanisms................... 179
7.6 General remarks about particle size formation in pipes......................... 180
Nomenclature.. 184
References.. 186

8 Acceleration induced droplet and bubble fragmentation 189
8.1 Critical *Weber* number.. 189
8.2 Fragmentation modes..200
8.3 Relative velocity after fragmentation...203
8.4 Breakup time...207
8.5 Particle production rate correlations 214
    8.5.1 Vibration breakup..214
    8.5.2 Bag breakup...214
    8.5.3 Bag and stamen breakup..216
    8.5.4 Sheet stripping and wave crest stripping following by
    catastrophic breakup...216
8.6 Droplets production due to highly energetic collisions..........................224
8.7 Acceleration induced bubble fragmentation 226
Nomenclature..230
References..232

9 Turbulence induced particle fragmentation and coalescence....................237
9.1. Homogeneous turbulence characteristics..237
9.2 Reaction of a particle to the acceleration of the surrounding
continuum 241
9.3 Reaction of particle entrained inside the turbulent
vortex – inertial range..243
9.4 Stability criterion for bubbles in continuum 244
9.5 Turbulence energy dissipation due to the wall friction.........................248
9.6 Turbulence energy dissipation due to the relative motion 250
9.7 Bubble coalescence probability 252
9.8 Coalescence probability of small droplets 257
Nomenclature..258
References..260

10 Liquid and gas jet disintegration 263
10.1 Liquid jet disintegration in pools 263
10.2 Boundary of different fragmentation mechanisms...............................266
10.3 Size of the ligaments...268
10.4 Unbounded instability controlling jet fragmentation 269
    10.4.1 No ambient influence 269
    10.4.2 Ambient influence 270
    10.4.3 Jets producing film boiling in the ambient liquid.........................273
    10.4.4 An alternative approach..275

  10.4.5 Jets penetrating two-phase mixtures...276
  10.4.6 Particle production rate ...277
 10.5. Jet erosion by high velocity gas environment...277
 10.6. Jet fragmentation in pipes...279
 10.7. Gas jet disintegration in pools ...280
 Nomenclature...283
 References ...286

11 Fragmentation of melt in coolant...289
 11.1 Introduction ...289
 11.2 Vapor thickness in film boiling ...291
 11.3 Amount of melt surrounded by continuous water...293
 11.4 Thermo-mechanical fragmentation of liquid metal in water...294
  11.4.1 External triggers...295
  11.4.2 Experimental observations ...300
  11.4.3 The mechanism of the thermal fragmentation...306
 11.5 Particle production rate during the thermal fragmentation ...322
 11.6 *Tang*'s thermal fragmentation model...324
 11.7 *Yuen*'s thermal fragmentation model ...327
 11.8 Oxidation ...327
 11.9 Superposition of thermal fragmentation ...328
  11.9.1 Inert gases...328
  11.9.2 Coolant viscosity increase ...329
  11.9.3 Surfactants...329
  11.9.4 Melt viscosity ...330
 Nomenclature...331
 References ...334

12 Nucleation in liquids...341
 12.1 Introduction ...341
 12.2 Nucleation energy, equation of *Kelvin* and *Laplace* ...342
 12.3 Nucleus capable to grow...344
 12.4 Some useful forms of the *Clausius-Clapeyron* equation, measures of superheating...345
 12.5 Nucleation kinetics ...348
  12.5.1 Homogeneous nucleation ...348
  12.5.2 Heterogeneous nucleation ...350
 12.6 Maximum superheat ...356
 12.7 Critical mass flow rate in short pipes, orifices and nozzles ...360
 12.8 Nucleation in the presence of non-condensable gases ...360
 12.9 Activated nucleation site density – state of the art...362
 12.10. Conclusions and recommendations...368
 Nomenclature...369
 References ...371

13 Bubble growth in superheated liquid 375
13.1 Introduction 375
13.2 The thermally controlled bubble growth 376
13.3 The *Mikic* solution 379
13.4 How to compute the mass source terms for the averaged
conservation equations? 382
    13.4.1 Non-averaged mass source terms 382
    13.4.2 The averaged mass source terms 384
13.5. Superheated steam 385
13.6 Diffusion controlled evaporation into mixture of gases inside the
bubble 386
13.7 Conclusions 387
Nomenclature 387
References 390
Appendix 13.1 Radius of a single bubble in a superheated liquid as a
function of time 392

14 Condensation of a pure steam bubble in a subcooled liquid 399
14.1 Introduction 399
14.2 Stagnant bubble 399
14.3 Moving bubble 401
14.4 Non-averaged source terms 406
14.5 Averaged source terms 407
14.6 Change of the bubble number density due to condensation 409
14.7 Pure steam bubble drifting in turbulent continuous liquid 410
14.8 Condensation from a gas mixture in bubbles surrounded by
subcooled liquid 413
    14.8.1 Thermally controlled collapse 413
    14.8.2 Diffusion controlled collapse 414
Nomenclature 415
References 419

15 Bubble departure diameter 421
15.1 How accurately can we predict bubble departure diameter for boiling? 421
15.2 Model development 423
15.3 Comparison with experimental data 429
15.4 Significance 432
15.5 Summary and conclusions 433
Nomenclature 434
References 435

16 How accurately can we predict nucleate boiling? 439
16.1 Introduction 439
16.2 New phenomenological model for nucleate pool boiling 444
    16.2.1 Basic assumptions 444

    16.2.2 Proposed model 446
    16.3 Data comparison 448
    16.4 Systematic inspection of all the used hypotheses 452
    16.5 Significance 453
    16.6 Conclusions 453
    Nomenclature.. 454
    References 456
    Appendix 16.1 State of the art of nucleate pool boiling modeling 459
    Appendix 16.2 Some empirical correlations for nucleate boiling................ 465

17 Heterogeneous nucleation and flashing in adiabatic pipes 467
    17.1 Introduction 467
    17.2 Bubbles generated due to nucleation at the wall............................ 468
    17.3 Bubble growth in the bulk 469
    17.4 Bubble fragmentation and coalescence.. 470
    17.5 Film flashing bubble generation in adiabatic pipe flow.................. 471
    17.6 Verification of the model 473
    17.9 Significance and conclusions.. 483
    Nomenclature.. 484
    References 486

18 Boiling of subcooled liquid.. 489
    18.1 Introduction 489
    18.2 Initiation of visible boiling on the heated surface........................... 489
    18.3 Local evaporation and condensation.. 490
        18.3.1 Relaxation theory 490
        18.3.2 Boundary layer treatment 493
    Nomenclature.. 495
    References 497

19 Natural convection film boiling 499
    19.1 Minimum film boiling temperature 499
    19.2 Film boiling in horizontal upwards-oriented plates 500
    19.3 Horizontal cylinder 502
    19.4 Sphere 502
    Nomenclature.. 502
    References 504

20 Forced convection boiling... 505
    20.1 Convective boiling of saturated liquid... 505
    20.2 Forced convection film boiling.. 507
        20.2.1 Tubes... 507
        20.2.2 Annular channel 510
        20.2.3 Tubes and annular channels 511
        20.2.4 Vertical flow around rod bundles... 511
    20.3 Transition boiling... 512

20.4 Critical heat flux 513
   20.4.1 The hydrodynamic stability theory of free convection DNB..........514
   20.4.2 Forced convection DNB and DO correlations................................517
   20.4.3 The 1995 look-up table...521
Nomenclature...521
References...523

21 Film boiling on vertical plates and spheres 527
21.1 Plate 527
   21.1.1 Introduction 527
   21.1.2 State of the art 528
   21.1.3 Problem definition...529
   21.1.4 Simplifying assumptions 530
   21.1.5 Energy balance at the vapor-liquid interface, vapor film
   thickness, average heat transfer coefficient...533
   21.1.6 Energy balance of the liquid boundary layer, layer thickness
   ratio 537
   21.1.7 Averaged heat fluxes..540
   21.1.8 Effect of the interfacial disturbances 542
   21.1.9 Comparison of the theory with the results of other authors...........543
   21.1.10 Verification using the experimental data......................................545
   21.1.11 Conclusions 546
   21.1.12 Practical significance..546
21.2 Sphere 547
   21.2.1 Introduction 547
   21.2.2 Problem definition...547
   21.2.3 Solution method 547
   21.2.4 Model 548
   21.2.5 Data comparison...557
   21.2.6 Conclusions 561
Nomenclature...561
References...564
Appendix 21.1 Natural convection at vertical plate......................................567
Appendix 22.2 Predominant forced convection only at vertical plate 567

22 Liquid droplets 569
22.1 Spontaneous condensation of pure subcooled steam – nucleation.........569
   22.1.1 Critical nucleation size 570
   22.1.2 Nucleation kinetics, homogeneous nucleation 573
   22.1.3 Droplet growth 575
   22.1.4 Self-condensation stop 577
22.2 Heat transfer across droplet interface without mass transfer 578
22.3 Direct contact condensation of pure steam on subcooled droplet 585
22.4 Spontaneous flashing of superheated droplet..587
22.5 Evaporation of saturated droplets in superheated gas 591

22.6 Droplet evaporation in gas mixture.. 594
Nomenclature.. 600
References 601

23 Heat and mass transfer at the film-gas interface 605
23.1 Geometrical film-gas characteristics.. 605
23.2 Convective heat transfer 607
    23.2.1 Gas side heat transfer 608
    23.2.2 Liquid side heat transfer due to conduction................................ 611
    23.2.3 Liquid side heat conduction due to turbulence 613
23.3 Spontaneous flashing of superheated film 621
23.4 Evaporation of saturated film in superheated gas 622
23.5 Condensation of pure steam on subcooled film 623
23.6 Evaporation or condensation in presence of non-condensable gases..... 624
Nomenclature.. 626
References 629

24 Condensation at cooled walls.. 631
24.1 Pure steam condensation... 631
    24.1.1 Onset of the condensation 631
    24.1.2 Condensation from stagnant steam (*Nusselt* 1916) at laminar liquid film.. 632
    24.1.3 Condensation from stagnant steam at turbulent liquid film (*Grigul* 1942) 633
24.2. Condensation from forced convection two-phase flow at liquid film... 634
    24.2.1 Down flow of vapor across horizontal tubes 634
    24.2.2 *Collier* correlation 635
    24.2.3 *Boyko* and *Krujilin* approach.. 635
    24.2.4 The *Shah* modification of the *Boyko* and *Krujilin* approach 636
24.3 Steam condensation from mixture containing non-condensing gases.... 636
    24.3.1 Computation of the mass transfer coefficient.............................. 638
Nomenclature.. 640
References 642

25 Discrete ordinate method for radiation transport in multi-phase computer codes 645
25.1 Introduction 645
    25.1.1 Dimensions of the problem 645
    25.1.2 Micro- versus macro-interactions.. 646
    25.1.3 The radiation transport equation (RTE) 646
25.2 Discrete ordinate method.. 647
    25.2.1 Discretization of the computational domain for the description of the flow 649
    25.2.2 Finite volume representation of the radiation transport equation ... 650
    25.2.3 Boundary conditions 656
25.3 Material properties.. 658

25.3.1 Source terms – emission from hot surfaces with known temperature 658
25.3.2 Spectral absorption coefficient of water 659
25.3.3 Spectral absorption coefficient of water vapor and other gases 663
25.4 Averaged properties for some particular cases occurring in melt-water interaction 663
25.4.1 Spherical cavity of gas inside a molten material 664
25.4.2 Concentric spheres of water droplets, surrounded by vapor, surrounded by molten material 665
25.4.3 Clouds of spherical particles of radiating material surrounded by a layer of vapor surrounded by water –*Lanzenberger*'s solution 669
25.4.4 Chain of infinite number of *Wigner* cells 685
25.4.5 Application of *Lanzenbergers*'s solution 686
Nomenclature 687
References 689

Index 691

26 Validation of multi-phase flow models (on CD attached to Vol. 1) 1
26.1 Introduction 3
26.2 Material relocation – gravitational waves (2D) 10
26.2.1 U-tube benchmarks 10
26.2.2 Gravitational 2D waves 15
26.3 Steady state single-phase nozzle flow 16
26.4 Pressure waves – single phase 17
26.4.1 Gas in a shock tube 17
26.4.2 Water in a shock tube 21
26.4.3 Pressure wave propagation in a cylinder vessel with free surface (2D) 22
26.5 2D: N_2 explosion in space filled previously with air 26
26.6 2D: N_2 explosion in space with internals filled previously with water 28
26.7 Film entrainment in pipe flow 32
26.8 Water flashing in nozzle flow 34
26.9 Pipe blow-down with flashing 38
26.9.1 Single pipe 38
26.9.2 Complex pipe network 42
26.10 Boiling, critical heat flux, post-critical heat flux heat transfer 43
26.11 Film boiling 50
26.12 Behavior of clouds of cold and very hot spheres in water 52
26.13 Experiments with dynamic fragmentation and coalescence 57
26.13.1 L14 experiment 57
26.13.2 L20 and L24 experiments 61
26.13.3 Uncertainty in the prediction of non-explosive melt-water interactions 62
26.13.4 Conclusions 63

26.14 L28, L31 experiment 64
26.15 PREMIX-13 experiment 69
26.16 PREMIX 17 and 18 experiments 75
26.17 RIT and IKE experiments 88
26.18 Assessment for detonation analysis 89
26.19 Examples of 3D capabilities 90
 26.19.1 Case 1. Rigid body steady rotation problem 90
 26.19.2 Case 2. Pure radial symmetric flow 92
 26.19.3 Case 3. Radial-azimuthal symmetric flow 94
 26.19.4 Case 4. Small-break loss of coolant 96
 26.19.5 Case 5. Asymmetric steam-water interaction in a vessel [65] 98
 26.19.6 Case 6. Melt relocation in a pressure vessel 102
26.20 General conclusions 104
References 104

1 Flow regime transition criteria

This Chapter presents a review of the existing methods for identification of flow patterns in two phase flow in pools, adiabatic and non-adiabatic channels, rod bundles and porous structures. An attempt is made to extend this information to be applicable in three-phase flow modeling. In addition the influence of the dynamic fragmentation and coalescence of the flow regimes is introduced.

1.1 Introduction

Transient multi-phase flows with temporal and spatial variation of the volumetric fractions of the participating phases can be represented by *sequences* of geometrical *flow patterns* that have some characteristic length scale. Owing to the highly random behavior of the flow in detail, the number of flow patterns needed for this purpose is very large. Nevertheless, this approach has led to some successful applications in the field of multi-phase flow modeling. Frequently modern mathematical models of *transient flows* include, among others, the following features:

1. Postulation of a limited number of idealized flow patterns, with transition limits as a function of local parameters for *steady state* flow (e.g., see Fig. 1.1);

2. Identification of one of the postulated idealized *steady state* flow patterns for each time step;

3. Computation of a characteristic *steady state length scale of the flow patterns* (e.g., bubble or droplet size) in order to address further constitutive relationships for interfacial heat, mass, and momentum transfer.

Various transfer mechanisms between mixture and wall, as well as between the velocity fields, depend on the flow regimes. This leads to the use of regime dependent correlations for modeling of the interfacial mass, momentum and energy transfer. The transfer mechanisms themselves influence strongly the flow pattern's appearance. That is why the first step of the coupling between the system PDEs and the correlations governing the transfer mechanisms is the flow regime identification.

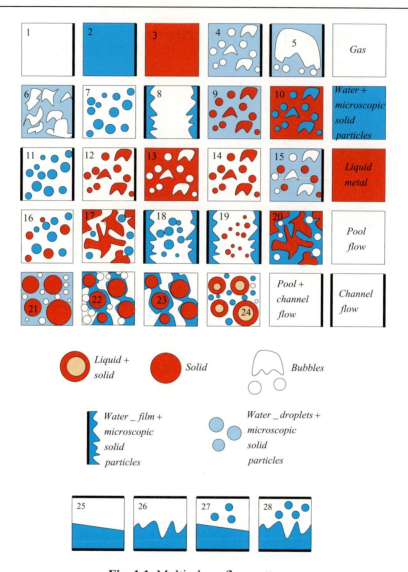

Fig. 1.1. Multi-phase flow patterns

We distinguish between flow patterns appearing in *pool* flow and in *channel* flow. In pool flow, $\gamma_v = 1$, there is no influence of the walls on the flow pattern. In channel flows characterized by $\gamma_v < 1$, however, this influence can be very strong resulting in patterns like film flow, slug flow etc.

Some flow patterns can be trivially identified by knowing only the values of the local volume fractions of the fields, α_l, and the consistency of the fields i.e. C_{li}, e.g. the single phase flows or flows consisting of three velocity fields with an initially postulated structure. For two interpenetrating velocity fields additional information is necessary to identify the flow pattern. There are analytical and experimental arguments for flow pattern identification which will be considered here. The emphasis of this Chapter is on how dynamics influence the transitions from one flow pattern into the other.

1.2 Pool flow

In pool flows we distinguish two main flow patterns: continuous liquid and continuous gas, and one intermediate between them. Next we discuss the conditions for existence of the bubble flow. We will realize that the dynamic fragmentation and coalescence will have an influence on the transition criterion through the bubble size.

Bubbly flow: The existence of the bubbly flow in pools is discussed in this section.

Non-oscillating particles: Consider equally sized spherical particles forming a rhomboid array. The average distance between the centers of two adjacent particles with diameter D_d and volumetric fraction α_d is then

$$\Delta \ell_d = D_d \left(\frac{\pi \sqrt{2}}{6 \alpha_d} \right)^{1/3}. \tag{1.1}$$

Here d stands for disperse. The *non-oscillating* particles will touch each other if

$$\Delta \ell_d = D_d. \tag{1.2}$$

This happens for a volume fraction of

$$\alpha_d = \frac{\pi}{6} \sqrt{2} \approx 0.74, \tag{1.3}$$

which is sometimes called in the literature the *maximum packing density* volume concentration. This consideration leads to the conclusion that bubble flow cannot exist for

$$\alpha_1 > 0.74, \tag{1.4}$$

and vice versa droplet flow cannot exist for

$$\alpha_1 < 0.26. \tag{1.5}$$

Oscillating particles: Oscillating particles will touch each other occasionally at larger average distance. This means that in strongly turbulent flows the existence of bubble flow should be expected to be limited by smaller volume fractions. In fact turbulent bubble flows in nature and technical facilities are observed up to

$$\alpha_1 < 0.25 \text{ to } 0.3, \tag{1.6}$$

Taitel et al. [19] (1980), *Radovich* and *Moissis* (1962), see in *Mishima* and *Ishii* [16] (1984). If the mean free path length of the oscillating particles is ℓ'_d the sphere of influence of this particle is $D_d + \ell'_d = D_d\left(1 + \ell'_d/D_d\right)$. In this case the particles may touch each other if

$$D_d\left(1 + \ell'_d/D_d\right) = D_d\left(\frac{\pi\sqrt{2}}{6\,\alpha_d}\right)^{1/3}, \tag{1.7}$$

or

$$\alpha_d = \frac{\pi}{6}\sqrt{2}/\left(1 + \Delta\ell'_d/D_d\right) < 0.74, \tag{1.8}$$

and consequently

$$\ell'_d/D_d > 0.44 \text{ to } 0.35. \tag{1.9}$$

The influence of the particle size on the flow regime transition: Brodkey [3] (1967) shows that bubbles with radii smaller than

$$D_{1,\text{solid like}} = 0.63\sqrt{\frac{2\sigma_2}{g\Delta\rho_{21}}} \approx 0.89\lambda_{RT}, \tag{1.10}$$

where λ_{RT} is the *Raleigh-Taylor* wavelength defined as follows

$$\lambda_{RT} = \sqrt{\frac{\sigma_2}{g\Delta\rho_{21}}}, \tag{1.11}$$

behave as a solid sphere and the coalescence is negligible. This argument was used by *Taitel* et al. in [19] (1980) to explain the existence of bubble flow in regions up to

$$\alpha_1 < 0.54, \tag{1.12}$$

if strong liquid turbulence destroys bubbles to dimensions

$$D_1(\varepsilon_2,...) < D_{1,\text{solid like}} \tag{1.13}$$

where ε_2 is the dissipation rate of the turbulent kinetic energy of the liquid.

Conclusions:

a) In the concept of modeling dynamic fragmentation and coalescence, bubble flow is defined if we have at least one bubble in the volume of consideration, Vol_{cell}, i.e.

$$n_1 Vol_{cell} > 1, \tag{1.14}$$

otherwise both phases are continuous. This is the trivial condition. Here n_1 is the number of bubbles per unit mixture volume.

b) For small bubble sizes, $D_1(\varepsilon_2,...) < D_{1,\text{solid like}}$, the bubbles behave as solid spheres and the coalescence probability is dramatically reduced. In this case bubble flow exists up to

$$\alpha_{1,B-Ch} = 0.54. \tag{1.15}$$

c) For larger bubble sizes, $D_1(\varepsilon_2,...) > D_{1,\text{solid like}}$, the transition between bubble and churn turbulent flow happens between

$$\alpha_{1,B-Ch} \approx 0.25 \text{ and } 0.54. \tag{1.16}$$

The size at which the lower limit holds is not exactly known. Assuming that this size is governed by a critical *Weber* number equal to 12 and using the bubble rise velocity in the pool as computed by *Kutateladze*

$$V_{1Ku} = \sqrt{2}\left[g\sigma_2(\rho_2-\rho_1)/\rho_2^2\right]^{1/4},$$

we obtain 6 times λ_{RT}. A linear interpolation between these two sizes gives

$$\alpha_{1,B-Ch} = 0.54 - 0.0567(D_1/\lambda_{RT} - 0.89). \tag{1.17}$$

The dependence of this transition criterion on the bubble size is remarkable. Modeling dynamic bubble size evolution gives different regime transition boundaries for different bubble sizes at the same gas volume fraction.

d) Churn turbulent flow exists between

$$0.54 < \alpha_1 < 0.74. \tag{1.18}$$

Note that the upper limit of the churn turbulent flow, $\alpha_{1,Ch-A} = 0.74$, seems to be a function of the *local Mach* number. The higher the local *Mach* number, the higher the upper limit due to the increasing turbulence. This consideration is supported by the position of the slip maximum as a function of the gas volume fraction in critical flow as measured by e.g. *Deichel* and *Winter* [5] (1990). The investigation of *Ginsberg* et al. [8] (1979) of flow behavior of volume-heated boiling pools shows that for fast transients the limit between bubble and churn turbulent flow and dispersed flow is higher. Thus one can assume that the upper limit is

$$\alpha_{1,Ch-A} = 0.74 + Ma(0.92 - 0.74) = 0.74 + 0.18Ma. \tag{1.19}$$

In accordance with this consideration, very slow flows do not have churn turbulent regimes and bubble flow goes directly into disperse droplet flow with increasing gas volume fraction.

1.3 Adiabatic flows

1.3.1 Channel flow – vertical pipes

For vertical pipe flows the following regime boundaries are defined as follows:

Bubble flow: As for the pool flow, in the concept of modeling dynamic fragmentation and coalescence, bubble flow is defined if we have at least one bubble in the volume of consideration, Vol_{cell}, i.e.

$$n_1 Vol_{cell} > 1, \tag{1.20}$$

otherwise both phases are continuous.

b) Bubble flow exists in the region

$$0 < \alpha_1 < \alpha_{1,bubble\ to\ slug} \quad \text{and} \quad D_h > D_{h,slug}. \tag{1.21}$$

where

$$D_{h,slug} = 19\frac{\Delta\rho_{21}}{\rho_2}\lambda_{RT}, \quad \textit{Taitel at al [19] (1980)}, \tag{1.22}$$

$$\alpha_{1,bubble\ to\ slug} = 0.54 \text{ for } D_1(\varepsilon_2,...) < 0.89\lambda_{RT}, \tag{1.23}$$

$$\alpha_{1,bubble\ to\ slug} = 0.54 - 0.0567(D_1/\lambda_{RT} - 0.89) \text{ for } 0.89\lambda_{RT} < D_1(\varepsilon_2,...) < 6\lambda_{RT}, \tag{1.24}$$

$$\alpha_{1,bubble\ to\ slug} \approx 0.25 \text{ for } D_1(\varepsilon_2,...) \geq 6\lambda_{RT} \tag{1.25}$$

Slug flow: Slug flow is defined as a train of large bubbles followed by mixtures of small bubbles and liquid or liquid only. The slug regime is never stationary. Slug flow exists if

$$D_h < D_{h,slug} \tag{1.26}$$

or if

$$D_h > D_{h,slug} \text{ and } \alpha_{1,bubble\ to\ slug} < \alpha_1 < \alpha_{1,slug\ to\ churn}. \tag{1.27}$$

This condition reflects the fact that slug flow can be transformed into churn turbulent flow if the gas volume fraction averaged over the entire pipe length is larger than those in the slug bubble section only. The gas volume fraction in the slug bubble section only is

$$\alpha_{1,slug\ to\ churn} = 1 - 0.813\left\{\frac{(C_0-1)|j|+0.35\,V_{TB}}{|j|+0.75\,V_{TB}b_1}\right\}^{0.75}, \tag{1.28}$$

Mishima and *Ishii* [16] (1984). The drift flux distribution coefficient for slug flow is

$$C_0 = 1.2, \tag{1.29}$$

the slug (*Taylor* bubble) raising velocity is

$$V_{TB} = \sqrt{\frac{\rho_2-\rho_1}{\rho_2}gD_h}, \tag{1.30}$$

the mixture volumetric flux is

$$j = \alpha_1 V_1 + (1-\alpha_1) V_2, \tag{1.31}$$

and

$$b_1 = \left(\frac{\rho_2 - \rho_1}{\eta_2^2 / \rho_2} g D_h^3 \right)^{1/18}. \tag{1.32}$$

The correlation contains the length of the *Taylor* bubble

$$\sqrt{2 \ell_{TB} \Delta \rho_{21} g / \rho_2} = |j| + 0.75 \, V_{TB} b_1. \tag{1.33}$$

The error for computing ℓ_{TB} is $\pm 100\%$. Thus, air-water flow in a pipe with diameter $D_h = 0.027 m$ at atmospheric conditions and phase volumetric flow rates of $\alpha_1 V_1 = 0.2$ to $2 \, m/s$, $(1-\alpha_1) V_2 = 0.2 m/s$, is a slug flow with characteristic slug length of $\ell_{TB} \approx 0.1$ to $0.5m$.

Churn turbulent flow: In accordance with *Mishima* and *Ishii* [16] (1984) churn turbulent flow exists under the following conditions:

$$\alpha_1 > \alpha_{1,\text{slug to churn}} \text{ and } \left[(D_h < D_{hc} \text{ and } V_1 < V_{11}) \text{ or } (D_h \geq D_{hc} \text{ and } V_1 < V_{12}) \right]. \tag{1.34}$$

Here

$$D_{hc} = \lambda_{RT} / \left[(\alpha_1 - 0.11)^2 N_{\eta 2}^{0.4} \right], \tag{1.35}$$

and

$$N_{\eta 2} = \eta_2 / \sqrt{\rho_2 \sigma_2 \lambda_{RT}} \tag{1.36}$$

is the viscous number. The first criterion is applied to flow reversal in the liquid film section along large bubbles,

$$D_h < D_{hc} \text{ and } V_1 > V_{11}, \tag{1.37}$$

where

$$V_{11} = (1 - 0.11/\alpha_1) V_{TB}. \tag{1.38}$$

In this case the flow reversal in the liquid film section along the large bubbles causes the transition.

The second criterion is applied to destruction of liquid slugs or large waves by entrainment or deformation

$$D_h \geq D_{hc} \text{ and } V_1 > V_{12}. \tag{1.39}$$

where

$$V_{12} = \frac{V_{2Ku}}{\alpha_1 N_{\eta 2}^{0.2}}, \tag{1.40}$$

and

$$V_{2Ku} = \left[g\sigma_2 (\rho_2 - \rho_1)/\rho_1^2 \right]^{1/4}, \tag{1.41}$$

is the *Kutateladze* terminal velocity for free falling droplets in gas. The correlation holds for low viscous flows $N_{\eta 2} < 1/15$ and relatively high liquid *Reynolds* number $\text{Re}_{23} > 1635$. In this case the churn flow bubble section following the slug disintegrates or the liquid waves and subsequent liquid bridges and slugs can be entrained as small droplets. This leads to the elimination of liquid slugs between large bubbles and to a continuous gas core. This is the criterion for transition from slug flow to annular-dispersed flow.

Annular film flow: The annular film flow is defined if

$$\alpha_1 > \alpha_{1,\text{slug to churn}} \text{ and } \left[(D_h < D_{hc} \text{ and } V_1 > V_{11}) \text{ or } (D_h \geq D_{hc} \text{ and } V_1 > V_{12}) \right]. \tag{1.42}$$

Annular film flow with entrainment: The annular film flow with entrainment is defined by *Kataoka* and *Ishi* [12] (1982) as follows

$$D_h > D_{hc} \text{ and } \alpha_1 > \alpha_{1,\text{slug to churn}} \text{ and } V_1 > V_{13} \text{ and }$$

$$\text{Re}_{2\delta} = \rho_2 V_2 4\delta_2 / \eta_2 > 160. \tag{1.43}$$

Here $\delta_2 = \frac{1}{2} D_h \left(1 - \sqrt{1-\alpha_2} \right)$ is the film thickness,

$$V_{13} = b_8 \frac{\sigma_2}{\eta_2 \alpha_1 \left(\frac{\rho_1}{\rho_2}\right)^{1/2}}, \qquad (1.44)$$

$$b_8 = N_{\eta 2}^{0.8}, \quad \text{for} \quad \text{Re}_{23} > 1635, \qquad (1.45)$$

$$b_8 = 11.78 \, N_{\eta 2}^{0.8} / \text{Re}_{23}^{1/3} \quad \text{for} \quad \text{Re}_{23} \leq 1635, \qquad (1.46)$$

$$\text{Re}_{23} = (1-\alpha_1)\rho_2 |V_{23}| D_h / \eta_2, \qquad (1.47)$$

$$V_{23} = (\alpha_2 V_2 + \alpha_3 V_3)/(1-\alpha_1). \qquad (1.48)$$

1.3.2 Channel flow – inclined pipes

Compared with the vertical flow the flow in horizontal pipes possess two additional flow patterns - stratified flow and stratified wavy flow. For the computation of the relative velocities and pressure drop for these flow pattern the work by *Mamaev* [13] (1969) is recommended. *Mamaev* et al. considered stratified flow possible for $Fr < Fr_{crit}$, where $Fr = \frac{(\rho w)^2 v_h}{g D_h}$, $v_h = X_1 v_1 + (1 - X_1) v_2$. The critical *Froud* number was obtained from experiments

$$Fr_{crit} = \left[\left(0.2 - \frac{2\cos\varphi}{\lambda_{fr}} \right) \middle/ (1-\beta)^2 \right] \exp(-2.5\beta),$$

where $\beta = X_1 v_1 / v_h$, and $\lambda_{fr} = \lambda_{fr}\left(\frac{\pi(1-\alpha_1)}{\pi - \theta} \frac{w_2 D}{v_2}, \frac{k}{D} \right)$ is the liquid site wall friction coefficient computed using the *Nikuradze* diagram. *Weisman* et al. [24] (1979), *Weisman* and *Kang* [25] (1981) and *Grawford* et al. [9] (1985) published a set of correlations for horizontal as well as vertical flows. Their correlations for horizontal flows are summarized at the end of this section.

Transition criteria are systematically elaborated by *Taitel* and *Dukler* [18] (1976), *Rouhani* et al. [17] (1983). The *Taitel* line of criteria development is presented here.

Stratified flow: Almost all results available in the literature provide a criterion for identification of the existence of the stratified flows based on the stability criterion Eq. (151b) from Chapter 2 in Volume 1 of this monograph

$$V_{1,stratified} - V_2 = \left[g\cos(\varphi - \pi/2)(\rho_2 - \rho_1)\left(\frac{\alpha_1}{\rho_1} + \frac{1-\alpha_1}{\rho_2}\right) \middle/ \frac{d\alpha_2}{d\delta_{2F}} \right]^{1/2}, \quad (1.49)$$

where δ_{2F} is the liquid thickness. φ is the angle defined between the upwards oriented vertical and the pipe axis – see Fig. 1.2. For a larger velocity difference the flow is no longer stratified and disintegrates into an intermittent flow like elongated bubble or slug or churn flow. For flow between two parallel plates

$$\frac{d\delta_{2F}}{d\alpha_2} = H, \quad (1.50)$$

where H is the distance between the two plates. For flows in a round tube, see Fig. 1.2, the angle θ with the origin the pipe axis is defined between the upwards oriented vertical and the liquid-gas-wall triple point as a function of the liquid volume fraction given by the equation

$$1 - \alpha_2 = \frac{\theta - \sin\theta\cos\theta}{\pi}. \quad (1.51)$$

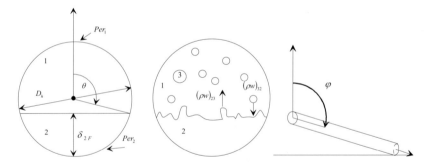

Fig. 1.2. Definition of the geometrical characteristics of the stratified flow

Having in mind that

$$\delta_{2F} = \frac{1}{2} D_h (1 + \cos\theta) \quad (1.52)$$

we obtain

$$\frac{d\alpha_2}{d\delta_{2F}} = \frac{d\alpha_2}{d\theta} \middle/ \frac{d\delta_{2F}}{d\theta} = \frac{4}{D_h} \frac{\sin\theta}{\pi}. \quad (1.53)$$

This criterion is in fact consistent with the *Kelvin-Helmholtz* gravity long wave theory -see *Milne Thomson* [14] (1968), or *Delhaye* [6], p. 90 (1981), or *Barnea* and *Taitel* [2] (1994).

Wallis and *Dobson* [23] (1973) compared the above equation with experimental data for channels with H ranging from 0.0254 to 0.305 m and corrected then by introducing a constant multiplier of 0.5. This result was in fact confirmed by *Mishima* and *Ishii* [15] in 1980. These authors obtained for low pressure $\frac{\alpha_1}{\rho_1} \gg \frac{1-\alpha_1}{\rho_2}$ the constant 0.487.

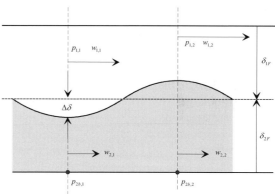

Fig. 1.3. The definition of the variables for the Hohannsen stability criterion

Hohannessen [10] (1972) considered the situation depicted in Fig. 1.3 and defined the transition of the stratified flow as equality of the static pressures at the bottom of the pipe $p_{2b,1}$ and $p_{2b,2}$ taking into account the change of the gas pressure due to cross section decrease by using the *Bernouli* equation. *Taitel* and *Dukler* [18] (1976), similarly to *Hohannessen* [10] (1972), equalized the buoyancy pressure increment required to create solution disturbance with finite amplitude to the increase of the gas dynamic pressure and obtained after linearization the following multiplier $\left(1-\frac{\delta_{2F}}{H}\right)$ again testing the result for low pressure. Defining the *Taylor* bubble velocity for inclined pipe with

$$V_{TB}^* = \left[\frac{\rho_2 - \rho_1}{\rho_2} D_h g \cos(\varphi - \pi/2)\right]^{1/2}, \qquad (1.54)$$

the criterion for a pipe is then

$$V_{1,stratified} - V_2 = \frac{1}{2}(1-\cos\theta)\left[\frac{g(\rho_2-\rho_1)\cos(\varphi-\pi/2)}{\frac{4}{D_h}\frac{\sin\theta}{\pi}}\left(\frac{\alpha_1}{\rho_1}+\frac{1-\alpha_1}{\rho_2}\right)\right]^{1/2}$$

(1.55)

or

$$\frac{V_{1,stratified} - V_2}{V_{TB}^*} = \frac{1}{4}(1-\cos\theta)\left[\frac{\pi}{\sin\theta}\rho_2\left(\frac{\alpha_1}{\rho_1}+\frac{1-\alpha_1}{\rho_2}\right)\right]^{1/2}.$$

(1.56)

This criterion is valid for gravity driven liquid flow. *Johnston* [11] (1985) compared the *Taitel* and *Dukler* result and found that the RHS of the above equation has to be multiplied with a factor ranging between 0.39 and 4 with 1 being a good choice. The accuracy of prediction varies from 2% for slow inclinations, 1/10, to 75% for 1/400 inclinations. *Anoda* et al. [1] (1989) confirmed the validity of the above equation for large diameter pipes (0.18m) and large pressures (3 to 7.3 MPa).

Bestion [4] (1990) reported data for stratification of horizontal flow for pressure range of 2 to 10 MPa. The data shows that if the liquid velocity is smaller than the bubble free rising velocity, $V_2 < V_{1Ku}$, the flow is stratified.

Stratified wavy flow: The surface of the liquid remains *smooth* if the gas velocity remains below some prescribed value. *Mamaev* et al. [13] reported in 1969 that waves started within $0.01w_2 \le w_1 \le 3.33w_2$ and are always there for $w_1 > 3.33w_2$. *Taitel* and *Dukler* [18] (1976) derived approximate expression for the gas velocity exciting waves

$$V_{1,wavy} - V_2 = \left[\frac{4\eta_2 g(\rho_2-\rho_1)\cos(\varphi-\pi/2)}{0.01\rho_2\rho_1 V_2}\right]^{1/2},$$

(1.57)

or after rearranging

$$\frac{V_{1,wavy} - V_2}{V_{TB}^*} = 20\left(\frac{\rho_2}{\rho_1}\right)^{1/2}\left(\frac{\rho_2 V_2 D_h}{\eta_2}\right)^{-1/2}.$$

(1.58)

For larger gas velocity the surface of the liquid is *wavy* (stratified wavy flow).

Annular flow – Fig.1.4: The first requirement for the flow to be annular is that the film volume fraction is

$$\alpha_2 < 0.24, \quad (1.59)$$

see in *Taitel* [20] (1990), p. 245. More information on the existence of the annular flow is contained in the *Weisman-Kang* flow map given below.

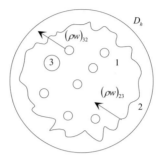

Fig. 1.4. Annular flow

Bubble flow: *Taitel* and *Dukler* [18] (1976) investigated gravity driven flow. The authors found the transition to dispersed bubble flow to occur if the liquid side share force due to turbulence equals the buoyancy force acting on the liquid. This results in the following criterion

$$V_{2,bubble} - V_2 = \left[\frac{4F}{b} \frac{\alpha_1 g (\rho_2 - \rho_1) \cos(\varphi - \pi/2)}{\rho_2 c_{2w}}\right]^{1/2} \quad (1.60)$$

where

$$c_{2w} = \frac{0.046}{\left(\dfrac{\rho_2 V_2 D_{h2}}{\eta_2}\right)^{1/5}}, \quad (1.61)$$

$$\frac{4F}{b} = 4\frac{\pi D_h^2/4}{D_h \sin\theta} = \frac{\pi}{\sin\theta} D_h. \quad (1.62)$$

F is the pipe cross section and b is the gas-liquid interface median if stratification is assumed. After rearrangement for pipe flow we obtain

$$V_{2,bubble} - V_2 = 8.26 \left[\frac{\alpha_1 D_h g(\rho_2 - \rho_1)\cos(\varphi - \pi/2)}{\sin\theta \rho_2}\right]^{1/2} \left(\frac{\rho_2 V_2 D_{h2}}{\eta_2}\right)^{1/10}, \quad (1.63)$$

or

$$\frac{V_{2,bubble} - V_2}{V_{TB}^*} = 8.26 \left(\frac{\alpha_1}{\sin\theta}\right)^{1/2} \left(\frac{\rho_2 V_2 D_{h2}}{\eta_2}\right)^{1/10}. \tag{1.64}$$

For smaller gas velocities the bubble flow exists. For larger gas velocity the flow is intermittent regime.

The *Weisman* and *Kang* [25] (1981) empirical flow map

Stratified-intermittent transition: *Weisman* et al. [24] (1979) defined the existence of the stratified flow if the liquid velocity is smaller than the prescribed value given below

$$\frac{\alpha_1 V_1}{V_{TB}} = 0.25 \left(\frac{\alpha_1}{\alpha_2} \frac{V_1}{V_{2,stratified}}\right)^{1.1}, \tag{1.65}$$

or

$$(\alpha V)_{2,stratified} = \left[0.25 V_{TB} (\alpha_1 V_1)^{0.1}\right]^{0.909}. \tag{1.66}$$

Stratified wavy flow: *Weisman* et al. [24] (1979) proposed the following correlation

$$\left(\frac{\lambda_{RT}}{D_h}\right)^{0.4} \left(\frac{\alpha_1 \rho_1 V_1 D_h}{\eta_1}\right)^{0.45} = 8\left(\frac{\alpha_1}{1-\alpha_1}\frac{V_1}{V_2}\right)^{0.16} \tag{1.67}$$

i.e.

$$(\alpha V)_{1,wavy} = \left\{101.6 \frac{\eta_1}{\rho_1 D_h} \frac{1}{\left[(1-\alpha_1)V_2\right]^{0.3556}} \left(\frac{D_h}{\lambda_{RT}}\right)^{0.889}\right\}^{1.55}. \tag{1.68}$$

For larger gas velocity the surface of the liquid is *wavy* (stratified wavy flow).

Transition to annular flow: The transition from stratified wavy to annular flow happens in convection regimes. This is the reason why the transition criterion is valid for horizontal as well for vertical flows. The transition conditions is defined by the *Weisman* et al. [24] (1979) correlation

$$1.9\left(\frac{\alpha_1}{1-\alpha_1}\frac{V_{1,annular}}{V_2}\right)^{1/8} = \left(\frac{\alpha_1 V_{1,annular}}{V_{2Ku}}\right)^{0.2} \left(\frac{\alpha_1 V_{1,annular}}{V_{TB}}\right)^{0.36} \tag{1.69}$$

i.e. if

$$\alpha_1 V_{1,annular} > 4.37 \frac{V_{2Ku}^{0.46} V_{TB}^{0.83}}{\left[(1-\alpha_1)V_2\right]^{0.287}}, \qquad (1.70)$$

where

$$V_{2Ku} = \left[g\sigma_2(\rho_2-\rho_1)/\rho_1^2\right]^{1/4}, \qquad (1.71)$$

the annular flow exists.

Transition to bubble flow: *Weisman* et al. [24] (1979) defined the existence of the dispersed bubble regime by comparing the pressure drop with a value obtained by data comparison as follows

$$\frac{(dp/dz)_{20}}{g(\rho_2-\rho_1)} = 2.89 \frac{\lambda_{RT}}{D_h}, \qquad (1.72)$$

where

$$\lambda_{RT} = \left[\sigma_{12}/(g\Delta\rho_{21})\right]^{1/2}, \qquad (1.73)$$

$$(dp/dz)_{20} = \frac{1}{2}\rho_2 \left(\frac{G}{\rho_2}\right)^2 \frac{\lambda_{R20}}{D_h}, \qquad (1.74)$$

is the pressure drop due to friction computed for liquid flow having mass flow rate equal to the mixture mass flow rate

$$G = \sum_{l=1}^{3} \alpha_l \rho_l w_l. \qquad (1.75)$$

If

$$(dp/dz)_{20} > 2.89 \frac{\lambda_{RT}}{D_h} g(\rho_2-\rho_1) \qquad (1.76)$$

bubble flow exists. Otherwise we have intermittent flow. Using the *Blasius* formula for the friction coefficient

$$\lambda_{R20} = \frac{0.316}{\left(\frac{GD_h}{\eta_2}\right)^{1/4}}, \tag{1.77}$$

the above criterion can be rewritten as follows

$$\frac{GD_h}{\eta_2} > 5.26 \left[\lambda_{RT} g \frac{D_h^2 \rho_2}{\eta_2^2} (\rho_2 - \rho_1) \right]^{4/7}. \tag{1.78}$$

Weisman and *Kang* [25] (1981) reported a criterion for existence of dispersed bubble flow for vertical and inclined flows defined as follows

$$\frac{\alpha_1 V_{1,bubble}}{V_{TB}} = 0.45 \left(\frac{j}{V_{TB}}\right)^{0.78} \left[1 - 0.65 \cos(\varphi - \pi/2)\right], \tag{1.79}$$

where $(\varphi - \pi/2)$ is the inclination angle. Solving with respect to $\alpha_1 V_1$ requires some iterations starting with initial value

$$\alpha_1 V_{1,bubble} = (1 - \alpha_2) V_2 / \left\{ \frac{1}{0.45 \, V_{TB}^{0.22} \left[1 - 0.65 \cos(\varphi - \pi/2)\right]} - 1 \right\}. \tag{1.80}$$

For smaller gas velocities the bubble flow exists. For larger gas velocity the flow is in intermittent regime.

Summary of the transition conditions: The decision procedure for flow pattern recognition is defined as follows

1) Compute from Eq.(1.78) $G_{dispersed}$. If $G > G_{dispersed}$ we have bubbly flow.
2) Compute $(\alpha V)_{1,annular}$ from Eq. (1.70). If $(\alpha V)_1 > (\alpha V)_{1,annular}$ we have annular flow.
3) Compute $(\alpha V)_{2,stratified}$ from Eq. (1.66). If $(\alpha V)_2 < (\alpha V)_{2,stratified}$ we have stratified flow.
4) Compute $(\alpha V)_{1,wavy}$ from Eq. (1.68). If $(\alpha V)_1 > (\alpha V)_{1,wavy}$ we have wavy flow.
5) Compute $(\alpha V)_{1,bubble}$ from Eq. (1.80). If $(\alpha V)_1 < (\alpha V)_{1,bubble}$ we have bubble flow.
6) In all other cases slug or plug flow is defined.

1.4. Heated channels

For heated channels the flow pattern maps for unheated flows have to be corrected after checking the heat transfer regime on the wall. If the heat flux on the wall exceeds the critical heat flux the inverted annular non-adiabatic flow occurs. There are investigations, which provide specific flow pattern transition criteria for non-adiabatic flow. Such an example is the study performed by *Doroschuk* et al. [7] (1982), whose results are summarized in Table 1.1.

Table 1.1. The *Doroschuk* et al. [7] (1982) flow pattern boundaries for heated flow in a vertical pipe. D_h=0.0088m.

The identification criteria are verified for the following conditions: $10^6 \leq p \leq 10^7 \, Pa$, $500 \leq G \leq 2500 \, kg/(m^2 s)$, $-0.2 \leq X_{1,eq} \leq 0.6$, $0 \leq \dot{q}''_w \leq 0.5 \times 10^6 \, W/m^2$, $X_{1,eq} = \frac{\sum X_i h_i - h'}{h'' - h'}$, X_1 is the mass flow concentration of the vapor velocity field 1.

One phase-to-subcooled nucleate boiling transition:

$$X_{1,eq} = -140 \left[\frac{\dot{q}''_w}{G(h''-h')} \right]^{1.1} \text{Re}_{20}^{0.2} \left(\frac{\rho_2}{\rho_1} \right)^{0.2}, \quad \text{Re}_{20} = GD_h/\eta_2,$$

Tarasova (1976), see in [7].

Bubble-to-slug flow transition: $X_{1,eq} = 0$

Slug-to-churn flow transition

$$X_{1,eq} = c_1 Fr_{20}^{-0.25} \left(\frac{\rho_2}{\rho_1} \right)^{-0.05} - 0.76 \times 10^2 \left[\frac{\dot{q}''_w}{G(h''-h')} \right]^{0.25} \left(\frac{\rho_2}{\rho_1} \right)^{0.5}, \quad Fr_{20} = \frac{G}{gD_h \rho_2^2}.$$

Churn-to-annular, dispersed flow transition:

$$X_{1,eq} = c_2 \, \text{Re}_{20}^{-0.5} \left(\frac{\rho_2}{\rho_1} \right)^{-0.5}$$

Strong disturbance on the film-to-small ripples transition:

$$X_{1,eq} = c_3 We_{20}^{-0.25} \left(\frac{\rho_2}{\rho_1} \right)^{-0.35}, \quad We_{20} = \frac{G^2 D_h}{\rho_2 \sigma_2}$$

Rippled film-to-micro film transition:

$$X_{1,eq} = c_4 \left(We_{20}/\text{Re}_{20}\right)^{-0.5} \left(\frac{\rho_2}{\rho_1}\right)^{0.25} \left(\frac{D_h}{0.008}\right)^{-0.25}$$

	c_1	c_2	c_3	c_4
tubes	0.05	130	3.5	0.11
annuli	0.08	120	3.0	0.07

1.5. Porous media

Tung and *Dhir* [21] analyzed non-boiling gas-liquid flow in porous media consisting of solid particles with uniform diameter D_3 assumed to form rhomboid arrays. The authors distinguish bubbly, annular and slug flow structure. Describing three-phase flow with solid particles requires also flow pattern transition criteria for the structure between the solid particles. That is why this information is of considerable interest for multi-phase flow modeling. The flow pattern criteria are functions of the gas volume fraction of the gas-liquid mixture only. For three velocity fields in which the third field consists of spheres the required gas-liquid void fraction is

$$\bar{\alpha}_1 = \frac{\alpha_1}{\alpha_1 + \alpha_2}. \tag{1.81}$$

Table 1.2. Flow regime limits as recommended by *Tung* and *Dhir* [21] (1990)

The **bubble flow** regime is divided into two sub regimes $\bar{\alpha}_1 \leq \bar{\alpha}_{10}$ and $\bar{\alpha}_{10} \leq \bar{\alpha}_1 \leq \bar{\alpha}_{11}$, where $\bar{\alpha}_{10}$ and $\bar{\alpha}_{20}$ are given by

$$\bar{\alpha}_{10} = \frac{\pi}{3} \frac{1-\bar{\alpha}_1}{\bar{\alpha}_1} \bar{D}_1 \left(1-\bar{D}_1\right) \left[6 \frac{\Delta \ell_3}{D_3} - 5\left(1+\bar{D}_1\right)\right] \quad \text{as long as} \quad \bar{\alpha}_{10} \geq 0$$

and

$$\bar{\alpha}_{11} = 0.6\left(1-\bar{D}_1^2\right) \quad \text{as long as} \quad \bar{\alpha}_{10} \geq 0.3,$$

where

$$D_1 = \min\left(\frac{1}{2} 2.7 \lambda_{RT}, \frac{1-\alpha_3}{\alpha_3} D_3\right)$$

is the bubble diameter

$$\bar{D}_1 = D_1/D_3,$$

$$\frac{\Delta \ell_3}{D_3} = \left(\frac{\pi \sqrt{2}}{6 \alpha_3} \right)^{1/3},$$

and $\Delta \ell_3$ is the average distance between two adjacent solid particles if they are assumed to be rhomboid arrays.

Pure **slug flow** occurs in the range of $\bar{\alpha}_{12} \leq \bar{\alpha}_1 \leq \bar{\alpha}_{13}$, where

$$\bar{\alpha}_{12} = \pi/6 \approx 0.52 \text{ and}$$

$$\bar{\alpha}_{13} = \pi\sqrt{2}/6 \approx 0.74.$$

Pure **annular flow** is assumed to occur in the range of $\bar{\alpha}_1 > \bar{\alpha}_{13}$.

1.6 Particles in film boiling

The particles in film boiling are surrounded by a film with dimensionless thickness

$$\delta_{1F}^* = \frac{\delta_{1F}}{D_3}, \qquad (1.82)$$

where $\delta_{1F}^* > 0$. The ratio of the volume of the sphere consisting of one particle and the surrounding film to the volume of the particle itself is

$$(\alpha_3 + \alpha_{1F})/\alpha_3 = (D_3 + 2\delta_{1F})^3 / D_3^3 = (1 + 2\delta_{1F}^*)^3. \qquad (1.83)$$

Therefore the vapor film volume fraction of the mixture is

$$\alpha_{1F} = \alpha_3 \left[(1 + 2\delta_{1F}^*)^3 - 1 \right]. \qquad (1.84)$$

Thus the condition that the films are not touching each other is

$$\alpha_3 + \alpha_{1F} = \alpha_3 \left(1 + 2\delta_{1F}^*\right)^3 < 0.25 \text{ to } 0.52 \qquad (1.85)$$

depending of the diameter of the film-particle system. The application of the criteria already derived for gas-liquid flows in porous structures is also used by using as a controlling void fraction the local gas volume fraction in the space outside the film-particles volumes,

$$\bar{\alpha}_1 = \frac{\alpha_1 - \alpha_{1F}}{\alpha_1 - \alpha_{1F} + \alpha_2}. \qquad (1.86)$$

The condition to have three-phase flow with continuous liquid and particles being in film boiling is then

$$\bar{\alpha}_1 < 0.74. \qquad (1.87)$$

This is a very important result. It simply demonstrates that particles in film boiling can be surrounded by much less continuous liquid mass than required in case of no film boiling.

1.7 Rod bundles

Bubble flow: *Venkateswararao* et al. [22] (1982) investigated two-phase flow in a vertical rod bundle and found that bubbles are seldom observed in the smallest gap between two rods. They migrate to the open area which exists between the rods. From this observation the authors recommend the upper limit of the bubble flow $\alpha_{1,\text{bubble to slug}}$ to be valid only for the part of the cross section inside the inscribed circle between the neighboring rods. The local void fraction in this circle

$$\alpha_{1,RB} = f\alpha_1 \qquad (1.88)$$

is f times grater then the averaged void fraction, where

$$f = \frac{\frac{4}{\pi}\left(\frac{\ell_{FR}}{D_{FR}}\right)^2 - 1}{\left(\sqrt{2}\frac{\ell_{FR}}{D_{FR}} - 1\right)^2} \qquad (1.89)$$

for a quadratic array, and

$$f = \frac{\dfrac{\sqrt{3}}{\pi}\left(\dfrac{\ell_{FR}}{D_{FR}}\right)^2 - \dfrac{1}{2}}{\left(2\dfrac{\sqrt{3}}{3}\dfrac{\ell_{FR}}{D_{FR}} - 1\right)^2} \qquad (1.90)$$

for a triangular array. Here ℓ_{FR} and D_{FR} are the pitch and the rod diameter, respectively. Thus, the comparison of $\alpha_{1,RB}$ with the limit, $\alpha_{1,\text{bubble to slug}}$, dictates whether bubble flow exists or not.

Slug flow: For slug flow in a vertical rod bundle there are no confining walls around the *Taylor* bubbles. The transition criteria in this case are quite different compared to the pipe flow. *Venkateswararao* et al. [22] (1982) observed *Taylor*-like bubbles occupying almost the entire free space in the cell. At the same time the number of cells occupied by *Taylor* bubbles increases and eventually the concentration of the occupied cells is great enough to cause coalescence $\alpha_{1,\text{slug coalescence}}$. For the quadratic rod array the authors recommend

$$\alpha_{1,\text{slug coalescence}} = \frac{\pi}{6\cos\theta}\frac{\ell_{FR}+D_{FR}}{2\ell_{FR}-D_{FR}}, \qquad (1.91)$$

where

$$\theta = \arcsin\frac{\ell_{FR}-D_{FR}}{\ell_{FR}+D_{FR}}. \qquad (1.92)$$

Similar is the expression for the triangular rod array.

Nomenclature

Latin

C_{li} mass concentration of species *i* inside the field *l*, *dimensionless*
C_0 drift flux distribution coefficient, *dimensionless*
D_d particle diameter (bubble, droplet, particle), *m*
$D_{1,solid\ like}$ bubbles with size less then this behave as a solid sphere, *m*
D_h hydraulic diameter, *m/s*
D_{FR} rod diameter, *m*
$Fr = \dfrac{(\rho w)^2 v_h}{g D_h}$, *Froud* number, *dimensionless*
$Fr_{20} = \dfrac{G}{g D_h \rho_2^2}$, *Froud* number, *dimensionless*
$G = \sum\limits_{l=1}^{3} \alpha_l \rho_l w_l$, mass flow rate, *kg/(m²s)*
g gravitational acceleration, *m/s²*
H distance between two parallel plates, *m*
h_l specific enthalpy of the velocity field *l*, *J/kg*
h'' saturated vapor specific enthalpy, *J/kg*
h' saturated liquid specific enthalpy, *J/kg*
$j = \alpha_1 V_1 + (1-\alpha_1) V_2$, mixture volumetric flux, *m/s*
ℓ_{FR} pitch diameter, *m*
ℓ_d^t mean free path length of oscillating particles, *m*
ℓ_{TB} length of the *Taylor* bubble, *m*
Ma local *Mach* number, *dimensionless*
$N_{\eta 2} = \eta_2 / \sqrt{\rho_2 \sigma_2 \lambda_{RT}}$
n_1 number of bubbles per unit mixture volume, *1/m³*
p pressure, *Pa*
$\dot{q}''_w$ heat flux from the wall into the flow, *W/m²*
$Re_{20} = G D_h / \eta_2$, Reynolds number, *dimensionless*
Vol_{cell} cell volume, *m³*
$V_{1Ku} = \sqrt{2} \left[g \sigma_2 (\rho_2 - \rho_1) / \rho_2^2 \right]^{1/4}$, *Kutateladze* bubble rise velocity in a pool, *m/s*
$V_{TB} = \sqrt{\dfrac{\rho_2 - \rho_1}{\rho_2} g D_h}$ slug (*Taylor* bubble) raising velocity, *m/s*

V_1 gas velocity, *m/s*
V_2 liquid velocity, *m/s*
$V_{1,stratified}$ gas velocity dividing the non-stratified from the stratified flow, *m/s*
$V_{2,stratified}$ liquid velocity dividing the non-stratified from the stratified flow, *m/s*
$V_{1,annular}$ gas velocity dividing the non-annular from the annular flow, *m/s*
$V_{1,wavy}$ velocity: for gas velocity larger than this velocity the surface of the liquid is *wavy* (stratified wavy flow), *m/s*
$V_{2,bubble}$ critical liquid velocity for transition into bubble flow, *m/s*
$We_{20} = \dfrac{G^2 D_h}{\rho_2 \sigma_2}$, *Weber* number, *dimensionless*
$X_{1,eq} = \dfrac{\sum X_l h_l - h'}{h'' - h'}$, equilibrium mass flow concentration of the vapor velocity field, *dimensionless*
X_1 mass flow concentration of the vapor velocity field 1, *dimensionless*
z axial coordinate, *m*

Greek

α_l local volume fractions of the fields *l*, *dimensionless*
α_d local volume fractions of the dispersed field *d*, *dimensionless*
$\alpha_{1,B-Ch}$ gas volume fraction that divides bubble and churn turbulent flow, *dimensionless*
$\alpha_{1,bubble\ to\ slug}$ gas volume fraction that divides bubble and slug flow, *dimensionless*
$\alpha_{1,slug\ to\ churn}$ gas volume fraction that divides slug and churn turbulent flow, *dimensionless*
β $= X_1 v_1 / v_h$, homogeneous gas mass fraction, *dimensionless*
γ_v volume porosity - volume occupied by the flow divided by the total cell volume, *dimensionless*
$\Delta \ell_d$ averaged distance between the centers of two adjacent particles, *m*
$\Delta \ell_3$ average distance between two adjacent particles if they are assumed to be rhomboid arrays, *m*
δ_{2F} liquid thickness, *m*
λ_{fr} friction coefficient, *dimensionless*
λ_{RT} *Rayleigh-Taylor* wavelength, *m*
ρ_1 gas density, *kg/m³*
ρ_2 liquid density, *kg/m³*
ρw mixture mass flow rate, *kg/(m²s)*

σ_2 viscous tension, *N/m*

ε_2 dissipation rate of the turbulent kinetic energy of the liquid, W/m^3

η_1 dynamic gas viscosity, *kg/(ms)*

η_2 dynamic liquid viscosity, *kg/(ms)*

v_h $= X_1 v_1 + (1 - X_1) v_2$, homogeneous specific volume, m^3/kg

θ angle with origin the pipe axis defined between the upwards oriented vertical and the liquid-gas-wall triple point, *rad*

φ angle between the upwards oriented vertical and the pipe axis, *rad*

References

1. Anoda Y, Kukita Y, Nakamura N, Tasaka K (1989) Flow regime transition in high-pressure large diameter horizontal two-phase flow, ANS Proc. 1989 National Heat Transfer Conference, 6-9 Aug., Philadelphia, Pennsylvania, ISBN: 0-89448-149-5, ANS Order Number 700143
2. Barnea D, Taitel Y (1994) Interfacial and structural stability of separated flow, Int. J. Multiphase Flow, vol 20 pp 387-414
3. Brodkey RS (1967) The phenomena of fluid motions, Addison-Wesley Press
4. Bestion D (1990) The physical closure lows in the CHATHRE code, Nuclear Engineering and Design, vol 124 pp 229-245
5. Deichel M, Winter ERF (1990) Adiabatic two-phase pipe flow at air-water mixtures under critical flow conditions, Int. J. Multiphase Flow, vol 16 no 3 pp 391-406
6. Delhaye JM (1981) Basic equations for two-phase flow, in Bergles AE. et al., Eds., Two-phase flow and heat transfer in power and process industries, Hemisphere Publishing Corporation, McGraw-Hill Book Company, New York
7. Doroschuk VE, Borevsky LY, Levitan LL (1982) Holographic investigation of steam-water flows in heated and unheated channels, 7th Int. Heat Transfer Conf. Munich TF15 pp 277-281
8. Ginsberg T, Jones OC Jr, Chen JC (Mid-Dec. 1979) Flow behavior of volume-heated boiling pools: implication with respect to transition phase accident conditions, Nuclear Technology, vol 46 pp 391-398
9. Grawford TJ, Weinberger CB, Weisman J (1985) Two-phase flow pattern and void fractions in downword flow. Part I: Steady state flow pattern, Int. J. Multiphase Flows, vol 11 no 2 pp 761-782. (1986) Part II: Void fractions and transient flow pattern, Int. J. Multiphase Flows, vol 12 no 2 pp 219-236
10. Hohannessen T (1972) Beitrag zur Ermittlung einer allgemeingültigen Stroemungsbilder-Karte, SIA Fachgruppe fuer Verfahrenstechnik, Tagung vom 14. Dez. 1972, ETH Zuerich
11. Johnston AJ (1985) Transient from stratified to slug regime in countercurrent flow, Int. J. Multiphase Flow, vol 11 no 1 pp 31-41
12. Kataoka I, Ishii M (July 1982) Mechanism and correlation of droplet entrainment and deposition in annular two-phase flow. NUREG/CR-2885, ANL-82-44
13. Mamaev WA, Odicharia GS, Semeonov NI, Tociging AA (1969) Gidrodinamika gasogidkostnych smesey w trubach, Moskva

14. Milne-Thomson LM (1968) Theoretical hydrodynamics, MacMillan & Co. Ltd., London
15. Mishima K, Ishii M (Dec. 1980) Theoretical prediction of onset of horizontal slug flow, Journal of Fluid Engineering, vol 102 pp 441-445
16. Mishima K, Ishii M (1984) Flow regime transition criteria for upward two-phase flow in vertical tubes, Int. J. Heat Mass Transfer, vol 27 no 5 pp 723-737
17. Rouhani SZ, Sohal MS (1983) Two-phase flow patterns: A review of research results, Progress in Nucl. Energy, vol 11 no 3 pp 219-259
18. Taitel Y, Dukler AE (Jan. 1976) A model for predicting flow regime transitions in horizontal and near horizontal gas-liquid flow, AIChE J., vol 22 no 1 pp 47-55
19. Taitel Y, Bornea D, Dukler AE (1980) Modeling flow pattern transitions for steady upward gas-liquid fow in vertical tubes, AIChE J., vol 26 no 3 p 345
20. Taitel Y (1990) Flow pattern transition in two - phase flow, Proc. 9 th Int. Heat Transfer Conf., Jerusalem, Israel, ed. by G. Hetstroni, Hemisphere, New York etc., vol 1 pp 237-254
21. Tung VX, Dhir VK (1990) Finite element solution of multi-dimensional two-phase flow through porous media with arbitrary heating conditions, Int. J. Multiphase Flow, vol 16 no 6 pp 985-1002
22. Venkateswararao P, Semiat R, Dukler AE (1982) Flow patter transition for gas-liquid flow in a vertical rod bundle, Int. J. Multiphase Flow, vol 8 no 5 pp 509-524
23. Wallis GB, Dobson JE (1973) The onset of slugging in horizontal stratified air-water flow, Int. J. Multiphase Flow, vol 1 pp 173-193
24. Weisman J, Duncan D, Gibson J, Grawford T (1979) Effect of fluid properties and pipe diameter on two - phase flow pattern in horizontal lines, Int. J. Multiphase Flow, vol 5 pp 437-462
25. Weisman J, Kang Y (1981) Flow pattern transition in vertical and upwardly inclined lines, Int. J. Multiphase Flow, vol 7 pp 271-291

2 Drag forces

2.1 Introduction

The pressure distribution around a particle moving in a continuum is non-uniform. Integrating the pressure distribution over the surface one obtains a resulting force that is different from zero. As shown in Vol. I, Chapter 6.2, the different spatial components of the integral correspond to different forces. One of these components is called drag force or form drag force. The averaging procedure over a family of particles gives some averaged force which can be used in the computational analysis based on coarse meshes in the space. The purpose of this section is to summarize the empirical information for computation of the drag forces in multiphase flow analysis.

Consider the discrete velocity field d surrounded by a continuum denoted by c. The force acting on a single particle multiplied by the number of particles per unit volume is

$$f_d^d = -\frac{\alpha_d}{\pi D_d^3/6} c_{cd}^d \frac{1}{2} \rho_{cd} |\Delta V_{cd}| \Delta V_{cd} \frac{\pi D_d^2}{4} = -\alpha_d \rho_{cd} \frac{1}{D_d} \frac{3}{4} c_{cd}^d |\Delta V_{cd}| \Delta V_{cd}. \quad (2.1)$$

Next we describe some experimentally observed effects influencing the *drag coefficient* c_{cd}^d.

It is well known from experimental observations that the drag coefficient depends on the radius, on the particle *Reynolds* number based on the absolute value of the relative velocity, and on whether the particle is solid, bubble, drop of pure liquid or drop of liquid with impurities of microscopic solid particles. Due to the internal circulation in the liquid or gas particle, its drag coefficient can be 1/3 of the corresponding drag coefficient of the solid particle with the same radius and *Reynolds* number. Small impurities hinder the internal circulation and lead to drag coefficients characteristic for solid particles.

Drag coefficients on deformed particles are 2 to 3 times larger than rigid sphere drag coefficients.

During the relative motion each particle deforms the continuum. *Increasing the concentration leads to increased resistance to the deformation in a restricted geometry.* This means that the mechanical cohesion of the family of particles with the surrounding continuum is stronger than the cohesion of the single particle moving with the same relative velocity. In other words, the *drag coefficient* of a *single* particle with a given radius and *Reynolds* number is *less* than the drag coefficient of a particular particle with the same radius and *Reynolds* number, belonging to a *family of particles* collectively moved through the continuum.

The velocity V_{wake} at any location δ behind a solid particle with diameter D_d, caused by its wake, e.g.

$$V_{wake} \approx \Delta V_{cd} / \left[0.2 + 0.24 \delta / D_d + 0.04 (\delta / D_d)^2 \right], \tag{2.2}$$

Stuhmiller et al. [30] (1989), can influence the interaction of the following particle with the continuum. For one and the same concentration and form of the particles, the less the compressibility of the particle, the grater the drag coefficient.

2.2 Drag coefficient for single bubble

The drag coefficients in bubbly flow are governed by three dimensionless numbers: The *Reynolds* number for a single bubble,

$$Re = D_1 \rho_2 |\Delta V_{12}| / \eta_2, \tag{2.3}$$

the *Eötvös* number,

$$Eo = (D_1 / \lambda_{RT})^2, \tag{2.4}$$

where

$$\lambda_{RT} = \left[\sigma / (g \Delta \rho_{21}) \right]^{1/2} \tag{2.5}$$

is the *Rayleigh-Taylor* instability wavelength and the *Morton* number

$$Mo = \frac{g \Delta \rho_{21}}{\sigma} \left(\frac{\eta_2^2}{\rho_2 \sigma} \right)^2. \tag{2.6}$$

Geary and *Rice* [11] observed that the aspect ratio of a bubble defined as the high (minor axis) divided by the breadth (major axis) varies

$$\frac{h}{b} = \begin{cases} 1 & \text{for } Ta < 1 \\ \{0.81 + 0.206 \tanh[2(0.8 - \log_{10} Ta)]\}^3 & \text{for } 1 \le Ta \le 39.8 \\ 0.24 \times 1.6 & \text{for } Ta > 39.8 \end{cases}$$

depending of the following dimensionless number

$$Ta = \frac{|\Delta V_{12}|}{\sqrt{g \lambda_{RT}}} \frac{D_1}{\lambda_{RT}}.$$

The data base for this correlation was within $1 \le Ta \le 50$.

Next we will consider four regimes of mechanical flow-bubble interaction.

1) For low relative velocities

$$Re \le 16, \qquad (2.7)$$

the force on a single particle depends linearly on the velocity difference, *Stokes* [29] (1880). This regime is called the *Stokes* regime. The drag coefficient for this regime is computed as follows

$$c_{21}^d = 24 / Re. \qquad (2.8)$$

2) We have a *viscous* regime (the force depends non-linearly on the velocity difference) if the following condition is satisfied

$$Re > 16. \qquad (2.9)$$

The drag coefficient for this regime is computed as follows

$$c_{21}^d = \frac{24}{Re}\left(1 + 0.1\, Re^{0.75}\right). \qquad (2.10)$$

This is the *Ishii* and *Zuber* [15] (1978) modification of the *Oseen* equation [22] (1910)

$$c_{21}^d = \frac{24}{Re}\left(1 + \frac{3}{16} Re\right) \qquad (2.11)$$

modified later by *Schiler* and *Nauman* [27] (1935) to

$$c^d_{21} = \frac{24}{Re}\left(1 + 0.15\, Re^{0.687}\right) \qquad (2.12)$$

and valid for $Re < 500$. For larger Reynolds number $5 < Re \leq 1000$ *Michaelides* [42] reported a correlation obtained by direct numerical simulation:

$$c^d_{cd} = \frac{2 - \eta_d/\eta_c}{2} c^d_{cd,0} + \frac{4\eta_d/\eta_c}{6 + \eta_d/\eta_c} c^d_{cd,2} \quad \text{for } 0 \leq \eta_d/\eta_c \leq 2,$$

$$c^d_{cd} = \frac{4}{\eta_d/\eta_c + 2} c^d_{cd,2} + \frac{\eta_d/\eta_c - 2}{\eta_d/\eta_c + 2} c^d_{cd,\infty} \quad \text{for } 2 < \eta_d/\eta_c < \infty,$$

where

$$c^d_{cd,0} = \frac{48}{Re}\left(1 + \frac{2.21}{\sqrt{Re}} - \frac{2.14}{Re}\right),$$

$$c^d_{cd,2} = 17\, Re^{-2/3},$$

$$c^d_{cd,\infty} = \frac{24}{Re}\left(1 + \frac{1}{6} Re^{2/3}\right),$$

with subscripts c used for continuum and d for disperse. The correlation is useful also for droplets in gas.

These two regimes are known in the literature as "undisturbed particles" because the distortions of the particles are negligible. The correlations for the undisturbed particles given above are valid for a slightly contaminated system, *Tomiyama* et al. [31, 32]. For pure liquids the drag coefficient is given by *Tomiyama* as 2/3 of that for slightly contaminated liquids. The "distorted particles" regime is characterized by a vortex system developing behind the particle, where the vortex departure creates a large wake region which distorts the particle itself and the following particles.

3) We have a *distorted* bubble regime (the single particle drag coefficient depends only on the particle radius and fluid properties, but not on the velocity or viscosity) if the following condition is satisfied

$$\frac{24}{Re}\left(1 + 0.1\, Re^{0.75}\right) \leq \frac{2}{3} D_1/\lambda_{RT} < \frac{8}{3}. \qquad (2.13)$$

The drag coefficient for this regime is computed as follows

$$c_{21}^d = \frac{2}{3}(D_1/\lambda_{RT}).$$ (2.14)

4) We have the regime of strongly deformed, *cap bubbles* if

$$D_1/\lambda_{RT} \geq 4.$$ (2.15)

Tomiyama et al. [32] proposed an empirical correlation for the drag coefficient and generalized the distorted and the cup bubble regimes as follows:

$$c_{21}^d = \frac{8}{3}\frac{Eo}{Eo+4}.$$ (2.16)

The data base for this correlation was reported to be for *Re* number up to 10^5 and for *Eo* number between 10^{-2} and 10^3, see Fig. 3 in [32].

In a later work *Tomiyama* [33] proposed a general correlation set for bubble drag coefficient:

For pure liquid

$$c_{21}^d = \max\left\{\min\left[\frac{16}{Re}(1+0.15Re^{0.687}),\frac{48}{Re}\right],\frac{8}{3}\frac{Eo}{Eo+4}\right\}.$$ (2.17)

For slightly contaminated liquid

$$c_{21}^d = \max\left\{\min\left[\frac{24}{Re}(1+0.15Re^{0.687}),\frac{72}{Re}\right],\frac{8}{3}\frac{Eo}{Eo+4}\right\}.$$ (2.18)

For contaminated systems

$$c_{21}^d = \max\left[\frac{24}{Re}(1+0.15Re^{0.687}),\frac{8}{3}\frac{Eo}{Eo+4}\right].$$ (2.19)

As noted by *Tomiyama* the system air-tap water may correspond to contaminated or slightly contaminated water. Water carefully distilled two or more times belongs to the pure liquid system. This correlations represents experimental data very well for $10^{-2} < Eo < 10^3$ and $10^{-3} < Re < 10^6$, and $10^{-14} < Mo < 10^7$.

2.3 Family of particles in continuum

In order to calculate the drag coefficient for a family of particles in continuum, *Ishii* and *Chawla* [16] (1979), among others, use the same relationship as for a single particle, changing properly only the *effective continuum viscosity* η_m (similarity assumption)

$$\frac{\eta_m}{\eta_c} = (1 - \frac{\alpha_d}{\alpha_{dm}})^{-2.5\alpha_{dm}\eta^*} \quad \eta^* = \frac{\eta_d + 0.4\eta_c}{\eta_d + \eta_c} \tag{2.20}$$

as a function of the volume concentration of the disperse phase and the *maximum packing* α_{dm} - see Table 2.1.

Table 2.1. Effective viscosity model

	Bubble in liquid	Drop in liquid	Drop in Gas	Solid Part. System
α_{dm}	≈ 1	≈ 1	0.26 to 1	≈ 0.62
η^*	0.4	≈ 0.7	1	1
$\frac{\eta_m}{\eta_c}$	$(1-\alpha_d)^{-1}$	$(1-\alpha_d)^{-1.75}$	$\approx (1-\alpha_d)^{-2.5}$ Brinkman [6] (1951) Roscoe [26], (1952)	$(1-\frac{\alpha_d}{0.62})^{-1.55}$

The experiments with liquid - liquid used for verification of the drag coefficient are summarized below:

- Water dispersed in mercury column with diameter 0.1 *m*. The water was introduced through a perforated plate with diameters of the nozzles 0.1 *m*. The volume fraction of water was varied between 0 and 0.5.

- Isobutanol water, isobutyl kentone water, loluen water column with diameter 0.0476 *m*. The dispersed liquid was introduced into the column with orifices having diameter 0.0016 to 0.0032 m.

- Kerosene water column with diameter 0.015 m and nozzle diameter 0.0015 *m* for introduction of the dispersed phase. The volumetric fraction of the dispersed phase varied between 0.05 and 0.4.

- Kerosene neptane column with diameter 0.032 *m*. The volumetric fraction of the dispersed phase varied between 0.05 and 0.7.

Next we discuss how to compute the drag coefficient for a *gas-liquid* system *in a pool* ($D_{hy} \gg D_1$) according to the recommendation of *Ishii* and *Chawla*, and extend their analysis to three-dimensional flow. In the three-phase case the forces have to be additionally corrected in order to take into account the influence of the third component. If we assume that the particles are completely surrounded by the continuum the correction for the influence of the third component can be easily introduced considering the bubbles and the continuous liquid as one mixture flowing through the fictitious channel volume fraction $\alpha_1 + \alpha_2$. The bubble concentration in this fictitious channel is

$$\alpha_d = \frac{\alpha_1}{\alpha_1 + \alpha_2}. \tag{2.21}$$

The assumption that the bubbles are completely surrounded by the continuum holds only if the size of the solid particles D_3 is considerably smaller than the size of the bubbles D_1. For other cases additional experimental information is necessary to describe completely this phenomenon.

1) We have the *Stokes regime* (the force on a single particle depends linearly on the velocity difference *Stokes* [29] (1880)) if the following condition is satisfied

$$Re < 16, \tag{2.22}$$

and

$$\frac{2}{3} D_1 / \lambda_{RT} < (1-\alpha_d)^{0.6} \frac{24}{Re}, \tag{2.23}$$

where

$$Re = D_1 \rho_2 |\Delta V_{12}| / \eta_m. \tag{2.24}$$

The drag coefficient is computed as follows

$$c_{21}^d = 24 / Re. \tag{2.25}$$

The drag force is therefore

$$\mathbf{f}_{21}^d = -(18\alpha_d \eta_m / D_1^2)(\mathbf{V}_2 - \mathbf{V}_1). \tag{2.26}$$

2) We have a viscous regime (the force depends non-linearly on the velocity difference) if the following condition is satisfied

$$Re > 16, \tag{2.27}$$

and

$$\frac{2}{3} D_1 / \lambda_{RT} < (1-\alpha_d)^{0.6} \frac{24}{Re}(1+0.1\, Re)^{0.75}. \tag{2.28}$$

The drag coefficient is computed as follows

$$c_{21}^d = \frac{24}{Re}(1+0.1\, Re^{0.75}), \tag{2.29}$$

which is as already mentioned the *Ishii* and *Zuber* [15] (1978) modification of *Schiler* and *Nauman* [27] (1935) valid for $Re < 500$. The drag force is therefore

$$\mathbf{f}_{21}^d = -(18\alpha_d \eta_m / D_1^2)(1+0.1\, Re^{0.75})(\mathbf{V}_2 - \mathbf{V}_1). \tag{2.30}$$

These two regimes are designated by *Ishii* and *Chawla* as "undisturbed particles" because the distortions of the particles are negligible. The "distorted particles" regime is characterized by a vortex system developing behind the particle, where the vortex departure creates a large wake region distorting the particle itself and the following particles.

3) We have a *distorted* bubble regime (the single particle drag coefficient depends only on the particle radius and fluid properties, but not on the velocity or viscosity) if the following condition is satisfied

$$(1-\alpha_d)^{0.6}\frac{24}{Re}(1+0.1Re^{0.75}) \leq \frac{2}{3} D_1 / \lambda_{RT} < \frac{8}{3}(1-\alpha_d)^{0.87}. \tag{2.31}$$

The three components of the drag coefficient computed as follows

$$c_{21}^d = \frac{2}{3}(D_1 / \lambda_{RT})\left(\frac{1+17.67 f^{6/7}}{18.67 f}\right)^2 ; \quad f = (1-\alpha_d)^{1.5}. \tag{2.32}$$

4) We have the regime of strongly deformed, *cap bubbles* if

$$\frac{2}{3} D_1 / \lambda_{RT} \geq \frac{8}{3}(1-\alpha_d)^{0.87}. \tag{2.33}$$

The drag coefficient for this regime is

$$c_{21}^d = \frac{8}{3}(1-\alpha_d)^2. \tag{2.34}$$

The drag force is therefore

$$\mathbf{f}_{21}^d = -\left[2\alpha_d(1-\alpha_d)^2 \rho_2 / D_1\right]|\mathbf{V}_2 - \mathbf{V}_1|(\mathbf{V}_2 - \mathbf{V}_1). \qquad (2.35)$$

For a flow in a pool this regime also exists for $\alpha_d > 0.3$.

The above described methods for computation of the bubble drag coefficient were verified as follows [15].

1. Air-water in column of 0.06 *m* diameter and bubble diameter of 0.00276 *m* in the region $\alpha_2 V_2$ = 0.44 to 1.03 *m/s*, $\alpha_1 V_1$ = 0.073 to 0.292 *m/s*, α_1 = 0.05 to 0.3. The error for the velocity measurements was reported of to be ± 5 %.

2. Nitrogen - kerosene-neptane column with diameter 0.032 *m* where α_1 was varied between 0 and 0.8. Air water bubble in column of 0.051 *m* bubble diameter varying between 0.002 to 0.004 *m*. The air volumetric fraction was varied between 0 and 0.35.

3. Air-water bubble flow with bubbles produced by nozzles having diameters of 0.00004 to 0.000078 *m*. The air volumetric fraction was varied between 0 and 0.34. Air-water bubble in column of 0.051 *m* bubble diameter varying between 0.002 to 0.004 *m*. The air volumetric fraction was varied between 0 and 0.18, for stagnant water and $\alpha_2 V_2$ = - 0.0541 *m/s*.

Beside the above mentioned regimes in pool flow, for flow in confined geometry there are additionally three kinds of interaction between gas and continuous liquid namely *churn turbulent, slug* and *film flows*. The identification of these regimes is discussed in Chapter 1.

5) For the *churn turbulent flow* the drag coefficient is calculated as for the previously discussed cap bubble regime.

6) For the computation of the drag coefficient for the axial direction in the *slug flow in a pipe Ishii* and *Chawla* propose

$$c_{21}^d = 9.8(1-\alpha_1)^3, \qquad (2.36)$$

where

$$D_{TB} = D_1 = 0.9 D_{hw}. \qquad (2.37)$$

The drag force is therefore

$$\mathbf{f}_{21}^{d} = -\left[7.35\alpha_d(1-\alpha_d)^3 \rho_2 / D_1\right]|\mathbf{V}_2 - \mathbf{V}_1|(\mathbf{V}_2 - \mathbf{V}_1). \tag{2.38}$$

This correlation was verified by experiments performed with:

1. Air low concentration solution of sodium sulfite Na_2SO_3 in water. The column diameter was 0.07 to 0.6 m and the nozzle inlet diameter was 0.00225 to 0.04 m. The air void fraction was varied between 0.02 and 0.03.

2. Air-water column with diameter 0.06 m, $\alpha_1 V_1 =$ 0.2198 to 0.367 m/s, $\alpha_2 V_2 =$ 0.44 to 1.03 m/s and $\alpha_1 =$ 0.1 to 0.35.

Note that slug flow does not exist in the pool flow.

2.4 Droplets-gas

Next we compute the drag coefficient for a *droplets-gas* system in a pool ($D_h \gg D_3$) according to the recommendation of *Ishii* and *Chawla*, and extending their analysis for three-dimensional flow. The effective viscosity for this case is

$$\frac{\eta_m}{\eta_1} = (1-\alpha_d)^{-2.5}, \tag{2.39}$$

Roscoe and *Brit* [25] (1952), *Brinkman* [7] (1952), where

$$\alpha_d = \frac{\alpha_3}{\alpha_1 + \alpha_3}. \tag{2.40}$$

1) The drag coefficient for the *Stokes* regime

$$Re < 1, \tag{2.41}$$

is computed as follows

$$c_{13}^d = \frac{24}{Re}; \ Re = D_3 \rho_1 \Delta V_{13} / \eta_m. \tag{2.42}$$

The drag force is therefore

$$\mathbf{f}_{13}^d = -(18\alpha_d \eta_m / D_3^2)(\mathbf{V}_1 - \mathbf{V}_3). \tag{2.43}$$

The drag coefficient for the viscous regime

$$1 \le Re < 1000 \tag{2.44}$$

is computed as follows

$$c_{13}^d = \frac{24}{Re}(1+0.1Re^{0.75}). \tag{2.45}$$

Therefore the drag force is

$$\mathbf{f}_{13}^d = -(18\alpha_d \eta_m / D_3^2)(1+0.1Re^{0.75})(\mathbf{V}_1 - \mathbf{V}_3). \tag{2.46}$$

3) The drag coefficient for *Newton*'s regime (for single particle - *Newton*)

$$Re \ge 1000 \tag{2.47}$$

is computed as follows

$$c_{13}^d = \frac{2}{3}(D_3/\lambda_{RT})\left(\frac{1+17.67 f^{6/7}}{18.67 f}\right)^2; \quad f = (1-\alpha_d)^3 \tag{2.48}$$

The drag force is therefore

$$\mathbf{f}_{21}^d = -\frac{1}{2}\alpha_d \rho_1 \frac{1}{\lambda_{RT}}\left(\frac{1+17.67 f^{6/7}}{18.67 f}\right)^2 |\mathbf{V}_1 - \mathbf{V}_3|(\mathbf{V}_1 - \mathbf{V}_3). \tag{2.49}$$

Again if solid particles participate in the flow we consider the gas-droplet flow as flowing in a fictitious channel with volume fraction of the all control volume and the volume fraction $\alpha_1 + \alpha_3$ of the droplets in this channel is

$$\alpha_d = \frac{\alpha_3}{\alpha_1 + \alpha_3}. \tag{2.50}$$

This assumption holds if the size of the solid particles, D_3, is much smaller then the size of the droplets, D_2. Otherwise one needs additional experimental information to perform the correction. As already mentioned, small impurities $C_{n3} > 0$, hinder the internal circulation of the droplets and lead to drag coefficients characteristic for the *solid particles in a gas*.

2.5. Solid particles-gas in presence of liquid. Solid particles-liquid in presence of a gas

Depending on the volumetric concentration of the macroscopic solid particles (the third velocity field) we distinguish the following cases:

1) The solid particles are touching each other in the control volume

$$\alpha_3 = \alpha_{dm}. \tag{2.51}$$

2) The solid particles are free in the flow

$$\alpha_3 < \alpha_{dm}. \tag{2.52}$$

In the second case we distinguish two sub cases:

2a) The volume fraction of the space among the particles if they were closely packed,

$$\alpha_2^* = \frac{1-\alpha_{dm}}{\alpha_{dm}} \alpha_3, \tag{2.53}$$

is smaller then the liquid volume fraction

$$\alpha_2^* < \alpha_2 \quad \left(\alpha_3 \leq \frac{\alpha_{dm}}{1-\alpha_{dm}} \alpha_2 \right). \tag{2.54}$$

2b) The volume fraction of the space among the particles if they were closely packed is larger then the liquid volume fraction

$$\alpha_3 > \alpha_2. \tag{2.55}$$

2.5.1 Solid particles: free particles regime

There is good experimental support for the two-phase regime: solid particle-gas, (2.31), or solid particles-liquid (2.32). We use further the notation d for discrete and c for continuous where $d = 3$. c can take values 1 or 2.

Following *Ishii* and *Chawla* in case of

$$\alpha_c + \alpha_d = 1 \tag{2.56}$$

2.5. Solid particles-gas in presence of liquid. Solid particles-liquid in presence of a gas

we have:

1) The *Stokes* regime is defined by

$$Re \leq 1. \tag{2.57}$$

The drag coefficient for this regime is

$$c_{cd}^d = 24/Re, \tag{2.58}$$

where

$$Re = D_d \rho_c |\Delta V_{cd}| / \eta_m, \tag{2.59}$$

$$\eta_m = \eta_c (1 - \frac{\alpha_d}{\alpha_{dm}})^{-1.55}, \tag{2.60}$$

$$\alpha_{dm} = 0.62. \tag{2.61}$$

The drag force is therefore

$$\mathbf{f}_{cd}^d = -(18\alpha_d \eta_m / D_d^2)(\mathbf{V}_c - \mathbf{V}_d). \tag{2.62}$$

Ishii and *Zuber* [15] (1978) compared the prediction of this equation with experimental data for glass particles with diameter $D_3 = 0.036$ *cm* in a tube with $D_h = 10$ *cm* filled with a glycerin - water column. In the region of $Re_\infty \approx 0.001$ to 0.06 and $\alpha_3 = 0$ to 0.45 the authors reported excellent agreement.

b) Viscous regime is defined by

$$1 \leq Re < 1000. \tag{2.63}$$

The drag coefficient for this regime is

$$c_{cd}^d = \frac{24}{Re}(1 + 0.1 Re^{0.75}). \tag{2.64}$$

Therefore the drag force is

$$\mathbf{f}_{cd}^d = -(18\alpha_d \eta_m / D_d^2)(1 + 0.1\ Re^{0.75})(\mathbf{V}_c - \mathbf{V}_d). \tag{2.65}$$

Note that for symmetric flow around a spheroid form with aspect ratio defined as

$$A = \text{semi-axes along the flow/semi-axes normal to the flow}, \qquad (2.66)$$

Militzer et al. [20] obtained

$$c_{cd}^d = \frac{4+A}{5}\left[\frac{24}{Re}(1+0.15Re^{0.687}) + \frac{0.42}{1+42500Re^{-1.16}}\right]$$

$$(1 + 0.00094 Re/A - 0.000754\,A\,Re + 0.0924/Re + 0.00276\,A^2). \qquad (2.67)$$

c) *Newton*'s regime is defined by

$$Re \geq 1000. \qquad (2.68)$$

The drag coefficient for this regime is

$$c_{cd}^d = 0.45\left(\frac{1+17.67 f^{6/7}}{18.67 f}\right)^2 ; \quad f = \sqrt{1-\alpha_d}\,\frac{\eta_c}{\eta_m}. \qquad (2.69)$$

The drag force is therefore

$$\mathbf{f}_{cd}^d = (0.3375\alpha_d \rho_c / D_d)\left(\frac{1+17.67 f^{6/7}}{18.67 f}\right)^2 |\mathbf{V}_c - \mathbf{V}_d|(\mathbf{V}_c - \mathbf{V}_d). \qquad (2.70)$$

Comparison with *Richardson* and *Zaki*'s data for $D_h = 0.062$ m, $D_s = 0.635$ cm for $\alpha_3 = 0$ to 0.4 shows a good agreement.

2.5.2 Solid particles in bubbly flow

If a heated solid sphere is falling in a liquid the boiling around the sphere changes the pressure field and consequently the drag coefficient. The investigations for estimation of the drag coefficients under boiling conditions for arbitrary velocity differences are not complete and future investigations are necessary. *Zvirin* et al. [39] (1990) published the following empirical method for computation of the drag coefficients in the limited region of *Reynolds* numbers and superheating:

$$c_{23}^d = c_{23,nb}^d \Big/ \left\{1.020 + 3.87\times 10^{-5}\left[T_3 - T'(p)\right]\right\}^2, \qquad (2.71)$$

where

2.5. Solid particles-gas in presence of liquid. Solid particles-liquid in presence of a gas

$$\log c_{23,nb}^d = -4.3390 + 1.589 \log Re_{32} - 0.154(\log Re_{32})^2, \quad (2.72)$$

is the drag coefficient measured for non-boiling conditions and

$$Re_{32} = D_3 \rho_2 |\Delta V_{32}| / \eta_2 \quad (2.73)$$

is the *Reynolds* number. The *Zvirin* et al. observations are valid in the following region $10^4 < Re_{32} < 3.38 \times 10^5$, $373 < T_3 < 973$ K.

For illustration if the temperature of the solid sphere is 973 K the drag coefficient is 8.8% lower than the value for the non-boiling case under atmospheric conditions and free falling. More complicated is the case if the free particles are part of a solid/liquid/gas mixture.

Consider first the bubble three-phase flow. As a first approximation we can assume that if the bubbles in the space among the particles are touching each other

$$\frac{\alpha_1}{\alpha_1 + \alpha_2} > 0.52 \quad (2.74)$$

the bubble three - phase flow cannot exist and vice versa if

$$\frac{\alpha_1}{\alpha_1 + \alpha_2} \le 0.52 \quad (2.75)$$

the bubble three-phase flow is defined. For the time being no experimental information is available to confirm the value 0.52. In any case if three-phase bubble flow is identified we distinguish two sub cases. If the volume fraction of the space among the particles if they were closely packed is smaller than the liquid fraction

$$\alpha_2^* < \alpha_2 \quad (2.76)$$

the theoretical possibility exists that the particles are carried only by a liquid. This hypotheses is supported if one considers the ratio of the free setting velocity in gas and liquid

$$\frac{w_{31\infty}}{w_{32\infty}} = \sqrt{\frac{\rho_3 - \rho_1}{\rho_3 - \rho_2} \frac{\rho_2}{\rho_1}} \gg 1. \quad (2.77)$$

We see that, due to the considerable differences between gas and liquid densities, the particles sink much faster in gas than in a liquid. Thus, most probably the solid particles are carried by the liquid. In this case the mixture can be considered as consisting of gas and liquid/solid continuum. The drag force between gas and solid is zero and the drag force between solid and liquid is computed for

$$\alpha_d = \alpha_3 / (\alpha_2 + \alpha_3) \tag{2.78}$$

and

$$\Delta \mathbf{V}_{dl} = \mathbf{V}_3 - \mathbf{V}_2 \tag{2.79}$$

If the volume fraction of the space among the closely packed particles is larger then the liquid volume fraction

$$\alpha_2^* > \alpha_2 \tag{2.80}$$

only

$$\alpha_3 - \frac{\alpha_{dm}}{1 - \alpha_{dm}} \alpha_2 = \alpha_3 (1 - \alpha_2 / \alpha_2^*) = \alpha_{31} \tag{2.81}$$

are surrounded by gas. So we can compute the drag and the virtual mass force between one single solid particle and gas as for a mixture

$$\alpha_d = \frac{\alpha_{31}}{\alpha_1 + \alpha_{31}} \tag{2.82}$$

namely

$$c_{13o}^d = c_{cd}^d (\alpha_d, \Delta \mathbf{V}_{dc}, ...) \tag{2.83}$$

$$c_{13o}^{vm} = c_{cd}^{vm} (\alpha_d, \Delta \mathbf{V}_{dc}, ...) \tag{2.84}$$

and multiply this force by the number of particles which are surrounded by gas having volumetric fraction α_{31}. The result is

$$c_{13}^d = \alpha_{31} c_{13o}^d, \tag{2.85}$$

$$c_{13}^{vm} = \alpha_{31} c_{13o}^{vm}. \tag{2.86}$$

The same has to be done for the calculation of the force between one single solid particle and liquid for the mixture $\alpha_2 + \alpha_{32}$,

2.5. Solid particles-gas in presence of liquid. Solid particles-liquid in presence of a gas

$$\alpha_d = \frac{\alpha_{32}}{\alpha_2 + \alpha_{32}}, \tag{2.87}$$

$$\Delta V_{dc} = V_3 - V_2, \tag{2.88}$$

namely

$$c_{23o}^d = c_{cd}^d(\alpha_d, \Delta V_{dc}, ...), \tag{2.89}$$

$$c_{23o}^{vm} = c_{cd}^{vm}(\alpha_d, \Delta V_{dc}, ...), \tag{2.90}$$

where

$$\alpha_{32} = \alpha_3 - \alpha_{31} = \alpha_3 \alpha_2 / \alpha_2^* \tag{2.91}$$

and multiply this force by the number of the particles which are surrounded by liquid α_{32}

$$c_{23}^d = \alpha_{32} c_{23o}^d, \tag{2.92}$$

$$c_{23}^{vm} = \alpha_{32} c_{23o}^{vm}. \tag{2.93}$$

In case that the bubbles in the space between the particles are touching each other

$$\frac{\alpha_1}{\alpha_1 + \alpha_2} > 0.52 \tag{2.94}$$

the more likely flow pattern is *three-phase disperse flow*. In this case the gas - liquid flow relative to the solid particles resembles two-phase gas liquid flow in a channel. Therefore the drag forces exerted by the solid particles should be larger than the drag forces exerted by the solid phase in the case of missing liquid. A possible correction of the drag force coefficients is

$$c_{13}^d = (1-\phi)c_{13o}^d, \tag{2.95}$$

$$c_{13}^{vm} = (1-\phi)c_{13o}^{vm}, \tag{2.96}$$

$$c_{23}^d = \phi\, c_{23o}^d, \tag{2.97}$$

$$c_{23}^{vm} = \phi \, c_{23o}^{vm}, \qquad (2.98)$$

where

$$\phi = \alpha_2 / (\alpha_1 + \alpha_2). \qquad (2.99)$$

2.5.3 Solid particles: dense packed regime

If over a period of time the volume fraction of the solid particles is

$$\alpha_{3a} = \alpha_{dm}, \qquad (2.100)$$

we have the constraint in the mass conservation equation

$$\frac{\partial}{\partial \tau}(\alpha_3 \rho_3) \leq 0. \qquad (2.101)$$

This means that the mass conservation equation for the solid particles has the form

$$\nabla \cdot (\alpha_3 \rho_3 \mathbf{V}_3 \gamma) = \mu_3 - \alpha_3 \rho_3 \frac{\partial \gamma_v}{\partial \tau} \qquad (2.102)$$

for

$$\tau = \tau_a + \Delta\tau, \; \alpha_{3a}(\tau_a) = \alpha_{dm} \; \text{if} \; \frac{\partial}{\partial \tau}(\alpha_3 \rho_3) \geq 0 \qquad (2.103)$$

and

$$\frac{\partial}{\partial \tau}(\alpha_3 \rho_3 \gamma_v) + \nabla \cdot (\alpha_3 \rho_3 \mathbf{V}_3 \gamma) = \mu_3 \qquad (2.104)$$

for

$$\tau = \tau_a + \Delta\tau, \; \alpha_{3a}(\tau_a) = \alpha_{dm} \; \text{if} \; \frac{\partial}{\partial \tau}(\alpha_3 \rho_3) < 0. \qquad (2.105)$$

In other words if the maximum packed density in a particular computational cell is reached no further mass accumulation in the cell is possible. In this case the forces between continuum and particles can be computed using the results of *Ergun* [10] (1952). *Ergun* generalized experimental results for pressure drop through packed beds ($\mathbf{V}_3 = 0$) for

2.5. Solid particles-gas in presence of liquid. Solid particles-liquid in presence of a gas

$$1 < \frac{1}{\alpha_3} \frac{\rho_1 D_3 |\Delta \mathbf{V}_{13}|}{\eta_1} \leq 4 \times 10^4 \qquad (2.106)$$

and correlated them with the following equation

$$\frac{\partial p}{\partial z} = -150 \left(\frac{\alpha_3}{1-\alpha_3} \right)^2 \frac{\eta_l}{D_3^2} V_l - 1.75 \frac{\alpha_3}{1-\alpha_3} \frac{\rho_l |\mathbf{V}_l|}{D_3} V_l \qquad (2.107)$$

where $l = 1$ or 2, $V_3 = 0$,

$$D_3 = \frac{Vol}{F_3} \frac{6}{\alpha_3}, \qquad (2.108)$$

F_3 is the total geometric surface of the solid particles in the volume Vol of the mixture consisting of phase 1 and phase 3.

Achenbach [1] reported a slightly modified *Ergun*'s equation

$$\frac{\partial p}{\partial z} = -160 \left(\frac{\alpha_3}{1-\alpha_3} \right)^2 \frac{\eta_l}{D_3^2} (V_l - V_3)$$

$$-0.6 \frac{\alpha_3}{1-\alpha_3} \frac{\rho_l |\Delta \mathbf{V}_{13}|}{D_3} (V_l - V_3) \left(\frac{1-\alpha_3}{\alpha_3} \frac{\eta_l}{\rho_l D_3 |\Delta \mathbf{V}_{13}|} \right)^{0.1} \qquad (2.109)$$

verified for $\frac{\alpha_3}{1-\alpha_3} \frac{\rho_l D_3 \Delta V_{13}}{\eta_l} \leq 10^5$, see in *Achenbach* or in [36].

The sum of the linear and the quadratic dependences of the forces on the velocity difference was first proposed by *Osborne Reynolds* (1900). The first term represents the viscous friction corresponding to *Poiseuille* flow (*Darcy*'s law) and the second - the turbulent dissipation.

Rewriting *Ergun*'s equation in terms of forces acting on phase 1 per unit mixture volume for $V_3 \neq 0$ we obtain

$$\alpha_1 \frac{\partial p}{\partial z} = -\alpha_1 150 \left(\frac{\alpha_3}{1-\alpha_3} \right)^2 \frac{\eta_l}{D_3^2} (V_l - V_3) - \alpha_1 1.75 \frac{\alpha_3}{1-\alpha_3} \frac{\rho_l |\Delta \mathbf{V}_{13}|}{D_3} (V_l - V_3) = f_{13}^d.$$

$$(2.110)$$

Comparing the above equation with the definition equation of the drag coefficient we obtain

$$c_{13}^d = \frac{200}{\dfrac{\alpha_1}{\alpha_3} \rho_1 \dfrac{D_3 |\Delta \mathbf{V}_{13}|}{\eta_1}} + \frac{7}{3}. \qquad (2.111)$$

Note that this expression was derived for pure solid - continuum flow, which means

$$\alpha_3 + \alpha_1 = 1. \qquad (2.112)$$

Expressions for drag coefficient in dense packed particles are obtained also by using the lattice Boltzmann method by *Koch* and *Hill* – see in [42] p.233,

$$c_{13}^d = c_{13,0}^d + \left[0.0673 + 0.212\alpha_3 + 0.0232(1-\alpha_3)^{-5} \right] \frac{\rho_1 D_3 |\Delta \mathbf{V}_{13}|}{\eta_1},$$

where

$$c_{13,0}^d = \frac{1 + 3(\alpha_3/2)^{1/2} + 2.11\alpha_3 \ln \alpha_3 + 16.14\alpha_3}{1 + 0.681\alpha_3 - 8.48\alpha_3^2 + 8.16\alpha_3^3}.$$

This correlation is reported to reproduce also the *Ergun*'s data. Note the contradiction for the influence of the Reynolds number comparing it with Eq. 2.111.

If the gas and the liquid are flowing around the densely packed particles the resistance forces between each of the phases l and the particles are considerably higher (see [38] (1936)). Part of the liquid, α_2^*, is bound between the particles by capillary force

$$\frac{\alpha_2^*}{1-\alpha_3} = \left[\frac{\alpha_3^2}{(1-\alpha_3)^3} \frac{\sigma_2 \cos \phi_{23}}{D_3^2 \rho_2 g} \right]^{0.264} / 22 \equiv \alpha_{2r}^*, \qquad (2.113)$$

Brown et al. [9] (1950), and remains for given pressure drop across the distance of consideration. It can normally be removed by evaporation rather then drainage. Here σ_2 is the surface tension and is θ_{23} the wetting contact angle between the liquid and particles (typically 0.8 for water/steel and 1 for UO$_2$/water or sodium. Usually the drag forces are corrected by means of the so called viscous and turbulent permeability coefficients, k_1^l, k_1^t respectively. *Lepinski* [18] (1984) uses the following form of the modified *Ergun* equation

2.5. Solid particles-gas in presence of liquid. Solid particles-liquid in presence of a gas

$$\frac{\partial p_1}{\partial z} + \rho_1 g + \frac{\alpha_1 \alpha_3^2}{(1-\alpha_3)^3} 150 \frac{\eta_1}{D_3^2 k_1^1}(w_1 - w_3)$$

$$+ \frac{\alpha_1^2 \alpha_3}{(1-\alpha_3)^3} 1.75 \frac{\rho_1}{D_3 k_1^t}|w_1 - w_3|(w_1 - w_3) = 0 \,. \tag{2.114}$$

Again rewriting this equation in term of forces acting on phase l per unit mixture volume we obtain

$$\alpha_1 \frac{\partial p_1}{\partial z} + \alpha_1 \rho_1 g + \frac{\alpha_1^2 \alpha_3^2}{(1-\alpha_3)^3} 150 \frac{\eta_1}{D_3^2 k_1^1}(w_1 - w_3)$$

$$+ \frac{\alpha_1^3 \alpha_3}{(1-\alpha_3)^3} 1.75 \frac{\rho_1}{D_3 k_1^t}|w_1 - w_3|(w_1 - w_3) = 0 \,. \tag{2.115}$$

Comparing with the definition equation for the drag force we find an expression defining the drag coefficients

$$c_{13}^d = \left(\frac{\alpha_1}{1-\alpha_3}\right)^3 \frac{200}{\frac{\alpha_1}{\alpha_3}\rho_1 \frac{D_3|\Delta V_{13}|}{\eta_1} k_1^1} + \frac{7}{3k_1^t} \,. \tag{2.116}$$

For $\alpha_1 = 1 - \alpha_3$ and k_1^1, $k_1^t \equiv 1$ this is exactly the expression resulting from the original *Ergun* equation.

The relative permeabilities are function of the effective saturation, α_{2eff}^*

$$k_1^1 = \left(1-\alpha_{2eff}^*\right)^3, \quad k_2^1 = \alpha_{2eff}^{*3}, \tag{2.117}$$

Brooks and *Corey* [8] (1966),

$$k_1^t = \left(1-\alpha_{2eff}^*\right)^5, \, k_2^t = \alpha_{2eff}^{*5}, \tag{2.118}$$

Reed [24] (1982) where

$$\alpha_{2eff}^{*} = \frac{\alpha_{2}^{*} - \alpha_{2r}^{*}}{1 - \alpha_{2r}^{*}} \quad \text{for } \alpha_{2}^{*} > \alpha_{2r}^{*}. \tag{2.119}$$

The true saturation is defined as

$$\alpha_{2}^{*} = \frac{\alpha_{2}}{1 - \alpha_{3}}. \tag{2.120}$$

In this case the virtual mass coefficient for the particles is

$$c_{23}^{vm} \approx \alpha_{2r}^{*}/\alpha_{3}. \tag{2.121}$$

Tung and *Dhir* [34] (1990) described the steady-state incompressible two-phase flow through porous media ("saturated" - all of the liquid is moving) modifying the *Ergun* equation

$$\alpha_{1} \frac{\partial p_{1}}{\partial z} + \alpha_{1} \rho_{1} g + \frac{\alpha_{1} \alpha_{3}^{2}}{(1 - \alpha_{3})^{3}} 150 \frac{\eta_{1}}{D_{3}^{2} \overline{k}_{1}^{1}} (w_{1} - w_{3})$$

$$+ \frac{\alpha_{1}^{2} \alpha_{3}}{(1 - \alpha_{3})^{4}} 1.75 \frac{\rho_{1}}{D_{3} \overline{k}_{1}^{t}} |w_{1} - w_{3}|(w_{1} - w_{3}) \pm f_{21}^{d} = 0, \tag{2.122}$$

where $l = 1, 2$. Comparing with the definition equation for the drag force we find expression defining the drag coefficients

$$c_{13}^{d} = \left(\frac{\alpha_{1}}{1 - \alpha_{3}}\right)^{3} \left[\frac{200}{\frac{\alpha_{1}}{\alpha_{3}} \rho_{1} \frac{D_{3} |\Delta \mathbf{V}_{13}|}{\eta_{1}}} \frac{1}{\alpha_{1} \overline{k}_{1}^{-1}} + \frac{7}{3} \frac{1}{(1 - \alpha_{3}) \alpha_{1} \overline{k}_{1}^{t}} \right]. \tag{2.123}$$

Tung and *Dhir* distinguish bubbly, annular and slug flow structure of the gas-liquid mixture. The different relative permeability multipliers are summarized in Table 2.2. In case of bubble flow the bubble diameter and flow regime limits are computed as shown in Chapter 1.

Table 2.2. Relative permeability multipliers as recommended by *Tung* and *Dhir* [34] (1990).

For all flow regimes, the liquid relative permeability multipliers *are given by*
$\overline{k}_2^l = \overline{k}_2^t = (1-\overline{\alpha}_1)^3$.
Gas relative permeability multipliers
a) Bubble flow, slug flow ($\overline{\alpha}_1 \leq \overline{\alpha}_{13}$): $\overline{k}_1^l = \left(\dfrac{\alpha_3}{1-\alpha_1}\right)^{4/3} \overline{\alpha}_1^3 \text{ and } \overline{k}_1^t = \left(\dfrac{\alpha_3}{1-\alpha_1}\right)^{2/3} \overline{\alpha}_1^3.$
b) Annular flow ($\overline{\alpha}_{14} \leq \overline{\alpha}_1 \leq 1$): $\overline{k}_1^l = \left(\dfrac{\alpha_3}{1-\alpha_1}\right)^{4/3} \overline{\alpha}_1^2 \text{ and } \overline{k}_1^t = \left(\dfrac{\alpha_3}{1-\alpha_1}\right)^{2/3} \overline{\alpha}_1^2.$

2.6. Annular flow

Film-wall force: The resisting force between wall and film per unit flow volume for *fully developed annular flow* is

$$f_{w2} = \frac{4}{D_h} \tau_{w2}, \qquad (2.124)$$

where

$$\tau_{w2} = \eta_2 \left[\frac{dw_2(r)}{dr}\right]_{r=R} = c_{w2} \frac{1}{2} \rho_2 w_2^2. \qquad (2.125)$$

For laminar flow,

$$Re_2 = \rho_2 w_2 \delta_2 / \eta_2 \leq 400. \qquad (2.126)$$

the assumption for a parabolic velocity profile leads to

$$c_{w2} = 4/Re_2, \qquad (2.127)$$

For turbulent flow,

$$Re_2 > 400, \tag{2.128}$$

the assumption for a 1/7 profile leads to

$$c_{w2} = 0.057 / Re_2^{1/4}. \tag{2.129}$$

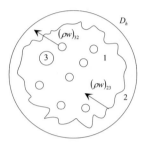

Fig. 2.1. Annular flow

Film-gas force: The gas resisting force per unit flow volume between film and gas is

$$f_1^d = -a_{12}\tau_{21} = -a_{12}c_{21}^d \frac{1}{2}\rho_1|w_2 - w_1|(w_2 - w_1)$$

$$= -\frac{2}{D_h}\sqrt{1-\alpha_2}\, c_{21}^d \rho_1|w_2 - w_1|(w_{2i} - w_1), \tag{2.130}$$

where a_{12} is the interfacial area density

$$a_{12} = \frac{4}{D_h}\sqrt{1-\alpha_2}. \tag{2.131}$$

Nikuradze (1933) proposed the following formula for computation of the friction coefficient in pipes

$$c_w = 0.005\left(1 + 300\frac{k_s}{D_h}\right), \tag{2.132}$$

over the range, $0.0001 \leq k_s/D_h \leq 0.03$ where k_s is the sand roughness grain size. *Wallis* [37] (1969) proposed to use as a roughness the film thickness which results the following equation

2.6. Annular flow

$$c_{21}^d = 0.005\left(1 + 300\frac{\delta_2}{D_h}\right) \approx 0.005(1 + 75\alpha_2),\tag{2.133}$$

where the ratio is approximately

$$\frac{\delta_{1F}}{D_h} \approx \frac{\alpha_2}{4}.\tag{2.134}$$

Hewitt and *Govan* used instead of the coefficient 0.005 the modified *Blasius* formula

$$c_{21}^d = \frac{0.079}{Re_{13}^{1/4}}(1 + 75\alpha_2)\tag{2.135}$$

where

$$Re_{13} = (\alpha_1\rho_1 w_1 + \alpha_3\rho_3 w_3)D_h/\eta_1.\tag{2.136}$$

Bharathan et al. proposed the following equation [4] (1978) which was *effectively for churn flow*

$$c_{21}^d = 0.005 + 14.14\,\alpha_2^{2.03}.\tag{2.137}$$

An alternative empirical correlation of the interfacial friction factor was proposed by *Bharathan* et al. [5] (1979)

$$c_{21}^d = 0.005 + A\delta^{*B},\tag{2.138}$$

where the constants A and B are given by

$$\log_{10} A = -0.56 + 9.07/D^*,\tag{2.139}$$

$$B = 1.63 + 4.74/D^*.\tag{2.140}$$

The dimensionless film thickness δ^* and diameter D^* are defined by

$$\delta^* = \delta_2/\lambda_{RT},\tag{2.141}$$

$$D^* = D_h/\lambda_{RT}.\tag{2.142}$$

Govan et al. [12] (1991) provided experimental data for effectively churn and annular flow for $\alpha_1 > 0.8$ and show that the *Bharathan* correlation gives much better agreement with the data whereas the *Wallis* correlation under predicts the data for $\alpha_1 < 0.92$. The values of the measured quantities $f^d/\Delta w^2$ vary between 40 kg/m^4 for $\alpha_1 \approx 0.8$ and ≈ 2.5 kg/m^4 for $\alpha_1 \approx 0.95$ for $\alpha_2 \rho_2 w_2 = 31.8$ to 47.7 kg/(m^{2}s).

Stephan and *Mayinger* [28] (1990) improved the *Barathan* et al. correlation based on high pressure experiments ($p = 6.7 \; 10^5$ to $13. \; 10^5$ Pa, $D_h = 0.0309$ m, $j_2 = 0.017$ to 0.0035 m/s, $j_1 = 5$ to 18 m/s) as follows

$$c_{21}^d = \frac{0.079}{Re_1^{1/4}}\left(1+115 \; \delta^{*B}\right), \tag{2.143}$$

where

$$Re_1 = \rho_1 w_1 D_h / \eta_1 , \tag{2.144}$$

$$B = 3.91/(1.8+3/D^*) . \tag{2.145}$$

An alternative correlation is proposed recently by *Ambrosini* et al. [2] (1991)

$$c_{21}^d = \frac{0.046}{Re_1^{1/5}}\left[1+13.8(\rho_1 V_1^2 D_h / \sigma)^{0.2} \; Re_1^{-0.6}(\delta_{2F} V_1^* / v_1 - 200\sqrt{\rho_1/\rho_2})\right],$$

$$\tag{2.146}$$

$$V_1^* = \sqrt{\tau_{12}/\rho_1} = V_1\sqrt{\frac{1}{2}c_{21}^d} , \tag{2.147}$$

$$\tau_{12} = \frac{1}{2}c_{21}^d \rho_1 V_1^2 , \tag{2.148}$$

which describes data in the region $c_{21}^d /(0.046/Re_1^{1/5}) \approx 1$ to 10 within $\pm 30\%$ error band.

Nigmatulin [22] (1982) obtained good agreement between the above expressions for c_{w2} and experimental data for $D_h = 0.0081$ to 0.0318 m and $Re_2 = 5$ to 1.5×10^5 for upstream air water and steam water flow. The experiments of *Nigmatulin* shows that between the film surface velocity and averaged film velocity the following relationship exists

$$w_{2i} = 2w_2 \text{ for } Re_2, \tag{2.149}$$

$$w_{2i} = 1.1w_2 \text{ for } Re_2. \tag{2.150}$$

In the same work the gas resisting force between film and gas is estimated to be

$$f_1^d = -a_{12}\tau_{21} = -a_{12}c_{21}^d \frac{1}{2}\rho_1 |w_{2i} - w_1|(w_{2i} - w_1)$$

$$= -\frac{2}{D_h}\sqrt{1-\alpha_2}c_{21}^d \rho_1 |w_{2i} - w_1|(w_{2i} - w_1), \tag{2.151}$$

and

$$c_{21}^d = 0.005\left[1 + 16\left(2\delta_2/D_h\right)^{1/2} + 1.5 \times 10^3 \left(2\delta_2/D_h\right)^2\right]. \tag{2.152}$$

Here δ_2 is the film thickness

$$\delta_2 = D_h(1 - \sqrt{1-\alpha_2})/2. \tag{2.153}$$

The f_{21}^d is computed in this way with an square mean error of 20%. The verification was performed by *Nigmatulin* with data for air-water and $p = 2.7 \times 10^5$ *Pa*, $D_h = 0.0318$ *m* and for steam - water and $p = (10 \text{ to } 100) \times 10^5$ *Pa* and $D_h = 0.0081$ and 0.0131 *m*.

Ueda [35] (1981) proposed the following correlation for the computation of the drag coefficient in annular flow

$$c_{21}^d = const \frac{1}{4}(2.85 - 2.10\alpha_1^{2.20})^4 \left[\frac{\rho_2 - \rho_1}{\rho_1} \frac{gD_{h21}}{(w_1 - w_2)^2}\right]^{0.7}$$

$$\times \left[\frac{\eta_1}{\eta_2} \frac{|w_1 - w_2|}{w_2}\right]^{0.1} \left[\frac{|w_1 - w_2|D_{h21}}{v_1}\right]^{-0.2} \tag{2.154}$$

where the

$$const \approx 2 \tag{2.155}$$

for horizontal flow and

$$D_{h21} = D_h \sqrt{1-\alpha_2}\,. \tag{2.156}$$

Some authors successfully correlated experimental data for interfacial friction coefficient using instead the relative velocity between film and gas, only the film velocity

$$f_{21}^d = -a_{12}\rho_2 c_{21}^{d*} \frac{1}{2} w_2 |w_2|\,. \tag{2.157}$$

Hugmark [13] (1973) correlated experimental data for film thickness of vertical upward flow as a function of film *Reynolds* number and summarized data for the relationships between drag coefficients and film thickness. Thus using this data base the relationship between liquid *Reynolds* number and drag coefficients is established. Later *Ishii* and *Gromles* correlated this relationship with the following correlation

$$\sqrt{c_{u1}^{d*}/4} = k\, Re_{2F}^m \tag{2.158}$$

where

$$Re_{2F} = \rho_2 w_2 4\delta_{2F}/\eta_2 \tag{2.159}$$

and k and m are given by

$$k = 3.73, \quad m = -0.47 \quad 2 < Re_{2F} < 100, \tag{2.160}$$

$$k = 1.962, \quad m = -1/3 \quad 100 < Re_{2F} < 1000, \tag{2.161}$$

$$k = 0.735, \quad m = -0.19 \quad 1000 < Re_{2F}\,. \tag{2.162}$$

The data base for the derivation of the correlation cover the range of

$$0.003 < \delta_{2F}/(D_h/2) < 0.1. \tag{2.163}$$

Careful evaluation of the interface friction factor taking into account the momentum pre-distribution due to entrainment and deposition done by *Lopes* and *Dukler* [19] (1986) shows that there is no unique relationships between c_{21}^d and Re_{2F}. The experimental data obtained by the *Lopes* and *Dukler* can be represented very well by the modified *Altstul* formula

$$c_{21}^d = \frac{1}{4}(3.331 \ln Re_1 - 33.582)^{-2}, \qquad (2.164)$$

for $4 \times 10^4 < Re_1 \leq 8.5 \times 10^4$, and

$$c_{12}^d = 0.014, \qquad (2.165)$$

for $8.5 \times 10^4 < Re_1 \leq 12 \times 10^4$, where

$$Re_1 = \rho_1 (V_1 - V_{2i}) D_{hc} / \eta_1, \qquad (2.166)$$

$$D_{hc} = D_h \sqrt{1 - \alpha_2}. \qquad (2.167)$$

Variation within $310 < Re_{23} = < 3100$, where

$$Re_{23} = \rho_2 (\alpha_2 V_2 + \alpha_3 V_3) D_h / \eta_2, \qquad (2.168)$$

does not influence the results.

2.7. Inverted annular flow

During some processes (e.g. reflooding phase of loss-of-coolant accidents in water cooled nuclear reactors) the temperature of the heated surface may exceed the maximum film boiling temperature. Thus the situation may occur where the superheated vapor flows up around the heated surface and a subcooled or saturated liquid jet. Since the positions of liquid and vapor phases are exactly opposite to the ones in normal annular flow, this flow pattern is called inverted annular flow. The drag force acting on a liquid in a fully developed inverted annular flow can be calculated in accordance to *Aritomi* et al. [3] (1990) as follows

$$f_{w2}^d = -a_{12w} \rho_1 c_{w2}^d \frac{1}{2} |w_1 - w_2|(w_1 - w_2) \qquad (2.169)$$

where

$$a_{12w} = \frac{4}{D_{hw}} \sqrt{a_2}, \qquad (2.170)$$

$$c_{w2}^d = 0.3164 / Re_{12}^{1/4}, \qquad (2.171)$$

$$Re_{12} = (w_1 - w_2)D_{hw}\sqrt{\alpha_2}/v_1.\tag{2.172}$$

2.8. Stratified flow in horizontal or inclined rectangular channels

Geometrical characteristics: Stratified flow may exists in regions with such relative velocities between the liquid and the gas which does not cause instabilities leading to slugging. Some important geometrical characteristics are specified here - see Fig. 2.2.

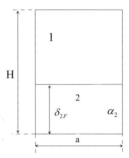

Fig. 2.2. Definition of the geometrical characteristics of the stratified flow

The perimeter of the pipe is then

$$Per_w = 2(a + H),\tag{2.173}$$

and the wetted perimeters for the gas and the liquid parts are

$$Per_{1w} = a + 2(H - \delta_{2F}) = a + 2\alpha_1 H,\tag{2.174}$$

$$Per_{2w} = a + 2\delta_{1F} = a + 2\alpha_2 H.\tag{2.175}$$

The gas-liquid interface median is then a, and the liquid level

$$\delta_{2F} = \alpha_2 H.\tag{2.176}$$

The hydraulic diameters for the gas and the liquid for computation of the pressure drop due to friction with the wall are therefore

$$D_{h1} = 4\alpha_1 F / Per_{1w} = \frac{4\alpha_1 aH}{a + 2H\alpha_1},\tag{2.177}$$

$$D_{h2} = 4\alpha_2 F / Per_{2w} = \frac{4\alpha_2 aH}{a + 2H\alpha_2}, \qquad (2.178)$$

and the corresponding *Reynolds* numbers

$$Re_1 = \frac{\alpha_1 \rho_1 w_1}{\eta_1} \frac{4\alpha_1 aH}{a + 2H\alpha_1}, \qquad (2.179)$$

$$Re_2 = \frac{\alpha_2 \rho_2 w_2}{\eta_2} \frac{4\alpha_2 aH}{a + 2H\alpha_2}. \qquad (2.180)$$

Here F is the channel cross section and Per_{1w} and Per_{2w} are the perimeters wet by gas and film, respectively.

If one considers the core of the flow the hydraulic diameter for computation of the gas-liquid friction pressure loss component is then

$$D_{h12} = 4\alpha_1 F / (Per_{1w} + a) = \frac{2\alpha_1 aH}{a + H\alpha_1} \qquad (2.181)$$

and the corresponding *Reynolds* number

$$Re_1 = \frac{\alpha_1 \rho_1 |w_1 - w_2|}{\eta_1} \frac{2\alpha_1 aH}{a + H\alpha_1}. \qquad (2.182)$$

The gas-wall, liquid-wall and gas-liquid interfacial area densities are

$$a_{1w} = \frac{Per_{1w}}{F} = \frac{a + 2\alpha_1 H}{aH}, \qquad (2.183)$$

$$a_{2w} = \frac{Per_{2w}}{F} = \frac{a + 2\alpha_2 H}{aH}, \qquad (2.184)$$

$$a_{12} = \frac{a}{F} = \frac{1}{H}. \qquad (2.185)$$

For the estimation of flow pattern transition criterion the following expression is sometimes required

$$\frac{d\alpha_2}{d\delta_{2F}} = \frac{1}{H}. \qquad (2.186)$$

Using the geometric characteristics and the *Reynolds* numbers the interfacial interaction coefficients can be computing by means of empirical correlations. The wall-gas and the wall-liquid interaction can be modeled as

$$f_1^{w\sigma,d} = -a_{1w}\tau_{w1} = a_{1w}c_{w1}^d \frac{1}{2}\rho_1 |w_1| w_1 = \frac{a+2\alpha_1 H}{aH} c_{w1}^d \frac{1}{2}\rho_1 |w_1| w_1, \qquad (2.187)$$

$$f_2^{w\sigma,d} = -a_{2w}\tau_{w2} = a_{2w}c_{w2}^d \frac{1}{2}\rho_2 |w_2| w_2 = \frac{a+2\alpha_2 H}{aH} c_{w2}^d \frac{1}{2}\rho_2 |w_2| w_2. \qquad (2.188)$$

The gas resisting force between film and gas is

$$f_1^{2\sigma,d} = -a_{12}\tau_{21} = -a_{12}c_{21}^d \frac{1}{2}\rho_1 |w_2 - w_1|(w_2 - w_1)$$

$$= -\frac{1}{H}c_{21}^d \frac{1}{2}\rho_1 |w_2 - w_1|(w_2 - w_1). \qquad (2.189)$$

For single-phase pipe flow the friction force per unit flow volume is equivalent to the pressure drop per unit length. The friction force can be then expressed either in terms of the drag coefficient or in terms of the friction coefficient

$$f_l^{w\sigma,d} = -a_{lw}\tau_{wl} = a_{lw}c_{wl}^d \frac{1}{2}\rho_l |w_l| w_l = \frac{a+2H}{aH}c_{wl}^d \frac{1}{2}\rho_l |w_l| w_l$$

$$= \frac{1}{2}\rho_l |w_l| w_l \frac{\lambda_{fr,wl}}{D_h} = \frac{1}{2}\rho_l |w_l| w_l \frac{a+2H}{4aH}\lambda_{fr,wl}. \qquad (2.190)$$

Obviously the relation between the drag coefficient c_{wl}^d and the friction coefficient $\lambda_{fr,wl}$ usually used in European literature is

$$c_{wl}^d = \frac{1}{4}\lambda_{fr,wl}. \qquad (2.191)$$

The friction coefficient in a pipe with a technical roughness can then be used to compute

$$\lambda_{fr,wl} = \lambda_{fr,wl}\left(\text{Re}_l = \rho_l w_l D_h / \eta_l, \ k/D_h\right), \qquad (2.192)$$

from the *Nukuradze* diagram given in *Idelchik*'s text book [14] (1975). k is the roughness of the pipe wall in m. In particular we will have

$$\lambda_{fr,w1} = \lambda_{fr,w1}(Re_1, \ k/D_{h1}), \tag{2.193}$$

$$\lambda_{fr,w2} = \lambda_{fr,w2}(Re_2, \ k/D_{h2}), \tag{2.194}$$

and

$$\lambda_{fr,12} = \lambda_{fr,12}(Re_{12}, \ k_{wave}/D_{h12}). \tag{2.195}$$

In the later case k_{wave} is the wave amplitude at the liquid interface.

Lee and *Bankoff* [17] (1982) measured interfacial shear stresses in a nearly horizontal stratified steam - saturated water flow. According to their results, the interfacial friction factor during the buildup of the roll waves can be expressed as follows

$$c_{21}^d = 0.012 + 2.694 \times 10^{-4} \left(\frac{Re_2^*}{1000}\right)^{1.534} (Re_1^* - Re_1^{**})/1000, \tag{2.196}$$

where

$$Re_1^{**} = 1.837 \times 10^5 / Re_2^{0.184} \tag{2.197}$$

represents the critical gas *Reynolds* number at which the transition to the roll wave regime take place. Here

$$Re_1^* = \frac{\alpha_1 \rho_1 w_1}{\eta_1} F/Per_1, \tag{2.198}$$

$$Re_2^* = \frac{\alpha_2 \rho_2 w_2}{\eta_2} F/Per_2. \tag{2.199}$$

F is the channel cross section and Per_1 and Per_2 are the wet perimeters of gas and film, respectively. Note the particular definition of the *Reynolds* numbers in the above correlation, instead of the hydraulic diameter the length scale is the cross section divided by the wet perimeter of the gas and film, respectively (the film *Reynolds* number was defined by *Lee* and *Bankoff* as mass flow rate per unit width of the film divided by the dynamic viscosity). The data base for Eq. (2.196) lies in the region $2000 \le Re_2 \le 12000$, $23000 \le Re_1 < 51000$.

2.9. Stratified flow in horizontal or inclined pipes

Geometrical characteristics: The geometric characteristics for round pipes are non linearly dependent on the liquid level, which makes the computation somewhat more complicated. Some important geometrical characteristics are specified here - see Fig. 2.3. The angle with the origin the pipe axis defined between the upwards oriented vertical and the liquid-gas-wall triple point is defined as a function of the liquid volume fraction by the equation

$$f(\theta) = -(1-\alpha_2)\pi + \theta - \sin\theta\cos\theta = 0. \tag{2.200}$$

The derivative

$$\frac{d\theta}{d\alpha_1} = \frac{\pi}{2\sin^2\theta} \tag{2.201}$$

will be used later. Having in mind that

$$\frac{df}{d\theta} = 2\sin^2\theta \tag{2.202}$$

the solution with respect to the angle can be obtained by using the *Newton* iteration method as follows

$$\theta = \theta_0 - \frac{f_0}{df/d\theta} = \theta_0 + \frac{(1-\alpha_2)\pi - \theta_0 + \sin\theta_0\cos\theta_0}{2\sin^2\theta_0} \tag{2.203}$$

where subscript *0* stands for the previous guess. We start with an initial value of $\pi/2$ [40]. *Biberg* proposed in 1999 [41] accurate direct approximation

$$\theta = \pi\alpha_2 + \left(\frac{3\pi}{2}\right)^{1/3}\left[1 - 2\alpha_2 + \alpha_2^{1/3} - (1-\alpha_2)^{1/3}\right] \tag{2.203b}$$

with an error less then $\pm 0.002 rad$ or

$$\theta = \pi\alpha_2 + \left(\frac{3\pi}{2}\right)^{1/3}\left[1 - 2\alpha_2 + \alpha_2^{1/3} - (1-\alpha_2)^{1/3}\right]$$

$$-\frac{1}{200}\alpha_2(1-\alpha_2)(1-2\alpha_2)\left\{1 + 4\left[\alpha_2^2 + (1-\alpha_2)^2\right]\right\}, \tag{2.203c}$$

with an error less then $\pm 0.00005 rad$.

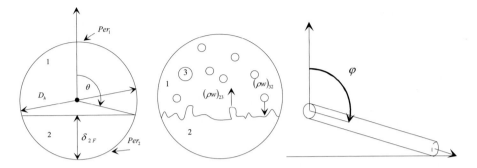

Fig. 2.3. Definition of the geometrical characteristics of the stratified flow

The perimeter of the pipe is then

$$Per_{1w} = \pi D_h, \qquad (2.204)$$

and the wetted perimeters for the gas and the liquid parts are

$$Per_{1w} = \theta D_h, \qquad (2.205a)$$

$$Per_{2w} = (\pi - \theta) D_h. \qquad (2.205b)$$

The gas-liquid interface median is then

$$b = D_h \sin(\pi - \theta) = D_h \sin \theta, \qquad (2.206)$$

and the liquid level

$$\delta_{2F} = \frac{1}{2} D_h (1 + \cos \theta). \qquad (2.207)$$

The hydraulic diameters for the gas and the liquid for computation of the pressure drop due to friction with the wall are therefore

$$D_{h1} = 4\alpha_1 F / Per_{1w} = \frac{\pi}{\theta} \alpha_1 D_h, \qquad (2.208)$$

$$D_{h2} = 4\alpha_2 F / Per_{2w} = \frac{\pi}{\pi - \theta} \alpha_2 D_h, \qquad (2.209)$$

and the corresponding *Reynolds* numbers

$$Re_1 = \frac{\alpha_1 \rho_1 w_1 D_h}{\eta_1} \frac{\pi}{\theta}, \qquad (2.210)$$

$$Re_2 = \frac{\alpha_2 \rho_2 w_2 D_h}{\eta_2} \frac{\pi}{\pi - \theta}. \qquad (2.211)$$

Here F is the channel cross section and Per_1 and Per_2 are the wet perimeters of gas and film, respectively.

If one considers the core of the flow the hydraulic diameter for computation of the gas-liquid friction pressure loss component is then

$$D_{h12} = 4\alpha_1 F / (Per_{1w} + b) = \frac{\pi}{\theta + \sin\theta} \alpha_1 D_h, \qquad (2.212)$$

and the corresponding *Reynolds* number

$$Re_{12} = \frac{\alpha_1 \rho_1 |w_1 - w_2| D_h}{\eta_1} \frac{\pi}{\theta + \sin\theta}. \qquad (2.213)$$

The gas-wall, liquid-wall and gas-liquid interfacial area density are

$$a_{1w} = \frac{Per_{1w}}{F} = \frac{4\alpha_1}{D_{h1}} = \frac{\theta}{\pi} \frac{4}{D_h}, \qquad (2.214)$$

$$a_{2w} = \frac{Per_{2w}}{F} = \frac{4\alpha_2}{D_{h2}} = \frac{\pi - \theta}{\pi} \frac{4}{D_h}, \qquad (2.215)$$

$$a_{12} = \frac{b}{F} = \frac{\sin\theta}{\pi} \frac{4}{D_h}. \qquad (2.216)$$

Ratel et al. [23] approximated this relation for smooth interface with

$$a_{12} \cong \frac{8}{\pi D_h} \sqrt{\alpha_2 (1 - \alpha_2)}. \qquad (2.217)$$

For the estimation of the flow pattern transition criterion the following expression is sometimes required

$$\frac{d\alpha_2}{d\delta_{2F}} = \frac{4}{D_h}\frac{\sin\theta}{\pi}. \tag{2.218}$$

The wall-gas and the wall-liquid forces per unit mixture volume can then be computed as follows

$$f_1^{w\sigma,d} = -a_{1w}\tau_{w1} = a_{1w}c_{w1}^d \frac{1}{2}\rho_1|w_1|w_1 = \frac{\theta}{\pi}\frac{4}{D_h}c_{w1}^d \frac{1}{2}\rho_1|w_1|w_1, \tag{2.219}$$

$$f_2^{w\sigma,d} = -a_{2w}\tau_{w2} = a_{2w}c_{w2}^d \frac{1}{2}\rho_2|w_2|w_2 = \frac{\pi-\theta}{\pi}\frac{4}{D_h}c_{w2}^d \frac{1}{2}\rho_2|w_2|w_2. \tag{2.220}$$

The gas resisting force between film and gas is

$$f_1^{2\sigma,d} = -a_{12}\tau_{21} = -a_{12}c_{21}^d \frac{1}{2}\rho_1|w_2-w_1|(w_2-w_1)$$

$$= -\frac{\sin\theta}{\pi}\frac{4}{D_h}c_{21}^d \frac{1}{2}\rho_1|w_2-w_1|(w_{2i}-w_1). \tag{2.221}$$

One choice for computation of the drag coefficient is the correlation by *Lee* and *Bankoff* [17] (1982).

Ratel and *Bestion* proposed in [33] the following empirical correlation

$$c_{21}^d = 6\times 10^{-8} Re_{10}^{1.75}\left(1+0.035 Re_{10}^{0.97} Fr^{*-1.3}\frac{\alpha_2^{0.44}}{\alpha_1^{0.5}}\right), \tag{2.222}$$

where

$$Re_{10} = \frac{\rho_1 w_1 D_h}{\eta_1}, \tag{2.223}$$

$$Fr^* = \frac{gD_h}{w_1^2}, \tag{2.224}$$

valid for the range $3.5\times 10^5 < Re_{10} < 7\times 10^5$. Unfortunately the authors retrieved the correlation with a computer code using in the momentum equations the erroneous term

$$\alpha_1 \alpha_2 (\rho_2 - \rho_1) g D_h \frac{\partial \alpha_1}{\partial z},$$

instead of the correct term

$$\alpha_1 \alpha_2 (\rho_2 - \rho_1) g \frac{F}{b} \frac{\partial \alpha_1}{\partial z},$$

where F is the channel cross section and b is the gas-liquid interface median.

For single-phase pipe flow the friction force per unit flow volume is equivalent to the pressure drop per unit length. The friction force can be then expressed either in terms of the drag coefficient c_{wl}^d or in terms of the friction coefficient $\lambda_{fr,wl}$

$$f_l^{w\sigma,d} = -a_{lw} \tau_{wl} = a_{lw} c_{wl}^d \frac{1}{2} \rho_l |w_l| w_l = \frac{4}{D_h} c_{wl}^d \frac{1}{2} \rho_l |w_l| w_l = \frac{1}{2} \rho_l |w_l| w_l \frac{\lambda_{fr,wl}}{D_h}. \quad (2.225)$$

Again as for rectangular channel we have

$$c_{wl}^d = \frac{1}{4} \lambda_{fr,wl} \quad (2.226)$$

and the procedure for computation of the forces applied to the rectangular channels can be used also for pipes.

Of course the approximation used in Eq. (2.187, 2.188) considers not the influence of the interfacial shear stress on the wall shear stress. Data comparison reported by *Biberg* [41] demonstrated a systematic under prediction of the liquid volume fraction in averaged with -5.5% if using this approach. By careful elaboration of an approximation for the turbulent velocity profile in both fields *Biberg* [41] come to the very important result

$$\frac{1}{\sqrt{\lambda_{fr,wl}}} = \frac{-1.8 \log \left[\frac{6.9}{\text{Re}_l^*} + \left(\frac{k}{3.7 D_{hl}^*} \right)^{1.11} \right]}{1 + 2\log(1 + b/Per_{lw}) \frac{w_1 - w_2}{w_1} \sqrt{\lambda_{fr,12}}}. \quad (2.227)$$

Here for free surface flow the equivalent hydraulic diameter is defined as

$$D_{hl}^* \approx \begin{cases} 8\delta_l/3 & \text{rectangular channel} \\ 4\alpha_l F/Per_{lw} & \text{pipe} \end{cases}, \qquad (2.228)$$

and for closed duct flow as

$$D_{hl}^* \approx \begin{cases} 4\delta_l/3 & \text{rectangular channel} \\ 4\alpha_l F/(Per_{lw}+b) & \text{pipe} \end{cases}. \qquad (2.229)$$

The Reynolds number $\text{Re}_l^* = \rho_l |w_l| D_{hl}^*/\eta_l$ is computed with the corresponding equivalent hydraulic diameter.

Nomenclature

Latin

a_{1w}	gas-wall interfacial area densities, *1/m*
a_{2w}	liquid-wall interfacial area densities, *1/m*
a_{12}	gas-liquid interfacial area densities, *1/m*
b	bubble breadth (major axis), *m*
c_{cd}^d	dispersed phase drag coefficient due to continuum action, *dimensionless*
c_{21}^d	bubble drag coefficient due to liquid action, *dimensionless*
c_{13}^d	particle drag coefficient due to gas action, *dimensionless*
$c_{23,nb}^d$	particle drag coefficient due to liquid action measured for non-boiling condition, *dimensionless*
c_{13o}^{vm}	single particle virtual mass coefficient due to gas action, *dimensionless*
c_{23o}^{vm}	single particle virtual mass coefficient due to liquid action, *dimensionless*
c_{13o}^d	single particle drag coefficient due to gas action, *dimensionless*
c_{23o}^d	single particle drag coefficient due to liquid action, *dimensionless*
c_{13}^{vm}	$=(1-\phi)c_{13o}^{vm}$, particles virtual mass coefficient due to gas action, *dimensionless*
c_{23}^{vm}	$=\phi\, c_{23o}^{vm}$, particles virtual mass coefficient due to liquid action, *dimensionless*
c_{13}^d	$=(1-\phi)c_{13o}^d$, particles drag coefficient due to gas action, *dimensionless*
c_{23}^d	$=\phi\, c_{23o}^d$, particles drag coefficient due to liquid action, *dimensionless*

c_{w1}^d	gas drag coefficient due to wall-gas interaction, *dimensionless*
c_{w2}^d	liquid drag coefficient due to wall-gas interaction, *dimensionless*
D_d	dispersed particle diameter, *m*
D_1	bubble diameter, *m*
D_3	particle diameter, *m*
D_{h1}	hydraulic diameters for the gas, *m*
D_{h2}	hydraulic diameters for the liquid, *m*
D_{h12}	hydraulic diameter for computation of the gas friction pressure loss component in a gas-liquid stratified flow, *m*
Eo	$= (D_1 / \lambda_{RT})^2$, Eötvös number, *dimensionless*
F	channel cross section, *m²*
$\mathbf{f}_{w1}$	gas resisting force between wall and gas per unit flow volume, *N/m³*
$\mathbf{f}_{w2}^d$	film resisting force between wall and film per unit flow volume, *N/m³*
$\mathbf{f}_{21}^d$	bubble force per unit flow volume due to liquid-gas interaction, *N/m³*
$\mathbf{f}_{13}^d$	particle force per unit flow volume due to liquid-gas interaction, *N/m³*
h	bubble high (minor axis), *m*
Mo	$= \dfrac{g \Delta \rho_{21}}{\sigma} \left(\dfrac{\eta_2^2}{\rho_2 \sigma} \right)^2$, Morton number, *dimensionless*
Per_w	perimeter of the rectangular channel, *m*
Per_{1w}	wetted perimeters for the gas, *m*
Per_{2w}	wetted perimeters for the liquid, *m*
p	pressure, *Pa*
Re	$= D_1 \rho_2 |\Delta V_{12}| / \eta_2$, Reynolds number for a single bubble, *dimensionless*
Re_{13}	$= (\alpha_1 \rho_1 w_1 + \alpha_3 \rho_3 w_3) D_h / \eta_1$, core Reynolds number, *dimensionless*
Re_1	$= \rho_1 w_1 D_h / \eta_1$, gas Reynolds number, *dimensionless*
Re_2	$= \rho_2 w_2 D_h / \eta_2$, liquid Reynolds number, *dimensionless*
Ta	$= \dfrac{|\Delta V_{12}|}{\sqrt{g \lambda_{RT}}} \dfrac{D_1}{\lambda_{RT}}$, *dimensionless*
V_{wake}	velocity behind a solid particle, *m/s*
$\mathbf{V}_1$	gas velocity, *m/s*
$\mathbf{V}_2$	liquid velocity, *m/s*
$\mathbf{V}_3$	droplet velocity, *m/s*
$w_{31\infty}$	particle sink velocity in gas, *m/s*
$w_{32\infty}$	particle sink velocity in liquid, *m/s*

w_1	axial gas velocity, m/s
w_2	axial liquid velocity, m/s
w_{2i}	axial liquid film interface velocity, *m/s*
z	axial coordinate, *m*

Greek

α_d	particle volume fraction, *dimensionless*
α_{dm}	volume concentration of the disperse phase and the maximum packing, *dimensionless*
α_3	particle volume fraction, *dimensionless*
α_{2eff}^*	effective saturation, *dimensionless*
γ	surface permeability - flow cross section divided by the overall cross section, *dimensionless*
∇	gradient
∇V_{cd}	velocity difference continuous minus disperse, *m/s*
∇V_{13}	velocity difference gas minus particle, *m/s*
$\Delta\tau$	time interval, s
∂	partial differential
δ	location, *m*
δ_{1F}	film thickness, *m*
ϕ	$=\alpha_2/(\alpha_1+\alpha_2)$, liquid volume fraction inside the gas liquid mixture within the three fluid mixture, *dimensionless*
λ_{RT}	$=\left[\sigma/(g\Delta\rho_{21})\right]^{1/2}$, *Rayleigh-Taylor* instability wavelength, *m*
η_m	effective continuum viscosity, *kg/(ms)*
η^*	viscosity coefficient, *dimensionless*
η_d	dispersed phase dynamic viscosity, *Pa s*
η_c	continuum dynamic viscosity, *Pa s*
k_1^l, k_1^t	viscous and turbulent permeability coefficients, respectively, *dimensionless*
$\lambda_{fr,wl}$	friction coefficient, *dimensionless*
μ_3	field 3 mass generation per unit mixture volume, *kg/(sm³)*
ρ_1	gas density, *kg/m³*
ρ_2	liquid density, *kg/m³*
ρ_3	particle density, *kg/m³*
σ_2	liquid-gas surface tension, *N/m*
τ	time, *s*

τ_a time at the old time level, s

τ_{w2} liquid share stress caused by the wall friction, N/m^2

τ_{w1} gas share stress caused by the wall friction, N/m^2

τ_{12} $= \dfrac{1}{2} c_{21}^d \rho_1 V_1^2$, gas side interfacial share stress at the liquid interface, N/m^2

θ_{23} wetting contact angle between the liquid and particles, rad

References

1. Achenbach E (1993) Heat and flow characteristics of packed beds, Experimental Heat Transfer, Fluid Mechanics and Thermodynamics, Kelleher M D et al. (eds) Elsevier
2. Ambrosini W, Andreussi P and Azzopardi BJ (1991) Int. J. Multiphase Flow, vol 17 no 4 pp 497-507
3. Aritomi M, Inoue A, Aoki S and Hanawa K (1990) Thermo- hydraulic behavior of inverted annular flow, NED vol 120 pp 281-291
4. Bharathan D, Richter HT and Wallis GB (1978) Air-water counter-current annular flow in vertical tubes, EPRI-NP-786
5. Bharathan D, Wallis GB, Richter HT (1979) Air - water countercurrent annular flow. EPRI NP - 1165, Electric Power Research Inst., Palo Alto, California
6. Brinkman HC (1951) J. Chem. Phys., vol 6 p 571
7. Brinkman HC (1952) J. Chem. Phys., vol 20 p 571
8. Brooks RH and Corey AT (1966) Properties of porous media affecting fluid flow, J. Irrig. and Drainage Div. Proc. ASChE, vol 92, IR2 pp 61
9. Brown GG et al. (1950) Unit operations, J. Wiley and Sons, Inc., New York, pp 210-228
10. Ergun S (1952) Fluid flow through packed columns, Chem. Eng. Prog., vol 48 no 2 pp 89-94
11. Geary NW and Rice RG (Oct. 1991) Circulation in bubble columns: correlations for distorted bubble shape, AIChE Journal, vol 37 no 10 pp 1593-1594
12. Govan AH, Hewitt GF, Richter HJ, and Scott A (1991) Flooding and churn flow in vertical pipes, Int. J. Multiphase Flow, vol 17 no 1 pp 27-44
13. Hugmark GA (Sept. 1973) Film thickness, entrainment and pressure drop in upward annular and dispersed flow, AIChEJ, vol 19 no 5 pp 1062-1065
14. Idelchik IE (1975) Handbook of hydraulic resistance, Second Edition, Hemisphere, Washington, translated from Russian in 1986
15. Ishii M and Zuber N (1978) Relative motion and interfacial drag coefficient in dispersed two - phase flow of bubbles, drops and particles, Paper 56 a, AIChE 71st Ann. Meet., Miami
16. Ishii M and Chawla TC (Dec.1979) Local drag laws in dispersed two-phase flow, NUREG/CR-1230, ANL-79-105
17. Lee SC and Bankoff SG (1983) Stability of steam - water countercurrent flow in an inclined channel. J. Heat Transfer, vol 105 pp 713-718
18. Lepinski RJ (Apr. 1984) A coolability model for postaccident nuclear reactor debries, Nucl. Technology, vol 65 pp 53-66

19. Lopes JCB and Dukler AE (Sept.1986) Droplet entrainment in vertical annular flow and its contribution to momentum Transfer, AIChE Journal, vol 32 no 9 pp 1500-1515
20. Militzer J, Kann J M, Hamdullahpur F, Amyotte P R, Al Towel A M (1998) Drag coefficients of axisymetric flow arround individual spheroidal particles, Powder Technol., vol 57 pp 193-195
21. Nigmatulin BI (1982) Heat and mass transfer and force interactions in annular - dispersed two - phase flow, 7th Int. Heat Transfer Conf., Munich, pp 337-342
22. Oseen CW (1910) Über die Stokessche Formel und über eine verwandte Aufgabe in der Hydrodynamik, Ark. F. Math. Astron. Och. Fys., vol 6 no 29
23. Ratel G and Bestion D (2000) Analysis with CHATHARE code of the stratified flow regime in the MERESA hot leg entrainment tests, 38th European Two Phase Group Meeting, Karlsruhe
24. Reed AW (Feb. 1982) The effect of channeling on the dryout of heated particulate beds immersed in a liquid pool, PhD Thesis, Massachusetts Institute of Technology, Cambridge
25. Roscoe R and Brit J (1952) Appl. Phys., vol 3 p 267
26. Roscoe R (1952) Brit. Appl. Phys., vol 3 p 26
27. Schiller L and Naumann AZ (1935) Z. Ver. Deut. Ing., vol 77 p 318
28. Stephan M and Mayinger F (1990) Countercurrent flow limitation in vertical ducts at high system pressure, Hetstroni G (ed), Proc. of The Ninth International Heat Transfer Conference, Jerusalem, Israel, vol 6 pp 47-52
29. Stokes GG (1880) Mathematical and physical papers, vol 1, Cambridge University Press, London
30. Stuhmiller JH, Ferguson RE and Meister CA (November 1989) Numerical simulation of bubble flow, EPRI Research Project Report NP-6557
31. Tomiyama A, Sakagushi T and Minagawa H (1990) Kobe University, private communication
32. Tomiyama A, Matsuoka T, Fukuda T and Sakaguchi T (April 3-7, 1995) A simple numerical method for solving an incompressible two/fluid model in a general curvilinear coordinate system, Proc. of The 2bd International Conference on Multiphase Flow '95 Kyoto, Kyoto, Japan, vol 2 pp NU-23 to NU-30
33. Tomiyama A (June 8-12, 1998) Struggle with computational bubble dynamics, Third International Conference on Multiphase Flow, ICMF 98, Lyon, France,
34. Tung VX and Dhir VK (1990) Finite element solution of multi-dimensional two-phase flow through porous media with arbitrary heating conditions, Int. J. Multiphase Flow, vol 16 no 6 pp 985-1002
35. Ueda T (1981) Two-phase flow - flow and heat transfer, Yokendo, Japan (in Japanese)
36. VDI-Wärmeatlas, VDI-Verlag, Düsseldorf (1991) 6. Aufl.
37. Wallis GB (1969) One-dimensional two-phase flow, New York: McGraw Hill
38. Wyckoff RD and Botset HG (1936) Physics, vol 73 p 25
39. Zvirin Y, Hewitt GF, Kenning DBR (1990) Boiling of free - falling spheres: Drag and heat transfer coefficients. Experimental Heat Transfer, vol 3 pp 185-214
40. Kolev NI (1977) Two-phase two-component flow (air-water steam-water) among the safety compartments of the nuclear power plants with water cooled nuclear reactors during lose of coolant accidents, PhD Thesis, Technical University Dresden
41. Biberg D (December 1999) An explicit approximation for the wetted angle in two-phase stratified pipe flow, The Canadian Journal of Chemical Engineering, vol 77 pp 1221-1224

42. Michaelides EE (March 2003) Hydrodynamic force and heat/mass transfer from particles, bubbles and drops – The Freeman Scholar Lecture, ASME Journal of Fluids Engineering, vol 125 pp 209-238

3. Friction pressure drop

3.1 Introduction

The drag force acting on the velocity field that is in contact with the wall can be calculated in different ways. One of them is to model the two-dimensional flow with fine resolution of the boundary layer, then to differentiate the velocity profile of the field being in a contact with the wall, and finally to compute the tangential stress at the wall. Due to the complexity of the flow pattern this is difficult to do for all flow patterns. That is why empirical methods are commonly used for practical applications.

3.2 Single-phase flow

The friction pressure loss computation for multi-phase flows uses the knowledge from the single-phase flow as a prerequisite. The pressure drop per unit length in a straight pipe with hydraulic diameter D_h, roughness k and length Δz is

$$\left(\frac{dp}{dz}\right)_R = \frac{1}{2}\rho|V|V\left(\frac{\lambda_R}{D_h} + \frac{\xi}{\Delta z}\right), \qquad (3.1)$$

where

$$\lambda_R = \lambda_R(\text{Re} = \rho V D_h / \eta, k / D_h) \qquad (3.2)$$

is the friction pressure loss coefficient. For technical roughness the *Nikuradze* diagram given in *Idelchik*'s text book [9] (1975) is recommend. The analytical approximations of this diagram are given below. First the minimum of λ_R is defined for $k = 0$ as follows:

(a) For laminar flow i.e. $Re < 2300$ the *Hagen* and *Poiseuille* law is valid

$$\lambda_R = 64 / Re. \qquad (3.3)$$

(b) For the intermediate region, $2300 < Re < 2818$, an interpolation between the *Hagen* and *Poiseuille* and the *Altshul* formula

$$\lambda_R = 0.028(Re/2300)^{2.667}, \qquad (3.4)$$

is necessary.

(c) For the turbulent region the *Altshul* formula

$$\lambda_R = 1/(1.8 \log Re - 1.64)^2, \qquad (3.5)$$

is recommend. For the case of $k > 0$ four *Reynolds* numbers are computed in order to define different regions of the *Nikuradze* diagram

$$Re_0 = 754 e^{0.0065/(k/D_h)}, \qquad (3.6)$$

$$Re_1 = 1160 \left(\frac{D_h}{k}\right)^{0.11}, \qquad (3.7)$$

$$Re_2 = 2090 \left(\frac{D_h}{k}\right)^{0.0635}, \qquad (3.8)$$

$$Re_3 = 560 D_h / k. \qquad (3.9)$$

The following approximations are recommended for the four regions of the *Nikuradze* diagram:

(a) For $Re < Re_0$ or $Re_0 > Re_1$, use the *Hagen* and *Poiseuille* formula;

(b) For $Re_0 < Re < Re_1$, use the *Samoilenko* formula

$$\lambda_R = 4.4 / \left(Re^{0.595} e^{0.00275 D_h / k}\right). \qquad (3.10)$$

(c) For $Re_1 < Re < Re_2$, interpolation between λ_{R2} and λ_{R3} is used

$$\lambda_R = \lambda_{R3} + (\lambda_{R2} - \lambda_{R3}) / e^{\left[0.0017(Re_2 - Re)\right]^2}, \qquad (3.11)$$

where for $k/D_h < 0.007$

$$\lambda_{R1} = 0.032, \qquad (3.12)$$

$$\lambda_{R2} = 7.244 / Re_2^{0.643}, \tag{3.13}$$

$$\lambda_{R3} = \lambda_{R1}, \tag{3.14}$$

and for $k/D_h \geq 0.007$

$$\lambda_{R1} = 0.0775 - 0.019/(k/D_h)^{0.286}, \tag{3.15}$$

$$\lambda_{R2} = 0.145(k/D_h)^{0.244}, \tag{3.16}$$

$$\lambda_{R3} = \lambda_{R1} - 0.0017. \tag{3.17}$$

(d) For $Re_2 < Re < Re_3$ the *Colebrook* and *Witte* formula is recommended

$$\frac{1}{\sqrt{\lambda_R}} = 1.74 - 2\log\left[\frac{1}{\sqrt{\lambda_R}}\frac{18.7}{Re} + 2(k/D_h)\right]. \tag{3.18}$$

One can avoid the solution of the above equation with respect to λ_R by using the following approximation

$$\lambda_R = \lambda_{R2}(Re/Re_2)^n, \tag{3.19}$$

where

$$n = \left[\log(\lambda_{R3}/\lambda_{R2})\right]/\left[\log(Re_3/Re_2)\right], \tag{3.20}$$

and

$$\lambda_{R2} = 7.244/Re_2^{0.643} \quad \text{for} \quad k/D_h < 0.007, \tag{3.21}$$

$$\lambda_{R2} = 0.145(k/D_h)^{0.244}. \tag{3.22}$$

Haland [20] proposed a very accurate explicit replacement of the *Colebrook* and *Witte* formula

$$\frac{1}{\sqrt{\lambda_R}} = -1.8\log\left[\frac{6.9}{Re} + \left(\frac{k}{3.7D_h}\right)^{1.11}\right]. \tag{3.18b}$$

which agrees with it to within ±1.5% for $4000 \le \mathrm{Re} \le 10^8$ and $k/D_h \le 0.05$.

(e) For $Re > Re_3$ the *Prandtl - Nikuradze* formula is recommended,

$$\lambda_R = 1/\left[1.74 + 2\log(0.5D_k/k)\right]^2. \tag{3.23}$$

Note that turbulent pipe flows with strong density boundary layer may change the pressure drop due to friction. For *supercritical fluids* in heated channels, the densyty in the boundary layer is smaller then the density in the bulk reducing the pressure drop, $\lambda_R = \lambda_{R,isotherm}\left[\rho_{wall}(T_{wall})/\rho_{bulk}(T_{bulk})\right]^{0.4}$, Kirilov et al. [21] (1990). This correlation is valid for $p/p_{cr} = 1.016 \div 1.22$, $\mathrm{Re} = 8\times10^4 \div 1.5\times10^6$.

3.3 Two-phase flow

The modeling of the two-phase friction pressure drop in a pipe started with the work by *Lockhart* and *Martinelli* [10] (1949). The authors defined the following auxiliary variables: The pressure gradient of the liquid flowing alone in the same tube

$$\left(\frac{dp}{dz}\right)_{R2} = \frac{1}{2}\frac{(\alpha_2\rho_2 V_2)^2}{\rho_2}\frac{\lambda_{R2}}{D_h}, \tag{3.24}$$

where

$$\lambda_{R2} = \lambda_{R2}(\mathrm{Re}_2 = \alpha_2\rho_2 V_2 D_h/\eta_2, \ k/D_h); \tag{3.25}$$

The pressure gradient of the gas flowing alone in the same tube

$$\left(\frac{dp}{dz}\right)_{R1} = \frac{1}{2}\frac{(\alpha_1\rho_1 V_1)^2}{\rho_1}\frac{\lambda_{R1}}{D_h} \tag{3.26}$$

where

$$\lambda_{R1} = \lambda_{R1}(\mathrm{Re}_1 = \alpha_1\rho_1 V_1 D_h/\eta_1, \ k/D_h); \tag{3.27}$$

The factor X_{LM}, called later the *Lockhardt* and *Martinelli* factor in honor of the authors,

$$X_{LM} = \phi_1^2/\phi_2^2 = \left(\frac{dp}{dz}\right)_{R2} \Big/ \left(\frac{dp}{dz}\right)_{R1} \qquad (3.28)$$

where the gas only two-phase friction multiplier is

$$\phi_1^2/\phi_2^2 = \left(\frac{dp}{dz}\right)_{R} \Big/ \left(\frac{dp}{dz}\right)_{R1} \qquad (3.29)$$

and the liquid only friction multiplier is

$$\phi_2^2 = \left(\frac{dp}{dz}\right)_{R} \Big/ \left(\frac{dp}{dz}\right)_{R2}. \qquad (3.30)$$

In case of the validity of the *Blasius* formula X_{LM} reduces to

$$X_{LM} = X_{tt} = \left(\frac{\rho_1}{\rho_2}\right)^{1/2} \left(\frac{\eta_2}{\eta_1}\right)^{0.1} \left(\frac{1-X_1}{X_1}\right)^{0.9}, \qquad (3.31)$$

see [8] (1983). Comparing with experimental data for air-water horizontal flow near to atmospheric pressure the authors found

$$\phi_1^2 = 1 + CX_{LM} + X_{LM}^2, \qquad (3.32)$$

$$\phi_2^2 = 1 + \frac{C}{X_{LM}} + \frac{1}{X_{LM}^2}, \qquad (3.33)$$

where C has the following values

$C = 20$ for turbulent liquid and turbulent gas,

$C = 12$ for laminar liquid and turbulent gas,

$C = 10$ for turbulent liquid and laminar gas, and

$C = 5$ for laminar liquid and laminar gas.

Other simple approximations can be found in the literature in the form $\phi_2^2 = 1/(1-\alpha_1)^m$, $m = 1.75$ to 2. *Ransom* et al. [15] (1988) recommended for C the following continuous approximation

$$C = \max\left\{2, \left(28 - 0.3\sqrt{G}\right)/\exp\left[\frac{\left(\log Y^* + 2.5\right)^2}{2.4 - 10^{-4} G} - 2\right]\right\}, \quad (3.34)$$

$$Y^* = \frac{\rho_1}{\rho_2}\left(\frac{\eta_2}{\eta_1}\right)^{0.2}. \quad (3.35)$$

Thus if the gas is the continuous phase

$$\left(\frac{dp}{dz}\right)_R = c_{1w}|V_1|V_1, \quad (3.36)$$

where

$$c_{1w} = c_{1w}^* \phi_1^2, \quad (3.37)$$

$$c_{2w} = 0, \quad (3.38)$$

$$c_{1w}^* = \frac{1}{2}\frac{(\alpha_1 \rho_1)^2}{\rho_1}\left(\frac{\lambda_{R1}}{D_h} + \frac{\xi}{\Delta z}\right) \quad (3.39)$$

and if the liquid is the continuous phase we have

$$\left(\frac{dp}{dz}\right)_R = c_{2w}|V_2|V_2, \quad (3.40)$$

where

$$c_{2w} = c_{2w}^* \phi_2^2, \quad (3.41)$$

$$c_{1w} = 0, \quad (3.42)$$

$$c_{2w}^* = \frac{1}{2}\frac{(\alpha_2 \rho_2)^2}{\rho_2}\left(\frac{\lambda_{R2}}{D_h} + \frac{\xi}{\Delta z}\right). \quad (3.43)$$

Later *Martinelli* and *Nelson* [11] (1948) introduced the following additional variables used successfully for description of friction pressure loss in two-phase steam-water flows at arbitrary parameter: The frictional pressure drop of the liquid following alone in the tube with flow rate equal to the total flow rate of the two phase flow

$$\left(\frac{dp}{dz}\right)_{R20} = \frac{1}{2}\frac{G^2}{\rho_2}\left(\frac{\lambda_{R20}}{D_h} + \frac{\xi}{\Delta z}\right), \qquad (3.44)$$

where

$$\lambda_{R20} = \lambda_{R20}\left(Re_{20} = D_h|G|/\eta_2, k/D_h\right), \qquad (3.45)$$

and the ratio

$$\phi_{20}^2 = \left(\frac{dp}{dz}\right)_R / \left(\frac{dp}{dz}\right)_{R20}, \qquad (3.46)$$

called later the *Martinelli-Nelson* multiplier. The relationship between ϕ_{20}^2 and ϕ_2^2 for the *Blasius* regime was analytically found

$$\phi_{20}^2 = (1-X_1)^{1.75}\phi_2^2. \qquad (3.47)$$

Later the *Martinelli-Nelson* method was provided with a more accurate database and approximation of ϕ_{20}^2 as a function of the local flow parameter. *Hewitt* makes the following proposal for the most accurate approximation of the available experimental data - see [8] (1982):

(a) In case of $\eta_2/\eta_1 < 1000$, the *Friedel* [6] (1979) correlation should be used:

$$\phi_{20}^2 = E + 3.24FH/(Fr^{0.045}We^{0.035}), \qquad (3.48)$$

where

$$E = (1-X_1)^2 + X_1^2 \frac{\rho_2}{\rho_1}\frac{\lambda_{10}}{\lambda_{20}}, \qquad (3.49)$$

$$\lambda_{10} = \lambda_{10}\left(\frac{\rho w D_h}{\eta_1}, \frac{k}{D_h}\right), \qquad (3.50)$$

$$\lambda_{20} = \lambda_{20}\left(\frac{\rho w D_h}{\eta_2}, \frac{k}{D_h}\right), \qquad (3.51)$$

$$F = X_1^{0.78}(1-X_1)^{0.24}, \tag{3.52}$$

$$H = \left(\frac{\rho_2}{\rho_1}\right)^{0.91}\left(\frac{\eta_1}{\eta_2}\right)^{0.19}\left(1-\frac{\eta_1}{\eta_2}\right)^{0.7}, \tag{3.53}$$

$$Fr = \frac{(G/\rho)^2}{gD_h}, \tag{3.54}$$

$$We = G^2 D_h /(\rho\sigma). \tag{3.55}$$

This correlation approximates 25 000 experimental points for vertical upwards co-current flow and for horizontal flow with 30 to 40% standard deviation for one- and two-component flow.

(b) In case of $\eta_2/\eta_1 \geq 1000$ and $G > 100$, the *Baroczy* correlation from 1965 modified by *Chisholm* [4] (1983) should be used

$$\phi_{2o}^2 = 1 + (Y^2 - 1)\left\{B[X_1(1-X_1)]^{0.875} + X_1^{1.75}\right\}. \tag{3.56}$$

Here B is computed as function of the dimensionless number

$$Y = \sqrt{\frac{\rho_2}{\rho_1}\left(\frac{\eta_1}{\eta_2}\right)^{0.25}} \tag{3.57}$$

as follows

$9.5 \geq Y$		$9.5 < Y < 28$		$Y \geq 28$	
$G \leq 500$	$B = 4.8$	$G \leq 600$	$B = \dfrac{520}{Y\sqrt{G}}$		$B = \dfrac{15000}{Y^2\sqrt{G}}$
$500 < G < 1900$	$B = 2400/G$	$G > 600$	$B = 21/Y$		
$1900 \leq G$	$B = 55/\sqrt{G}$				

(c) In the case of $\eta_2/\eta_1 \geq 1000$ and $G \leq 100$, the *Martinelli* correlation is recommended.

3.4 Three-dimensional flow in a porous structure

For friction pressure loss computation in porous three-dimensional structures one should bear in mind that the friction coefficient c has three components corresponding to the three directions of the c. m. velocity components. The hydraulic diameters for each of the three space directions are defined at the locations in the control volume, in which the space velocity components are defined. Together with the wall roughness they are input parameters for friction pressure loss calculation.

In case of cross flow in a vertical rod bundle the total pressure drop can be calculated as follows. We denote with S_{BE} the averaged distance between the rods in triangular array arrangement and with D_{BE} the outer rod diameter. In the distance Δr,

$$\Delta r / \left(\frac{\sqrt{3}}{2} S_{BE} \right)$$

rods are placed. *Gunter* and *Shaw* [7] (1945) proposed a correlation for computation of the pressure drop per single row. Across all the distance Δr the frictional pressure drop coefficient proposed by *Gunter* and *Shaw* has to be multiplied by

$$\Delta r / \left(\frac{\sqrt{3}}{2} S_{BE} \right).$$

Thus the friction pressure drop coefficient is

$$\lambda_{Ru} = \frac{b}{\mathrm{Re}_r^m} \left(\frac{S_{BE}}{\sqrt{3}} \right) \left(\frac{D_{hu}}{S_{BE}} \right)^{0.4} \Delta r / \left(\frac{\sqrt{3}}{2} S_{BE} \right) \tag{3.58}$$

where

$$D_{hu} = (2\sqrt{3} S_{BE}^2 - \pi D_{BE}^2)/(\pi D_{BE}), \tag{3.59}$$

$$\mathrm{Re}_r = D_{hu} \left| G \right|_{lu} / \eta_2, \tag{3.60}$$

$b = 180$, $m = 1$ for laminar flow, $b = 2.26$, $m = 0.145$ for turbulent flow.

3.5 Heated channels

The above algorithm is valid strictly for an *adiabatic* flow. For a *boiling* flow in a heated channel, the application of the above correlations is justified, if there is no *sub cooled boiling* with a net steam volume fraction different from zero. In the last case the friction pressure drop *increases* several times compared to the adiabatic case with the same mass flow rate. This cannot be taken into account by the above mentioned methods. For the estimation of the pressure loss in case of sub cooled boiling it is worth to paying attention to *Nigmatulin's* works [13], relying on the several successful comparisons with experimental data. For the computation of the friction pressure gradient

$$\left(\frac{dp}{dz}\right)_R = \frac{4}{D_h}\tau \qquad (3.61)$$

in the direction z, *Avdeev* proposes the method described in [2, 3] (1983, 1986). *Staengl* and *Mayinger* [17] (1989) recommend for the sub cooled boiling the use of *Friedel*'s correlation together with their correlation for prediction of the drift flux parameter given in Chapter 4. For the pressure loss they reported a standard deviation of ± 25%.

Zheng et al. [19] (1991) proposed a new empirical correlation for prediction of the frictional pressure drop per unit length for heated channels in the form

$$\left(\frac{dp}{dz}\right)_R = \frac{\lambda_R}{D_h}\frac{1}{2}\rho|V|V, \qquad (3.62)$$

where the friction coefficient is computed as follows

$$\lambda_R = \lambda_R(Re = D_h|G|/\eta_{eff}, k/D_h), \qquad (3.63)$$

where the effective viscosity is

$$\eta_{eff}/\eta_2 = (1-X_1)^{3.67} + \frac{\eta_1}{\eta_2}X_1^{2.43} + 3.2727\times10^6\left(\frac{p}{p_{cr}}\right)^{2.48}$$

$$\times\left(1-\frac{p}{p_{cr}}\right)^{9.07}We^{-0.97}X_1^{0.001}(1-X_1)^{0.4} \qquad (3.64)$$

Here the *Weber* number

$$We = G^2 La/(\rho_1\sigma) \qquad (3.65)$$

is computed using the *Laplace* constant

$$La = \sqrt{2}\lambda_{RT},\qquad(3.66)$$

and modified in the case of rough pipes

$$We_{rough} = \frac{\lambda_{1,rough}}{\lambda_{2,smooth}} We.\qquad(3.67)$$

This correlation was compared with 7172 data in the region $D_h = 0.003$ to $0.1 m$, $p = 1$ to 200 bar, $G = 290$ to 10 000 $kg/(m^2 s)$. The comparison shows that 70 to 90 % of the data are within the ± 30 % error band and presents some improvement of with respect to the *Friedel*'s correlation for mass flow rates below 4560 $kg/(m^2 s)$. This correlation is valid for heat transfer regimes without boiling crisis. In the case that the local heat transfer is associated with dry heated surface the authors recommend to use the following two-phase multiplier for Eq. (3.46)

$$\phi_{20}^2 = \left(\frac{\eta_1}{\eta_2}\right)^{0.2}\left(\frac{\rho_1}{\rho_2}\right)^{0.8}\left[1+X_1(\frac{\rho_2}{\rho_1}-1)\right]^{1.8}\qquad(3.68)$$

For heated pipes with internal structure promoting swirls and droplet deposition at the wall in all heat transfer regimes the authors proposed to use the *Martinelli - Nelson* method with the following two-phase friction multiplier

$$\phi_{20}^2 = 1+X_1\left(\frac{\rho_2}{\rho_1}\frac{\lambda_1}{\lambda_2}-1\right)+2.351X_1^{2.637}(1-X_1)^{0.565}\frac{\rho_2}{\rho_1}.\qquad(3.69)$$

The correlation was compared with 1018 data for $D_h = 0.01325$ and 0.02181 m, rib height of 0.000775 and 0.0011 m, incline 56.9 and 48.7°, vertical pipes, $p = 20$ to 210 bar, $G = 300$ to 1500 $kg/(m^2 s)$. 85 % of the data are in the 30 % error band.

Knowing the frictional pressure loss we can compute the *dissipation rate of the kinetic energy of turbulence* the field l per unit volume of the mixture by multiplying the friction pressure gradients by the corresponding velocities in each direction, add the thus-obtained products, and obtain

$$\alpha_l\rho_l\varepsilon_l = \left[u\left(\frac{dp}{dr}\right)_R + v\left(\frac{1}{r^\kappa}\frac{dp}{d\theta}\right)_R + w\left(\frac{dp}{dz}\right)_R\right]_l\qquad(3.70)$$

This is an important characteristic for a flow in a confined geometry. It contains information about the *characteristic size and frequency of the turbulent eddies*, which are of interest for the flow structure identification as well as for the heat and mass transfer modeling.

3.6 Three-phase flow

There are only a few investigations on frictional pressure drop in the three-phase flows. For liquid - solid two phase pressure drop due to friction *Sakagushi* et al. [16] (1990) correlated their own experimental data for vertical flow within an error band of ± 2% as follows

$$\left(\frac{dp}{dz}\right)_R = \frac{1}{2}\rho_2 w_2^2 \frac{\lambda_{R2}}{D_h} \phi^2 \tag{3.71}$$

where

$$\lambda_{R2} = \lambda_{R2}\left(\frac{k}{D_h}, \frac{\rho_2 w_2 D_h}{\eta_2}\right), \tag{3.72}$$

$$\phi^2 = 1 + \min\left[120, \ 5.28\times10^3\left(0.08 - \frac{D_3}{D_h}\right)\left(\frac{w_2}{\Delta w_{23\infty}}\right)^{-2.8}\alpha_3\right], \tag{3.73}$$

for $\frac{D_3}{D_h} \leq 0.08$ and

$$\phi^2 = 1 \tag{3.74}$$

for $\frac{D_3}{D_h} > 0.08$.

For liquid - solid - gas flow *Tomiyama* [18] (1990) approximate with the following correlation the experimental data obtained in the *Sakagushi*'s laboratory for gas phase in a form of a bubbles $\alpha_1 < 0.3$ with standard deviation of ±10%

$$\left(\frac{dp}{dz}\right)_R = \frac{1}{2}\rho_2(\alpha_2 w_2)^2 \frac{\lambda_{R2}}{D_h}\left(1 + 350\sqrt{\frac{\alpha_1}{\text{Re}_2 \ Fr_2}} / \left[0.98(1-\alpha_1)(1-\alpha_3)^{3.8}\right]\right), \tag{3.75}$$

where the *Blasius* formula for computation of the liquid only friction coefficient

$$\lambda_{R2} = 0.3164 / Re_2^{1/4}, \qquad (3.76)$$

with

$$Re_2 = \alpha_2 \rho_2 w_2 D_h / \eta_2, \qquad (3.77)$$

was used and

$$Fr_2 = \frac{(\alpha_2 w_2)^2}{gD_h}. \qquad (3.78)$$

The data belong to the following region $0 \leq \alpha_1 w_1 \leq 0.121$, $0.488 \leq \alpha_2 w_2 \leq 1.02$, $0 \leq \alpha_3 w_3 \leq 0.0375$, $p \approx 10^5$, $\rho_3 = 2400 \div 2640$, $0.0209 \leq D_h \leq 0.0504$, $0.00115 \leq D_3 \leq 0.00416$.

For higher void fraction $\alpha_1 > 0.3$ Minagawa [12] (1990) observed a flow pattern similar to the slug flow in the two-phase flow. It is remarkable that for both regimes, bubble and slug flow, the particles are completely surrounded by liquid. For the liquid - solid - gas slug flow *Minagawa* proposed the following correlation.

$$\left(\frac{dp}{dz}\right)_R = \phi^2 \left(\frac{dp}{dz}\right)_{32} \qquad (3.79)$$

where

$$\phi^2 = 1 + \frac{const}{X} + \frac{1}{X^2}, \qquad (3.80)$$

$$X^2 = \frac{(dp/dz)_{32}}{(dp/dz)_{10}}, \qquad (3.81)$$

$$\left(\frac{dp}{dz}\right)_{32} = \phi_{32}^2 \left(\frac{dp}{dz}\right)_2, \qquad (3.82)$$

$$\phi_{32}^2 = 1 + 400\alpha_3 / \left\{ \left[\left(1 + \frac{D_3}{D_h / 0.038}\right)^{3.62} \right] \left(\frac{w_2}{\Delta w_{32\infty}}\right)^{2.8} \right\}, \qquad (3.83)$$

$$\left(\frac{dp}{dz}\right)_{10} = \frac{1}{2}\rho_1(\alpha_1 w_1)^2 \frac{\lambda_{10}}{D_h}, \tag{3.84}$$

$$\lambda_{10} = \lambda_{10}\left(\frac{k}{D_h}, \frac{\alpha_1 \rho_1 w_1 D_h}{\eta_1}\right), \tag{3.85}$$

$$\left(\frac{dp}{dz}\right)_2 = \frac{1}{2}\rho_2 w_2^2 \frac{\lambda_{10}}{D_h}, \tag{3.86}$$

$$\lambda_2 = \lambda_2\left(\frac{k}{D_h}, \frac{\rho_2 w_2 D_h}{\eta_2}\right) \tag{3.87}$$

D_h	D_3	const	$\overline{\Delta}\%$
0.0209	0.00114	57.7	18.3
	0.00257	40.1	18.1
	0.00417	49.8	20.1
	0.00296 (AL)	52.4	23.0
0.0307	0.00114	49.1	11.7
	0.00257	55.9	21.3
	0.00417	47.9	11.8
	0.00296 (AL)	50.9	28.8
0.0507	0.00114	113	20.5
	0.00257	105	30.0

Nomenclature

Latin

C	constant
c_w	modified friction coefficient, kg/m^3
D_{BE}	outer rod diameter, m
D_h	hydraulic diameter, m
$\left(\dfrac{dp}{dz}\right)_R$	pressure drop per unit length, Pa/m
$\left(\dfrac{dp}{dz}\right)_{R1}$	pressure gradient of the gas flowing alone in the same tube, Pa/m
$\left(\dfrac{dp}{dz}\right)_{R2}$	pressure gradient of the liquid flowing alone in the same tube, Pa/m

Nomenclature

Fr $= \dfrac{(G/\rho)^2}{gD_h}$, total flow *Froud* number, *dimensionless*

G mass flow rate, $kg/(m^2 s)$

k roughness, m

La $= \sqrt{2}\lambda_{RT}$, *Laplace* constant, m

p pressure, N/m^2

p_{cr} critical pressure, N/m^2

Re *Reynolds* number, *dimensionless*

S_{BE} averaged distance between the rods in triangular array arrangement, m

V velocity, m/s

We $= G^2 La /(\rho_1 \sigma)$, modified *Weber* number for smooth pipe, *dimensionless*

We_{rough} $= \dfrac{\lambda_{1,rough}}{\lambda_{2,smooth}} We$, modified *Weber* number for rough pipe, *dimensionless*

We $= G^2 D_h /(\rho \sigma)$, total flow *Weber* number, *dimensionless*

X_{LM} *Lockhardt* and *Martinelli* factor, *dimensionless*

X_l l field mass fraction, *dimensionless*

z axial coordinate, m

Greek

α_l l field volume fraction, *dimensionless*

Δz pipe length, m

Δr distance, m

ε_l dissipation rate of the kinetic energy of turbulence the field l per unit volume of the mixture, W/m^3

ϕ_1^2 gas only two phase friction multiplier, *dimensionless*

ϕ_2^2 liquid only friction multiplier, *dimensionless*

η_{eff} effective dynamic viscosity, $kg/(ms)$

λ_{R10} friction coefficient for the total mass flow rate considered as gas, *dimensionless*

λ_{R20} friction coefficient for the total mass flow rate considered as liquid, *dimensionless*

λ_R friction pressure loss coefficient, *dimensionless*

ξ local pressure loss coefficient, *dimensionless*

τ share stress, N/m^2

ρ_l l field density, kg/m^3

ρ density, kg/m^3

Subscripts

1   gas
2   liquid
3   field 3

References

1. Abuaf N, Zimmer GA, Saha P (June 1981) A study of non equilibrium flashing of water in a converging diverging nozzle, vol 1 Experimental, vol 2 Modeling, NUREG/CR-1864, BNL-NUREG-51317
2. Avdeev AA (1983) Hydrodynamics of turbulent bubble two phase mixture, High Temperature Physic, vol 21 no 4 pp 707-715, in Russian
3. Avdeev AA (1986) Application of the Reynolds analogy to the investigation of the surface boiling in forced convection, High Temperature Physics, vol 24 no 1 pp 111 - 119, in Russian.
4. Chisholm D (1983) Two-phase flow in pipelines and heat exchanger, George Godwin, London and New York, p 110
5. Ergun S (1952) Fluid flow through packed columns, Chem. Eng. Prog. vol 48 no 2 pp 89-94
6. Friedel L (1979) New friction pressure drop correlations for upward, horizontal, and downward two - phase pipe flow. Presented at the HTFS Symposium, Oxford, September 1979 (Hoechst AG Reference No. 372217/24 698)
7. Gunter A Y, Shaw WA (1945) A general correlation of friction factors for various types of surfaces in cross flow, ASME Trans., vol 57 pp 643-660
8. Hetstroni G (1982) Handbook of multiphase systems. Hemishere Publ. Corp., Washington etc., McGraw-Hill Book Company, New York etc.
9. Idelchik IE (1975) Handbook of hydraulic resistance, Second edition, Hemisphere, Washington, 1986, translation o a Russian edition
10. Lockhart RW, Martinelli RC (1949) Proposed correlation of data for isothermal two-phase, two- component flow in pipes, Chem. Eng. Prog., vol 45 no 1 pp 39-48
11. Martinelli RC, Nelson DB (1948) Prediction of pressure drop during forced circulation boilng of water, Trans. ASME, vol 70, p 695
12. Minagawa H (1990) Pressure drop for liquid - gas - solid - slug flow, Kobe University, private communication
13. Nigmatulin BI (1982) Heat and mass transfer and force interactions in annular - dispersed two - phase flow, 7th Int. Heat Transfer Conf., Munich, pp 337-342
14. Palazov V (31. July 1991) Prediction of single-phase friction pressure losses in a converging-diverging nozzle, Private communication
15. Ransom VH et al. (1988) RELAP5/MOD2 Code Manual Volume 1: Code Structure, System Models, and Solution Methods, NUREG/CR-4312 EGG-2396, rev 1, pp 209-216

16. Sakagushi T, Minagawa H, Tomyama A, Shakutsui H (1989) Characteristics of pressure drop for liquid - solid two - phase flow in vertical pipes, Reprint from Memories of the faculty of engineering, Kobe University, no 36 pp 63-90
17. Staengel G, Mayinger F (March 23-24, 1989) Void fraction and pressure drop in subcooled forced convective boiling with refrigerant 12, Proc. of 7th Eurotherm Seminar Thermal Non-Equilibrium in Two-Phase Flow, Roma, pp 83-97
18. Tomyama A, Sakagushi T, Minagawa H (1990) Kobe University, private comunication
19. Zheng Q et al. (1991) Druckverlust in glatten und innenberippten Verdampferrohren, Wärme- und Stoffübertragung, vol 16 pp 323-330
20. Haland SE (1983) Simple and explicit formulas for the friction factor in turbulent pipe flow, J. Fluids Eng., vol 98 pp 173-181
21. Kirilov PL, Yur'ev YuS an Bobkov VP (1990) Handbook of thermal-hydraulic calculations, in Russian, Energoatomizdat, Moscow, Russia, pp 130-132

4. Diffusion velocities for algebraic slip models

4.1 Introduction

Historically stationary bubble flows or particle flows in pipes are investigated by measuring the volumetric flows of gas and liquid among the other flow characteristics. This measurements are easily recomputed in term of relative cross section averaged volume fluxes or relative velocities. The obtained results for the relative velocities are frequently compared with the bubble free rise velocity or particle free fall velocity. We consider next bubbles and solid particles in a continuum as particles and their free rising and free settling velocities in stationary continuum will be called simple free particle velocities. It was found that the free particle velocity is a function of the particle size. The next interesting finding was that the cross section averaged relative velocity differs from the free particle velocity which can be explained by the non-uniformity of the velocity and volume concentration profiles depending on the particle-continuum density ratio, orientation of the pipe flow and local parameters. Much later careful measurements provided a data base for the velocity and concentration profiles in pipes. The obtained experimental information is usually generalized by empirical correlations. Some of them are purely empirical while others are based on sound mathematical principles of averaging. The correlations are very useful inside the measurement data banks for steady state flow. In the past usually mixture momentum equations are used for computation of the mixture momentum and the redistribution of the momentum was simulated by using this empirical information. Later this approach was extended to transient pipe flows and then to 3D flows. This approach is very useful if the simulated processes are slowly changing. For fast processes the instant momentum redistribution of the mixture momentum leads to non-adequate process description especially in cases of very strong interfacial heat and mass transfer. For this case the complete set of momentum equations has to be integrated and the drag coefficients have to be provided. Usually the drag coefficients measured for steady flows are used in the latter approach. If one has a method to compute the drag coefficients from the correlation for steady state relative velocities the collected correlation data bank is then a reliable data base also for the drag coefficients. In this chapter we will provide first an approximation for computation of steady state drag coefficients from relative velocity measurements for single particles and for flows. Then we will present a collection of correlation for different cases. The correlation can be used as already mentioned

a) for steady state flows in addition to the mixture momentum equation completing the description of the mechanical interaction,
b) for slow transient flows in addition to the mixture momentum equation completing the description of the mechanical interaction,
c) for strong transient flows in form of drag coefficients used in the separated momentum equations.

In pool flows with low particle concentration the drag coefficients for single particles are the proper choice. For a cloud of particles the drag forces have to be modified taking into account the increased adhesion to the continuum. One should never forget that the drag forces for single particles are only the first choice for pipe flows. The better choice is the effective drag force based on drift flux models.

4.2 Drag as a function of the relative velocity

4.2.1 Wall force not taken into account

We start with the simplified momentum equations for the continuous and the disperse phase, denoted with c and d, respectively, neglecting compressibility, interfacial mass transfer, the spatial acceleration and the viscous terms

$$\alpha_c \rho_c \frac{\partial w_c}{\partial \tau} + \alpha_c \nabla p + \alpha_c \rho_c g - f_d^d - f_d^{vm} = 0, \tag{4.1}$$

$$\alpha_d \rho_d \frac{\partial w_d}{\partial \tau} + \alpha_d \nabla p + \alpha_d \rho_d g + f_d^d + f_d^{vm} = 0, \tag{4.2}$$

where the drag force per unit mixture volume

$$f_d^d = -\alpha_d \rho_c \frac{3}{4} \frac{c_d^d}{D_d} |\Delta w_{cd}| (w_c - w_d), \tag{4.3}$$

and the virtual mass force per unit mixture volume

$$f_d^{vm} = -\alpha_d \rho_c c_d^{vm} \frac{\partial}{\partial \tau} (w_c - w_d) \tag{4.4}$$

are functions of the relative velocities. All parameters in the above equations are cross section averaged. We multiply Eq. (4.1) by $\alpha_d \rho_d$, Eq. (4.2) by $\alpha_c \rho_c$ and subtract the thus obtained equations. The result is

$$\alpha_c \rho_c \alpha_d \rho_d \frac{\partial}{\partial \tau}(w_c - w_d) + (\alpha_c \alpha_d \rho_d - \alpha_d \alpha_c \rho_c)\nabla p - (\alpha_c \rho_c + \alpha_d \rho_d)(f_d^d + f_d^{vm}) = 0.$$
(4.5)

Dividing the thus obtained equation by $\alpha_c \rho_c \alpha_d \rho_d$ and replacing the forces defined by Eqs. (4.3) and (4.4) we obtain

$$(1 + bc_d^{vm})\frac{\partial}{\partial \tau}\Delta w_{cd} + (\frac{1}{\rho_c} - \frac{1}{\rho_d})\nabla p + b\frac{3}{4}\frac{c_d^d}{D_d}|\Delta w_{cd}|\Delta w_{cd} = 0,$$
(4.6)

or

$$\frac{\partial}{\partial \tau}\Delta w_{cd} + a\Delta w_{cd} = a\Delta w_{cd,\tau \to \infty},$$
(4.7)

where

$$a = b\frac{3}{4}\frac{c_d^d}{D_d}|\Delta w_{cd}|/(1 + bc_d^{vm}),$$
(4.8)

$$b = \frac{\alpha_c \rho_c + \alpha_d \rho_d}{\alpha_c \rho_d},$$
(4.9)

$$\Delta w_{cd,\tau \to \infty} = -(\frac{1}{\rho_c} - \frac{1}{\rho_d})\nabla p /(b\frac{3}{4}\frac{c_d^d}{D_d}|\Delta w_{cd}|).$$
(4.10)

We see from the Eq. (4.6) that the effect of the virtual mass force is to increase the effective particle inertia. For initial condition

$$\tau = 0,$$
(4.11)

$$\Delta w_{cd} = \Delta w_{cd,o},$$
(4.12)

the analytical solution of the above equation for constant pressure gradient and for $\tau = \Delta \tau$ is

$$\Delta w_{cd} = \Delta w_{cd,\tau \to \infty} + (\Delta w_{cd,o} - \Delta w_{cd,\tau \to \infty})e^{-a\Delta \tau}.$$
(4.13)

We see that for

$$\Delta \tau \to \infty,$$
(4.14)

$$\Delta w_{cd} \to \Delta w_{cd,\tau\to\infty} \qquad (4.15)$$

the velocity difference approaches the steady state velocity difference.

Equation (4.13) is valid for all direction. For the steady state case and vertical flow Eq. (4.1) reduces to

$$\nabla p = -\rho_c g + f_d^d / \alpha_c . \qquad (4.16)$$

Substituting Eq. (4.16) into Eq. (4.10) we obtain

$$\frac{\alpha_c + \alpha_d}{\alpha_c} \rho_c \frac{3}{4} \frac{c_d^d}{D_d} \Delta w_{cd,\tau\to\infty}^2 = g(\rho_d - \rho_c), \qquad (4.17)$$

and

$$f_d^d = -\frac{\alpha_c \alpha_d}{\alpha_c + \alpha_d} g(\rho_d - \rho_c). \qquad (4.18)$$

The latter equation is the well-known forces balance equation for a free falling sphere for $\alpha_c \to 1$, $\alpha_d \to 0$.

Note that Eq. (4.17) is valid also if we take into account the wall friction force.

For the vertical flow the drag coefficient can be computed from Eq. (4.17) resulting in

$$c_d^d = \frac{4}{3} D_d g \frac{|\rho_d - \rho_c|}{\rho_c} \frac{\alpha_c}{\alpha_c + \alpha_d} / \Delta w_{cd\infty}^2 . \qquad (4.19)$$

As already mentioned in the introduction, the steady state drag coefficient can be used also in the transient solution – Eq.(4.13). The coefficient a then takes the form

$$a = \frac{(\alpha_c \rho_c + \alpha_d \rho_d)|\rho_d - \rho_c|}{\rho_c \rho_d (\alpha_c + \alpha_d)} g / \left[\Delta w_{cd\infty} (1 + bc_d^{vm}) \right]. \qquad (4.20)$$

Remember that in this case $\Delta w_{cd\infty}$ is the steady state free settling velocity for the family of solid spheres or the free rising velocity for a family of bubbles.

The usual method for deriving the terminal speed of a spherical particle falling (or rising) under gravity is to consider the balance between *buoyancy and drag forces*. Employing the drag coefficient for a particle in an infinite medium $c_{d,single}^d$, we have

$$\frac{4}{3}\pi r_d^3 g |\Delta\rho|_{cd} = c_{d,single}^d \pi r_d^2 \frac{1}{2} \rho_c \Delta w_{cd\infty,single}^2 \qquad (4.21)$$

or

$$c_{d,single}^d = \frac{4}{3} D_d g \frac{|\Delta\rho|_{cd}}{\rho_c} \bigg/ \Delta w_{cd\infty,single}^2. \qquad (4.22)$$

$\Delta\rho$ is the density difference. For the case of a drag coefficient not depending on the bubble form the above equation reduces to

$$\Delta w_{cd\infty,single} = const \sqrt{D_d g \frac{|\Delta\rho|_{cd}}{\rho_c}}. \qquad (4.23)$$

It is fundamental velocity scaling parameter in bubbly and slug flow. For slug flow the slug bubble diameter is comparable with the hydraulic diameter and therefore

$$\Delta w_{cd\infty,single} = const \sqrt{D_h g \frac{|\Delta\rho|_{cd}}{\rho_c}}, \qquad (4.24)$$

see *Dimitrescu* [8] (1943). For large bubbles where the size is comparable with the *Rayleigh-Taylor* instability length

$$\lambda_{RT} = \sqrt{\frac{\sigma_2}{g(\rho_2 - \rho_1)}}, \qquad (4.25)$$

$$\Delta w_{cd\infty,single} = const \sqrt{\sqrt{\frac{\sigma_2}{g(\rho_2 - \rho_1)}} g \frac{|\Delta\rho|_{cd}}{\rho_c}} = const \left(\frac{\sigma g \Delta\rho_{cd}}{\rho_c^2}\right)^{1/4}, \qquad (4.26)$$

see *Kutateladze*. More information about the constants will be given later on in this Chapter. Equation (4.22) is in fact Eq. (4.19) for disappearing concentration of the dispersed phase. Equation (4.19) is more general. Eliminating the group

$$\frac{4}{3} D_d g \frac{|\Delta\rho|_{cd}}{\rho_c} = c^d_{d,single} \Delta w^2_{cd\infty,single} \tag{4.27}$$

from Eqs. (4.19) and (4.22) results in

$$c^d_d = c^d_{d,single} \frac{\alpha_c}{\alpha_c + \alpha_d} \bigg/ \left(\frac{\Delta w_{cd\infty}}{\Delta w_{cd\infty,single}} \right)^2 . \tag{4.28}$$

This result demonstrates that the single particle drug coefficient has to be modified in order to obtain an effective drag coefficient for clouds of mono-disperse particles. *Richardson* and *Zaki* [27] proposed in 1954 to correlate experimental data with the function

$$\frac{\Delta w_{cd\infty}}{\Delta w_{cd\infty,single}} = \alpha_c^n . \tag{4.29}$$

where *n* is depending on the particle-continuum system and local parameters. Using Eq. (4.29) Eq. (4.28) takes the form

$$c^d_d = \frac{c^d_{d,single}}{(\alpha_c + \alpha_d)\alpha_c^{2n-1}} . \tag{4.30}$$

Note that the drag coefficient may also depend on the relative velocity. This makes iteration necessary to compute the drag coefficient. To avoid this, the *Wallis* collection of correlations given in the Section 4.3 is recommended.

4.2.2 Wall forces taken into account

We consider a steady state, one-dimensional, fully developed flow, consisting of two velocity fields designated with *c* and *d* without any mass sources. The continuous field wets the pipe wall. After inserting the pressure gradient from the mixture momentum equation

$$\frac{dp}{dz} + \rho g + f_{wc} = 0 \tag{4.31}$$

into the momentum equation for the continuous velocity field

$$\alpha_c \left(\frac{dp}{dz} + \rho_c g \right) + f_{wc} + \alpha_d \rho_c \frac{3}{4} \frac{c^d_d}{D_d} |\Delta w_{cd}|(w_c - w_d) = 0 \tag{4.32}$$

POSTKARTE AUS DEM HARENBERG-KALENDER OSTSEEKÜSTE 2008

Rapsblüte am Leuchtturm Staberhuk auf Fehmarn

Foto Fan & Mross

and solving with respect to c_d^d we obtain

$$c_d^d = \frac{4}{3} D_d \frac{\alpha_c \Delta \rho_{cd} g + f_{wc}}{\rho_c |\Delta w_{cd}|(w_d - w_c)}. \qquad (4.33)$$

Thus, using correlations for the wall friction force f_{wc} and a drift flux correlation (or other type of correlations) for computing of the relative velocity Δw_{dc}, we can easily compute from the above equation the drag coefficient c_d^d.

4.3 Two velocity fields

4.3.1 Single bubble terminal velocity

Wallis [35] (1974) approximated the experimental observations of a number of authors for the terminal speed of bubble by the algorithm presented in Table 4.1. The dimensionless terminal velocity

$$V^* = \Delta w_{dc,\infty} \left(\frac{\rho_c^2}{\eta_c g \Delta \rho} \right)^{1/3} \qquad (4.34)$$

is approximated as a function of the dimensionless bubble size

$$r^* = \frac{D_d}{2} \left(\frac{\rho_c g \Delta \rho}{\eta_c^2} \right)^{1/3}, \qquad (4.35)$$

and the *Archimedes* number

$$Ar = \sqrt{\frac{\sigma^3 \rho_c^2}{\eta_c^4 g \Delta \rho}}. \qquad (4.36)$$

In a later work *Wallis* et al. [36] (1976) provided for this regions also the corresponding drag coefficients which are also summarized in Table 4.1. They are function of the *Reynolds* and *Weber* numbers defined as follows

$$Re_d = D_d \rho_c \Delta w_{dc} / \eta_c, \qquad (4.37)$$

$$We_d = D_d \rho_c \Delta w_{dc}^2 / \sigma. \qquad (4.38)$$

4. Diffusion velocities for algebraic slip models

Table 4.1. The terminal speed of bubble V^* [35] (1974), and the corresponding drag coefficients c_1^d [36] (1976)

Region	Range	V^*	c_1^d	Range
1	$r^* < 1.5$	$r^{*2}/3$	$\dfrac{16}{Re_d}$	$Re_d < 2.25$
2A	$1.5 \leq r^* < \min(13.4, r^*_{2A \to 3})$	$0.408 r^{*1.5}$	$\dfrac{13.6}{Re_d^{0.8}}$	$2.25 \leq Re_d < 2068$, $We_d < 4$
2D	$13.4 \leq r^* < r^*_{2D \to 3}$	$\dfrac{1}{9} r^{*2}$	$\dfrac{73}{Re_d}$	$Re_d \geq 2068$, $We_d < 4$
3	$r^*_{2D \to 3} \leq r^* < r^*_{3D \to 4}$	$Ar^{1/3}\sqrt{2/r^{*2}}$	$We_d = 4$	
4	$r^*_{3D \to 4} \leq r^* < r^*_{4 \to 5}$	$\sqrt{2} Ar^{1/6}$	$\dfrac{We_d}{3}$	$We_d < 8$
5	$r^*_{4D \to 5} \leq r^*$	$\sqrt{r^*}$	$\dfrac{8}{3}$	

The boundary of the ranges are defined as follows

$$r^*_{2A \to 3} = \sqrt{\frac{\sqrt{2}}{0.408}} Ar^{1/6} = 1.862 Ar^{1/6}, \qquad (4.39)$$

$$r^*_{2D \to 3} = \left(9\sqrt{2}\right)^{2/5} Ar^{2/15} = 2.77 Ar^{2/15}, \qquad (4.40)$$

$$r^*_{3D \to 4} = Ar^{1/3}, \qquad (4.41)$$

$$r^*_{4 \to 5} = 2 Ar^{1/3}. \qquad (4.42)$$

The first range is for sufficiently small bubbles (D_1 less than 0.0005m), where *the viscous forces dominate inertia* forces and the rise velocity can be predicted from the theory of the "*creeping flow*" as long as the interface remains spherical.

In the second region the bubbles behave approximately as solid particles.

In Region 3 *the shape of the bubbles departs significantly from sphericity*, they move in a helical or zig-zag path, and the velocity decreases as the "effective radius", r_1, increases. The effective radius is the radius of a sphere which would have the same volume as the dispersed globule.

In the Region 4 (bubble diameter approximately 0.001 to 0.02m) the terminal bubble velocity

$$\Delta w_{dc,\infty} = Ku \, V_{Ku} \qquad (4.43)$$

is independent on size and is proportional to the *Kutateladze* terminal velocity

$$V_{Ku} = \left[g \frac{\Delta \rho_{cd}}{\rho_c} \left(\frac{\sigma}{g \Delta \rho_{cd}} \right)^{1/2} \right]^{1/2} = \left(\frac{\sigma g \Delta \rho_{cd}}{\rho_c^2} \right)^{1/4} \qquad (4.44)$$

The coefficient is

$$Ku = \sqrt{2} \, . \qquad (4.45)$$

The inclination of the pipe can be taken into account by taking into account only the component of the gravity acceleration being parallel to the pipe axis

$$V_{Ku}^* = \left(\frac{\sigma g \cos \varphi \Delta \rho_{cd}}{\rho_c^2} \right)^{1/4} . \qquad (4.46)$$

where φ is the angle between the positive flow direction and the upwards directed vertical. The *Kutateladze* terminal velocity is frequently used as a scaling factor for correlating the so called weighted mean drift velocity in the drift flux theory for bubbly flow.

The collective motion of bubbles in this region for $\alpha_1 < 0.25$ to 0.30 is called by *Ishii* and *Chawla* motion of distorted particles. The collective motion of bubbles in this region for $\alpha_1 > 0.25$ to 0.30 is named churn-turbulent flow.

In the Region 5 the bubbles are very large assuming *spherical cap* shape and a flat base. In many publications such a bubble is named *Taylor* bubble. Both viscous and surface tension forces can be neglected and the rise velocity is given by a balance between form-drag and buoyancy. The analytical result derived by *Davis* and *Taylor* for this regime is equivalent to

$$\Delta w_{dc,\infty} \equiv V_{TB} = \sqrt{D_h g \Delta \rho_{cd} / \rho_c} \, . \qquad (4.47)$$

The inclination of the pipe can be taken into account by taking into account only the component of the gravity acceleration being parallel to the pipe axis

$$\Delta w_{dc,\infty} \equiv V_{TB}^* = \sqrt{D_h g \cos \varphi \Delta \rho_{cd} / \rho_c} \, . \qquad (4.48)$$

where φ is the angle between the positive flow direction and the upwards directed vertical. The *Taylor* terminal velocity is frequently used as a scaling factor for correlating the so called weighted mean drift velocity in the drift flux theory for slug flows. Note that the ratio

$$\frac{V_{Ku}^*}{V_{TB}^*} = \left(\frac{\sigma}{D_h^2 g \cos\varphi \Delta\rho_{cd}} \right)^{1/4} = \sqrt{\frac{\lambda_{RT}}{D_h}} . \qquad (4.49)$$

The collective motion of bubbles in this regime in channels is named *slug flow*. For vessels with diameters much larger than 40 λ_{RT} the slug bubbles can not be sustained due to the interfacial instability and they disintegrate to cap bubbles, *Kataoka* and *Ishii* - see in [20] (1987).

The rise velocity of a group of bubbles is less then the terminal rise velocity of one single bubble due to of the mutual interference of the bubbles. *Zuber* and *Findlay* [37] propose the following relationship

$$\Delta w_{dc\infty} = \Delta w_{dc\infty,single} \alpha_c^{n-1} . \qquad (4.50)$$

where $n = 3$ for Regions 1 and 2, $n = 5/2$ for churn-turbulent flow. For the churn-turbulent flow *Ishii* [15,16] (1977) proposes $n = 2.75$. For the same region *Clark* and *Flemmer* [4] (1985) measured in a vertical 0.1 m - diameter pipe $n = 1.702$. The differences between the different authors can be explained by the differences of the pipe diameters for which the data are obtained, which influence the distribution profile of the void fraction. *Wallis* [34], p.178, (1969) proposes the following relationship

$$n = 4.7 \frac{1 + 0.15 Re_{d\infty}^{0.687}}{1 + 0.253 Re_{d\infty}^{0.687}} , \qquad (4.51)$$

where

$$Re_{d\infty} = \rho_c D_d \Delta w_{dc,\infty} / \eta_c . \qquad (4.52)$$

Thus the relationship (4.30) is slightly modified

$$c_{cd}^d = \frac{c_{cd,\infty}^d}{(\alpha_d + \alpha_c)\alpha_c^{2n-3}} . \qquad (4.53)$$

If for the computational simulation of pool bubbly flow the used discretization mesh size is in the order of magnitude of a few bubble diameters one can use the above method.

4.3.2 Single particle terminal velocity

Consider a family of solid particles with a representative volume median diameter D_d, which are further denoted as *discrete field d* moving in a continuum liquid, denoted further as a *continuum c*. The mixture flows in a channel with a hydraulic diameter D_h. In this case the maximal volumetric fraction which can be occupied by solid spheres is

$$\alpha_d \leq \alpha_{d\max} \quad (\alpha_{d\max} = \pi/6 \cong 0.52). \tag{4.54}$$

To describe the *relative velocity of the solid particles* we use a similar approach to that in Section 3.1 - first estimation of the terminal speed of a single particle in an infinite continuum, and thereafter using it for the estimation of the terminal speed of a group of particles. Again the result is used to compute the terminal speed. The data are approximated by the following set of correlations – see *Wallis* [35] (1974):

$$V^* = \frac{2}{9} r^{*2} \qquad r^* < 1.5 \text{ (entirely viscous flow)} \tag{4.55}$$

$$V^* = 0.307\ r^{*1.21} \qquad 1.5 \leq r^* < 10 \quad \text{(inertia forces become important)} \tag{4.56}$$

$$V^* = 0.693\ r^{*0.858} \qquad 10 \leq r^* < 36 \quad \text{(reflects the effect of vortex shading)} \tag{4.57}$$

$$V^* = 2.5\ \sqrt{r^*} \qquad 36 \leq r^* \quad \text{(fully-developed turbulent wake)}. \tag{4.58}$$

The falling velocity of a *group of particles* is less then the terminal falling velocity of one single particle due to the mutual interference of the particles. For this case *Richardson* and *Zaki* [27], p.47 proposed in 1954 to use in Eqs. (4.29, 4.30) the following exponents

$$n = 4.65 + 19.5\ D_d/D_h, \qquad Re_d < 0.2, \tag{4.59}$$

$$n = (4.35 + 17.5\ D_d/D_h)\ Re_d^{-0.03}, \qquad 0.2 \leq Re_d < 1, \tag{4.60}$$

$$n = (4.45 + 18.0\ D_d/D_h)\ Re_d^{-0.1}, \qquad 1 \leq Re_d < 200, \tag{4.61}$$

$$n = 4.45 \, Re_d^{-0.1}, \qquad 200 \leq Re_d < 500, \qquad (4.62)$$

$$n = 2.39, \qquad 500 < Re_d. \qquad (4.63)$$

Rowe [28] found a convenient empirical equation for the estimation of the *Richardson-Zaki* exponent

$$n = 2.35 \frac{2 + 0.175 Re_{d\infty}^{3/4}}{1 + 0.175 Re_{d\infty}^{3/4}} \qquad (4.64)$$

representing the data with standard deviation of ± 0.02.

Zwirin et al. [38] (1989) experimentally observe that the free settling velocity of particles in liquid, having averaged temperature greater than the liquid saturation temperature, is higher than the free settling velocity of particles with temperature lower than the saturated temperature

$$V/V_{T<T'} = 1.02 + 3.87 \, 10^{-5} \, (T - T') \qquad (4.65)$$

for $373 < T < 973.14 \, K$.

4.3.3 Cross section averaged bubble rise velocity in pipes – drift flux models

4.3.3.1 Basics

Computing the weighted mean velocity defined by

$$w_d = \frac{\frac{1}{A}\int_A \alpha_{d,local} w_{d,local} dA}{\frac{1}{A}\int_A \alpha_{d,local} dA}, \qquad (4.66)$$

which appears in the cross section averaged conservation equations results in the expression

$$w_d = C_0 j + V_{dj}^*. \qquad (4.67)$$

j is the volumetric flux of the mixture which is equivalent to the center of volume velocity of the mixture. Here the *distribution parameter*

$$C_0 = \frac{\frac{1}{A}\int_A \alpha_{d,local} j_{local} dA}{\left(\frac{1}{A}\int_A j_{local} dA\right)\left(\frac{1}{A}\int_A \alpha_{d,local} dA\right)}, \quad (4.68)$$

and the weighted mean drift velocity

$$V_{dj}^* = \frac{\frac{1}{A}\int_A \alpha_{d,local} V_{dj,local} dA}{\left(\frac{1}{A}\int_A \alpha_{d,local} dA\right)}, \quad (4.69)$$

are averages over the cross section of the channel. The local drift velocity of the dispersed phase is defined with respect to the total volume flux

$$V_{dj,local} = w_{d,local} - j. \quad (4.70)$$

This expression has been known in the literature as the *Zuber* and *Findlay* model since 1965– see *Zuber* and *Findlay* [37]. A large number of publication exists on correlating experimental data with the so called distribution parameter C_0 and the weighted mean drift velocity V_{dj}^*. We will give a summary reviewing this literature in this chapter for gas liquid flows.

4.3.3.2 Some useful relationships

Having the definition equation (4.44) the mean gas and liquid velocities are easily computed as a function of the mixture mass flow rate ρw, the densities and the local volume fractions

$$w_c = \frac{\rho w(1-\alpha_d C_o) - \alpha_d \rho_d V_{dj}^*}{(1-\alpha_d)\left[\rho w(1-\alpha_d C_o) + \alpha_d \rho_d C_o\right]}, \quad (4.71)$$

$$w_d = \frac{\rho w - (1-\alpha_d C_o)\rho_c w_c}{\alpha_d \rho_d}. \quad (4.72)$$

The relative velocity is therefore

$$\Delta w_{dc} = \frac{(1-C_o)w_c - V_{dj}^*}{1-\alpha_d C_o}. \quad (4.73)$$

Defining the averaged volumetric flow concentration

$$\dot{\alpha}_d = \frac{j_d}{j}, \qquad (4.74)$$

the drift flux equation takes the form

$$\frac{\dot{\alpha}_d}{\alpha_d} = C_0 + \frac{V_{dj}^*}{j}, \qquad (4.75)$$

or

$$\alpha_d = \frac{\dot{\alpha}_d}{C_0 + \dfrac{V_{dj}^*}{j}}. \qquad (4.76)$$

The velocity ratio takes the form

$$\frac{w_d}{w_c} = \frac{1-\alpha_d}{\dfrac{1}{C_0 + V_{dj}^*/j} - \alpha_d}. \qquad (4.77)$$

The above relations are very useful for comparison of information from different publications on this issue.

Zuber and *Findlay* [37] clearly demonstrated that the reason for success of correlating data is in using flow regime dependent correlations. As we will show in the following literature review there are also the so called general drift flux correlations, some of them containing smooth change over the flow pattern boundaries.

Having the drift flux parameter we compute easily from Eqs. (4.59) through (4.61) the field velocities and the relative velocity, respectively. If separated momentum equations are used the cloud drag coefficient can be easily computed by using Eqs. (4.19) and (4.61).

4.3.3.3 Bubble or churn-turbulent flow

Zuber and *Findlay* [37] (1965) demonstrated that for adiabatic flows the data of several authors can be generalized with the following expressions

$$w_d = C_0 j + 1.53\, V_{Ku}. \qquad (4.78)$$

For bubble flow or churn-turbulent flow the distribution parameter is

$$C_0 = 1.2 . \tag{4.79}$$

The distribution parameter may take values less then 1 for wall void picking profiles and values up to 1.5 for strong center line picking of the void fraction. For slug flow observed for values of

$$\alpha_1 \approx 0.26 \tag{4.80}$$

the recommended expression is

$$w_d = C_0 j + 0.35 \, V_{TB} . \tag{4.81}$$

The correlations are valid for co-current up flow. It seems that for small diameters

$$D_h < 19.1 \left(\frac{\sigma}{g \cos \varphi \Delta \rho_{cd}} \right)^{1/2} , \tag{4.82}$$

or

$$D_h / \lambda_{RT} < 19.5 \tag{4.83}$$

the slug flow is always the expected regime. Note that *Kataoka* and *Ishii* [20] used instead of 19.5 the value of 40.

Coddington and *Macian* [5] compared in the year 2000 a set of empirical correlation for $0.1 \le p \le 15 MPa$, $1 \le G \le 2000 kg/(m^2 s)$, channel length between 1.7 and 3.7 m, for pipes and rod bundles with rod diameters 9.5, 10.7, 12.2, 12.3 mm, and hydraulic diameters between 4 and 13 mm, with uniform and variable axial heat distribution, subcooling between 118 and 0 K, and heat fluxes between 5 and 3377 kW/m^2. The results for the mean absolute error and the standard deviation of the absolute error are given in Table 4.2.

This very informative study confirms once again the sound physical basics of the *Zuber* and *Findlay* correlation.

Lellouche [17] (1974) correlated the void cross section distribution parameter with

$$C_0 = 1/\left[A_o + (1 - A_o) \alpha_1^{B_o} \right], \tag{4.84}$$

where

Table 4.2. Wide range void fraction correlation

Correlation	Year	Data Source	Aver. err.	Stand. dev. in %
Zuber-Findlay	1965	Tube	−0.025	11.4
Ishii	1977	Tube	0.048	12.6
Gardner	1980	Tube	0.056	11.1
Liao, Parlos and Grifith	1985	Tube	0.028	9.4
Takeochi	1992	Tube	0.040	8.3
Sun	1980	RB+Tube	−0.041	11.4
Jowitt	1981	RB	0.057	11.6
Sonnenburg	1989	RB+Tube	0.049	9.7
Toshiba	1989	RB	0.019	10.3
Dix	1971	RB	−0.010	9.2
Bestion	1985	RB+Tube	0.018	8.8
Chexal-Lellouche	1992	RB+Tube	−0.017	7.8
Inoue	1993	RB	0.003	8.3
Maier and Coddington	1996	RB	−0.002	7.1

RB: Rod Bundle

$k_1 = 0.833$,

$A_o = k_1 + (1 - k_1) \, p/22115000$,

$B_o = \dfrac{1}{1 - A_0}(1 + 1.57 \rho_1 / \rho_2)$,

and the weighted mean drift velocity with

$$V_{1j}^* = 1.41 \, V_{Ku} \,. \tag{4.85}$$

The correlation is valid for co-current up flow.

Clark and *Flemmer* [4] (1985) reported a correlation for bubble flow in vertical tube with hydraulic diameter $D_h = 0.1m$. The void cross section distribution parameter is correlated taking into account the void fraction dependence which was already discussed by *Zuber* but not established for practical applications: For up flows

$$C_0 = 0.934(1 + 1.42\alpha_1), \tag{4.86}$$

and for down flow with

$$C_0 = 1.521(1-3.67\alpha_1),\qquad(4.87)$$

The weighted mean drift velocity is correlated again as proposed by *Zuber* $V_{1j} = 1.53\, V_{Ku}$, for both cases.

Ishii [15] (1977) reported a correlation for bubble flow ($w_2 > 0.5$) in a vertical tube. For a round tube the void cross section distribution parameter is correlated with

$$C_0 = 1.2 - 0.2\sqrt{\rho_1/\rho_2},\qquad(4.88)$$

for a rectangular channel with

$$C_0 = 1.35 - 0.35\sqrt{\rho_1/\rho_2},\qquad(4.89)$$

and the weighted mean drift velocity is correlated with

$$V_{1j}^* = \sqrt{2}\,(1-\alpha_1)^{1.75}\, V_{Ku},\qquad(4.90)$$

for both cases. For churn-turbulent bubbly flow *Ishii* uses the *Zuber-Findlay* expression with constant, $\sqrt{2}$, $V_{1j}^* = \sqrt{2}\, V_{Ku}$. For slug flow Ishii uses the *Zuber-Findlay* expression $V_{1j}^* = 0.35\, V_{TB}$.

In a later work *Kataoka* and *Ishii* [20] (1987) investigated pool flows. They used the same void cross section distribution parameter for a round tube and a rectangular channel as before but introduced viscosity dependence in the weighted mean drift velocity as follows. For the low viscous case characterized by

$$N\eta_2 \le 2.25 \times 10^{-3},\qquad(4.91)$$

$$N\eta_2 = \eta_2 / \sqrt{\rho_2 \sigma_2 \lambda_{RT}}\quad\text{(Viscous number)},\qquad(4.92)$$

and small diameters

$$D_h^* \le 30,\qquad(4.93)$$

where

$$D_h^* = D_h / \lambda_{RT},\qquad(4.94)$$

the weighted mean drift velocity is correlated with

$$V_{1j}^* = 0.0019 D_h^{*0.809} \left(\rho_1/\rho_2\right)^{-0.157} N\eta_2^{-0.562} V_{Ku}. \tag{4.95}$$

For large diameters

$$D_h^* > 30, \tag{4.96}$$

the weighted mean drift velocity is correlated with

$$V_{1j}^* = 0.03 \left(\rho_1/\rho_2\right)^{-0.157} N\eta_2^{-0.562} V_{Ku}. \tag{4.97}$$

For the high viscous case characterized by

$$N\eta_2 > 2.25 \times 10^{-3}, \tag{4.98}$$

the weighted mean drift velocity is reported for large diameters

$$D_h^* > 30, \tag{4.99}$$

and is

$$V_{1j}^* = 0.92 \left(\rho_1/\rho_2\right)^{-0.157} V_{Ku}. \tag{4.100}$$

The accuracy of the void fraction prediction reported by the authors is ±20% for $\alpha_1 w_1 = 0.1$ to 2.5, $D_h = 0.01$ to 0.6, $p = (1$ to $25)10^5$ Pa for air-water, $p = (1$ to $182)10^5$ Pa for steam-water.

4.3.3.4 Bubble flow in an annular channel

Staengl and *Mayinger* [33] (1989) reported for this case the following correlation the void cross section distribution parameter with

$$C_0 = \dot{\alpha}_1 \left[1 + 1.409 Fr^{-0.01}\left(1 - \frac{h'-h_2}{h''-h'}\right)^{0.164}\left(\frac{\rho_1}{\rho_2}\right)^{0.694}\left(\frac{1-X_1}{X_1}\right)^{0.864}\left(1 - \frac{p}{p_{cr}}\right)^{0.124}\right], \tag{4.101}$$

where

$$\dot{\alpha}_1 = \frac{1}{1 + \frac{1-X_1}{X_1}\frac{\rho_1}{\rho_2}}, \tag{4.102}$$

$$Fr = \frac{G^2}{gD_h\rho_2^2}, \qquad (4.103)$$

$$X_1 = \frac{\alpha_1\rho_1 w_1}{G}, \qquad (4.104)$$

and the weighted mean drift velocity with

$$V_{1j}^* = 1.18\, V_{Ku}. \qquad (4.105)$$

The correlation is obtained for subcooled boiling of CCl_2F_2 in a channel with D_h = 0.014, for the parameter region $10 \le T' - T_2 \le 50\ K$, $12\ 10^5 \le p \le 40\ 10^5\ Pa$, $500 \le G \le 3000\ kg/(m^2s)$.

4.3.3.5 Slug flow in a tube

As already mentioned the free rising Taylor bubble velocity is used as a velocity scale for the weighted mean drift velocity for slug flow in a tube.

Dimitresku [8] (1943) correlated the void cross section distribution parameter with

$$C_0 = 1.2, \qquad (4.106)$$

and the weighted mean drift velocity with

$$V_{1j}^* = 0.351\, V_{TB}^*. \qquad (4.107)$$

The correlation was validated for

$$\frac{\rho_2 j D_h}{\eta_2} > 8000, \qquad (4.108)$$

and volume of *Taylor* bubble $> (0.4\ D_h)^3$. *Zuber* and *Findlay* [37] (1965) confirmed the data for $D_h \le 0.05$. For larger diameters, $D_h > 0.05$, *Delhaye* et al. [7] (1981) proposed to use

$$V_{1j}^* = 0.56\, V_{TB}^*. \qquad (4.109)$$

Wallis [34] (1969) observed the dependence on the bubble to pipe radius as follows

$$V_{1j}^* = V_{TB} \tag{4.110}$$

for

$$D_1 < D_h/8, \tag{4.111}$$

and

$$V_{1j}^* = 1.13\, V_{TB} e^{-D_1/D_h} \tag{4.112}$$

for

$$D_h/8 < D_1 < 0.6. \tag{4.113}$$

Kuroda - see in [29] (1987) correlated the weighted mean drift velocity with

$$V_{1j}^* = \left\{ 0.35 - \frac{0.25}{\left[\left(\sqrt{Bo} - 1.9\right)/2.12\right]^{2.67} + 1} \right\} V_{TB}^* \tag{4.114}$$

where

$$Bo = \rho_2 g D_h^2 / \sigma. \tag{4.115}$$

Bendiksen [1] (1985) investigated vertical flow within $0 < 4/Eo < 0.6$ and correlated the weighted mean drift velocity with

$$V_{1j}^* = 0.344 \frac{1 - 0.96 e^{-0.0165 Eo}}{\left(1 - 0.52 e^{-0.0165 Eo}\right)^{3/2}} \sqrt{1 + \frac{20}{Eo}\left(1 - \frac{6.8}{Eo}\right)}\, V_{TB}^* \tag{4.116}$$

where

$$Eo = \frac{g \Delta \rho_{21} D_h^2}{\sigma} \tag{4.117}$$

is the pipe *Eötvös* number. *Bendiksen* observed that the void cross section distribution parameter is different for laminar

$$C_0 = 2.29 \left[1 - \frac{20}{Eo}\left(1 - e^{-0.0125 Eo}\right) \right], \tag{4.118}$$

and turbulent flow

$$\frac{\log Re + 0.309}{\log Re - 0.743}\left[1 - \frac{2}{Eo}\left(3 - e^{-0.025 Eo}\log Re\right)\right], \qquad (4.119)$$

where

$$Re = \rho_2 j D_h / \eta_2. \qquad (4.120)$$

Bendiksen observed that for large *Froud* numbers $Fr > 3.5$ where $Fr = j/V_{TB}$ there is no strong influence of the inclination on the distribution parameter $C_o \approx$ 1.19 to 1.2., but for $Fr < 3.5$ the distribution parameter may vary within 20% $C_o \approx$ 1. to 1.2.

4.3.3.6 Annular flow

Ishii [15] (1977) correlated the void cross section distribution parameter for annular flow with

$$C_0 = 1 + \frac{1 - \alpha_1}{\alpha_1 + \left[\frac{1 + 75(1 - \alpha_1)}{\sqrt{\alpha_1}}\frac{\rho_1}{\rho_2}\right]^{1/2}}, \qquad (4.121)$$

and the weighted mean drift velocity with

$$V_{1j}^* = 8.16(C_o - 1)(1 - \alpha_1)^{1/2} V_{TB}. \qquad (4.122)$$

For

$$|j_1| < V_{TB}\left(\frac{1}{C_0} - 0.1\right) \qquad (4.123)$$

Ishii reported the existence of churn turbulent flow.

Delhaye et al. [7] (1981) correlated the void cross section distribution parameter with

$$C_0 = 1, \qquad (4.124)$$

and the weighted mean drift velocity with

$$V_{1j}^* = 23\sqrt{\frac{\eta_2 \alpha_2 w_2}{\rho_1 D_h} \frac{\rho_2 - \rho_1}{\rho_2}}.\qquad(4.125)$$

4.3.3.7 Full-range drift-flux correlation

There are successful attempts to generalize data for all flow regimes in a pipe flow by the so called full-range drift flux correlations. Three of the best examples will be given below.

The Holmes correlation from 1981: Holmes [13] (1981) correlated the void cross section distribution parameter with

$$C_0 = 1/\left[A_o + (1-A_o)\alpha_1^{B_o}\right].\qquad(4.126)$$

Here

$$A_o = 1 - 0.328\,(1 - 1.5228\,10^{-3}\,p)F^2,$$

$$B_o = B_1 B_2 / \left[1 + F(B_2 - 1)\right],$$

$$B_1 = 2.94 + 7.763\,10^{-4}\,p + 9.702\,10^{-6}\,p^2,$$

$$B_2 = \sqrt{\rho_2 / \rho_1},$$

$$G_1 = 251.7 + 232.7\,p^{1/2} + 18.39\,p,$$

$$G_2 = -4.8 + 6650\,\rho_1/\rho_2 + 7.62\,p^{1/2} + 14.20\,p,$$

$$G_{HS} = 1/\sqrt{\left|x_1/G_2^2 + (1-x_1)/G_1^2\right|},$$

$$F = 1/e^{20683.9\,|G|G_{HS}}.$$

The weighted mean drift velocity was correlated with

$$V_{1j}^* = \frac{(1-\alpha_1 C_o)C_o K(\alpha_1)}{\sqrt{\rho_1/\rho_2}\,\alpha_1 C_o + 1 - \alpha_1 C_o} V_{Ku}.\qquad(4.127)$$

Here

$0 \leq \alpha_1 \leq \alpha_1^*$	$K(\alpha_1) = 1.53 C_o$
$\alpha_1^* \leq \alpha_1 \leq \alpha_2^*$	$K(\alpha_1) = \dfrac{\dfrac{1.53}{C_o}(\alpha_2^* - \alpha_1)^2 + Ku(\alpha_1^* - \alpha_1)^2}{(\alpha_2^* - \alpha_1)^2 + (\alpha_1 - \alpha_1^*)^2}$
$\alpha_2^* \leq \alpha_1 \leq 1$	$K(\alpha_1) = Ku$

$Ku = w_1 / V_{Ku}$	(Kutateladze number) $K(Ku)$
≤ 2	0
4	1
10	2.1
14	2.5
20	2.8
28	3.0
≥ 50	3.2

The transition limits are given as follows: α_1^* for bubble-to-film flow and α_2^* for slug-to-annular flow. The limits are set as follows $\alpha_1^* = 0.18$, $\alpha_2^* = 0.45$.

The Chexal et al. correlation from 1989: Chexal et al. [2] (1989) correlated the void cross section distribution parameter with

$$C_o = FrC_{ov} + (1-Fr)C_{oh}\left[1 + \alpha_1^{0.05}(1-\alpha_1)^2\right]. \qquad (4.128)$$

The Froud number is computed as follows

$Re_1 > 0$

$Fr = (\pi - 2\varphi^*)/\pi$ for $\left(0 \leq \varphi^* \leq \pi/2\right)$.

$Re_1 < 0$

$Fr = 1$ if $\varphi^* < \dfrac{8}{9}\dfrac{\pi}{2}$,

$Fr = (\pi/2 - \varphi^*)/(\dfrac{1}{9}\dfrac{\pi}{2})$ for $(\dfrac{8}{9}\dfrac{\pi}{2} < \varphi^* \leq \dfrac{\pi}{2})$.

φ^* = pipe orientation angle measured from vertical ($0 \leq \varphi \leq \pi/2$). $\varphi^* = 0$ for vertical pipe, $\varphi^* = \pi/2$ for horizontal pipe. Here C_{ov} is valid for vertical flow. C_{oh}

for horizontal co current flow and is defined as for vertical flow but using absolute values of the volume fluxes j_1 and j_2.

$Re_1 > 0$

$C_{ov} = C_{ov}^+$

where

$C_{ov}^+ = L / \left[A_o + (1 - A_o) \alpha_1^{B_o} \right]$

$Re_1 < 0$

$C_{ov} = \max \left[C_{ov}^+, \frac{V_{1j}^o (1-\alpha_1)^{0.2}}{|j_1| + |j_2|} \right]$

where

$B_o = (1 + 1.57 \rho_1 / \rho_2)/(1 - k_1)$,

$L = \left[1 - \exp(-C_1 \alpha_1) \right] / \left[1 - \exp(-C_1) \right]$,

$C_1 = 4 p_{kr}^2 / \left[p(p_{kr} - p) \right]$,

$A_o = k_1 + (1 - k_1)(\rho_1 / \rho_2)^{1/4}$,

$k_1 = \min(0.8, A_1)$,

$A_1 = 1 / \left[1 + \exp(-Re/60000) \right]$,

$Re = Re_1$ if $Re_1 > Re_2$ or $Re_1 < 0$,

$Re = Re_2$ if $Re_1 \leq Re_2$.

The local gas superficial *Reynolds* number is defined as follows

$Re_1 = \alpha_1 \rho_1 w_1 D_h / \eta_1$.

The local liquid superficial *Reynolds* number is defined as follows

$Re_2 = \alpha_2 \rho_2 w_2 D_h / \eta_2$.

The weighted mean drift velocity was correlated by *Chexal* et al. as follows.

$$V_{1j}^* = FrV_{1jv} + (1-Fr)V_{1jh} \qquad (4.129)$$

for co-current up flow.

$$V_{1j}^* = FrV_{1jv} + (Fr-1)V_{1jh}$$

for co current down flow. V_{1jh} considers only co-current horizontal flow ($j_1 > 0$, $j_2 > 0$) and is evaluated using the same equation as for the vertical flow using positive values of j_1 and j_2.

$$V_{1jv} = V_{1jv}^o C_9,$$

$$V_{1jv}^o = 1.41 C_2 C_3 C_4 \ V_{Ku}.$$

Computation of the C_2 factor:

$$\rho_2/\rho_1 > 18,$$

$$C_5 = \sqrt{150\rho_1/\rho_2}, \ C_6 = C_5/(1-C_5),$$

$C_5 \geq 1$, $C_2 = 1$,

$C_5 < 1$, $C_2 = 1/[1-\exp(-C_6)]$,

$$\rho_2/\rho_1 \leq 18,$$

$$C_2 = 0.4757 \left[\ln(\rho_2/\rho_1)\right]^{0.7}.$$

Computation of the C_4 factor:

$D_2 = 0.09144$ (normalizing diameter),

$$C_7 = (D_2/D_h)^{0.6}, \ C_8 = C_7/(1-C_7),$$

$C_7 \geq 1$, $C_4 = 1$,

$C_7 < 1$, $C_4 = 1/[1-\exp(-C_8)]$.

Computation of the C_3 and C_9 factors:

Co-current up flow ($Re_1 > 0$, $Re_2 > 0$)

$$C_9 = (1-\alpha_1)^{k_1},$$

$$C_3 = \max\left[0.5,\ 2e^{(-|Re_2|/60000)}\right],$$

Co-current down flow ($Re_1 < 0$, $Re_2 < 0$)

$$C_9 = \min\left[0.7, (1-\alpha_1)^{0.65}\right],$$

$$C_3 = 2(C_{10}/2)^{B_2},$$

where

$$C_{10} = 2\exp\left[\frac{|Re_2|}{350000}\right]^{0.4} -1.75|Re_2|^{0.03}\exp\left[-\left(\frac{D_1}{D_h}\right)^2 \frac{|Re_2|}{50000}\right] + \left(\frac{D_1}{D_h}\right)^{0.25}|Re_1|^{0.001},$$

and $D_1 = 0.0381$ (normalizing diameter),

$$B_2 = 1/\left[1+0.05\left(\frac{|Re_2|}{350000}\right)^{0.4}\right].$$

Counter-current flow ($Re_1 > 0$, $Re_2 < 0$) : Two solution for the void fraction, α_{11}, α_{12}., The desired void fraction, α_1 known a priori, is used to select the appropriate C_3 as follows: for

$$\alpha_1 = \max(\alpha_{11}, \alpha_{12})$$

$$C_3 = 2(C_{10}/2)^{B_2},$$

for

$$\alpha_1 = \min(\alpha_{11}, \alpha_{12}),$$

$$C_3 = \min\left\{2\left(\frac{C_{10}}{2}\right)^{B_2}, \ 2\left[\left(\frac{C_{10}}{2}\right)^{B_2}\left(\frac{j_2}{j_2^*}\right) + \left(1 + \frac{|Re_2|}{60000}\right)\left(1 - \frac{j_2}{j_2^*}\right)\right]\right\}.$$

j_2^* is j_2 on the counter-current flow limiting (CCFL) line corresponding to j_1 and is calculated using

$$C_3 = \min\left\{2(C_{10}/2)^{B_2}\right\},$$

$$C_9 = (1-\alpha_1)^{k_1}.$$

On CCFL line $dj_1/d\alpha_1 = 0$.

Kawanishi et al. [21] (1990) compared the prediction of α_1 with 1353 data points for $p < 180 \ 10^5$, $D_h \leq 0.61$. The error band was ±30.7%.

The Kawanishi et al. correlation from 1990: Kawanishi et al. [21] (1990) correlated the void cross section distribution parameter with

$j \leq -3.5$ $\quad C_0 = 1.2 - 0.2\sqrt{\rho_1/\rho_2}$, *Ishii* [15] (1977) for round tube,

(4.130)

$-3.5 < j \leq -2.5$ $\quad C_0 = 0.9 + 0.1\sqrt{\rho_1/\rho_2} - 0.3(1 - \sqrt{\rho_1/\rho_2})(2.5 + j)$ (4.131)

$-2.5 < j < 0$ $\quad C_0 = 0.9 + 0.1\sqrt{\rho_1/\rho_2}$ (4.132)

$0 \leq j$ $\quad C_0 = 1.2 - 0.2\sqrt{\rho_1/\rho_2}$, *Ishii* [15] (1977) for round tube,

(4.133)

The weighted mean drift velocity was correlated with

$j \leq 0$ $\quad V_{1j}^* = \sqrt{2}V_{Ku}$, $\quad$ *Zuber* and *Findlay* [37] (4.134)

$0 < j < 0.24$ $\quad$ Interpolate between V_{1j} for $j \leq 0$ and V_{1j} for $j = 0.24$.

(4.135)

$j \geq 0.24$

$p \leq 15 \; 10^5$

$D_h \leq 0.05$ $\qquad V_{1j}^* = 0.35 V_{TB}$, $\quad$ Zuber and Findlay [37] $\qquad$ (4.136)

$0.05 < D_h \leq 0.46$ $\qquad V_{1j}^* = 0.52 V_{TB}$, $\hfill$ (4.137)

$0.46 < D_h$ $\qquad V_{1j}^* = 0.52 V_{TB, D_h = 0.46}$, $\hfill$ (4.138)

$15 \times 10^5 < p \leq 180 \times 10^5$

$D_h \leq 0.02 \; m$ $\qquad V_{1j}^* = 0.35 V_{TB}$, $\quad$ Zuber and Findlay [37] $\qquad$ (4.139)

$0.02 < D_h \leq 0.24$ $\qquad$ Interpolate between
V_{1j}^* for $D_h = 0.02$ and V_{1j}^* for $D_h = 0.24$. $\qquad$ (4.140)

$0.24 < D_h \leq 0.46$ $\qquad V_{1j}^* = 0.048 \sqrt{\dfrac{\rho_2}{\rho_1}} V_{TB}$ $\hfill$ (4.141)

$0.46 < D_h$ $\qquad V_{1j}^* = 0.048 \sqrt{\dfrac{\rho_2}{\rho_1}} V_{TB}$ for $D_h = 0.46$. $\hfill$ (4.142)

Kawanishi et al. [21] (1990) compared the prediction of α_1 with 1353 data points for $p < 180 \times 10^5$, $D_h \leq 0.61$. The error band was $\pm 16.8\%$ which makes the correlation more accurate than that of *Chexal* et al. for this region.

Liao, Parlos and Grifith (1985) correlation: The *Liao, Parlos* and *Grifith* (1985) correlation taken from *Coddington* and *Macian* [5] consists of two regions: For bubbly flow

$$j_2 > 2.34 - 1.07 \left(\frac{g \sigma \Delta \rho_{21}}{\rho_2^2} \right)^{1/4}, \qquad (4.143)$$

$$C_0 = 1, \qquad (4.144)$$

$$V_{1j}^* = 1.53 (1 - \alpha_1)^2 V_{Ku}. \qquad (4.145)$$

For churn turbulent flow

$$C_0 = \left\{1.2 - 0.2\sqrt{\frac{\rho_1}{\rho_2}}\left[1 - \exp(-18\alpha_1)\right]\right\}, \tag{4.146}$$

which is the *Ishii* [15] (1977) correlation for around tube modified by *Liao* et al. [18] with the multiplier $1 - \exp(-18\alpha_1)$. For the weighted mean drift velocity the gas density was used in the denominator in the *Kutateledze* number (scaling with the free falling droplet velocity)

$$V_{1j}^* = 0.33\left(\frac{g\sigma\Delta\rho_{21}}{\rho_1^2}\right)^{1/4}. \tag{4.147}$$

For annular flow $|j_1| > V_{TB}\left(\frac{1}{C_0} - 0.1\right)$, the authors used the *Ishii* [15] (1977) results as follows: For the distribution parameter

$$C_0 = 1 + \frac{1-\alpha_1}{\alpha_1 + \left[\frac{1+75(1-\alpha_1)}{\sqrt{\alpha_1}}\frac{\rho_1}{\rho_2}\right]^{1/2}}, \tag{4.148}$$

and for the weighted mean drift velocity

$$V_{1j}^* = 8.16(C_o - 1)(1-\alpha_1)^{1/2} V_{TB}. \tag{4.149}$$

Gardner [10] correlation (1980):

$$\frac{\alpha_1}{\sqrt{1-\alpha_1}} = 11.2\left(\frac{j_1}{V_{ku}}\rho_1 w_2^2 \sqrt{\frac{g\Delta\rho_{21}}{\sigma_2^3}}\right)^{2/3}. \tag{4.150}$$

Maier and Coddincton correlation (1986): The best fit of the data is given by the *Maier* and *Coddincton* [23] 1986 correlation

$$C_0 = 1.0062 + 2.57 \times 10^{-9} p, \tag{4.151}$$

$$V_{1j}^* = 0.8 - 1.23 \times 10^{-7} p + 5.63 \times 10^{-15} p^2$$

$$+ G\left(1.05 \times 10^{-3} - 8.81 \times 10^{-11} p + 6.73 \times 10^{-19} p^2\right). \tag{4.152}$$

4.3.3.8 Horizontal stratified flow

Fraß and *Wiesenberg* [9] obtained a correlation for stratified horizontal flow valid for high pressures 40 to 160 *bar*:

$$S = \left(0.38 Fr_2^{-3/2} + 1.2\right)\left(\frac{X_1}{1-X_1}\frac{\rho_2}{\rho_1}\right)^{0.76}, \qquad (4.153)$$

where

$$Fr_2 = \frac{(\rho w)^2 v_2}{g D_h}. \qquad (4.154)$$

4.3.3.9 Mamaev et al. method for inclined pipes

Stratified flow: For the computation of the relative velocities and pressure drop for this flow pattern the work by *Mamaev* - see Ch. 1 Ref. [13] (1969) is recommended.

$$\alpha_1 = 1 - A^{0.4} \quad \text{for} \quad 0 \le A \le 0.18, \qquad (4.155)$$

$$\alpha_1 = 0.615(1-A) \quad \text{for} \quad 0.18 < A \le 1, \qquad (4.156)$$

where

$$A = \lambda_{fr}(1-\beta)^2 Fr_{crit}/(-2\cos\varphi). \qquad (4.157)$$

Note that *Mamaev* et al. considered stratified flow possible for $Fr_h < Fr_{crit}$, where $Fr_h = \frac{(\rho w)^2 v_h^2}{g D_h}$, $v_h = X_1 v_1 + (1-X_1)v_2$. The critical *Froud* number was obtained from experiments

$$Fr_{crit} = \left[\left(0.2 - \frac{2\cos\varphi}{\lambda_{fr}}\right)\bigg/(1-\dot{\alpha}_1)^2\right]\exp(-2.5\dot{\alpha}_1), \qquad (4.158)$$

where $\dot{\alpha}_1 = X_1 v_1 / v_h$, and $\lambda_{fr} = \lambda_{fr}\left(\frac{\pi(1-\alpha_1)}{\pi-\theta}\frac{w_2 D}{v_2}, \frac{k}{D_h}\right)$ is the liquid site wall friction coefficient computed using the *Nikuradze* diagram.

Slug flow, churn-turbulent flow, or bubbly flow: For the computation of the relative velocities and pressure drop for this flow pattern the work by *Mamaev* see Ch. 1 Ref. [13] (1969) is again recommended. The volume fraction is then

$$\alpha_1 = 0.81\dot{\alpha}_1 \left[1 - \exp\left(-2.2 Fr_{crit}^{1/2}\right)\right], \tag{4.159}$$

where the required variables are computed as in the previous section. This correlation goes together with the friction pressure drop correlation obtained by the authors

$$\left(\frac{dp}{dz}\right)_{fr} = \lambda_{fr} \frac{1}{2D} \left[\alpha_1 \rho_1 w_1^2 + (1-\alpha_1)\rho_2 w_2^2\right] \Phi_{2o}, \tag{4.160}$$

where

$$\lambda_{fr} = \lambda_{fr} \left[\frac{X_1 w_1 + (1-X_1)w_2}{X_1/\rho_1 + (1-X_1)/\rho_2} \frac{w_2 D_h}{v_2}, \frac{k}{D_h}\right], \tag{4.161}$$

$$\Phi_{2o} = \left\{1 + 0.78\dot{\alpha}_1 \left[1 - \exp\left(-2.2 Fr_{crit}^{1/2}\right)\right] - 0.22\dot{\alpha}_1 \left[1 - \exp\left(-15\rho_2/\rho_1\right)\right]\right\} / (1-\dot{\alpha}_1). \tag{4.162}$$

4.3.4 Cross section averaged particle sink velocity in pipes – drift flux models

In this case we calculate the drift flux parameter using the appropriate correlation from Table 4.3.

Table 4.3. Drift flux parameters for a mixture of liquid and solid particles ($\alpha_{d\max} < \pi/6$)

Concentration (distribution) parameter C_o	Weighted mean drift velocity
	$V_{dj}^* = -const V_T$, where $V_T = \sqrt{gD_h \dfrac{\rho_d - \rho_c}{\rho_c}}$
	$const = ...$
Covier and *Aziz* [3] (1972)	
1.2	1.77 (*Newton* regime, $C^d = 0.44$)

Oedjoe and *Buchanan* [25] (1966) for flow of water slurries of coal, D_d = 0.062 to 0.0318 and gravel, D_d = 0.0047 to 0.0138

1.2   0.88

Wallis [34] (1969)

1.2 $V_{dj}^* = (1-\alpha_d)^{2.39} V_{dj,\alpha_d=0}^*$

Sakaguchi at al. [32] (1990), $0.8 \leq C_o \leq 1.4$

$$[1 + \frac{-1.39 \frac{j_{c,w_d=0}}{j_{cd}} + 2.48 \frac{D_d}{D_{hcd}} + 0.488}{\left(\frac{\alpha_d}{0.01}\right)^{1.2} + 1}] \quad V_{dj}^* = -C_o j_{c,w_d=0}$$

$$[1.213 - 0.294(w_c - w_d)/\Delta w_{cd\infty}]$$

where

$$j_{c,w_d=0} = \Delta w_{cd\infty} \frac{\alpha_c^m}{\alpha_d^n},$$

$$m = \left[-0.457 \left(\frac{D_d}{D_h}\right)^{0.7} + 1.31\right]/\alpha_d^{1.2},$$

$$n = -0.797 \frac{D_d}{D_h} + 0.732,$$

$$\Delta w_{cd} = \sqrt{\frac{4}{3} \frac{D_d}{c_{cd}^d} \frac{\rho_d - \rho_c}{\rho_c} g},$$

$$c_{dc}^d = \frac{24}{Re_{dc}}\left(1 + 0.15 Re_{dc}^{0.687}\right)$$

$$+ \frac{0.42}{1 + 42500 Re_{dc}^{-1.16}}.$$

For $\alpha_d < 0.005$ one should use $C_o = 1$ and $j_{c,w_d=0} = \alpha_c \Delta w_{cd\infty}$. The experimental data base includes diameters 0.00114, 0.00257 and 0.00417 m for aluminum ceramic particles with density 2270, 2380 and 2400, respectively, and diameter 0.00296 for aluminum particles with density 2640. The pipe diameters are 0.0209, 0.0306, 0.0508. The solid volumetric fraction was $0.005 < \alpha_3 < 0.1$, the superficial liquid velocity $0.193 < j_c < 1.51$, and the superficial solid particle velocity $0.000450 < j_d < 0.0407$.

We see from Table 4.3 that *Covier* and *Azi*s and *Oedjoe* and *Buchman* used as a weighted mean drift velocity the free-setting velocity of single solid particles. *Wallis* introduced the correction for the influence of neighbor particles in a confined flow.

Sakagushi et al. [31, 30] rewrite the original *Zuber* drift flux correlation (4.67) in terms of the velocities

$$w_d = \frac{C_o(1-\alpha_d)w_c}{1-\alpha_d C_o} + \frac{V_{dj}^*}{1-\alpha_d C_o}, \qquad (4.163)$$

calculate the liquid flux for which the solid particle velocity, w_d, is equal to zero

$$j_{d,w_d=0} = (1-\alpha_d)w_{c,w_d=0} = -\frac{V_{dj}^*}{C_o} \qquad (4.164)$$

called suspension volumetric flux of the liquid phase, and rewrite the *Zuber* drift flux correlation in terms of the suspension volumetric flux of the liquid

$$w_d = C_o(j - j_{d,w_d=0}). \qquad (4.165)$$

The analog to Eq. (4.165) in terms of the suspension volumetric flux of the liquid phase is

$$\frac{\Delta w_{cd}}{\Delta w_{cd\infty}} = [(1-C_o)\frac{w_c}{\Delta w_{cd\infty}} + C_o \frac{j_{d,w_d=0}}{\Delta w_{cd\infty}}]/(1-\alpha_d C_o) \qquad (4.166)$$

Sakagushi et al. proposed instead of the use the *free settling velocity* as a main parameter in the drift flux description, to use the *suspension volumetric* flux of the liquid phase. The relationship between this velocity and the *weighted mean drift velocity* is given with Eq. (4.164). The experimental data of *Sakagushi* are correlated by means of the correlations given in Table 4.3. They are obtained by using the *fast closing valves* method to measure the volumetric fraction of each phase. The experimental data for the particles volumetric fraction are approximated with an index of deviation of +10.6 and −8.61%. Comparing with the data of other authors, *Sakagushi* shows that they are correlated with an index of deviation +57.3 and −311%. Using *Sakagushi*'s correlation from Table 4.3 we can rewrite Eq. (4.166) in the following form

$$\frac{\Delta w_{cd}}{\Delta w_{cd\infty}} = [(1-C_o)\frac{w_c}{\Delta w_{cd\infty}} + C_o \frac{\alpha_c^m}{\alpha_d^n}]/(1-\alpha_d C_o) \qquad (4.167)$$

4.4 Slip models

The critical two phase is co-current flow. Some authors used in this case successfully slip correlation describing the gas-to-liquid velocity ratio S as a function of the local parameters. In what follows we give some of them.

Kolev [22] (1982) propose the following form of empirical correlation

$$S = w_1/w_2 = 1 + \frac{(n+1)^{n+1}}{n^n} X_1 (1-X_1)^n \left[(\rho_2/\rho_1)^m - 1\right]. \qquad (4.168)$$

Here m determines the magnitude of the maximum of S and n the position at the X_1 coordinate where this maximum occurs

$$S = 1 + \frac{X_1}{X_{1m}} \left(\frac{1-X_1}{1-X_{1m}}\right)^{\frac{1-X_{1m}}{X_{1m}}} \left[(\rho_2/\rho_1)^m - 1\right] \qquad (4.169)$$

where

$$S = S_{max} = (\rho_2/\rho_1)^m \qquad (4.170)$$

for

$$X_1 = X_{1m}. \qquad (4.171)$$

This relationship gives the maximum possible slip ratios for gas mass flow concentration greater than 0.5, and divides the regions of experimentally observed and not observed slip ratios for X_1 less than 0.5. The measurements for critical air - water made by *Deichsel* and *Winter* [6] (1990) for $T = 293$ K and $p = (1.65, 2.65, 3.37, 4.15)$ 10^5 are represented successful using the above equation and $X_{1m} \approx 0.1$, $m \approx (0.18, 0.17, 0.16, 0.14)$, respectively.

One year later *Petry* [26] (1983) reported a similar relationship correlating data for two-phase flow of R12 in capillary tubes

$$\log S = 1/[A (\log X_1)^2 + B \log X_1 + C], \qquad (4.172)$$

where $A = 6.3014$, $B = 25.5632$, $C = 23.3784$. This correlation gives $S_m = 5.5$ at $X_{1m} = 0.013$. Similar to those authors a correlation was proposed by *Michaelides* and *Parikh* [24] (1982)

$$S = 1 + \frac{X_1}{X_{1m}} \exp\left[(X_{1m} - X_1)/X_{1m}\right]\left[(\rho_2/\rho_1)^m - 1\right] \qquad (4.173)$$

where

$$m = 1/2, \qquad (4.174)$$

$$X_{1m} = 1/C = 0.1. \qquad (4.175)$$

C was experimentally observed in the region 7 to 15. A probably more general form is

$$S = 1 + \left(\frac{X_1}{X_{1m}}\right)^\kappa \exp\kappa\left[(X_{1m} - X_1)/X_{1m}\right]\left[(\rho_2/\rho_1)^m - 1\right] \qquad (4.176)$$

where

$$\kappa > 1. \qquad (4.177)$$

The fact that there is a maximum of the $S(X_1)$ function is confirmed by *Grieb* [11] (1989) showing the change of the flow pattern from continuous liquid to continuous gas with liquid entrainment. *Hug* and *Loth* [14] (1992) proposed a new semi-empirical correlation for S_2

$$S_2 = (\rho_2/\rho_1) 2X_1(1-X_1)/\left\{2X_1(1-X_1) + \left[1 + 4X_1(1-X_1)\left(\frac{\rho_2}{\rho_1} - 1\right)\right]^{1/2} - 1\right\}. \qquad (4.178)$$

This simple correlation gives a comparable result to the very complicated and extensively tested EPRI correlation [2] (1989) and approximate reasonably air - water data for mass flow rates $G = 50$ to 1330 $kg/(m^2s)$ at atmospheric pressure.

Deichsel and *Winter* [6] (1990) found indirect experimental evidence that in the critical cross section the local critical flow at every point of the cross section is nearly homogeneous. They found that there is a profile of the gas volumetric fraction which can be described as follows

$$\alpha_1(r) = \alpha_{1,r=0} - (\alpha_{1,r=0} - \alpha_{1,r=R_h})(r/R_h)^n, \quad (4.179)$$

where

$$\alpha_{1,r=0} = 1 - (1-\alpha_1)^{3.1}, \quad (4.180)$$

$$\alpha_{1,r=R_h} = 1 - (1-\alpha_1)^{\alpha_1/3}, \quad (4.181)$$

$$n = 1.015/(1.015 - \alpha_1). \quad (4.182)$$

The authors assumed that at each point of the critical cross section the flow velocity is equal to the sound velocity and computed the slip as the ratio of the cross section averaged gas velocity to the cross section averaged liquid velocity

$$S = \frac{1-\alpha_1}{\alpha_1} \left[\int_A a_h(\alpha_1, p)\alpha_1(r)dA \right] / \left\{ \int_A a_h(\alpha_1, p)[1-\alpha_1(r)]dA \right\}. \quad (4.183)$$

4.5 Three velocity fields – annular dispersed flow

Consider the mixture of gas, film and droplets. The relative droplets velocity with respect to the gas in accordance with *Kataoka* and *Ishii* [19] (1982)

$$\Delta w_{13} = w_1 - w_3 = (D_3/4)\left\{[g(\rho_3 - \rho_1)]^2/(\eta_1\rho_1)\right\}^{1/3} (1-\alpha_d)^{1.5}, \quad (4.184)$$

depends on the volume averaged droplet size. Here

$$\alpha_d = \alpha_3/(1-\alpha_2) \quad (4.185)$$

is the volume *concentration of droplets in the gas-droplets mixture*. In annular-dispersed flow, most droplets are in a wake regime due to their relatively small size. That is why one can consider approximately the flow consisting of a core having a volume fraction and a density of

$$\alpha_{core} = \alpha_3 + \alpha_1, \quad (4.186)$$

$$\rho_{core} = \alpha_d\rho_3 + (1-\alpha_d)\rho_1, \quad (4.187)$$

respectively, and a film. This allows us to compute the drift flux parameter for an annular flow using *Ishii*'s correlation [15] (1977)

$$C_{ocore} = 1 + \alpha_2 / \left[\alpha_{core} + \sqrt{(1 + 75\alpha_2) \rho_{core} / \left(\sqrt{\alpha_{core}} \rho_2 \right)} \right], \quad (4.188)$$

$$V_{corej} = (C_{ocore} - 1) \sqrt{(\rho_2 - \rho_{core}) D_{hy} g \alpha_2 / (0.015 \rho_2)}, \quad (4.189)$$

and thereafter to compute the film velocity

$$w_2 = \frac{\rho w (1 - \alpha_{core} C_{ocore}) - \alpha_{core} \rho_{core} V_{corej}}{(1 - \alpha_{core})[\rho_2 (1 - \alpha_{core} C_{ocore}) + \alpha_{core} \rho_{core} C_{ocore}]}. \quad (4.190)$$

Having the relative velocity Δw_{13}, we compute the gas velocity from the definition of the c.m. velocity, namely

$$w_1 = (\rho w - \alpha_2 \rho_2 w_2 + \alpha_3 \rho_3 \Delta w_{13}) / (\alpha_1 \rho_1 + \alpha_3 \rho_3). \quad (4.191)$$

Having the actual field velocities and the c.m. velocity, the diffusion velocities are easily obtained from the definition equations.

In case of dispersed flow (*gas and particles only*) the gas velocity is

$$w_1 = (\rho w + \alpha_3 \rho_3 \Delta w_{13}) / (\alpha_1 \rho_1 + \alpha_3 \rho_3) \quad (4.192)$$

Similarly we can model the flow if the gas phase or if the second field carry out solid particles. In both cases the particle diameter is an input parameter.

4.6 Three-phase flow

Depending on the volumetric concentration of the macroscopic solid particles (the third velocity field) we distinguish the following cases:

1. The *solid particles are touching each other* in the control volume

$$\alpha_3 = \alpha_{dm}. \quad (4.193)$$

In this case we suppose that the *friction* between wall and particles and among the particles itself *obstructs their movement*

$$V_3 = 0. \quad (4.194)$$

In this case it is not possible to have simultaneously a droplet field. In the free space between the particles, a mixture of liquid and gas can flow. It is convenient in such cases to simply correct the volumetric porosities, the permeabilities, and the hydraulic diameters, and to describe the flow of gas-liquid mixture through the free space.

2. The *solid particles are free in the flow* and the volume fraction of the space between them if they were closely packed,

$$\alpha_2^* = \frac{1-\alpha_{dm}}{\alpha_{dm}}\alpha_3, \tag{4.195}$$

is *smaller* than the liquid volume fraction

$$\alpha_2^* < \alpha_2. \tag{4.196}$$

In this case the question arises how the particles are distributed in the mixture. The ratio of the free settling velocities in gas and liquid

$$\sqrt{\frac{\rho_3-\rho_1}{\rho_3-\rho_2}\frac{\rho_2}{\rho_1}}$$

gives an idea how to answer this question. We see that, due to considerable differences between gas and liquid densities, the *particles sink much faster in gas than in the liquid*. Thus most probably the solid particles are carried out by the liquid, which corresponds to the experimental observations. In this case one can consider the flow consisting of gas, droplets, and a fictitious liquid velocity field consisting of the liquid field and the macroscopic solid particles field, having a density of

$$\rho_2' = (\alpha_2\rho_2 + \alpha_3\rho_3)/\alpha_2' \tag{4.197}$$

volume concentration in the mixture of

$$\alpha_2' = \alpha_2 + \alpha_3. \tag{4.198}$$

mixture mass flow rate and velocity of the fictitious mixture of the velocity fields 2 and 3, V'_2. Further our task reduces to the estimation of the momentum redistribution between the components 2 and 3, which means calculation the velocities V_2 and V_3. For this purpose we use either

$$V_2 = V_2' - \frac{\alpha_3\rho_3}{\alpha_2\rho_2 + \alpha_3\rho_3}\Delta V_{23}, \tag{4.199}$$

$$V_3 = V_2 - \Delta V_{23} \tag{4.200}$$

estimating ΔV_{23} empirically, or a proper correlation from Table 4.3, and the proposal made by *Giot* see in [12] (1982) and *Sakagushi* [30, 38] (1987,1988). Both authors assume that the particles behave as they are moving in a fictitious channel with hydraulic diameter

$$D_{hcd} = D_h \sqrt{1 - \alpha_1 - \alpha_3} \tag{4.201}$$

filled with liquid, a local volumetric particles concentration of

$$\alpha_c = \frac{\alpha_2}{\alpha_2 + \alpha_3}, \tag{4.202}$$

and a density of the continuous and discrete phase

$$\rho_c = \rho_2, \tag{4.203}$$

$$\rho_d = \rho_3. \tag{4.204}$$

3. The *solid particles are free in the flow*

$$\alpha_3 > \frac{\alpha_{dm}}{1 - \alpha_{dm}} \alpha_2 \tag{4.205}$$

and the volume fraction of the space between them if they were closely packed is larger then the liquid volume fraction

$$\alpha_2^* > \alpha_2. \tag{4.206}$$

This means that part of the particles, α_{31}

$$\alpha_3 - \frac{\alpha_{dm}}{1 - \alpha_{dm}} \alpha_2 = \alpha_3 (1 - \alpha_2 / \alpha_2^*) \tag{4.207}$$

are surrounded by gas. Probably in this case the particles surrounded by gas will settle down and will be carried out mechanically by the other particles carried out by the liquid. The question arises how to compute their velocity. As a first approximation one can correct the density of the particles surrounded by the liquid, with the weight of the rest of the particles

$$\rho_3^* = \rho_3 \alpha_3 / \left(\frac{\alpha_{dm}}{1-\alpha_{dm}} \alpha_2 \right) = \rho_3 \alpha_2^* / \alpha . \tag{4.208}$$

and proceed as in the previous case. But this question has to be clarified experimentally, which has not been done as far as the author knows.

Nomenclature

Latin

Ar	$= \sqrt{\dfrac{\sigma^3 \rho_c^2}{\eta_c^4 g \Delta \rho}}$	Archimedes number, *dimensionless*
Bo	$= \rho_2 g D_h^2 / \sigma$, Bond number, *dimensionless*	
C_0	distribution parameter	
c_l^d	drag coefficient acting on the field *l*, *dimensionless*	
$c_{d,single}^d$	drag coefficient for a particle in an infinite medium, *dimensionless*	
Eo	$= \dfrac{g \Delta \rho_{21} D_h^2}{\sigma} = \left(D_h / \lambda_{RT} \right)^2$, pipe *Eötvös* number, *dimensionless*	
Fr	$= \dfrac{G^2}{g D_h \rho_2^2}$, square of *Froud* number based on the liquid density, *dimensionless*	
Fr_h	$= \dfrac{(\rho w)^2 v_h^2}{g D_h}$, square of *Froud* number based on the mixture density, *dimensionless*	
f_l^d	drag force acting on the field *l* per unit mixture volume, N/m^3	
f_l^{vm}	virtual mass force acting on the field *l* per unit mixture volume, N/m^3	
g	gravity acceleration, m/s^2	
h	specific enthalpy, J/kg	
j	volumetric flux of the mixture - equivalent to the center of volume velocity of the mixture, m/s	
$N\eta_2$	$= \eta_2 / \sqrt{\rho_2 \sigma_2 \lambda_{RT}}$, liquid viscous number, *dimensionless*	
p	pressure, Pa	
Re_d	$= D_d \rho_c \Delta w_{dc} / \eta_c$, *Reynolds* number for dispersed particle surrounded by continuum, *dimensionless*	
S	$= w_1 / w_2$, slip (velocity ratio), *dimensionless*	

Nomenclature

r^* $= \dfrac{D_d}{2}\left(\dfrac{\rho_c g \Delta \rho}{\eta_c^2}\right)^{1/3}$, bubble size, *dimensionless*

T temperature, *K*

V^* $= \Delta w_{dc,\infty}\left(\dfrac{\rho_c^2}{\eta_c g \Delta \rho}\right)^{1/3}$, terminal velocity, *dimensionless*

V_{Ku} $= \left(\dfrac{\sigma g \Delta \rho_{cd}}{\rho_c^2}\right)^{1/4}$, *Kutateladze* terminal velocity of dispersed particle in continuum, *m/s*

V_{TB}^* $= \sqrt{D_h g \cos\varphi \Delta \rho_{cd}/\rho_c}$, *Taylor* terminal velocity, *m/s*

V_{dj}^* weighted mean drift velocity, *m/s*

We_d $= D_d \rho_c \Delta w_{dc}^2/\sigma$, *Weber* number for dispersed particle surrounded by continuum, *dimensionless*

v_h $= X_1 v_1 + (1-X_1)v_2$, homogeneous mixture specific volume, *m³/kg*

w_l cross section averaged axial velocity of field *l*, *m/s*

w_d weighted mean velocity, *m/s*

X_1 $= \dfrac{\alpha_1 \rho_1 w_1}{G}$, gas mass flow concentration, *dimensionless*

Greek

α_l volume fraction of field *l*, *dimensionless*

$\dot{\alpha}_d$ $= \dfrac{j_d}{j}$, averaged volumetric flow concentration of the field *d*, *dimensionless*

$\dot{\alpha}_1$ $= \dfrac{1}{1+\dfrac{1-X_1}{X_1}\dfrac{\rho_1}{\rho_2}}$, averaged volumetric flow concentration of the gas, *dimensionless*

α_{core} $= \alpha_3 + \alpha_1$, core volume fraction: droplet + gas, *dimensionless*

α_{dm} particle volume fraction at which the solid particles are touching each other in the control volume, *dimensionless*

Δw_{cd} velocity difference: continuum minus disperse, *m/s*

$\Delta w_{cd\infty}$ steady state free settling velocity for the family of solid spheres or the free rising velocity for a family of bubbles, *m/s*

$\Delta w_{cd,o}$ velocity difference at zero time, *m/s*

$\Delta w_{cd,\tau\to\infty}$ steady state velocity difference, *m/s*

$\Delta \rho_{cd}$ density difference: continuum minus disperse, kg/m^3
Φ_{2o} two phase friction multiplier, *dimensionless*
φ angle between the positive flow direction and the upwards directed vertical, *rad*
λ_{RT} $= \sqrt{\dfrac{\sigma_2}{g(\rho_2 - \rho_1)}}$, *Rayleigh-Taylor* instability length, *m*
ρ_l density of field *l*, kg/m^3
ρ_{core} $= \alpha_d \rho_3 + (1 - \alpha_d)\rho_1$, core density: droplet + gas, kg/m^3
ρw mixture mass flow rate, $kg/(m^2 s)$
σ gas-liquid surface tension, N/m
η dynamic viscosity, $kg/(ms)$

Subscripts

1   field 1
2   field 2
3   field 3
c   continuum
d   disperse
cr   critical
core   droplets + gas in annular flow

Superscripts

$'$ saturated liquid
$''$ saturated vapor

References

1. Bendiksen KH (1985) On the motion of long bubbles in vertical tubes, Int. J. Multiphase Flow, vol 11 pp 797 - 812
2. Chexal B, Lellouche G, Horowitz J and Healzer J (Oct. 10-13, 1989) A void fraction correlation applications, Proc. of the Fourth International Topical Meeting on Nuclear Reactor Thermal- Hydraulics. Mueller U, Rehme K, Rust K, (eds) G Braun Karlsruhe Karlsruhe, vol 2 pp 996-1002
3. Covier GW and Azis K (1972) The flow of complex mixtures in pipes, Von Nostrand Reinhold, p 469
4. Clark NN and Flemmer RL (1985) Predicting the holdup in two-phase bubble up flow and down flow using the Zuber and Findlay drift flux model, AIChE J., vol 31 no 3 March, pp 500-503

5. Coddington P and Macian R, (2000) A study of the performance of void fraction correlations used in the context of drift-flux two phase flow models, Internet publication of the Proceedings of Trends in Numerical and Physical Modeling for Industrial Multiphase Flows, 27th - 29th September, Corse, France
6. Deichsel M and Winter ERF (1990) Adiabatic two-phase pipe flow at air-water mixtures under critical flow conditions, Int. J. Multiphase Flow, vol 16 no 3 pp 391-406
7. Delhaye JM, Giot M, and Riethmueller ML (1981) Thermodynamics of two-phase systems for industrial design and nuclear engineering. Hemisphere Publ. Corp., Mc Graw-Hill Book Company
8. Dimitresku DT, (1943) Stroemung an einer Luftblase im senkrechten Rohr, Z. angw. Math. Mech., vol 23 no 3 pp 139-149
9. Fraß F and Wiesenberger J (September 1976) Druckverlust und Phasenverteilung eines Dampf/Wasser-Gemisches in horizontaldurchströmten Rohr, BWK, vol 28 no 9
10. Gardner C G (1980) Fractional vapor content of a liquid pool through which vapor is bubbled, Int. Journal of Multiphase Flow, vol 6 pp 399-410
11. Grieb G (June 1989) New slip correlation of forced convection two - phase flow, Nucl. Energy, vol 28 no 3 pp 155-160
12. Hetstroni G (1982) Handbook of multiphase systems. Hemisphere Publ. Corp., Washington etc., McGraw-Hill Book Company, New York etc.
13. Holmes JA (13-15 Sept. 1981) Description of the drift flux model in the LOCA-Code RELAP-UK. Heat and Fluid Flow in Water Reactor Safety. Manchester,
14. Hug R and Loth J (Jan.-March 1992) Analytical two phase flow void prediction method, J. Thermophysics, vol 6 no 1 pp 139-144
15. Ishii M (1977) One dimensional drift-flux model and constitutive equations for relative motion between phases in various two-phase flow regimes. ANL-77-47 Argone National Laboratory, Argone
16. Ishii M and Mishima K (Dec. 1980) Study of two-fluid model and interfacial area. NUREG/CR-1873, ANL-80-111
17. Lellouche GS (1974) A model for predicting two-phase flow, BNL-18625
18. Liao LH, Parlos A and Grifith P (1985) Heat transfer, carryover and fall back in PWR steam generators during transients, NUREG/CR-4376, EPRI NP-4298
19. Kataoka I and Ishii M (July 1982) Mechanism and correlation of droplet entrainment and deposition in annular two-phase Flow. NUREG/CR-2885, ANL-82-44
20. Kataoka I and Ishii M (1987) Drift flux model for large diameter pipe and new correlation for pool void fraction, Int. J. Heat and Mass Transfer, vol 30 no 9 pp 1927-1939
21. Kawanishi K, Hirao Y and Tsuge A (1990) An Experimental Study on Drift Flux Parameters for Two-Phase Flow in Vertical Round Tubes, NED, vol 120 pp 447-458
22. Kolev NI (1982) Modeling of transient non equilibrium, non homogeneous systems, Proc. of the seminar "Thermal Physics 82 (Thermal Safety of VVER Type Nuclear Reactors)", Karlovy Vary, May 1982, Chechoslovakia, vol 2 pp 129-147 (in Russian).
23. Maier D and Coddington P (1986) Validation of RETRAN-03 against a wide range of rod bundle void fraction data, ANS Transactions, vol 75 pp 372-374
24. Michaelidies E and Parikh S (1982) The prediction of critical mass flux by the use of Fanno lines, Nucl. Eng. Des., vol 71 pp 117-124
25. Oedjoe D and Buchman BH (1966) Trans. Inst. Chem. Eng., vol 44 no 10 p 364
26. Petry G (1983) Two - phase flow of R12 in capillary tubes under the critical flow state. Ph. D. Thesis, Technical University of Munich, Germany

27. Richardson JF and Zaki WN (1954) Sedimentation and fluidization: Part I, Trans. Instn. Chem. Eng., vol 32 pp 35-53
28. Rowe PN (1987) A convenient empirical equation for the estimation of the Richardson-Zaki exponent, Chem. Eng. Science, vol 42 no 11 pp 2795-2796
29. Sakagushi T, Minagawa H, Kato Y, Kuroda N, Matsumoto T, and Sohara K (1987) Estimation of in-situ volume fraction of each phase in gas-liquid-solid three-phase flow, Trans. of JSME, vol 53 no 487 pp 1040-1046
30. Sakaguchi T et al. (1986) Volumetric fraction of each phase in gas-liquid and liquid-solid two-phase flow, Memories of the faculty of engineering, Kobe University, no 33 pp 73-102
31. Sakaguchi T, Minagawa H, and Sahara K (March 22-27, 1987) Estimation of volumetric fraction of each phase in gas-liquid-solid three-phase flow, Proc. of the ASME-JSME Thermal Engineering Joint Conference, Honolulu, Hawaii, pp 373-380
32. Sakaguchi T, Minagawa H, Tomiama A, and Sjakutsui H (Jan.1990) Estimation of volumetric fraction in liquid - solid two - phase flow, Trans. JSME, vol 56 no 521 pp 5-10 (in Japanese)
33. Staengel G and Mayinger F (March 23-24, 1989) Void fraction and pressure drop in subcooled forced convective boiling with refrigerant 12, Proc. of 7th Eurotherm Seminar Thermal Non-Equilibrium in Two-Phase Flow, Roma pp 83-97
34. Wallis GB (1969) One dimensional two-phase flow, New York: McGraw Hill
35. Wallis GB (1974) The terminal speed of single drops or bubbles in an infinite medium, Int. J. Multi phase Flow, vol 1 pp 491-511
36. Wallis GB, Richter H J, and Kuo J T (Dec.1976) The separated flow model of two-phase flow, EPRI NP-275
37. Zuber N and Findlay JA (Nov. 1965) Average volumetric concentration in two-phase flow systems, Trans. ASME, J.of Heat Transfer, vol 84 no 4 pp 453-465
38. Zwirin Y, Hewitt GF and Kenning DB (1989) Experimental study of drag and heat transfer during boiling on free falling spheres, Heat and Technology, vol 7 no 3-4 pp 13-23

5. Entrainment in annular two-phase flow

5.1 Introduction

Entrainment is a process defined as mechanical mass transfer from the continuous liquid velocity field into the droplet field. Therefore entrainment is only possible if there is a wall in the flow, i.e. in channel flow or from the surfaces in pool flows. The surface instability on the film caused by the film-gas relative velocity is the reason for droplets formation and their entrainment.

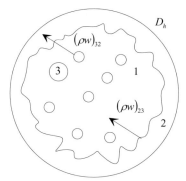

Fig. 5.1. Annular flow

The entrainment is quantitatively described by models for the following characteristics:

(1) Identification of the conditions when the entrainment starts;

(2) The mass leaving the film per unit time and unit interfacial area, $(\rho w)_{23}$, or the mass leaving the film and entering the droplet field per unit time and unit mixture volume, μ_{23}

$$\mu_{23} = a_{12}(\rho w)_{23} \qquad (5.1)$$

where a_{12} is the interfacial area density, i.e. the surface area between gas and film per unit mixture volume;

(3) Size of the entrained droplets.

Note that there is no general correlation for entrainment and deposition up to now. In what follows we summarize some results presented in the literature for quantitative modeling of these processes.

5.2 Some basics

Taylor, see [3] or [23] obtained the remarkable result for the interface averaged entrainment velocity

$$u_{23} = const \left(\rho_1/\rho_2\right)^{1/2} \Delta V_{12} \lambda_m^* f_m^* \qquad (5.2)$$

where λ_m^*, defined by

$$2\pi\lambda_m^* = \lambda_m \rho_1 \Delta V_{12}^2 / \sigma , \qquad (5.3)$$

and f_m^*, defined by

$$2 f_m^* = f_m / \left[\left(\rho_1/\rho_2\right)^{1/2} \rho_1 \Delta V_{12}^3 / \sigma\right], \qquad (5.4)$$

are the dimensionless wavelength and the frequency of the fastest growing of the unstable surface perturbation waves which are complicated functions of the *Taylor* number based on relative velocity,

$$Ta_{12} = \left(\rho_2/\rho_1\right)\left[\sigma/\left(\eta_2 \Delta V_{12}\right)\right]^2, \qquad (5.5)$$

where $f_m^* \leq \sqrt{3}/9$ and $0.04 \leq \lambda_m^* f_m^* \leq \sqrt{3}/6$ for $10^{-5} \leq Ta_{12} \leq 10^4$ and $\lambda_m^* f_m^* = \sqrt{3}/6$ for $Ta_{12} > 10^4$. Comparison with experimental data for engine spray where $\Delta V_{12} = V_2$, indicates that $\frac{1}{2}/\left(const\ \lambda_m^* f_m^*\right)$ is of order of 7, see *Bracco* [4]. This means

$$u_{23} \approx 14 \left(\rho_1/\rho_2\right)^{1/2} \Delta V_{12}. \qquad (5.6)$$

Following *Schneider* et al. [22] (1992) and assuming that (a) the surface entrainment velocity is equal to the liquid side surface friction velocity,

$$u_{23} = u_2^{*1\sigma} = \left(\tau_2^{1\sigma}/\rho_2\right)^{1/2}, \tag{5.7}$$

and (b) the liquid side surface shear stress is equal to the vapor side shear stress due to the gas flow

$$\tau_2^{1\sigma} = c_{12}^d \frac{1}{2}\rho_1 \Delta V_{12}^2, \tag{5.8}$$

we obtain a result

$$u_{23} = \left(\frac{c_{12}^d}{2}\right)^{1/2} \left(\rho_1/\rho_2\right)^{1/2} \Delta V_{12}, \tag{5.9}$$

which is close to *Taylor*'s solution.

The internal turbulent pulsation can also cause surface entrainment not influenced by the gas environment directly. Mathematically it can be expressed as

$$u_{23} = u_2^{*1\sigma} + u_2' = \left(c_{12}^d \frac{1}{2}\rho_1 \Delta V_{12}^2/\rho_2\right)^{1/2} + cV_2. \tag{5.10}$$

Here the interface averaged fluctuation velocity $u_2' = cV_2$ is proportional to the jet velocity. The proportionality factor can be of the order of 0.1, see *Faeth* [5]. With the *Braco*'s constant this expression results in

$$u_{23} = u_2^{*1\sigma} + u_2' = 14\left(\frac{\rho_1}{\rho_2}\right)^{1/2} \Delta V_{12} + 0.1\, V_2. \tag{5.11}$$

5.3 Correlations

This Section contains a set of empirical correlation used in the literature for modeling the entrainment mass flow rate.

The *Nigmatulin* correlation: *Nigmatulin* [16, 17] (1982) analyzed experimental data for entrainment for non-heated flow in the region $3 < We_{12} < 10$, $3 \times 10^3 < (Re_{2\delta} = \rho_2 w_2 \delta_2/\eta_2) < 1.5 \times 10^4$ and $10 < \rho_2/\rho_1 < 100$ and for heated flow with G = 500 to 2000 kg/(m²s), w_1 = 3 to 40 m/s and $\dot{q}_{w2}''$ = 0.1 to 3.6 MW/m² and obtained the following

$$\frac{(\rho w)_{23}}{\alpha_2 \rho_2 w_2} = 29 \frac{We_{Zeichik}}{Re_2} \left(\frac{\rho_2}{\rho_1}\right)^{1/2} \left(\frac{\delta_{2F}}{D_h}\right), \tag{5.12}$$

where the shear stress $\tau_2^{1\sigma}$ in the *Weber* number is computed as given in Chapter 2. The data are reproduced by the *Nigmatulin* correlation with an error of $\pm 20\%$. Nigmatulin reported also information about the equilibrium entrained mass fraction of the total liquid mass, E_∞, defined as follows

$$1 - E_\infty = 3.1\, Y^{-0.2} \quad \text{for} \quad Y \leq 8000 \tag{5.13}$$

$$1 - E_\infty = 1700\, Y^{-0.9} \quad \text{for} \quad Y > 8000 \tag{5.14}$$

where

$$Y = c_1 c_2 (\rho_{core}/\rho_2)(\eta_2/\eta_1)^{0.3} Fr_1 Eo, \tag{5.15}$$

$$Fr_1 = V_1^2/(gD_h), \tag{5.16}$$

$$Eo = (D_h/\lambda_{RT})^2, \tag{5.17}$$

$$\lambda_{RT} = (\sigma/g\Delta\rho_{21})^{1/2}, \tag{5.18}$$

$$c_1 = \Delta\rho_{21}/\rho_1 \quad \text{for} \quad \Delta\rho_{21}/\rho_1 \leq 10, \tag{5.19}$$

$$c_1 = 1 \quad \text{for} \quad \Delta\rho_{21}/\rho_1 > 10, \tag{5.20}$$

$$c_2 = 0.15 \quad \text{for} \quad Eo^{1/2} \geq 20, \tag{5.21}$$

$$c_2 = 1 \quad \text{for} \quad Eo^{1/2} < 20. \tag{5.22}$$

The data base for this correlation is: (a) air-water, p = 0.18 to 0.3 *MPa*, G = 90 to 2140 $kg/(m^2 s)$, X_1 = 0.08 to 0.8, V_1 = 16 to 80 m/s; (b) steam - water, p = 2 to 16 *MPa*, G = 253 to 4000 $kg/(m^2 s)$, X_1 = 0.08 to 0.9. This correlation gives the best agreement compared to other correlations existing up to 1995.

The *Zaichik* inception model: *Zaichik* et al. [29] provided in 1998 the following criterion for the inception of the entrainment

$$\frac{We_{Zeichik}}{Re_{2F}} = \frac{1}{4} \frac{\tau_2^{1\sigma} \eta_2}{\sigma_2 \rho_2 w_2} = 2.5 \times 10^{-5} + \frac{6 \times 10^{-4}}{(Re_{2F} - 160)^{0.6}} \tag{5.23}$$

which represents excellent data in the region $160 < Re_{2F} < 2000$.

The *Kataoka* and *Ishii* model: *Kataoka* and *Ishii* [11] (1982) found that entrainment starts if the following condition is fulfilled

$$Re_{2F} > Re_{2Fc}. \qquad (5.24)$$

Then the authors distinguish two entrainment regimes depending on whether the droplet field is under-entrained or over-entrained with respect to the equilibrium condition. The entrained mass fraction of the total liquid mass flow that defines the boundary between these two regimes is defined by the following correlation

$$E_\infty = \tanh\left(7.25 \times 10^{-7} We_{Ishii}^{1.25} Re_{23}^{1/4}\right). \qquad (5.25)$$

The above correlation for E_∞ was verified with data in the region $2\times 10^4 < We_{Ishii}^{1.25} Re_{23}^{1/4} < 8\times 10^6$, $1 < p < 4$ bar, $0.0095 < D_h < 0.032$ m, $320 < Re_2 < 6400$, $\alpha_1 V_1 < 100$ m/s. In this region of parameters the maximum value of the dimensionless entrainment $(\rho w)_{23} D_h / \eta_2$ was ≈ 20. The *under-entrained regime* (entrance section and smooth injection of liquid as a film causing excess liquid in the film compared to the equilibrium condition) is defined by

$$Re_2 > Re_{2\infty}. \qquad (5.26)$$

In this regime the entrainment mass flow rate is described by the following correlation which was published in [12] in 1983

$$(\rho w)_{23} = \frac{\eta_2}{D_h} \left[\begin{array}{l} 0.72 \times 10^{-9} Re_{23}^{1.75} We_{Ishii} (1-E_\infty)^{0.25} \left(1 - \frac{E}{E_\infty}\right)^2 \\ + 6.6 \times 10^{-7} \left(Re_{23} We_{Ishii}\right)^{0.925} (1-E)^{0.185} \left(\frac{\eta_1}{\eta_2}\right)^{0.26} \end{array} \right]. \qquad (5.27)$$

For the *over-entrained regime* (entrainment is caused by shearing-off of roll wave crests by gas core flow) defined by

$$Re_2 \le Re_{2\infty} \qquad (5.28)$$

the entrainment mass flow rate is correlated by *Kataoka* and *Ishii* [12] as follows:

$$(\rho w)_{23} = \frac{\eta_2}{D_h} 6.6 \times 10^{-7} \left(Re_{23} We_{Ishii} \right)^{0.925} (1-E)^{0.185} \left(\frac{\eta_1}{\eta_2} \right)^{0.26}. \qquad (5.29)$$

Lopez de Bertodano et al. [15] recommended in 1998 to replace the correlation by *Kataoka* and *Ishii* with the following correlation

$$(\rho w)_{23} = \frac{\eta_2}{D_h} \left\{ 4.47 \times 10^{-7} \left[\left(Re_{2F} - Re_{2Fc} \right) We_{Lopez} \right] \right\}^{0.925} \left(\frac{\eta_1}{\eta_2} \right)^{0.26}. \qquad (5.30)$$

This correlation gives the same results as the *Kataoka* and *Ishii* correlation for low pressure and relative low mass flow rates but better agreement with the data for high pressure and large gas mass flow rates.

The *diffusion droplet deposition rate* has been investigated by several authors and reliable data are available. The most frequently used in the literature is *Paleev's* correlation [19] (1966), which is used by *Kataoka* and *Ishii* slightly modified

$$(\rho w)_{32} = (\eta_3 / D_h) 0.022 (\eta_1 / \eta_3)^{0.26} (Re_{23} - Re_2)^{0.74} \qquad (5.31)$$

for existing droplets, i.e. for $\alpha_3 > 0.0001$. The *Kataoka - Ishii* model was based on data in the following region $277 < Re_{23} < 5041$, $1414 < We_{Ishii} < 9602$, $0.0095 < D_h < 0.032$.

The *Whalley* et al. model: *Whalley* et al. [27] (1974) propose for the entrainment and deposition mass flow rates

$$(\rho w)_{32} = f \frac{\alpha_1}{\alpha_1 + \alpha_3} \rho_3, \qquad (5.32)$$

$$(\rho w)_{23} = f \left(\frac{\alpha_1}{\alpha_1 + \alpha_3} \rho_3 \right)_{eq}, \qquad (5.33)$$

respectively, where $\frac{\alpha_3}{\alpha_1 + \alpha_3} \rho_3$ is the homogeneous droplet concentration in kg/m^3 in the vapor core. The equilibrium droplet concentration

$$\left(\frac{\alpha_3}{\alpha_1+\alpha_3}\rho_3\right)_{eq} = 186.349\tau^{*2} + 0.185919\tau^* - 0.0171915 \qquad \tau^* \leq 0.047$$
(5.34)

$$\left(\frac{\alpha_3}{\alpha_1+\alpha_3}\rho_3\right)_{eq} = 96.6903\tau^{*2} + 10.55840\tau^* - 0.3097050 \qquad 0.047 < \tau^* \leq 0.100$$
(5.35)

$$\left(\frac{\alpha_3}{\alpha_1+\alpha_3}\rho_3\right)_{eq} = 51.6429\tau^{*2} + 27.13020\tau^* - 1.6586300 \qquad 0.100 < \tau^* \leq 0.300$$
(5.36)

$$\left(\frac{\alpha_3}{\alpha_1+\alpha_3}\rho_3\right)_{eq} = 145.8329\tau^{*2.13707} \qquad 0.300 < \tau^*$$
(5.37)

was experimentally correlated with the dimensionless group

$$\tau^* = \tau_{12}\delta_2/\sigma \quad \textit{Hutchinson and Whalley} [9] (1973). \tag{5.38}$$

Here, δ_2 is the film thickness, σ the surface tension and τ the interfacial stress. The interfacial stress is given by the following equations

$$\tau_{12} = \frac{\lambda_{R12}}{8}\rho_{core}j_{core}^2 \quad \textit{Wallis} [28] (1970) \tag{5.39}$$

where

$$j_{core} = \alpha_1 w_1 + \alpha_3 w_3 \tag{5.40}$$

is the core (gas + droplet) superficial velocity,

$$\rho_{core} = (\alpha_1\rho_1 w_1 + \alpha_3\rho_3 w_3)/j_{core} \tag{5.41}$$

is the core density,

$$\lambda_{R12} = \lambda_{R1}(1 + 360\delta_2/D_h) \tag{5.42}$$

is the gas-film friction coefficient,

$$\lambda_{R1} = 0.3164 / \text{Re}_{core}^{1/4} \tag{5.43}$$

is the friction coefficient computed using the core *Reynolds* number defined as follows

$$\text{Re}_{core} = (\alpha_1 \rho_1 w_1 + \alpha_3 \rho_3 w_3) D_h / \eta_1 \tag{5.44}$$

and the film thickness δ_2 is evaluated with the triangular relationship

$$\left(\frac{dp}{dz}\right)_2 = \frac{4\delta_2}{D_h} \left(\frac{dp}{dz}\right)_{Tph} \tag{5.45}$$

in accordance with *Turner* and *Wallis* [25] (1965), where

$$\left(\frac{dp}{dz}\right)_{Tph} \approx 4\tau_{12}/D_h \tag{5.46}$$

is the two-phase friction pressure drop per unit length in the core,

$$\left(\frac{dp}{dz}\right)_2 = \frac{1}{2}\rho_2 w_2^2 \lambda_{R2} / D_h \tag{5.47}$$

is the friction pressure drop per unit length in the film. Here the film friction coefficient

$$\lambda_{R2} = \lambda_{R2}(\text{Re}_2, k/D_h) \tag{5.48}$$

is estimated using *Hewitt's* analysis [7] (1961). The deposition coefficient, f, is estimated by *Katto's* [10] (1984) correlation given in Table 5.1 in the next Section. Since the amount of deposition is equal to that of entrainment at the equilibrium state, the same "f" is used in Eqs. (5.32) and (5.33).

The *Sugawara* model: For the range of application $p = 2.7 \times 10^5$ to 90×10^5 Pa and *Reynolds* number $\text{Re}_1 = 3 \times 10^4$ to 7×10^5 *Sugawara* [20] (1990) proposes the following entrainment correlation:

$$(\rho w)_{23} = 1.07 \left(\frac{\tau_{12}\delta_{2eq}}{\sigma}\right)\left(\frac{V_1\eta_2}{\sigma}\right)\left(\frac{\rho_2}{\rho_1}\right)^{0.4}, \tag{5.49}$$

where

$$\delta_{2eq} = k_2 \quad \text{for} \quad Re_1 > 1\ 10^5, \tag{5.50}$$

$$\delta_{2eq} = k_2 (2.136 \log_{10} Re_1 - 9.68) \quad \text{for} \quad Re_1 \leq 1\ 10^5, \tag{5.51}$$

is the wavelength and

$$k_2 = 0.57 \delta_2 + 21.73\ 10^3 \delta_2^2 - 38.8\ 10^6 \delta_2^3 + 55.68\ 10^9 \delta_2^4 \tag{5.52}$$

is hydrodynamic equivalent wave roughness.

The entrainment correlation was verified in conjunction with the deposition correlation given in Table 6.1 in the next Section. The applicability of this equation is limited in the region pressure: $p = 1 \times 10^5$ to 70×10^5 Pa, $Re_1 = \dfrac{V_1 D_h}{v_1} = 10^4$ to 10^6 and $\dfrac{\alpha_3}{\alpha_1 + \alpha_3} \dfrac{\rho_3}{\rho_1}$ from 0.04 to 10.

The *Ueda* model: *Ueda* [26] (1981) proposed a simple model for the entrainment based on his experiments with air-water and air - alcohol

$$(\rho w)_{23} = 3.54\ 10^{-3} U^{0.57} \quad \text{for} \quad 120 < U < 5000 \tag{5.53}$$

where

$$U = \frac{\tau_{12}}{\sigma_2} \left(\frac{\alpha_2 w_2}{\sigma_2} \right)^{0.6}. \tag{5.54}$$

For $U < 120$ linear interpolation to $(\rho w)_{23} = 0$ at $U = 0$ can be used.

The *Hewit* and *Gowan* model: A model was proposed by *Hewitt* and *Govan* [8] (1989). For *entrainment* the authors recommend the *Govan* et al. correlation [6] (1988)

$$(\rho w)_{23} = 0, \quad \text{for} \quad Re_2 < Re_{2\infty} \tag{5.55}$$

$$(\rho w)_{23} = 5.75 \times 10^{-5} \alpha_1 \rho_1 w_1 \left[(Re_2 - Re_{2\infty})^2 \frac{\eta_2^2}{D_h \sigma} \frac{\rho_2}{\rho_1^2} \right]^{0.316}, \tag{5.56}$$

for

142   5. Entrainment in annular two-phase flow

$$Re_2 \geq Re_{2\infty},\qquad(5.57)$$

and

$$1 < (Re_2 - Re_{2\infty})^2 \frac{\eta_2^2}{D_h \sigma} \frac{\rho_2}{\rho_1^2} < 10^7,\qquad(5.58)$$

where

$$Re_{2\infty} = \exp(5.8504 + 0.4249 \frac{\eta_1}{\eta_2}\sqrt{\frac{\rho_2}{\rho_1}}),\ \textit{Owen}\ \text{and}\ \textit{Hewitt}\ [18]\ (1987),\qquad(5.59)$$

is the local equilibrium film *Reynolds* number. For example for $Re_{2\infty} \approx 459$ air/water flow at atmospheric pressure $(\rho w)_{23}/(\alpha_1 \rho_1 w_1)$ takes values $\approx 5.75 \times 10^5$ to 9.37×10^{-3}. As shown later by *Schadel* et al. [21] (1990) there is some dependence of $Re_{2\infty}$ on the gas velocity and tube diameter not taken into account in Eq. (5.56). For deposition the authors used the correlation given in Table 5.1 in the next Section.

The *Schadel* et al. data: Recently additional data for the local equilibrium film *Reynolds* number as a function of the gas velocity for three - different pipe diameters have been provided by *Schadel* et al. [21] (1990) - see Table 5.1.

Table 5.1. Local equilibrium film *Reynolds* number as a function of the gas velocity. Parameter pipe diameter. *Schadel* et al. [21] (1990).

$D_h = 0.0254$, $c_{23} = 1.2425\ 10^{-4}$ ms/kg

w_1	32	55	68	80	89	99	115
$Re_{2\infty}$	259	242	212	212	212	212	126

$D_h = 0.0420$, $c_{23} = 1.09\ 10^{-4}$ ms/kg

w_1	19.5	36.5	53.0	72.0
$Re_{2\infty}$	306	314	235	243

$D_h = 0.05715$, $c_{23} = 1.0825\ 10^{-4}$ ms/kg

w_1	125	33	41	49
$Re_{2\infty}$	451	451	276	276

$c_{23} = 1.175\ 10^{-4}$ ms/kg (averaged for all data)

Schadel et al. summarized their data for entrainment by the following correlation

$$(\rho w)_{23} = c_{23} w_1 (\rho_1 \rho_2)^{1/2} \eta_2 (Re_2 - Re_{2\infty}) \qquad (5.60)$$

or

$$(\rho w)_{23} = c_{23} \frac{\eta_2}{D_h} \rho_1 w_1 D_h \left(\frac{\rho_2}{\rho_1}\right)^{1/2} (Re_2 - Re_{2\infty}), \qquad (5.61)$$

where the coefficient c_{23} is given in the Table 5.1. Not that this coefficient is not dimensionless. For the deposition the authors recommend the correlation given in Table 5.1. *Schadel* at al. correlations for entrainment and deposition are based on data for air/water vertical flow in the following region: $0.0254 \leq D_h \leq 0.0572$, $20 < w_1 < 120 \ m/s$, $12 < \alpha_2 \rho_2 w_2 < 100 \ kg/(m^2 s)$.

Tomiyama and Yokomizo [24] (1988) found experimentally that behind spacer grids in nuclear reactor cores, the value of f increases with the increase of the pulsation components of the gas velocity.

Note that all of the entrainment models are proposed with a counterpart deposition model.

5.4 Entrainment increase in boiling channels

There are two important characteristics of the annular flow in boiling channels. The steam mass flow generated at the wall surface contributes substantially to the fragmentation of the liquid, and creates instabilities in the film with characterizing for the bubble departures frequencies. In addition, the diminishing of the film to zero called dry out (DO) causes important changes of the behavior of the boiling systems. It reduces the steam production and increase the wall temperature which for imposed constant heat flux my cause melting of the surface and its failure.

Kodama and *Kataoka* [31] reported in 2002 a dimensional correlation for the net entrainment rate due to the bubble break up with accuracy up to a constant that have to be derived from experiments

$$\left(\rho w\right)_{23\_bubbles\_breakup} = const \frac{\dot{q}''_{w2}}{\Delta h \rho''} \exp\left[-\frac{\delta_2 \sqrt{\tau_{2w}/\rho_2}}{158.7(\eta''/\eta')^{2.66} \, 30 w_2}\right]. \qquad (5.62)$$

Milashenko et al. [30] reported in 1989

$$\mu_{23\_bubbles\_breakup} = \frac{1.75}{\pi D_h} \alpha_2 \rho_2 w_2 \left(\dot{q}''_{w2} 10^{-6} \frac{\rho''}{\rho'}\right)^{1.3}. \qquad (5.63)$$

Regarding the dry out film thickness *Groeneveld* [32] reported that in all of his dry out experiments a complete drying of the film was observed. Trying to compute dry out heat flux by using three fluid models I found in [35] that the uncertainty of the entrainment and deposition models does not allow very accurate prediction of the location of the dry out unless some assumption is made for the limiting film thickness. Haw to estimate it is still not clear. There are authors proposing to use some Weber number stability criterion in a form

$$\delta_{2,crit} = We_{12,crit} \frac{\sigma_{12}}{\rho_1 \Delta w_{KH}^2}, \qquad (5.64)$$

[33], where $We_{12,crit} \sim const$ and Δw_{KH} is the *Kelvin-Helmholtz* relative velocity causing the self-amplifying instability, or

$$\delta_{2,crit} = 0.16 \left[\frac{\sigma_{12}}{\rho_2 (g \rho_2/\eta_2)^2}\right]^{0.2}, \qquad (5.65)$$

see in [34]. Final resolution of this problem remains to be done. It is important to note, that the accuracy of the mechanistic prediction of the location of the dry out depends on the accuracy of the used correlations. Therefore, only increasing the accuracy of the entrainment and deposition correlation will improve the accuracy of the mechanistic dry out predictions.

5.5 Size of the entrained droplets

Kataoka at al. [13] (1983) correlated the droplet size after the entrainment with a large experimental data base for air-water at low pressure $p \approx 1.2$ *bar*: $10 < Re_2 < 9700$, $2.5 < Re_1 < 17 \times 10^4$, $3 < We_{31} < 20$. Note that the dimensionless numbers used below

$$Re_1 = \alpha_1 \rho_1 V_1 D_h / \eta_1, \qquad (5.66)$$

$$Re_2 = \alpha_2 \rho_2 V_2 D_h / \eta_2, \qquad (5.67)$$

and

$$We_{31} = \rho_1 (\alpha_1 V_1)^2 D_{3E} / \sigma, \qquad (5.68)$$

are based on superficial velocities. The final correlation for the median particle size in an always observed log-normal distribution is

$$We_{31} = 0.01 Re_1^{2/3} \left(\frac{\rho_1}{\rho_2}\right)^{-1/3} \left(\frac{\eta_1}{\mu_2}\right)^{2/3}, \qquad (5.69)$$

with the ratio of the maximum to median size $D_{3,\max} / D_{3E} = 3.13$. We_{31} takes values between 3 and 20 in the considered region and is accurate within $\pm 40\%$. In contrast to pool flow, the walls influence the processes in the channel. For this reason the averaged stable droplet diameter in channel flow depends on the hydraulic diameter of the channel, D_h. Note the difference between the *Kataoka* et al. correlation and the proposed ones by other authors, e.g. the correlation proposed by *Azzopardi* et al. [2] (1980)

$$We_{31} = 1.9 Re_1^{0.1} \left(\frac{\rho_1}{\rho_2}\right)^{0.6} \left(\frac{\rho_1 (\alpha_1 V_1)^2 D_h}{\sigma}\right)^{2/3}, \qquad (5.70)$$

or those proposed by *Ueda* [26] (1981)

$$\frac{\rho_1 V_1^2 D_{3E}}{\sigma} = 0.68 \left(\sigma^{0.1} D_h^{-1/4}\right) \frac{\rho_1 V_1 D_h}{\eta_1} \left(\frac{\rho_1 V_1^2 D_h}{\sigma}\right)^{0.4}, \qquad (5.71)$$

(not dimensionless) verified in the region $3000 < \rho_1 V_1 D_h / \eta_1 < 50000$.

Recently *Ambrosini* et al. [1] (1991) proposed a new correlation for the droplet size in pipes in the presence of film at the wall

$$\frac{D_{3\infty}}{\delta_2} = 22 \left(\frac{\sigma}{\rho_1 c_{12}^d V_1^2 \delta_2}\right)^{0.5} \left(\frac{\rho_1}{\rho_2}\right)^{0.83} \exp\left[0.6 \alpha_3 D_h / D_{3\infty} + 99 / (\rho_1 V_1^2 D_h / \sigma)\right], \qquad (5.72)$$

where δ_2 is the film thickness, c_{12}^d the gas-film drag coefficient. Here 1 stands for gas, 2 for film and 3 for droplets. The data correlated are in the region $\alpha_2\rho_2 V_2 + \alpha_3\rho_3 V_3 \approx 40$ to $140 \ kg/(m^2 s)$, $V_2 \approx 22$ to $67 \ m/s$, and the results are in the region of $D_2 \approx 0.025$ to $0.2 \ mm$. Note that Eq. (5.72) attempts to correlate integral results of simultaneously happening fragmentation and coalescence, which - as shown in Section 7.6 - is not correct.

Nomenclature

Latin

a_{12}	interfacial area density, i.e. the surface area between gas and film per unit mixture volume, m^2/m^3
c_{12}^d	vapor side shear stress coefficient at the liquid surface due to the gas flow, *dimensionless*
c_1	$= \Delta\rho_{21}/\rho_1$
const	constant, *dimensionless*
D_h	hydraulic diameter, *m*
$\left(\dfrac{dp}{dz}\right)_2$	friction pressure drop per unit length in the film, *Pa/m*
$\left(\dfrac{dp}{dz}\right)_{Tph}$	two phase friction pressure drop per unit length in the core, *Pa/m*
E	$= \dfrac{\alpha_3\rho_3 w_3}{\alpha_2\rho_2 w_2 + \alpha_3\rho_3 w_3}$, mass fraction of the entrained liquid, entrainment, *dimensionless*
E_∞	equilibrium mass fraction of the entrained liquid, entrainment, equilibrium entrainment, *dimensionless*
Eo	$=(D_h/\lambda_{RT})^2$, Eötvös number, *dimensionless*
Fr_1	$= V_1^2/(gD_h)$, gas *Froud* number, *dimensionless*
f	deposition coefficient, *dimensionless*
f_m^*	frequency of the fastest growing of the unstable surface perturbation waves, *dimensionless*
G	mass flow rate, $kg/(m^2 s)$
j_{core}	$= \alpha_1 w_1 + \alpha_3 w_3$ core (gas + droplet) superficial velocity, *m/s*
k_2	film wavelength, *m*
p	pressure, *Pa*

Re_1 $= \alpha_1 \rho_1 |w_1| D_h / \eta_1$, local gas film *Reynolds* number based on the hydraulic diameter, *dimensionless*

$Re_{1,strat}$ $= \dfrac{\alpha_1 \rho_1 |w_1 - w_2| D_h}{\eta_1} \dfrac{\pi}{\theta + \sin\theta}$, local gas film *Reynolds* number based on the gas hydraulic diameter for stratified flow, *dimensionless*

Re_2 $= \alpha_2 \rho_2 |w_2| D_h / \eta_2$, local film *Reynolds* number based on the hydraulic diameter, *dimensionless*

$Re_{2\delta}$ $= \rho_2 w_2 \delta_2 / \eta_2$, local film *Reynolds* number based on the liquid film thickness, *dimensionless*

$Re_{2\infty}$ $= Re_{23}(1 - E_\infty)$, local equilibrium film *Reynolds* number based on the hydraulic diameter, *dimensionless*

Re_{23} $= \rho_2(1-\alpha_1)|w_{23}|D_h / \eta_2$, total liquid *Reynolds* number, *dimensionless*

Re_{2F} $= \rho_2 w_2 4\delta_{2F} / \eta_2$, local film *Reynolds* number, *dimensionless*

Re_{2Fc} $= 160$, critical local film *Reynolds* number, *dimensionless*

Re_{core} $= (\alpha_1 \rho_1 w_1 + \alpha_3 \rho_3 w_3) D_h / \eta_1$, core *Reynolds* number, *dimensionless*

Ta_{12} $= (\rho_2/\rho_1)[\sigma/(\eta_2 \Delta V_{12})]^2$, *Taylor* number based on relative velocity, *dimensionless*

$\dot{q}''_{w2}$ heat flux, MW/m^2

U $= \dfrac{\tau_{12}}{\sigma_2}\left(\dfrac{\alpha_2 w_2}{\sigma_2}\right)^{0.6}$, group used in the *Ueda* correlation

$u_2^{*1\sigma}$ $= (\tau_2^{1\sigma}/\rho_2)^{1/2}$, liquid side surface friction velocity, m/s

u'_2 $= cV_2$ interface averaged fluctuation velocity, m/s

u_{23} interface averaged entrainment velocity, m/s

V_1 gas velocity, m/s

V_2 jet velocity, continuum liquid velocity, m/s

$V_{1,wave}$ for gas velocity larger then $V_{1,wave}$ the horizontal pipe flow is no longer stratified with smooth surface, m/s

$V_{1,stratified}$ for gas velocity larger then $V_{1,stratified}$ horizontal pipe flow is no longer stratified, m/s

We_{Ishii} $= \dfrac{\rho_1(\alpha_1 w_1)^2 D_h}{\sigma_2}\left(\dfrac{\rho_2 - \rho_1}{\rho_1}\right)^{1/3}$, *Weber* number for the *Ishii* entrainment correlation, *dimensionless*

We_{Lopez} $= \dfrac{\rho_1 w_1^2 D_h}{\sigma_2}\left(\dfrac{\rho_2 - \rho_1}{\rho_1}\right)^{1/2}$, *Weber* number for the *Lopez* et al. correlation, *dimensionless*

$We_{Zeichik} = \dfrac{\tau_2^{1\sigma}\delta_{2F}}{\sigma_2}$, Weber number for the *Zeichik* et al. correlation, *dimensionless*

$We_{12} = \dfrac{\rho_1 D_h \Delta V_{12}^2}{\sigma_2}$, Weber number, *dimensionless*

$We_{31} = \rho_1(\alpha_1 V_1)^2 D_{3E}/\sigma$, Weber number, *dimensionless*

$w_{23} = \dfrac{\alpha_2 w_2 + \alpha_3 w_3}{1-\alpha_1}$, center of volume velocity of film and droplet together, *m/s*

w_1 axial cross section averaged gas velocity, *m/s*

w_2 axial cross section averaged film velocity, *m/s*

w_3 axial cross section averaged droplets velocity, *m/s*

X_1 gas mass flow divided by the total mass flow, *dimensionless*

$Y = c_1 c_2 (\rho_{core}/\rho_2)(\eta_2/\eta_1)^{0.3} Fr_1 Eo$

Greek

α_1 gas volume fraction, *dimensionless*

α_2 film volume fraction, *dimensionless*

α_3 droplets volume fraction, *dimensionless*

ΔV_{12} relative velocity, *m/s*

$\Delta w_{12} = w_1 - w_2$, relative velocity, *m/s*

$\delta_{2F} = D_h(1-\sqrt{1-\alpha_2})/2$,   film thickness in annular flow, *m*

δ_{2eq} equilibrium thickness, *m*

η_1 gas dynamic viscosity, *kg/(ms)*

η_2 liquid dynamic viscosity, *kg/(ms)*

λ_{R12} gas-film friction coefficient, *dimensionless*

λ_{R1} friction coefficient, *dimensionless*

$\lambda_{R2} = \lambda_{R2}(\text{Re}_2, k/D_h)$, film friction coefficient, *dimensionless*

λ_m^* wavelength of the fastest growing of the unstable surface perturbation waves, *dimensionless*

$\lambda_{RT} = (\sigma/g\Delta\rho_{21})^{1/2}$, *Rayleigh-Taylor* wavelength, *m*

μ_{23} mass leaving the film and entering the droplet field per unit time and unit mixture volume, *kg/(m³s)*

ρ_1 gas density, *kg/m³*

ρ_2 liquid density, *kg/m³*

ρ_{core} $=(\alpha_1\rho_1 w_1 + \alpha_3\rho_3 w_3)/j_{core}$, core density, kg/m^3
$(\rho w)_{23}$ entrainment mass flow rate, mass leaving the film per unit time and unit interfacial area, $kg/(m^2 s)$
$(\rho w)_{32}$ deposition mass flow rate, mass leaving the droplet field per unit time and unit interfacial area and deposed into the film, $kg/(m^2 s)$
σ surface tension, N/m
$\tau_2^{|\sigma}$ liquid side surface shear stress
τ_{12} interfacial stress, N/m^2
τ^* $=\tau_{12}\delta_2/\sigma$, interfacial stress, *dimensionless*
θ angle with origin of the pipe axis defined between the upwards oriented vertical and the liquid-gas-wall triple point, *rad*

References

1. Ambrosini W, Andreussi P, Azzopardi BJ (1991) A physical based correlation for drop size in annular flow, Int. J. Multiphase Flow, vol 17 no 4 pp 497-507
2. Azzopardi BJ, Freeman G, King D J (1980) Drop size and deposition in annular two-phase flow, UKAEA Report AERE-R9634
3. Batchelor GK (Editor) (1958) Collected works of Taylor GI, Cambridge Univ. Press, Cambridge, MA
4. Bracco FV (Feb. 25 - March 1, 1985) Modeling of engine sprays, Proc. International Congress & Exposition Detroit, Michigan, pp 113-136
5. Faeth GM (April 3-7 1995) Spray combustion: A review, Proc. of The 2nd International Conference on Multiphase Flow '95 Kyoto, Kyoto, Japan
6. Govan AH, Hewitt GF, Owen DG, Bott TR (1988) An improved CHF modelling code, 2nd UK National Heat Transfer Conference, Glasgow
7. Hewitt GF (1961) Analysis of annular two phase, application of the Dukler analysis to vertical upward flow in a tube, AERE-R3680
8. Hewitt GF, Govan AH (March 23-24, 1989) Phenomenological modelling of non-equilibrium flows with phase change, Proc. of 7th Eurotherm Seminar Thermal Non-Equilibrium in Two-Phase Flow, Roma, pp 7-27
9. Hutchinson P, Whalley PB (1973) Possible caracterization on entrainment in annular Flow, Chem. Eng. Sci., vol 28 p 974
10. Katto Y (1984) Prediction of critical heat flux for annular flow in tubes taking into account of the critical liquid film thickness concept, Int. J. Heat Mass Transfer, vol 27 no 6 pp 883-890
11. Kataoka I, Ishii M (July 1982) Mechanism and correlation of droplet entrainment and deposition in annular two-phase flow. NUREG/CR-2885, ANL-82-44
12. Kataoka I, Ishii M (March 20-24, 1983) Entrainment and deposition rates of droplets in annular two-phase flow, ASME-JSME Thermal Engineering Joint Conference Proceedings, Honolulu, Hawaii, vol 1 pp 69-80, Eds. Yasuo Mori, Wen-Jei Yang
13. Kataoka I, Ishii M, Mishima K (June 1983) Transaction of the ASME, vol 5 pp 230-238

14. Lopes JCB, Dukler AE (Sept. 1986) Droplet entrainment in vertical annular flow and its contribution to momentum transfer, AIChE Journal, vol 32 no 9 pp 1500-1515
15. Lopez de Bertodano MA, Assad A, Beus S (1998) Entrainment rate of droplets in the ripple-annular regime for small vertical ducts, Third International Conference on Multiphase Flow, ICMF'98, Lyon, France, June 8-12, CD Proceedings
16. Nigmatulin BI (1982) Heat and mass transfer and force interactions in annular - dispersed two - phase flow, 7-th Int. Heat Transfer Conference Munich, pp 337 - 342
17. Nigmatulin BI, Melikhov OI, Khodjaev ID (April 3-7, 1995) Investigation of entrainment in a dispersed-annular gas-liquid flow, Proc. of The 2bd International Conference on Multiphase Flow '95 Kyoto, Japan, vol 3 pp P4-33 to P4-37
18. Owen GD, Hewitt GF (1987) An improved annular two-phase flow model, 3rd BHRA Int. Conf. on Multiphase Flow, The Hague
19. Paleev II, Filipovich BS (1966) Phenomena of liquid transfer in two-phase dispersed annular flow. Int. J. Heat Mass Transfer, vol 9 p 1089
20. Sugawara S (1990) Droplet deposition and entrainment modeling based on the three-fluid model, Nuclear Engineering and Design, vol 122 pp 67-84
21. Schadel SA, Leman GW, Binder JL, Hanratty TJ (1990) Rates of atomization and deposition in vertical annular flow, Int. J. Multiphase Flow, vol 16 no 3 pp 363-374
22. Schneider JP, Marchiniak MJ, Jones BG (Sept. 21-24, 1992) Breakup of metal jets penetrating a volatile liquid, Proc. of the Fifth Int. Top. Meeting On Reactor Thermal Hydraulics NURETH-5, vol 2 pp 437-449
23. Taylor GI (1963) Generation of ripples by wind blowing over a viscous fluid, in "The scientific papers of Sir Geoffrey Ingham Taylor", Cambridge University Press, vol 3 Ch 25 pp 244-254
24. Tomiyama A, Yokomiyo O (February 1988) Spacer - effects on film flow in BWR fuel bundle, J. Nucl.Sc.Techn., Atomic Energy Society of Japan, vol 25, no 2, pp 204-206
25. Turner JM, Wallis GB (1965) An analysis of the liquid film annular flow, Dartmouth Colege, NYO-3114-13
26. Ueda T (1981) Two-phase flow - flow and heat transfer, Yokendo, Japan, in Japanese
27. Whalley PB et al. (1974) The calculation of critical heat flux in forced convection boiling, Proc. 5th Int. Heat Transfer Conf., Tokyo, vol 4 pp 290-294
28. Wallis GB (1970) Annular two-phase flow, Part I, A Simple Theory, J. Basic Eng., vol 92 p 59
29. Zaichik LI, Nigmatulin BI, Aliphenkov VM (June 1998) Droplet deposition and film atomization in gas-liquid annular flow, Third International Conference on Multiphase Flow, ICMF'98, Lion, France
30. Milashenko VI, Nigmatulin BI, Petukhov VV and Trubkin NI (1989) Burnout and distribution of liquid in evaporative channels of various lengths. Int. J. Multiphase Flow vol 15 no 3 pp 393-402
31. Kodama S and Kataoka I (April 14-18, 2002) Study on analytical prediction of forced convective CHF in the wide range of quality, Proceedings of ICONE10, 10TH International Conference on Nuclear Engineering Arlington, VA, USA, Paper nr ICONE10-22128
32. Groeneveld DC (2001) Private communication
33. Lovell TW (1977) The effect of scale on two phase countercurrent flow flooding in vertical tubes, Masters Thesis, Thayer School of Engineering, Dartmouth College
34. Borkowski JA and Wade NL eds. (1992) TRAC-BF1/MOD1, Models and correlations

35. Kolev NI (Sept. 1991) A three-field model of transient 3D multi-phase, three-component flow for the computer code IVA3, Part 2: Models for the interfacial transport phenomena. Code Validation. KfK 4949, Kernforschungszentrum Karlsruhe

6. Deposition in annular two-phase flow

6.1 Introduction

Droplets in the gas core may follow the turbulent gas pulsation depending on their size. Therefore some kind of turbulent diffusion from regions of high droplet concentrations into the region with smaller concentration is possible if the droplets are small enough. The deposition of droplets on the walls is defined as a process of transfer of droplets from the gas bulk flow to the wall leading to increase of the film wall thickness. If the gas core is turbulent there is in any case droplet deposition on the wall.

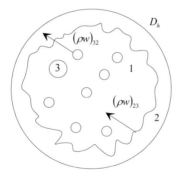

Fig. 6.1. Annular flow

6.2 Analogy between heat and mass transfer

Historically they are several empirical attempts in the literature to model this process. Most of the reported methods rely on the analogy between heat and mass transfer. The droplet mass flow rate is defined as a droplet diffusion process

$$(\rho w)_{32} = K_3 \left(\frac{\alpha_3}{\alpha_1 + \alpha_3} \rho_3 - 0 \right) = K_3 \frac{\alpha_3}{\alpha_1 + \alpha_3} \rho_3. \tag{6.1}$$

Here $(\rho w)_{32}$ is the deposition mass flow rate in $kg/(m^2 s)$. K_3 is commonly reported in the literature as a droplet mass transfer coefficient. $\frac{\alpha_3}{\alpha_1 + \alpha_3} \rho_3$ is the mass concentration of droplets in the droplet-gas mixture per unit core volume (droplet+gas), i.e. in kg/m^3. This quantity is called sometimes in the literature the droplet mass density. The starting point is the well-known *Colburn* correlation for heat transfer in a pipe

$$Nu = \frac{hD_h}{v} = 0.023 \left(\frac{VD_h}{v} \right)^{0.8} Pr^{1/3}. \tag{6.2}$$

The analogy between heat and mass transfer gives

$$\frac{K_1 D_h}{D_3^t} = 0.023 \left(\frac{\rho_1 V_1 D_h}{\eta_1} \right)^{0.8} \left(\frac{v_1^t}{D_3^t} \right)^{1/3} \tag{6.3}$$

Here v^t is cinematic turbulent viscosity of gas and D_3^t is turbulent diffusion constant of the droplets through the gas. Solving with respect to the droplet mass transfer coefficient and dividing by the gas velocity results in

$$\frac{K_1}{V_1} = 0.023 \left(\frac{V_1 D_h}{v_1} \right)^{-0.2} \left(\frac{v_1^t}{D_3^t} \right)^{-2/3} \frac{v_1^t}{v_1}. \tag{6.4}$$

Assuming $K_3 \approx K_1$, $v_1^t \approx v_1$, $D_3^t \approx D_1^t$ and $\frac{v_1^t}{D_3^t} \approx Pr$ and substituting in Eq. (6.1) we obtain

$$(\rho w)_{32} = 0.023 \, V_1 \, Re_1^{-1/5} \, Pr_1^{-2/3} \frac{\alpha_3}{\alpha_1 + \alpha_3} \rho_3 = K_3 \frac{\alpha_3}{\alpha_1 + \alpha_3} \rho_3, \tag{6.5}$$

where

$$Re_1 = \frac{\rho_1 V_1 D_h}{\eta_1} \tag{6.6}$$

is the gas *Reynolds* umber. The analogy eliminated the need to estimate the turbulent diffusion constant of the droplets through the gas

$$D_3^t = \frac{1}{3}\ell_3 u_3',\qquad(6.7)$$

which is associated with the estimation of the mean free path ℓ_3 in which the droplet changes its direction by colliding with a turbulent eddy and of the droplet pulsation velocity u_3'. The analogy is also applicable only for very small particles, which follow immediately the gas pulsations. Table 6.1 gives different empirical correlations and the region of validation of each correlation.

6.3 Fluctuation mechanism in the boundary layer

Next we consider the fluctuation mechanism in the boundary layer more carefully. The number of droplets striking the film interface per unit time and unit surface is

$$\dot{n}_{32}'' = \frac{1}{4}\frac{n_3}{\alpha_1+\alpha_3}u_3'.\qquad(6.8)$$

Here n_3 is the number of the particles per unit mixture volume (gas + droplet + film), $\dfrac{n_3}{\alpha_1+\alpha_3}$ is the number of the particles per unit core volume (gas +droplet), and u_3' is the particle fluctuation velocity. The droplet mass deposited into the film per unit surface and unit time is then

$$(\rho w)_{32} = \rho_3 \dot{n}_{32}''\frac{\pi}{6}D_3^3 = \frac{1}{4}\rho_3\frac{n_3}{\alpha_1+\alpha_3}\frac{\pi}{6}D_3^3\,u_3' = \frac{1}{4}u_3'\frac{\alpha_3}{\alpha_1+\alpha_3}\rho_3$$

$$= K_{32}\frac{\alpha_3}{\alpha_1+\alpha_3}\rho_3.\qquad(6.9)$$

Thus we see the physical meaning of the deposition mass transfer coefficient

$$K_{32} = \frac{1}{4}u_3'.\qquad(6.10)$$

Matsuura et al. [9] assumed that the droplet fluctuation velocity is proportional to the gas fluctuation velocity in the gas boundary layer

$$u'_3 \approx f \, u'_1, \tag{6.11}$$

that the gas fluctuation velocity is 90% of the gas friction velocity

$$u'_1 = 0.9 V_1^*. \tag{6.12}$$

The interfacial shear stress

$$\tau_{12} = \frac{1}{2} c_{21}^d \rho_1 V_1^2 \tag{6.13}$$

and the gas friction velocity at the interface

$$V_1^* = \sqrt{\tau_{12}/\rho_1} = V_1 \sqrt{\frac{1}{2} c_{21}^d}, \tag{6.14}$$

can be estimated by using the *Blasius* formula

$$c_{12}^d = 0.057 / \mathrm{Re}_1^{1/4}. \tag{6.15}$$

For the turbulent gas boundary layer at the interface we have

$$c_{12}^d = 0.057 / \mathrm{Re}_1^{1/4}. \tag{6.16}$$

Consequently

$$(\rho w)_{32} = f \, 0.038 \, \frac{\alpha_3}{\alpha_1 + \alpha_3} \rho_3 V_1 \, \mathrm{Re}_1^{-1/8}. \tag{6.17}$$

The missing information in the considerations by *Matsuura* et al. [9] is the fact that the capability of the droplets to follow the gas turbulent pulsations depends on their size. This relation was provided by *Zaichik* et al. [21].

6.4 *Zaichik*'s theory

Zaichik et al. [21] succeeded for the first time to derive expression based on the turbulence fluctuation theory in dispersed flows

$$(\rho w)_{32} = \rho_3 \frac{\dot{n}''_{32}}{\alpha_1 + \alpha_3} \frac{\pi}{6} D_3^3 = \frac{\alpha_3}{\alpha_1 + \alpha_3} \rho_3 V_1^* \frac{\dot{n}''_{32} \frac{\pi}{6} D_3^3}{\alpha_3 V_1^*}$$

$$= \frac{\alpha_3}{\alpha_1+\alpha_3}\rho_3 V_1^* \frac{0.247}{\left(1+0.54\Delta\tau_{13}^{*1/3}\right)\left[1+\left(\frac{\Delta\tau_{13}^*}{1+0.54\Delta\tau_{13}^{*1/3}}\right)^{1/2}\right]}. \tag{6.18}$$

Here

$$\Delta\tau_{13} = \frac{\rho_3 D_3}{18\eta_1} \tag{6.19}$$

is the *Stokes* relaxation time and

$$\Delta\tau_{13}^* = \Delta\tau_{13}\frac{D_3}{2}/V^*. \tag{6.20}$$

Substituting the *Blasius* expression for the gas friction velocity into the *Zaichik* formula we obtain

$$(\rho w)_{32} = 0.042\frac{\alpha_3}{\alpha_1+\alpha_3}\rho_3 V_1 \operatorname{Re}_1^{-1/8} \times \frac{1}{\left(1+0.54\Delta\tau_{13}^{*1/3}\right)\left[1+\left(\frac{\Delta\tau_{13}^*}{1+0.54\Delta\tau_{13}^{*1/3}}\right)^{1/2}\right]}.$$

$$\tag{6.21}$$

Comparing with the *Masuura* et al. expression we see that the function f is about

$$f = \frac{1}{\left(1+0.54\Delta\tau_{13}^{*1/3}\right)\left[1+\left(\frac{\Delta\tau_{13}^*}{1+0.54\Delta\tau_{13}^{*1/3}}\right)^{1/2}\right]}. \tag{6.22}$$

6.5 Deposition correlations

Comparing with the correlation in Table 6.1 we see that *Nigmatulin* properly correlate in 1982 the data by using $(\rho w)_{32} \propto \alpha_3\rho_3 V_1 \operatorname{Re}_1^{-1/8}$. The *Paleev* and *Filippovich* correlation contains the dependence $(\rho w)_{32} \propto (\alpha_3\rho_3 w_3)^{3/4}$ which is close to the theoretical one but does not have the slight dependence on the *Reynolds* number and on the particle size explicitly. The *Lopes* and *Dukler* correlation from 1986 possesses the proper dependence $(\rho w)_{32} \propto V_1^*$ by taking into account the dependence on the particle size.

Table 6.1. Deposition correlations

Paleev and *Filippovich* [13] (1966):

$$(\rho w)_{32} = 0.022\, \rho_1 V_1\, \text{Re}_1^{-1/4} \left(\frac{\alpha_3}{\alpha_1} \frac{\rho_3}{\rho_1} \right)^{0.74}$$

validated for air - water mixture at $p \approx 10^5\, Pa$, $0.1 < \alpha_3 \rho_3 < 1.3$, $3 \times 10^4 < \text{Re}_1 < 8.5 \times 10^4$. Sometimes [21] this correlation is used in slightly modified form

$$(\rho w)_{32} = \frac{\eta_3}{D_h} 0.022 \left(\frac{\eta_1}{\eta_3} \right)^{1/4} \left(\frac{\alpha_3 \rho_3 w_3 D_h}{\eta_3} \right)^{3/4}.$$

Sugawara [14] (1990):

$$(\rho w)_{32} = 0.009\, \text{Pr}_1^{-2/3}\, V_1\, \text{Re}_1^{-1/5} \left(\frac{\alpha_1}{\alpha_1 + \alpha_3} \frac{\rho_3}{\rho_1} \right)^{-0.5} \frac{\alpha_1}{\alpha_1 + \alpha_3} \rho_3$$

validated for steam - water flows in the region $1 \times 10^5 < p < 70 \times 10^5\, Pa$, $0.04 < \frac{\alpha_1}{\alpha_1 + \alpha_3} \frac{\rho_3}{\rho_1} < 10$, $10^4 < \text{Re}_1 < 10^6$.

Nigmatulin [10] (1982):

$$(\rho w)_{32} = 0.008 \frac{\alpha_3}{\alpha_1 + \alpha_3} \rho_3 V_3 \text{Re}_1^{-0.12} \left(\frac{\alpha_1}{\alpha_1 + \alpha_3} \right)^{-0.16} f$$

where

$f = \sigma^*$ for $\sigma^* > 1$,

$f = \sigma^{*1/2}$ for $\sigma^* \leq 1$,

$$\sigma^* = 0.16 \frac{\sigma}{V_1 \sqrt{\eta_1 \eta_2}} \left(\frac{\rho_1}{\rho_2} \right)^{1/4}.$$

Validated for $p = (10\ \text{to}\ 100) \times 10^5\, Pa$, $D_h = 0.013\, m$, $0.2 \leq \sigma^* \leq 10$, $0.01 \leq \rho_1/\rho_2 \leq 0.1$, $10^4 < \text{Re}_1 < 10^5$. Error for $(\rho w)_{32}/\alpha_3 \rho_3 V_3$: +35%, -40%

Katto [6] (1984):

$$(\rho w)_{32} = 0.405\sigma^{0.915} \frac{\alpha_1}{\alpha_1 + \alpha_3} \rho_3 \text{ for } \sigma < 0.0383$$

$$(\rho w)_{32} = 9.480 \times 10^4 \sigma^{4.7} \frac{\alpha_1}{\alpha_1 + \alpha_3} \rho_3 \text{ for } \sigma \geq 0.0383$$

Owen and *Hewit* [12] (1987):

$$(\rho w)_{32} = \frac{0.18}{\sqrt{\frac{\rho_1 D_h}{\sigma}}} \frac{\alpha_1}{\alpha_1 + \alpha_3} \rho_3, \text{ for } \frac{\alpha_1}{\alpha_1 + \alpha_3} \frac{\rho_3}{\rho_1} < 0.3$$

$$(\rho w)_{32} = \frac{0.083}{\sqrt{\frac{\rho_1 D_h}{\sigma}}} \left(\frac{\alpha_1}{\alpha_1 + \alpha_3} \frac{\rho_3}{\rho_1} \right)^{-0.65} \frac{\alpha_1}{\alpha_1 + \alpha_3} \rho_3 \text{ for } \frac{\alpha_1}{\alpha_1 + \alpha_3} \frac{\rho_3}{\rho_1} \geq 0.3$$

validated for air-genklane, fluorheptane, air-water, steam-water, and $p < 110 \times 10^5$ Pa.

Lopes and *Ducler* [7] (1986):

$$(\rho w)_{32} = K_3 \frac{\alpha_3}{\alpha_1 + \alpha_3} \rho_3 V_1^*$$

where

$$V_1^* = \sqrt{\tau_{12}/\rho_1}$$

is the friction velocity,

$$\tau_{12} = \frac{1}{2} \frac{c_{12}^d}{4} \rho_1 (V_1 - V_{2i})^2$$

is the interfacial shear stress

$$c_{12}^d = (3.331 \ln Re_1 - 33.582)^{-2} \text{ for } 4 \times 10^4 < Re^1 < 8.5 \times 10^4$$

and

$c_{12}^d = 0.056$ for $8.5 \times 10^4 < Re_1 < 12 \times 10^4$,

is the approximation of the data of *Lopes* and *Dukler* for the friction coefficient,

$$Re_1 = \rho_1 (V_1 - V_{2i}) D_{hc} / \eta_1$$

is the gas Reynolds number and

$$D_{hc} = D_h \sqrt{1 - \alpha_2}$$

is the gas core diameter. The experimental data for the modified mass transfer coefficient K_3 are a unique function of the dimensionless particle relaxation time

$$\tau^+ = \frac{1}{18} \frac{\rho_3}{\rho_1} \left(\frac{\rho_1 V_1^* D_3}{\eta_1} \right)^2$$

in the range $10^{-2} < \tau^+ < 10^6$ the data are represented by

$K_3 = 2 \times 10^{-5}$ for $\tau^+ < 0.2$ (*Brownian* diffusion),

$K_3 = 4.93 \times 10^{-4} (\tau^+)^{1.99}$ for $0.2 < \tau^+ < 25$,

$K_3 = 0.3$ for $25 < \tau^+$.

Schadel et al. [15] (1990)

$$(\rho w)_{32} = \frac{0.034 \, \alpha_3 \rho_3 V_3}{D_h^{0.6}} \quad \text{for} \quad \frac{\alpha_3 \rho_3 V_3}{\alpha_1 V_1} < \frac{0.078}{D_h^{0.6}}$$

$$(\rho w)_{32} = \frac{0.021}{D_h^{0.6}} \quad \text{for} \quad \frac{\alpha_3 \rho_3 V_3}{\alpha_1 V_1} \geq \frac{0.078}{D_h^{0.6}}$$

$0.0254 < D_h < 0.05715$ m, $19.5 < V_1 < 115$ m/s, vertical air / water flow.

The mass entering the film per unit time and unit mixture volume is

$$\mu_{32} = a_{12} (\rho w)_{32} . \tag{6.23}$$

For channels with defined hydraulic diameter, D_h, the interfacial area density is

$$a_{12} = \frac{4}{D_h}\sqrt{1-\alpha_2} \ . \tag{6.24}$$

Nomenclature

Latin

a_{12} interfacial area density, i.e. the surface area between gas and film per unit mixture volume, m^2/m^3

c_{12}^d vapor side shear stress coefficient at the liquid surface due to the gas flow, *dimensionless*

c_1 $= \Delta p_{21} / \rho_1$

$const$ constant, dimensionless

D_h hydraulic diameter, m

$\left(\dfrac{dp}{dz}\right)_2$ friction pressure drop per unit length in the film, Pa/m

$\left(\dfrac{dp}{dz}\right)_{Tph}$ two-phase friction pressure drop per unit length in the core, Pa/m

E $= \dfrac{\alpha_3 \rho_3 w_3}{\alpha_2 \rho_2 w_2 + \alpha_3 \rho_3 w_3}$, mass fraction of the entrained liquid, entrainment, *dimensionless*

E_∞ equilibrium mass fraction of the entrained liquid, entrainment, equilibrium entrainment, *dimensionless*

Eo $= (D_h / \lambda_{RT})^2$, Eötvös number, *dimensionless*

Fr_1 $= V_1^2 /(gD_h)$, gas *Froud* number, *dimensionless*

f deposition coefficient, *dimensionless*

f_m^* frequency of the fastest growing of the unstable surface perturbation waves, *dimensionless*

G mass flow rate, $kg/(m^2 s)$

j_{core} $= \alpha_1 w_1 + \alpha_3 w_3$ core (gas + droplet) superficial velocity, m/s

k_2 film wavelength, m

p pressure, Pa

Re_1 $= \alpha_1 \rho_1 |w_1| D_h / \eta_1$, local gas film *Reynolds* number based on the hydraulic diameter, *dimensionless*

$Re_{1,strat} = \dfrac{\alpha_1 \rho_1 |w_1 - w_2| D_h}{\eta_1} \dfrac{\pi}{\theta + \sin\theta}$, local gas film *Reynolds* number based on the gas hydraulic diameter for stratified flow, dimensionless

$Re_2 = \alpha_2 \rho_2 |w_2| D_h / \eta_2$, local film *Reynolds* number based on the hydraulic diameter, dimensionless

$Re_{2\delta} = \rho_2 w_2 \delta_2 / \eta_2$, local film *Reynolds* number based on the liquid film thickness, dimensionless

$Re_{2\infty} = Re_{23}(1 - E_\infty)$, local equilibrium film *Reynolds* number based on the hydraulic diameter, dimensionless

$Re_{23} = \rho_2 (1-\alpha_1) |w_{23}| D_h / \eta_2$, total liquid *Reynolds* number, dimensionless

$Re_{2F} = \rho_2 w_2 4\delta_{2F} / \eta_2$, local film *Reynolds* number, dimensionless

$Re_{2Fc} = 160$, critical local film *Reynolds* number, dimensionless

$Re_{core} = (\alpha_1 \rho_1 w_1 + \alpha_3 \rho_3 w_3) D_h / \eta_1$, core *Reynolds* number, dimensionless

$Ta_{12} = (\rho_2/\rho_1)\left[\sigma/(\eta_2 \Delta V_{12})\right]^2$, *Taylor* number based on relative velocity, dimensionless

$\dot{q}''_{w2}$ heat flux, MW/m^2

$U = \dfrac{\tau_{12}}{\sigma_2}\left(\dfrac{\alpha_2 w_2}{\sigma_2}\right)^{0.6}$, group used in the *Ueda* correlation

$u_2^{*1\sigma} = (\tau_2^{1\sigma}/\rho_2)^{1/2}$, liquid side surface friction velocity, m/s

$u_2' = cV_2$ interface averaged fluctuation velocity, m/s

u_{23} interface averaged entrainment velocity, m/s

V_1 gas velocity, m/s

V_2 jet velocity, continuum liquid velocity, m/s

$V_{1,wave}$ for gas velocity larger then $V_{1,wave}$ the horizontal pipe flow is no longer stratified with smooth surface, m/s

$V_{1,stratified}$ for gas velocity larger then $V_{1,wave}$ horizontal pipe flow is no longer stratified, m/s

$We_{Ishii} = \dfrac{\rho_1(\alpha_1 w_1)^2 D_h}{\sigma_2}\left(\dfrac{\rho_2 - \rho_1}{\rho_1}\right)^{1/3}$, *Weber* number for the *Ishii* entrainment correlation, dimensionless

$We_{Lopez} = \dfrac{\rho_1 w_1^2 D_h}{\sigma_2}\left(\dfrac{\rho_2 - \rho_1}{\rho_1}\right)^{1/2}$, *Weber* number for the *Lopez* et al. correlation, dimensionless

$We_{Zeichik} = \dfrac{\tau_2^{1\sigma} \delta_{2F}}{\sigma_2}$, Weber number for the *Zeichik* et al. correlation, *dimensionless*

$We_{12} = \dfrac{\rho_1 D_h \Delta V_{12}^2}{\sigma_2}$, Weber number, *dimensionless*

$w_{23} = \dfrac{\alpha_2 w_2 + \alpha_3 w_3}{1 - \alpha_1}$, center of volume velocity of film and droplet together, m/s

w_1 axial cross section averaged gas velocity, m/s
w_2 axial cross section averaged film velocity, m/s
w_3 axial cross section averaged droplets velocity, m/s
X_1 gas mass flow devided by the total mass flow, *dimensionless*
Y $= c_1 c_2 (\rho_{core} / \rho_2)(\eta_2 / \eta_1)^{0.3} Fr_1 Eo$

Greek

α_1 gas volume fraction, *dimensionless*
α_2 film volume fraction, *dimensionless*
α_3 droplets volume fraction, *dimensionless*
ΔV_{12} relative velocity, m/s
Δw_{12} $= w_1 - w_2$, relative velocity, m/s
δ_{2F} $= D_h (1 - \sqrt{1 - \alpha_2})/2$, film thickness in annular flow, m
δ_{2eq} equilibrium thickness, m
η_1 gas dynamic viscosity, $kg/(ms)$
η_2 liquid dynamic viscosity, $kg/(ms)$
λ_{R12} gas-film friction coefficient, *dimensionless*
λ_{R1} friction coefficient, *dimensionless*
λ_{R2} $= \lambda_{R2}(Re_2, k/D_h)$, film friction coefficient, *dimensionless*
λ_m^* wavelength of the fastest growing of the unstable surface perturbation waves, *dimensionless*
λ_{RT} $= (\sigma / g \Delta \rho_{21})^{1/2}$, *Rayleigh-Taylor* wavelength, m
μ_{23} mass leaving the film and entering the droplet field per unit time and unit mixture volume, $kg/(m^3 s)$
ρ_1 gas density, kg/m^3
ρ_2 liquid density, kg/m^3
ρ_{core} $= (\alpha_1 \rho_1 w_1 + \alpha_3 \rho_3 w_3)/j_{core}$, core density, kg/m^3

$(\rho w)_{23}$ entrainment mass flow rate, mass leaving the film per unit time and unit interfacial area, $kg/(m^2 s)$

$(\rho w)_{32}$ deposition mass flow rate, mass leaving the droplet field per unit time and unit interfacial area and deposited into the film, $kg/(m^2 s)$

σ surface tension, N/m

$\tau_2^{1\sigma}$ liquid side surface shear stress

τ_{12} interfacial stress, N/m^2

τ^* $= \tau_{12} \delta_2 / \sigma$, interfacial stress, *dimensionless*

θ angle with origin of the pipe axis defined between the upwards oriented vertical and the liquid-gas-wall triple point, *rad*

References

1. Govan AH, Hewitt GF, Owen DG, and Bott TR (1988) An improved CHF modeling code, 2nd UK National Heat Transfer Conference, Glasgow
2. Hewitt GF (1961) Analysis of annular two phase, application of the Dukler analysis to vertical upward flow in a tube, AERE-R3680
3. Hewitt GF and Gowan AH (March 23-24, 1989) Phenomenological modeling of non-equilibrium flows with phase change, Proc. of 7th Eurotherm Seminar Thermal Non-Equilibrium in Two-Phase Flow, Roma, pp 7-27
4. Hutchinson P and Whalley PB (1973) Possible caracterization on entrainment in annular flow, Chem. Eng. Sci., vol 28 p 974
5. Kataoka I and Ishii M (July 1982) Mechanism and correlation of droplet entrainment and deposition in annular two-phase flow. NUREG/CR-2885, ANL-82-44
6. Katto Y (1984) Prediction of critical heat flux for annular flow in tubes taking into account of the critical liquid film thickness concept, Int. J. Heat Mass Transfer, vol 27 no 6 pp 883-890
7. Lopes JCB and Ducler AE (Sept.1986) Droplet entrainment in vertical annular flow and its contribution to momentum transfer, AIChE Journal, vol 32 no 9 pp 1500-1515
8. Lopez de Bertodano MA, Assad A and Beus S (June 8-12, 1998) Entrainment rate of droplets in the ripple-annular regime for small vertical ducts, Third International Conference on Multiphase Flow, ICMF'98, Lyon, France, CD Proceedings
9. Matsuura K, Kataoka I and Serizawa A (April 23-27, 1995) Prediction of droplet deposition rate based on Lagrangian simulation of droplet behavior, Proc. Of the 3rd JSME/ASME Joint International Conference on Nuclear Engineering, Kyoto, Japan, vol 1 pp 105-109
10. Nigmatulin BI (1982) Heat and mass transfer and force interactions in annular - dispersed two - phase flow, 7-th Int. Heat Transfer Conference Munich, pp 337 - 342
11. Nigmatulin BI, Melikhov OI and Khodjaev ID (April 3-7, 1995) Investigation of entrainment in a dispersed-annular gas-liquid flow, Proc. of The 2bd International Conference on Multiphase Flow '95 Kyoto, Japan, vol 3 pp P4-33 to P4-37
12. Owen GD and Hewitt GF (1987)An improved annular two-phase flow model, 3rd BHRA Int. Conf. on Multiphase Flow, The Hague

13. Paleev II and Filipovich BS (1966) Phenomena of liquid transfer in two-phase dispersed annular flow. Int. J. Heat Mass Transfer, vol 9 p 1089
14. Sugawara S (1990) Droplet deposition and entrainment modeling based on the three-ffluid model, Nuclear Engineering and Design, vol 122 pp 67-84
15. Schadel SA, Leman GW, Binder JL, and Hanratty TJ (1990) Rates of atomization and deposition in vertical annular flow, Int. J. Multiphase Flow, vol 16 no 3 pp 363-374
16. Tomiyama A and Yokomiyo O (February 1988) Spacer - effects on film flow in BWR fuel bundle, J. Nucl. Sc. Techn., Atomic Energy Society of Japan, vol 25 no 2 pp 204-206
17. Turner JM and Wallis GB (1965)An analysis of the liquid film annular flow, Dartmouth Colege, NYO-3114-13
18. Ueda T (1981) Two-phase flow - flow and heat transfer, Yokendo, Japan (in Japanese)
19. Wallis G B (1970) Annular two-phase flow, Part I, A simple theory, J. Basic Eng., vol 92 p 59
20. Whalley PB et al. 4 (1974) The calculation of critical heat flux in forced convection boiling, Proc. 5th Int. Heat Transfer Conf., Tokyo, vol 4 pp 290-294
21. Zaichik LI, Nigmatulin BI and Alipchenko VM (June 8-12, 1998) Droplet deposition and film atomization in gas-liquid annular flow, Third International Conference on Multiphase Flow, ICMF'98, Lyon, France, CD Proceedings

7. Introduction to fragmentation and coalescence

7.1 Introduction

As already mentioned in Section 1.1, transient multiphase flows with temporal and spatial variation of the volumetric fractions of the participating phases can be represented by *sequences* of geometrical *flow patterns* that have some characteristic length scale. Owing to the highly random behavior of the flow in detail, the number of flow patterns needed for this purpose is very large. Nevertheless, this approach has led to many successful applications in the field of multiphase flow modeling. Frequently modern mathematical models of *transient flows* include, among others, the following features:

1. Postulation of a limited number of idealized flow patterns, with transition limits as a function of local parameters for *steady state* flow (e.g., see Fig. 1.1);

2. Identification of one of the postulated idealized *steady state* flow patterns for each time;

3. Computation of a characteristic *steady state length scale of the flow patterns* (e.g., bubble or droplet size) in order to address further constitutive relationships for interfacial heat, mass, and momentum transfer.

A steady state length scale should not be used for highly transient processes. *Mechanical disintegration* of fluid occurs in the flow over a *finite time*. In many transient processes, the characteristic time constant may be comparable with the time scale of the macroscopic process under consideration. There is a class of multi-phase flows for which the prediction of the transient flow pattern length scale is crucial. An example is the mathematical description of the interaction between molten metal and water or between a cold liquid and a hot liquid that is at a temperature much higher than the saturation temperature of the cold liquid. The mathematical description of violent explosions that occur during such interactions is possible only if the *dynamic fragmentation and coalescence* modeling corresponds to the real physics. Thus there is a need to model the continuous fragmentation and coalescence dynamics for multi-phase flows. Moreover, an adequate theory to predict the length scale of the flow structure should automatically provide information for the flow pattern identification. As already mentioned the opposite is the widespread practice today.

A possible formalism to model dynamic fragmentation and coalescence is the following. Describe multi-phase flows by means of three mutually interacting velocity fields, three abstract fluids having their own temperature and velocity. Define the correspondence between the abstract fluids and the real physical world: (1) the first fluid, $l = 1$, as a gas; (2) the second fluid, $l = 2$, as a liquid; and (3) the third fluid, $l = 3$, as another liquid. Write the local volume and time average mass conservation equation for each velocity field – compare with Eq. (1.62) in Vol.1,

$$\frac{\partial}{\partial \tau}(\alpha_l \rho_l) + \nabla \cdot (\alpha_l \rho_l \mathbf{V}_l) = \mu_l. \tag{7.1}$$

Here α_l is the averaged volumetric fraction, ρ_l is the density, $\mathbf{V}_l$ is the velocity vector and μ_l is the mass source density. Define the particle number density n_l as a volume-averaged number of discrete particles of the fluid l (e.g., bubbles or droplets) per unit flow volume, and write the local volume and time-averaged conservation equation for this quantity neglecting diffusion effects – compare with Eq. (1.109) in Vol.1,

$$\frac{\partial}{\partial \tau}(n_l \gamma_v) + \nabla \cdot (\mathbf{V}_l n_l) = \gamma_v \left(\dot{n}_{l,kin} - \dot{n}_{l,coal} + \dot{n}_{l,sp} \right)$$

$$\equiv \gamma_v n_l \left\{ \frac{1}{n_l} \dot{n}_{l,kin} + f_{l,sp} - \frac{1}{2} \left[\left(f_{d,col}^s + f_{d,col}^{no} \right) P_{d,coal}^{no} + f_{d,col}^o P_{d,coal}^o \right] - f_{l,spectrum\_cut} \right\}$$

for $\alpha_l \geq 0$. \hfill (7.2)

Here $\dot{n}_{l,sp}$ is the particle production term, that is the increase in the number of particles per unit time and unit mixture volume due to mechanical fragmentation such as splitting (subscript *sp*), and $\dot{n}_{l,coal}$ is the particle sink term, that is, the decrease, in the number of particles due to mechanical coalescence (subscript coal). $\dot{n}_{l,kin}$ is the number of particles generated or lost per unit time and unit mixture volume due to evaporation and/or condensation.

The next important step in the modeling is the estimation of the length scale of the particles needed for computation of mass, momentum and energy transport between the velocity fields, given the number of particles per unit mixture volume, n_d, and their volume fraction α_d in the flow mixture. It is possible to proceed in different ways, but one of them seems to be easy and practicable: We assume that the dispersed phase is *locally mono-disperse*, that is, that all particles in a control volume have the same size. In this case the product of the particle number density,

n_d, and the volume of single particles, $\pi D_d^3/6$, assuming spherical shapes with constant diameter D_d, gives the volume fraction α_d, and therefore

$$D_d = \left(\frac{6}{\pi}\frac{\alpha_d}{n_d}\right)^{1/3}. \tag{7.3}$$

As already shown in [9] and the references given there, this method can be used to describe the dynamic evolution of the length scale of a multi-phase structure. The success of the application depends on whether appropriate experimental information is available and, if it is available, on whether this information can be generalized to provide the source terms in Eq. (7.2).

A further sophistication of the theory needs the *multi-group approach* already successfully exercised in neutron physics in the last 50 years with a detailed description of the dynamic interaction among groups of particles. This very ambitious task is outside the scope of this work. A compromise between complexity and a practicable approach is shown in [2], where a two-group approach for *macro-* and *microscopic* liquid metal particles in water-gas flows was used. Other promising examples for bubbly flow are reported in [24, 1, 12].

In Chapters 7 to 9 we concentrate our attention on the available knowledge for computation of the particle production term $\dot{n}_{l,sp}$, and the particle sink term $\dot{n}_{l,coal}$. The number of particles born per unit time and unit mixture volume, $\dot{n}_{l,kin}$, due to evaporation and condensation is subject to the nucleation theory and will be discussed later.

The experimental observations useful for development of the fragmentation or coalescence model are classified here as follows:

1. Identification of the kind of process leading to fragmentation or coalescence;
2. Measurements of the size of the final products of the fragmentation or coalescence;
3. Definition of the final state of the fragmentation or coalescence process and measurement of the duration of the process.

The quantitatively estimated characteristics are functions of the local flow parameters and initial conditions. These characteristics are used to model the production and sink rates. These models are then used in the range of parameters for which they are valid in macroscopic fluid models.

The following discussion will be restricted to mono-disperse particles in a single computational cell. We will also discuss the implication of this assumption.

7.2 General remarks about fragmentation

There are different but partially overlapping processes leading to disintegration of the continuum and the formation of dispersed particles or leading to the disintegration of unstable droplet and the formation of finer particles as shown in Fig. 1.4 in Chapter 1:

The main characteristics of each fragmentation process are

1. the stable particle diameter after the fragmentation is finished, $D_{d\infty}$, and
2. the duration of fragmentation, $\Delta \tau_{br}$.

It is very important to note that $D_{d\infty}$ is defined here for situations without particle coalescence.

After the fragmentation process we assume that

1. the total mass of the particles is the same as before the fragmentation,
2. all newly formed particles are at the same temperature.

Knowing $D_{d\infty}$ we compute the particle number density after the fragmentation process using assumption 1,

$$n_{d\infty} = \alpha_d / \left(\pi D_{d\infty}^3 / 6 \right) \tag{7.4}$$

and finally the time-averaged production rate

$$\dot{n}_{d,sp} \approx \frac{n_{d\infty} - n_d}{\Delta \tau_{br}} = \frac{n_d}{\Delta \tau_{br}} \left[\left(\frac{D_d}{D_{d\infty}} \right)^3 - 1 \right]. \tag{7.5}$$

The production rate $\dot{n}_{d,sp}$ can be written in the form

$$\dot{n}_{d,sp} = n_d f_{d,sp}, \tag{7.6}$$

where

$$f_{d,sp} = \left[\left(\frac{D_d}{D_{d\infty}}\right)^3 - 1\right] \Big/ \Delta\tau_{br} , \qquad (7.7)$$

is the *fragmentation frequency* of a single particle. If $D_d = D_{d\infty}$, the fragmentation frequency is zero and therefore the production rate $\dot{n}_{d,sp}$ is zero.

If the process is considered over a time interval $\Delta\tau$ (e.g., computational time step) that is larger than the fragmentation period $\Delta\tau_{br}$, the $\Delta\tau_{br}$ in Eq. (7.5) should be replaced by $\Delta\tau$, which in fact means that during the time interval $\Delta\tau$ the particle fragmentation is completed and stable conditions are reached. Only if $\Delta\tau \gg \Delta\tau_{br}$, is it justified to use the steady state flow pattern length scale instead of the transient one.

For volumes with zero convective net flux and no coalescence, the steady-state value of n_d approaches $n_{d\infty}$, that is, D_d approaches $D_{d\infty}$. The situation changes if coalescence effects are included.

Obviously, we need $D_{d\infty}$ and $\Delta\tau_{br}$ to compute the average mono-disperse particle production rate $\dot{n}_{d,sp}$. The quantitative description of the above discussed 16 modes of fragmentation concentrates on providing information for $D_{d\infty}$ and $\Delta\tau_{br}$ for the particular conditions and geometry.

7.3 General remarks about coalescence

Fragmentation is one of the processes controlling the particle size. Not less important is the collision and coalescence of particles in a continuum. The collision is caused by existence of *spatial velocity differences among the particles themselves*. This spatial relative velocity is caused by different factors, e.g. non-linear trajectory of the particles, turbulent fluctuations etc. Not each collision leads necessarily to coalescence. Thus *modeling particle agglomeration means modeling of collision and coalescence mechanisms*. In the following we discuss a simple formalism for the mathematical description of particle collision and coalescence and the available formalized empirical information needed for practical application of the theory.

7.3.1 Converging disperse field

Because of various geometrical obstacles, boundary conditions or variety of interactions inside the flow the particles may change their velocity in magnitude and direction. Consider a cloud of particles moving from a center radially outwards. None of the particles will touch the other. In this case there is no collision. The condition for collision in a flow of dispersed particles without oscillation over the mean velocity values is that they are coming together. Mathematically it can be expressed as follows: It is only if the relative velocity

$$V^{rel} = \left| \max\left[0, \, \mathbf{V}_d\left(\mathbf{r}+\Delta\mathbf{r}\right) - \mathbf{V}\left(\mathbf{r}\right)\right] \right|, \tag{7.8}$$

is negative, collisions and therefore agglomerations may take place. Here $\mathbf{r}$ is the position vector of the point having particle velocity vector $\mathbf{V}_d$. In this case

$$\nabla \mathbf{V}_d < 0. \tag{7.9}$$

Usually, for a practical computation, $\Delta V_{dd}^{no} = V^{rel}$ should be averaged across the computational cell so that we have

$$\Delta V_{dd}^{no} \approx \left[\left(\Delta x \frac{\partial u}{\partial x}\right)^2 + \left(\Delta y \frac{\partial v}{\partial y}\right)^2 + \left(\Delta z \frac{\partial w}{\partial z}\right)^2\right]^{1/2}. \tag{7.10}$$

As in the molecular kinetic theory, the average distance between collisions, or the mean free path ℓ_{col}, is given by the ratio of the distance of collisions $V^{rel}\Delta\tau$ and the collision frequency along this distance

$$\ell_{col} \approx const \frac{1}{n_d \pi D_d^2}. \tag{7.11}$$

It is recommended to resolve this distance by the computation because along this distance the interaction happens and the velocity of the particles changes as follows

$$\Delta V_{dd}^{no} \approx \ell_{col} \nabla \mathbf{V}_d. \tag{7.12}$$

7.3.2 Analogy to the molecular kinetic theory

Usually the agglomerated particles per unit time and unit volume are defined as

$$\frac{dn_d}{d\tau} = \dot{n}_{d,coal} = -f_{d,coal} n_d / 2, \tag{7.13}$$

where $f_{d,coal}$ is the coalescence frequency of single particles with the dimension s^{-1}. $\dot{n}_{d,coal}$ is the instant coalescence rate. The number of particles remaining after a time interval $\Delta\tau$ per unit volume is easily obtained by integration of Eq. (7.13)

$$\int_{n_d}^{n_{d,\tau+\Delta\tau}} d\ln n_d = -\frac{1}{2}\int_{\tau}^{\tau+\Delta\tau} f_{d,coal} d\tau, \qquad (7.14)$$

or

$$n_{d,\tau+\Delta\tau} = n_d e^{-f_{d,coal}\Delta\tau/2}. \qquad (7.15)$$

The time-averaged coalescence rate is therefore

$$\dot{n}_{d,coal} = \frac{n_d - n_{d,\tau+\Delta\tau}}{\Delta\tau} = n_d\left(1 - e^{-f_{d,coal}\Delta\tau/2}\right)/\Delta\tau. \qquad (7.16)$$

The coalescence frequency is defined as the product of the collision frequency and the coalescence probability, $f_{d,coal}^p$,

$$f_{d,coal} = f_{d,col} f_{d,coal}^p, \qquad (7.17)$$

which expresses the fact that not each collision leads necessarily to coalescence. To model the coalescence means to find adequate physical models for $f_{d,col}$ and $f_{d,coal}^p$, which is the purpose of this Section.

We start with the analogy to molecular kinetic theory and discuss the differences resulting from the different nature of the droplet agglomeration compared with the random molecular collision.

Postulate a hypothetical particle-continuum mixture in which the particles are colliding at random with the following properties (see *Rohsenow* and *Choi* [18] (1961), p.487):

1. The particles are hard spheres, resembling billiard balls, having diameter D_d and mass m.

2. The particles exert no forces on each other except when they collide.

3. The collisions are perfectly elastic and obey the classical conservation laws of momentum and energy.

4. The particles are uniformly distributed through the gas. They are in a state of continuous motion and are separated by distances that are large compared with their diameter D_d.

5. All directions of particle velocity fluctuations are equally probable. The speed (magnitude of the velocity) of particles can have any value between zero and infinity.

One usually singles out a particle as it travels in a straight path from one collision to the next: its speed and direction of motion changes with each collision. Imagine that at a given instant all particles but the one in question are frozen in position and this particle moves with an averaged speed V^{rel}. At the instant of collision, the center-to-center distance of the two particles is D_d. The collision cross section of the target area of the particle is $\frac{1}{4}\pi D_d^2$. In time $\Delta\tau$ the moving particle sweeps out a cylindrical volume of the length $V^{rel}\Delta\tau$ and the cross section $\frac{1}{4}\pi D_d^2$. Any particle whose center is in this cylinder will be struck by the moving particle. The number of collisions in the time $\Delta\tau$ is

$$n_d \frac{1}{4}\pi D_d^2 V^{rel} \Delta\tau, \qquad (7.18)$$

where n_d is the number of particles present per unit volume, assumed to be uniformly distributed in space. The collision frequency of a single particle is defined as the number of collisions per unit time

$$f_{d,col} = n_d \frac{1}{4}\pi D_d^2 V^{rel} = \frac{3}{2}\frac{\alpha_d}{D_d} V^{rel}. \qquad (7.19)$$

By multiplying the collision frequency of a *single* particle, $f_{d,col}$, with the number of particles per unit mixture volume we obtain the total collision frequency, i.e. the number of collisions per unit mixture volume and per unit time

$$\dot{n}_{d,col} = f_{d,col} n_d = n_d^2 \frac{\pi}{4} D_d^2 V^{rel}. \qquad (7.20)$$

Assuming all particles move at *averaged space* velocities V^{rel} (*Clausius*) the result is

$$f_{d,col} = \frac{1}{0.75} n_d \frac{\pi}{4} D_d^2 V^{rel} = \frac{1}{0.75} \frac{3}{2} \frac{\alpha_d}{D_d} V^{rel}. \qquad (7.21)$$

If the particles are assumed to possess a *Maxwellian relative speed distribution* the result is

$$f_{d,col} = \frac{1}{\sqrt{2}} n_d \frac{\pi}{4} D_d^2 V^{rel} = \frac{1}{\sqrt{2}} \frac{3}{2} \frac{\alpha_d}{D_d} V^{rel}. \qquad (7.22)$$

The assumptions 1 and 3 do not hold for real liquid droplet collision because droplets are deformable, elastic and may agglomerate after random collisions. The collision frequency is not an independent function of the coalescence probability. The functional relationship is not known. That is why some authors correct the collision frequency with a constant less than one estimated by comparison with experiments, e.g. *Rosenzweig* et al. [19] (1980) give for relatively low V^{rel} and non-oscillatory coalescence *const* = 0.0001. *Howarth* [6] (1967) obtained a modified form of Eq. (7.22), which includes additionally the multiplier $8/\sqrt{3\alpha_d}$, namely

$$f_{d,col} = \frac{1}{\sqrt{2}} \frac{6\alpha_d}{D_d} V^{rel} 8/\sqrt{3\alpha_d} = \frac{\sqrt{24\alpha_d}}{D_d} V^{rel}. \qquad (7.23)$$

The dependence of the collision frequency on $\approx \alpha_d^m$ in the above equation is confirmed by experiments as follows. For droplets - *Howarth* [6] (1967) $\approx \alpha_d^{0.6}$, for liquid - liquid droplets *Coulaloglu* and *Tavlarides* [3] (1976), *Madden* [14] (1962), *Komasawa* et al. [11] (1968) $\approx \alpha_d^{0.45}$, and for bubbles *Sztatecsny* et al. [21] (1977) $\approx \alpha_d^{0.6}$.

Similar considerations can be repeated for dispersed particles having diameters D_{d1}, D_{d2} and particle densities n_{d1}, n_{d2}. The result is

$$\dot{n}_{d,col} = n_{d1} n_{d2} \frac{\pi}{4} \left(\frac{D_{d1} + D_{d2}}{2} \right)^2 V^{rel}, \qquad (7.24)$$

see the pioneer work of *Smoluchowski* [22] in 1918.

Hibiki and *Ishii* [5] used in 1999 the *Loeb* [13] notation from 1927 of Eq. (7.19) modifying it as follows

$$f_{d,col} = \frac{F_{col}}{4(1-\alpha_d)} V^{rel}, \qquad (7.25)$$

where

$$F_{col} \approx n_d 4\pi D_d^2 \qquad (7.26)$$

is the interface available for collision per unit mixture volume. $1-\alpha_d$ takes into account the reduction of the volume available for collisions. *Hibiki* and *Ishii* [5] (1999) modified the above relation to

$$f_{d,col} = \frac{F_{col}}{4(\alpha_{d,\max} - \alpha_d)} V^{rel} = \frac{\alpha_d}{\alpha_{d,\max} - \alpha_d} \frac{6}{D_d} V^{rel} \qquad (7.27)$$

by introducing the maximum allowable void fraction $\alpha_{d,\max} = 0.52$ for existence of bubbly flow. This modification ensures that the collision frequency increases to infinite if the bubble volume concentration reaches the maximum packing concentration.

Two spherical particles with initial size D_{d0} possess surface energy of $E_{\sigma 0} = 2\pi D_{d0}^2 \sigma$. After coalescence the new spherical particle has a size $D_d = 2^{1/3} D_{d0}$, and therefore less surface energy, $E_\sigma = \pi D_d^2 \sigma$. The difference is $\Delta E_\sigma = E_{\sigma 0} - E_\sigma = 2\pi D_{d0}^2 \left(1 - 1/2^{1/3}\right)$. This energy has to be supplied by some kinetic energy associated with the particles before the collision. The kinetic energy forcing the two colliding bubbles to coalesces in this case is there virtual mass kinetic energy, $\Delta E_k = 2\left[\frac{1}{2}\rho_c \left(V^{rel}\right)^2\right]\frac{1}{2}\frac{\pi D_{d0}^3}{6}$. For droplets this is simply the kinetic energy of the droplets before collision $\Delta E_k = 2\left[\frac{1}{2}\rho_d \left(V^{rel}\right)^2\right]\frac{\pi D_{d0}^3}{6}$.

The theoretical minimum for V^{rel} required to produce bubble coalescence is computed by equalizing the change of surface energy of the both particle to the kinetic energy lost during the coalescence of particles $\Delta E_\sigma = \Delta E_k$, and therefore

$$\frac{\rho_c D_{d0} \left(V^{rel}\right)^2}{\sigma} = 24\left(1 - 1/2^{1/3}\right) = 4.95 \qquad (7.28)$$

or

$$V^{rel} > \left(\frac{4.95\sigma}{\rho_c D_{d0}}\right)^{1/2}. \tag{7.29}$$

For droplets we have,

$$\frac{\rho_d D_{d0}\left(V^{rel}\right)^2}{\sigma} = 12\left(1 - 1/2^{1/3}\right) = 2.476 \tag{7.30}$$

or

$$V^{rel} > \left(\frac{2.476\sigma}{\rho_d D_{d0}}\right)^{1/2}. \tag{7.31}$$

Comparing Eqs. (7.29) with (7.31) we realize that at the same relative velocities bubbles will coalesce less frequently then droplets having the same size.

The nature of V^{rel} depends

(a) on the turbulent fluctuation of the particles,

(b) on the difference of the relative velocities caused by the differences of the particle size, and

(c) on the non-uniform velocity field.

Even using an average particle size, the second and the third components may differ from zero. We call the coalescence caused by reasons (a), (b), and (c), *oscillatory*, *spectral*, and *non-oscillatory* coalescence, respectively. While for the oscillatory coalescence the driving force moving the particles is exerted from the oscillating turbulent eddies and therefore pushing out the continuum between two particles and moving the particles apart (under some circumstances before they coalesce), for the spectral and the non-oscillatory coalescence the forces leading to collisions inevitably act towards coalescence – for contracting particle free path length. Therefore we have for the probability of the oscillatory coalescence

$$P^o_{d,coal} \leq 1, \quad d = \text{bubbles, droplets,} \tag{7.32}$$

and for non-oscillatory and spectral coalescence in the case of contracting particle free path length

$$P^{no}_{d,coal} = P^s_{d,coal} = 1, \quad d = \text{bubbles,} \tag{7.33}$$

$$-\infty < P^{no}_{d,coal} = P^{s}_{d,coal} \le 1, \quad d = \text{droplets}. \tag{7.34}$$

The last condition reflects the fact that for high droplet - droplet relative velocities the splitting of the resulting unstable droplet is possible.

7.4 Superposition of different droplet coalescence mechanisms

Let us rewrite the final expression for the coalescence frequency for the average particle diameter D_d

$$f_{d,coal} = f^{s}_{d,col} P^{s}_{d,coal} + f^{no}_{d,col} P^{no}_{d,coal} + f^{o}_{d,col} P^{o}_{d,coal}$$

$$= 4.9 \frac{\sqrt{\alpha_d}}{D_d} \left(\Delta V^{s}_{dd} P^{s}_{d,coal} + \Delta V^{no}_{dd} P^{no}_{d,coal} + V'_d P^{o}_{d,coal} \right)$$

$$= 4.9 \frac{\sqrt{\alpha_d}}{D_d} \left[\left(\Delta V^{s}_{dd} + \Delta V^{no}_{dd} \right) P^{no}_{d,coal} + V'_d P^{o}_{d,coal} \right] \tag{7.35}$$

where

$$P^{no}_{d,coal} = 0, \quad \Delta V^{s}_{dd} + \Delta V^{no}_{dd} < \left(\frac{12\sigma}{\rho_d D_d} \right)^{1/2}, \tag{7.36}$$

$$P^{no}_{d,coal} = f\left(\Delta V^{s}_{dd} + \Delta V^{no}_{dd} \right) \approx \Delta m_{d1} / m_{d2}, \tag{7.37}$$

$$P^{o}_{d,coal} \approx 0, \quad \text{for} \quad \Delta \tau_{col} / \Delta \tau_{coal} < 1, \tag{7.38}$$

$$P^{o}_{d,coal} \approx 0.032 \left(\Delta \tau_{col} / \Delta \tau_{coal} \right)^{1/3}, \quad \text{for} \quad \Delta \tau_{col} / \Delta \tau_{coal} \ge 1, \tag{7.39}$$

$$\Delta \tau_{col} / \Delta \tau_{coal} = 1.56 \left(\frac{D_d \sigma_d}{3\rho_d + 2\rho_c} \right)^{1/2} / V'_d. \tag{7.40}$$

The first term in Eq. (7.35) takes into account the coagulation caused by differences in the particle size leading to a relative movement between the particles, the second takes into account the spatial change of the particle velocities, also responsible for dramatic coagulation in stagnation points, and the third term takes into account the turbulence induced coagulation. This method is very easy to implement in computer codes.

7.5 Superposition of different bubble coalescence mechanisms

The method already described for particles is also valid for bubble agglomeration. The frequency of coalescence of a single bubble is

$$f_{d,coal} = f^s_{d,col} P^s_{d,coal} + f^{no}_{d,col} P^{no}_{d,coal} + f^o_{d,col} P^o_{d,coal}$$

$$= 4.9 \frac{\sqrt{\alpha_d}}{D_d} \left[\left(\Delta V^s_{dd} + \Delta V^{no}_{dd} \right) P^{no}_{d,coal} + V'_d P^o_{d,coal} \right], \qquad (7.41)$$

where

$$P^{no}_{d,coal} = 0, \quad \Delta V^s_{dd} + \Delta V^{no}_{dd} < \left(\frac{24\sigma}{\rho_c D_d} \right)^{1/2}, \qquad (7.42)$$

$$P^{no}_{d,coal} = 1, \quad \Delta V^s_{dd} + \Delta V^{no}_{dd} \geq \left(\frac{24\sigma}{\rho_c D_d} \right)^{1/2}, \qquad (7.43)$$

$$P^o_{d,coal} \approx 0, \quad \text{for} \quad \Delta\tau_{col} / \Delta\tau_{coal} < 1, \qquad (7.44)$$

$$P^o_{d,coal} \approx 0.032 \left(\Delta\tau_{col} / \Delta\tau_{coal} \right)^{1/3}, \quad \text{for} \quad 1 \leq \Delta\tau_{col} / \Delta\tau_{coal} \leq 3.149, \qquad (7.45)$$

$$\Delta\tau_{col} / \Delta\tau_{coal} = 3.35 \times 10^{-13} \frac{\sigma_c^2}{\eta_c \rho_c \left(D_d V'_d \right)^3}. \qquad (7.46)$$

The first term in Eq. (41) takes into account the coagulation caused by differences in the bubble size leading to relative bubble movement, the second takes into account the spatial change of the bubble velocities also responsible for dramatic coagulation in stagnation points, and the third term takes into account the turbulence induced coagulation.

7.6 General remarks about particle size formation in pipes

Droplets can be formed from a continuous liquid in channels due to different reasons:

- Transition from churn turbulent flow in which the liquid is the continuous phase into a disperse regime in which the liquid is disintegrated;
- Distortion of a jet in inverted annular flow due to hydrodynamic instability;
- Disintegration of hydrodynamic unstable drops into smaller droplets due to acceleration induced fragmentation as will be discussed in Chapter 8;
- Droplet entrainment from a film attached at the wall.

Let us now examine more precisely the basic physics of the particle size formation in pipes with film on the wall where entrainment and deposition of droplets takes place, see *Kolev* [8].

As discussed in Chapter 5 the entrainment is quantitatively described by models for the following characteristics:

(1) Identification of the conditions when the entrainment starts;

(2) The mass leaving the film per unit time and unit interfacial area, $(\rho w)_{23}$, or the mass leaving the film and entering the droplet field per unit time and unit mixture volume, μ_{23}

$$\mu_{23} = a_{12}(\rho w)_{23} \qquad (7.47)$$

where a_{12} is the interfacial area density, i.e. the surface area between gas and film per unit mixture volume;

(3) Size of the entrained droplets.

For channels with defined hydraulic diameter the interfacial area density is

$$a_{12} = \frac{4}{D_h}\sqrt{1-\alpha_2} . \qquad (7.48)$$

The particle production rate in case of entrainment is

$$\dot{n}_{23} = \mu_{23} / \left(\rho_2 \frac{\pi}{6} D_{3E}^3\right) = \frac{6}{\pi} \frac{4}{D_h} \sqrt{1-\alpha_2} \left(\rho w\right)_{23} / \left(\rho_2 D_{3E}^3\right) . \qquad (7.49)$$

While the entrainment transports droplets from the film into the gas bulk flow, the deposition reverses this process and transports droplets from the bulk region into the film. Entrainment takes place only if certain conditions are fulfilled. In contrast the deposition is controlled by turbulent fluctuations (a kind of macroscopic turbulent diffusion) and lift forces due to velocity gradients and takes place under all conditions. While entrainment is quantitatively described by the above-mentioned three models, the deposition is described only by a model for the mass flow rate entering the film, $(\rho w)_{32}$ - see Chapter 6. The mass entering the film per unit time and unit mixture volume is

$$\mu_{32} = a_{12}(\rho w)_{32}, \qquad (7.50)$$

and the particle sink term for velocity field $l = 3$ is

$$\dot{n}_{32} = \mu_{32} / \left(\rho_3 \frac{\pi}{6} D_3^3 \right) = \frac{6}{\pi} \frac{4}{D_h} \sqrt{1-\alpha_2} \, (\rho w)_{32} / \left(\rho_3 D_3^3 \right). \qquad (7.51)$$

Thus, the net particle production rate due to entrainment and deposition is

$$\dot{n}_{23} - \dot{n}_{32} = \frac{6}{\pi} \frac{4}{D_h} \sqrt{1-\alpha_2} \left[(\rho w)_{23} / \left(\rho_2 D_{3E}^3 \right) - (\rho w)_{32} / \left(\rho_3 D_3^3 \right) \right], \qquad (7.52)$$

where D_{3E} is the volume-averaged size of the droplets leaving the film. Obviously a proper description of the particle production rate needs proper prediction of the entrainment and deposition mass flow rate and the entrainment diameter of the droplets.

Finally, we write the conservation equation of the droplets, Eq. (1.109), in case of three-fluid gas-film-droplet flow neglecting the turbulent diffusion

$$\frac{\partial n_3}{\partial \tau} + V_3 \nabla n_3 = \dot{n}_{3,sp} - \dot{n}_{3,coal} + \dot{n}_{23} - \dot{n}_{32}. \qquad (7.53)$$

Replacing the source terms by their equals, we obtain finally

$$\frac{\partial n_3}{\partial \tau} + V_3 \nabla n_3 = n_3 \left\{ \left[\left(\frac{D_3}{D_{3\infty}} \right)^3 - 1 \right] / \Delta \tau_{br} - f_{3,coal} / 2 \right\}$$

$$+ \frac{6}{\pi} \frac{4}{D_h} \sqrt{1-\alpha_2} \left[(\rho w)_{23} / \left(\rho_2 D_{3E}^3 \right) - (\rho w)_{32} / \left(\rho_3 D_3^3 \right) \right]. \qquad (7.54)$$

Obviously, for large hydraulic diameters $D_h \to \infty$, i.e. for pool flow, entrainment and deposition do not influence the particle number density.

For flow with constant droplet velocity $V_3 \approx const$, the droplet mass conservation equation simplifies to

$$\rho_3 \left(\frac{\partial \alpha_3}{\partial \tau} + V_3 \nabla \alpha_3 \right) = \mu_{23} - \mu_{32} = \frac{4}{D_h} \sqrt{1-\alpha_2} \left[(\rho w)_{23} - (\rho w)_{32} \right]. \quad (7.55)$$

Having in mind the relationship between particle number density, volume fraction of droplets and the droplet diameter

$$\frac{\pi}{6} D_3^3 = \alpha_3 / n_3, \quad (7.56)$$

we can write

$$\alpha_3 \left(\frac{\partial D_3}{\partial \tau} + V_3 \nabla D_3 \right) = \frac{D_3}{3} \left[\frac{\partial \alpha_3}{\partial \tau} + V_3 \nabla \alpha_3 - \frac{\pi}{6} D_3^3 \left(\frac{\partial n_3}{\partial \tau} + V_3 \nabla n_3 \right) \right]. \quad (7.57)$$

Substituting into the above equation the right hand side (RHS) from Eqs. (7.54) and (7.55) one obtains the differential equation governing the particle size in the steady state flow

$$\alpha_3 \left(\frac{\partial D_3}{\partial \tau} + V_3 \nabla D_3 \right) = \frac{D_3}{3} \left[\begin{array}{c} \frac{4}{D_h} \sqrt{1-\alpha_2} (\rho w)_{23} \left(1 - \frac{\rho_3}{\rho_2} \frac{D_3^3}{D_{3E}^3} \right) / \rho_3 \\ -\alpha_3 \left\{ \left[\left(\frac{D_3}{D_{3\infty}} \right)^3 - 1 \right] / \Delta \tau_{br} - f_{3,coal}/2 \right\} \end{array} \right]. \quad (7.58)$$

Some plausible conclusions are immediately drawn from the above equation:

1. Deposition does not influence the particle size because the deposition terms cancel.

2. Entrainment, particle disintegration in gas, and coalescence are the three mechanisms influencing the particle size.

3. If the entrained droplets have a size equal to the droplet size of the core, i.e. $D_{3E} \approx D_3$, the entrainment does not influence the particle size because of

$$\lim_{D_{3E} \to D_3} \left(1 - \frac{\rho_3}{\rho_2} \frac{D_3^3}{D_{3E}^3}\right) \to 0. \tag{7.59}$$

4. There is an exact relationship between the local droplet number density, i.e. local droplet diameter, for known volume fraction α_3 and the three participating mechanical phenomena, (a) droplet fragmentation, (b) coalescence of droplets, and (c) entrainment, which is not independent of the initial and boundary conditions.

> Consequently, for steady state flow it cannot be expected to obtain an exact correlation for the local droplet diameter, which depends only on the local conditions but not on the initial conditions.

Therefore correlations modeling droplet diameter in steady state pipe flows as a function of the local parameters only but not of the steady state boundary conditions should be used very carefully.

Obviously the need to increase the accuracy of the dynamic models for prediction of droplet sizes in pipes leads necessarily to the need of increasing accuracy of the models of the above-mentioned four mechanical phenomena.

For the case of the fully developed steady state flow Eq. (7.58) reduces to

$$D_3 = \left[\frac{\frac{4}{D_h}\sqrt{1-\alpha_2}\,(\rho w)_{23}/\rho_3 + \alpha_3\left(1/\Delta\tau_{br} + f_{3,coal}/2\right)}{\frac{4}{D_h}\sqrt{1-\alpha_2}\,(\rho w)_{23}/\left(\rho_2 D_{3E}^3\right) + \alpha_3/\left(\Delta\tau_{br} D_{3\infty}^3\right)} \right]^{1/3}, \tag{7.60}$$

and for the case of the fully developed steady state flow without entrainment we obtain

$$D_3 = D_{3\infty}\left(1 + \Delta\tau_{br} f_{3,coal}/2\right)^{1/3} > D_{3\infty}. \tag{7.61}$$

One should not be surprised that the final effective droplet size is larger than the stable size under the local flow conditions because the coalescence acts towards increasing it.

7. Introduction to fragmentation and coalescence

Note that the identification of the velocity fields as disperse require *initialization of the particle length scale* equal to some typical geometrical length scale of the confined channel, for example, the local hydraulic diameter.

Nomenclature

Latin

a_{12} interfacial area density, i.e. the surface area between gas and film per unit mixture volume, $1/m$

D_h hydraulic diameters, m

$D_d = \left(\dfrac{6\,\alpha_d}{\pi\,n_d}\right)^{1/3}$ particle diameter assuming spherical shapes, m

$D_{d\infty}$ stable particle diameter after the fragmentation is finished, m

D_{3E} equivalent diameter of the entrained droplet, m

$F_{col} \approx n_d 4\pi D_d^2$, interface available for collision per unit mixture volume, m^2

$f_{d,sp}$ fragmentation frequency of a single particle, s^{-1}

$f_{d,coal} = f_{d,col} f_{d,coal}^P$, coalescence frequency of single particles, s^{-1}

$f_{d,col}$ collision frequency, s^{-1}

$f_{d,coal}^P$ coalescence probability, s^{-1}

ℓ_{col} average distance between collisions, or the mean free path, m

n_l particle number density: volume-averaged number of discrete particles of the fluid l (e.g., bubbles or droplets) per unit flow volume, m^{-3}

$n_{d\infty} = \alpha_d / \left(\pi D_{d\infty}^3 / 6\right)$, particle number density after the fragmentation is finished, m^{-3}

n_{d1}, n_{d2} particle densities of the two groups of particle with diameters D_{d1} and D_{d2}, respectively, m^{-3}

$\dot{n}_{l,sp}$ particle production term - increase in the number of particles per unit time and unit mixture volume due to mechanical fragmentation such as splitting (subscript sp), $m^{-3} s^{-1}$

$\dot{n}_{l,coal}$ particle sink term - the decrease in the number of particles due to mechanical coalescence (subscript $coal$), instant coalescence rate, $m^{-3} s^{-1}$

$\dot{n}_{l,kin}$ number of particles generated or lost per unit time and unit mixture volume due to evaporation and/or condensation, $m^{-3} s^{-1}$

$\dot{n}_{23}$ particle production rate in case of entrainment, $m^{-3} s^{-1}$

$\dot{n}_{32}$	particle sink term in case of deposition, $m^{-3}s^{-1}$
$P^o_{d,coal}$	≤ 1, probability of the oscillatory coalescence, *dimensionless*
$P^{no}_{d,coal}$	probability of the non oscillatory coalescence, *dimensionless*
$P^s_{d,coal}$	probability of the spectral coalescence, *dimensionless*
$\mathbf{r}$	position vector
$\mathbf{V}_l$	velocity vector, *m/s*
$\mathbf{V}_d$	particle velocity vector
V^{rel}	$= \left| \max\left[0, \mathbf{V}_d(\mathbf{r}+\Delta\mathbf{r}) - \mathbf{V}_d(\mathbf{r})\right]\right|$, particle velocity difference at two different points at the same time, *m/s*

Greek

α_l	averaged volumetric fraction, *dimensionless*
$\alpha_{d,\max}$	$=0.52$, maximum allowable void fraction for existence of bubbly flow, *dimensionless*
Δm_{d1}	mass change of the target droplet *d1* after colliding with the projectile droplet *d2* having mass m_{d2}, see Chapter 8 Section 6, *kg*
m_{d2}	mass of the single particle belonging to the second group of particles, *kg*
ΔV^s_{dd}	artificial particle-particle relative velocity resulting from the assembly averaging of the relative velocities, *m/s*
ΔV^{no}_{dd}	$= V^{rel}$, particle-particle velocity difference averaged over the computational cell volume, *m/s*
V'	fluctuation particle velocity component, *m/s*
$\Delta \tau_{br}$	duration of fragmentation, *s*
$\Delta \tau_{col}$	duration of the contact of the particles due to collision, *s*
$\Delta \tau_{coal}$	time interval necessary to push out the medium between two colliding particles, sometimes called coalescence time, *s*
$\Delta \tau$	time interval, *s*
ρ_l	density, *kg/m³*
$(\rho w)_{23}$	entrainment mass flow rate, mass leaving the film per unit time and unit interfacial area, *kg/(m²s)*
$(\rho w)_{32}$	deposition mass flow rate entering the film, *kg/(m²s)*
μ_l	mass source density, *kg/(m³s)*
μ_{23}	$= a_{12}(\rho w)_{23}$, mass entering the film per unit time and unit mixture volume, *kg/(m³s)*

μ_{32} $= a_{12}(\rho w)_{32}$ mass entering the film per unit time and unit mixture volume, $kg/(m^3 s)$

σ liquid-gas surface tension, N/m

Subscripts

l	= 1 first fluid – gas
	= 2 second fluid – liquid
	= 3 third fluid - another liquid
d	disperse
c	continuous
23	from 2 to 3
32	from 3 to 2

References

1. Antal SP, Ettorre SM, Kunz RF, Podowski MZ (27th-29th September, 2000) Development of a next generation computer code for the prediction of multi-component multi-phase flows, Internet publication of the Proceedings of Trends in Numerical and Physical Modeling for Industrial Multiphase Flows, Course, France
2. Chen X, Yuen WW, Theofanous TG (1995) On the constitutive description of micro-interactions concept in steam explosions, Proceedings of the Seventh International Topical Meeting on Nuclear Reactor Thermal Hydraulics NURETH-7, New York, USA, NUREG/CP-0142
3. Coulaloglu CA, Tavlarides LL (1976) Drop size distribution and coalescence frequencies of liquid - liquid dispersions in flow vessels, A. I. Ch. E., vol 22 no 2 pp 289-297
4. Griffith L (1943) A theory of the size distribution of particles in a comminuted system, Canadian J. Research, vol 21 no 6 pp 57-64
5. Hibiki T, Ishii M (August 15-17, 1999) Interfacial area transport of air-water bubbly flow in vertical round tubes, CD proc. of the 33rd Nat. Heat Transfer Conf., Albuquerque, New Mexico
6. Howarth W J (1967) Measurement of coalescence frequency in an agitated tank, A. I. Ch. E. J. , vol 13 no 5 pp 1007-1013
7. Kataoka I, Ishii M, Mishima K (June 1983) Transaction of the ASME, vol 105 pp 230-238
8. Kolev NI (1993) Fragmentation and coalescence dynamics in multi-phase flows, Experimental Thermal and Fluid Science, vol 6 pp 211 - 251
9. Kolev NI (October 3-8, 1999) Verification of IVA5 computer code for melt-water interaction analysis, Part 1: Single phase flow, Part 2: Two-phase flow, three-phase flow with cold and hot solid spheres, Part 3: Three-phase flow with dynamic fragmentation and coalescence, Part 4: Three-phase flow with dynamic fragmentation and coalescence – alumna experiments, Proc. of the Ninth International Topical Meeting on Nuclear Reactor Thermal Hydraulics (NURETH-9), San Francisco, California

10. Kolomentzev AI, Dushkin AL (1985) Vlianie teplovoj i dinamiceskoj neravnovesnosty faz na pokasatel adiabatj v dwuchfasnijh sredach, TE 8, pp 53-55
11. Komasawa S et al (February 1968) Behavior of reacting and coalescing dispersed phase in a stirred tank reactor, Journal of Chemical Engineering of Japan, vol 1 no 1 pp 208-211
12. Lo S (27th-29th September, 2000) Application of population balance to CFD modeling of gas-liquid reactors, Internet publication of the Proceedings of Trends in Numerical and Physical Modeling for Industrial Multiphase Flows, Course, France.
13. Loeb LB (1927) The kinetic theory of gases, Dover, New York
14. Madden AJ (1962) Coalescence frequencies in agitated liquid-liquid systems, A. I. Ch. E. J., vol 8 no 2 pp 233-239.
15. Mugele RA, Evans HD (1951) Droplet size distribution in sprazs, Ing. Eng. Chem., vol 43 pp 1317-1324
16. Nukiama S, Tanasawa Y (1938) Trans. Soc. Mech. Engrs. (Japan), vol 4 no 14 p 86
17. Pilch M, Erdman CA, Reynolds A B (Aug. 1981) Acceleration induced fragmentation of liquid drops, Charlottesville, VA: Department of Nucl. Eng., University of Virginia, NUREG/CR-2247
18. Rohsenow WM, Choi H (1961) Heat, mass, and momentum transfer, Prentice - Hall, Inc., Engelwood Cliffs, New Jersey
19. Rosenzweig AK, Tronov VP, Perguschev LP (1980) Coaliszencija kapel vody v melkodispersnyh emulsijach tipa voda v nevti, Journal Pricladnoj Chimii, no 8 pp 1776-1780
20. Rosin P, Rammler E (1933) Laws governing the fineness of powdered coal, J. Inst. Fuel, vol 7 pp 29-36
21. Sztatecsny K, Stöber K, Moser F (1977) Blasenkoaleszenz und - zerteilung in einem Rührkessel, Chem. Ing. Techn., vol 49 no 2 p 171
22. Smoluchowski M (1918) Versuch einer mathematischen Theorie der Koagulationskinetik kolloider Lösungen, Zeitschrift für Physikalische Chemie, Leipzig, Band XCII pp 129-168
23. Sauter J (1929) NACA Rept. TM-518
24 Tomiyama A (June 8-12, 1998) Struggle with computational bubble dynamics, Third International Conference on Multiphase Flow, ICMF 98, Lyon, France
25. Wallis GB (1969) One-dimensional two-phase flow, McGraw-Hill, New York

8. Acceleration induced droplet and bubble fragmentation

8.1 Critical *Weber* number

Consider pool flow, that is, flow without any wall influence. Fluid particles in multiphase mixtures experience forces acting to destroy them and forces acting to retain their initial form. The *hydrodynamic stability* limit is usually described by the ratio of the forces acting to destroy the particles, the shear forces $t\pi D_d^2$, where **t** is the tangential force per unit surface, and the forces acting to retain the particle form, for example, surface tension forces $\sigma_d \pi D_d$ (see Figs. 8.1 and 8.2),

$$We_d = \text{Shear force/Surface tension force} = tD_d^2 / (\sigma_d D_d) = tD_d / \sigma_d. \quad (8.1)$$

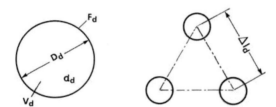

Fig. 8.1. (a) Particle size. (b) Average distance between adjacent particles if they form a rhomboid array

This ratio is called the *Weber* number in honor of *Heinrich Weber* (1842-1913), who used it first. At the end of fragmentation, the *Weber* number is

$$We_{d\infty} = tD_{d\infty} / \sigma_d \quad (8.2)$$

There are two considerations leading to a theoretical upper and lower limit of $We_{d\infty}$:

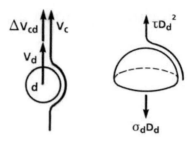

Fig. 8.2. Hydrodynamic stability affecting forces

(a) *Triebnigg* [53] (1929) was the first who equated the surface tension force to the shear force acting on the particles

$$\sigma_d \pi D_{d\infty} \approx c_{cd}^d \frac{1}{2} \rho_c (V_c - V_{d\infty})^2 \pi D_{d\infty}^2 / 4 \qquad (8.3)$$

and obtained for the upper limit of the *Weber* number

$$We_{d\infty} \equiv \frac{\rho_c (V_c - V_{d\infty})^2}{\sigma_d / D_{d\infty}} \approx 8 / c_{cd}^d . \qquad (8.4)$$

The drag coefficient for a solid sphere in the turbulent regime, c_{cd}^d, is ≈ 0.4, and therefore the upper limit is

$$We_{d\infty} \approx 20 . \qquad (8.5)$$

Note that Eq. (8.5) does not reflect the experimentally observed dependence of $We_{d\infty}$ on c_{cd}^d i.e. $We_{d\infty}$ on $Re_d = \rho_c D_{d0} \Delta V_{cd} / \eta_c$, where $\Delta V_{cd} = V_c - V_d$, - see *Lane* [38] (1951), *Hanson* [22] (1963). *Hinze* [24] (1949) shows that during the fragmentation process the particle changes shape and experiences a considerably higher drag; therefore, in nature $We_{d\infty}$ is less than 20.

b) The *Kelvin - Helmholtz* (KH) stability analysis provides the information necessary for the estimation of the lower limit of the *Weber* number. The *Kelvin - Helmholtz* instability is caused by the relative motion of two continuous phases - see in *Chandrasekhar* [9] (1981). The most unstable wavelength is (for gas as a continuous phase)

$$\delta_{d,KH} \approx 3\pi (1 + \rho_c / \rho_d) \sigma_d / (\rho_c \Delta V_{cd\infty}^2) . \qquad (8.6)$$

-see Fig. 8.3.

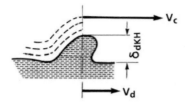

Fig. 8.3. The most unstable wavelength during relative motion of two continuous phases

If the entrained particle in this process has a size approximately equal to the most unstable wavelength

$$D_{d\infty} \approx \delta_{d,KH}, \qquad (8.7)$$

the critical *Weber* number should be

$$We_{d\infty} \equiv \frac{\rho_c (V_c - V_{d\infty})^2}{\sigma_d / D_{d\infty}} \approx 3\pi (1 + \rho_c / \rho_d) \qquad (8.8)$$

which e.g. for water droplets in gas means

$$We_{d\infty} \geq 9.52. \qquad (8.9)$$

At very high relative velocities *Pilch* et al. [42] indicated that inside the droplet the liquid boundary layer with thickness

$$\lambda_{bk} = 2\pi D_d \left(\frac{3 \operatorname{Re}_d}{2\pi} \right)^{-1/2} \qquad (8.10)$$

forms and has stabilizing effect acting against the *Kelvin - Helmholtz* instabilities. This was the reason for the recommendation made recently by *Dinh* et al. [13] (1997) to limit the prediction of Eq. (8.8) by the water boundary layer thickness.

192   8. Acceleration induced droplet and bubble fragmentation

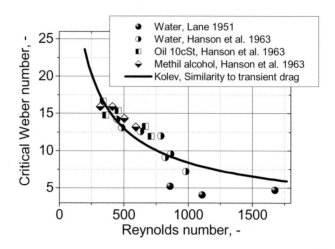

Fig. 8.4. Variation of the critical *Weber* number $We_{d\infty}$ for suddenly applied relative velocities with *Reynolds* number Re_d

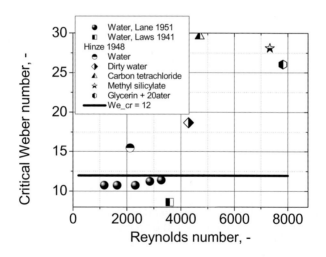

Fig. 8.5. Variation of the critical *Weber* number $We_{d\infty}$ for gradually applied relative velocities with *Reynolds* number Re_d. 1-Water, *Lane* 1951; 2-Water, *Laws* 1941; 3-7 *Hinze* 1948: Water; 4-Dirty water, 5-Carbon tetrachloride, 6-Methyl salicylate, 7-Glycerin+20 % water

Haas [21] (1964) provided experimental data for mercury droplets in air in the region

$$2000 < \mathrm{Re}_d = \rho_c D_d \Delta V_{cd} / \eta_c < 17600 \tag{8.11}$$

and found that below $We_{cd} \approx 5.2$ there is no breakup, and above $We_{cd} \approx 6$, there is. *Haas* found

$$We_{d\infty} \approx 5.6 \tag{8.12}$$

valid for experiments with initial *Weber* numbers ranging from 3.36 to 37.5, a region characterized by bag breakup. This seems to be the lower limit for the critical *Weber* number.

In fact, the experimental observations of several authors for low viscosity liquids provide a value of $We_{d\infty}$ of about

$$5 < We_{d\infty} < 20 \tag{8.13}$$

with the most commonly used value being

$$We_{d\infty} \approx 12, \tag{8.14}$$

as shown in Fig. 8.5.

Within the margin $5 < We_{d\infty} < 20$ there is a dependence on the *Reynolds* number not considered into the above mentioned approaches. *Sarjeant* [48] (1979) summarized the data of *Hinze* [25] (1949), *Lane* [38] (1951), *Hanson* et al. [22] (1963), see Figs. 8.4 and 8.5, and found that for suddenly applied relative velocities within $\mathrm{Re}_d = 300$ and 10^5 where

$$\mathrm{Re}_d = \rho_c D_d \Delta V_{cd} / \eta_c, \tag{8.15}$$

the critical *Weber* number varies as given in Fig. 8.4. The data given in Fig. 8.4 indicate proportionality with the transient drag. Comparing with the approximation of the lowest boundary of the transient drag for accelerating spheres as computed by *Brauer* [6] (1992) we obtain

$$We_{d\infty} = 55\left(\frac{24}{\mathrm{Re}_d} + \frac{20.1807}{\mathrm{Re}_d^{0.615}} - \frac{16}{\mathrm{Re}_d^{2/3}}\right) \quad \text{for} \quad 200 < \mathrm{Re}_d < 2000, \tag{8.16}$$

and

$$We_{d\infty} \approx 5.48 \quad \text{for} \quad 2000 < \mathrm{Re}_d. \tag{8.17}$$

The value of $Re_d \approx 200$ is chosen to take into account the observation of *Schröder* and *Kitner* [49] (1965) reporting that a droplet oscillates only in the presence of a vortex tail behind the droplet, which requires a *Reynolds* number of 200 at least. This is the explanation why an upper limit of the critical *Weber* number is observed experimentally. The computer simulation reported by *Brauer* [6] (1992) also supports this observation for flow around an accelerated sphere.

For gradually applied relative velocities *Taylor* [52] (1949) provides a theoretical analysis leading to the critical *Weber* number that is about $\sqrt{2}$ times greater than the critical *Weber* number for suddenly applied relative velocities which is not definitely confirmed from experiments. For a free falling droplet in a gravitational field that can be considered as gradual application of relative velocity, the drag force in Eq. (8.3) is equal to the buoyancy force

$$c_{cd}^d \frac{1}{2} \rho_c \left(V_c - V_{d\infty} \right)^2 \pi D_{d\infty}^2 / 4 = \left(\pi D_{d\infty}^3 / 6 \right) g \Delta \rho_{dc} . \tag{8.18}$$

Therefore

$$D_{d\infty}^2 g \Delta \rho_{dc} / \sigma_d = 6 \tag{8.19}$$

or having in mind that the terminal velocity of a large drop is about

$$\Delta V_{cd} \approx \left(\sqrt{2} \text{ to } 1.7 \right) \left(\frac{\sigma_d g \Delta \rho_{dc}}{\rho_c^2} \right)^{1/4} , \tag{8.20}$$

$$We_{d\infty} \approx 4.8 \text{ to } 7.1 . \tag{8.21}$$

Magarevey and *Taylor* [40] (1956) observed that free falling water droplets with artificially produced initial diameter of 12 to 20 *mm* disintegrate before reaching the "terminal velocity" corresponding to the initial drop size which should reduce the value estimated by means of Eq. (8.19). Equation (8.19) can be written in somewhat different form

$$\left(D_{d\infty} / \lambda_{RT} \right)^2 = 6 , \tag{8.22}$$

where

$$\lambda_{RT} = \left[\sigma_d / \left(g \Delta \rho_{dc} \right) \right]^{1/2} \tag{8.23}$$

is the scale of the *Rayleigh - Taylor* instability wavelength for the case where gas and liquid are interpenetrating due to gravity. The above described approach considers the stability limit of the free falling liquid globules.

There is another point of view to approach this limit, namely, to consider the transition from churn turbulent bubble flow into dispersed flow – see left Fig. 8.6a.

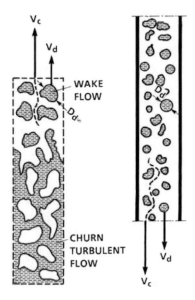

Fig. 8.6. a) Transition from churn-turbulent to dispersed wake. b) Droplet fragmentation in channels

For churn turbulent two - phase flow, the relative velocity between gas and liquid is nearly independent of the dispersed particle size, and is given by *Ishii* [28] (1977)

$$|\Delta V_{13}| \approx \sqrt{2} \left(\sigma_3 g |\Delta \rho_{13}| / \rho_1^2 \right)^{1/4}. \tag{8.24}$$

Here 1 stands for gas and 3 stands for liquid. For the wake flow regime of droplets in a gas, the relative velocity is

$$|\Delta V_{13}| \approx \left[g \Delta \rho_{13}^2 / (\eta_1 \rho_1) \right]^{1/3} D_3 / 4, \tag{8.25}$$

see in *De Jarlais* et al. [12] (1986). Obviously at the transition between both regimes the relative velocity should be the same. *De Jarlais* et al. use this argument and after solving both equations for D_3 they obtain the expression

$$D_3 = D_{d\infty} = 4\left[2\sigma_d/(g\Delta\rho_{dc})\right]^{1/2}\left\{\left[\rho_c\sigma_d\sqrt{\sigma_d/(g\Delta\rho_{dc})}\right]^{1/2}/\eta_c\right\}^{-1/3} = 4\sqrt{2}\lambda_{RT}N_{\eta d}^{1/3},$$
(8.26)

where $d \equiv 3$, $c \equiv 1$ and λ_{RT} is the *Rayleigh - Taylor* wavelength as defined with Eq. (8.23).

$$N_{\eta d} = \eta_c/(\rho_c\sigma_d\lambda_{RT})^{1/2} = Ar^{-1/2}$$
(8.27)

is the viscosity number. The experimental observation of the authors confirms that this expression approximates the maximum droplet size. Relative velocities larger than those predicted by Eq. (8.25) leads to further fragmentation governed by the fragmentation mechanism for pool flow as discussed before.

Comparing with experimental data, *Ruft* [47] (1977) found that the constant $4\sqrt{2}$ in Eq. (8.26) should be replaced by 20. The complete algorithm proposed by *Ruft* for free falling droplets representing the data in Fig. 8.7 is

(i) $\left(\dfrac{\Delta\rho_{dc}}{\rho_c}\right)^2 \Big/ N_{\eta d}^4 < 10^7$, $D_{d\infty} = 20 N_{\eta d}^{1/3} \lambda_{RT}$, $\left(We_{d\infty} \approx 40 N_{\eta d}^{2/3}\right)$; (8.28)

(ii) $10^7 \leq \left(\dfrac{\Delta\rho_{dc}}{\rho_c}\right)^2 \Big/ N_{\eta d}^4 < 10^9$, $D_{d\infty} \approx 3\lambda_{RT}$, $\left(We_{d\infty} \approx 6\right)$; (8.29)

(iii) $10^9 \leq \left(\dfrac{\Delta\rho_{dc}}{\rho_c}\right)^2 \Big/ N_{\eta d}^4 < 3\times 10^9$ interpolation between (ii) and (iv); (8.30)

(iv) $3\times 10^9 \leq \left(\dfrac{\Delta\rho_{dc}}{\rho_c}\right)^2 \Big/ N_{\eta d}^4$, $D_{d\infty} \approx 3.9\lambda_{RT}$, $\left(We_{d\infty} \approx 7.75\right)$. (8.31)

The 28 experimental data are in the region

$$90 \leq \left(\dfrac{\Delta\rho_{dc}}{\rho_c}\right)^2 \Big/ N_{\eta d}^4 < 10^{18}.$$
(8.32)

In this region the uncertainty is

$$3 \leq D_{d\infty}/\lambda_{RT} < 10.$$
(8.33)

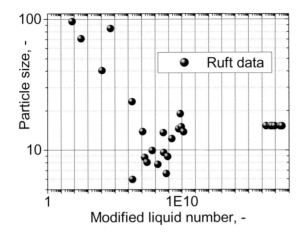

Fig. 8.7. Dimensionless maximum droplet diameter $\left(D_{d,\max}/\lambda_{RT}\right)^2$ as a function of a modified liquid number $\left(\Delta\rho_{dc}/\rho_c\right)^2/N_{\eta d}^4$

Using Eq. (8.20) the *Ruft* approximation can be expressed in terms of *Weber*'s number as given in the brackets. Thus *Ruft*'s recommendations should be considered in any case as an upper bound of the particle size for gradually applied relative velocity.

Brodkey [7] (1967) and *Gelfand* et. al. [17] (1973) approximated the dependence of the critical *Weber* number on the viscosity, see Fig. 8.8, experimentally observed by *Haas* [21] (1964), *Hanson* et al. [22] (1963), *Hinze* [26] (1955), *Hassler* [23] (1971) by multiplying the critical *Weber* number for water by

$$1+\left(1.077 \text{ to } 1.5\right) On_d^{(1.6 \text{ to } 1.64)} \tag{8.34}$$

where the *Ohnesorge* number is defined as follows

$$On_d = \eta_d/\sqrt{\rho_d D_d \sigma_d} = We_d^{1/2}/Re_d . \tag{8.35}$$

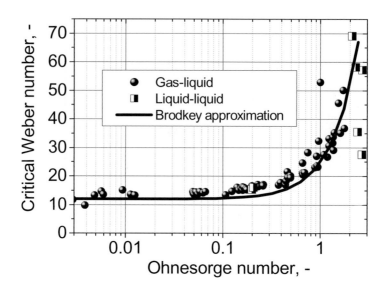

Fig. 8.8. Critical Weber number as a function of the *Ohnesorge* number (viscosity). Data for gas - liquid by *Hanson* et al. [22], *Haas* [21], *Hassler* [23], *Hinze* [26]. Data for liquid-liquid by *Li* and *Folger*, [39]. Approximation by *Brodkey* [7]

Summary: Thus, if

$$We_d > We_{d\infty} \quad \text{and} \quad Re_d > 200.$$

a drop exposed on relative velocity is unstable. For *suddenly* applied relative velocity the recommended critical *Weber* number is

$$We_{d\infty} = 55\left(\frac{24}{Re_d} + \frac{20.1807}{Re_d^{0.615}} - \frac{16}{Re_d^{2/3}}\right)\left[1 + 1.077 On_d^{1.64}\right], \tag{8.36}$$

for $200 < Re_d < 2000$ and

$$We_{d\infty} = 5.48\left[1 + 1.077 On_d^{1.64}\right], \tag{8.37}$$

for $2000 \le Re_d$. As the experiments performed by *Hsiang* and *Faeth* [27] show, there is no fragmentation if

$On_d > 4$, \hfill (8.38)

see Fig. 8.8.

For *gradually* applied relative velocity the critical *Weber* number is expected to be larger than the critical *Weber* number for *suddenly* applied relative velocity.

The decision which of both regimes should be taken, depends on the ratio of the velocity relaxation time, $\Delta\tau_{\Delta V}$, to the breakup time $\Delta\tau_{br}$. If $\Delta\tau_{\Delta V} \gg \Delta\tau_{br}$ the condition for suddenly applied relative velocity should be used. If $\Delta\tau_{\Delta V} \ll \Delta\tau_{br}$ the condition for gradually applied relative velocity is more appropriate. The estimation of $\Delta\tau_{\Delta V}$ and $\Delta\tau_{br}$ will be discussed later. Experimental data are needed in order to establish the behavior of the unstable droplet between the above discussed two extreme cases.

For free falling drops the method proposed by *Ruft* is recommended.

Important remarks about the size initialization by flow pattern transition:

The acceleration induced droplet fragmentation in channel flow, Fig. 8.6b, is preceded usually by the transition from the structure with continuous liquid to the structure with continuous gas - see Fig. 8.6a. This situation is easily identified if one *keeps a record of the flow pattern during the computational analysis*. Thus, if during the integration time step the liquid was continuous at the old time level and is getting dispersed at the new time level, the initial median size of the new born family of unstable drops is $D_{d\infty}$ computed as follows:

- In case of transition from churn turbulent to annular-disperse two-phase flow, the mean size of the new born unstable droplets is well prescribed by Eq. (8.26) or Eq. (8.28).

- The transition from inverted annular flow into dispersed flow is visualized by jet disintegration in a pipe. For estimation of the initial size of the drops born the formalism proposed by *De Jarlais* et al. given Chapter 9 is recommended.

- If the relative velocity between gas and the new born droplets is larger than the critical one, the unstable droplets follow the dynamic fragmentation mechanism already described for pool flow.

8.2 Fragmentation modes

The sequences of pattern of deforming and fragmenting particles can be classified in groups called fragmentation modes. The qualitative description may vary depending on the different author's views but some important *common* features are confirmed by many authors. We use further the following verbal description of the different modes:

1. In the *vibration breakup* the droplet is excited by the gas vortex tail behind the drop to oscillate with its natural oscillation frequency and eventually to decompose into two or more large fragments.

2. *Bag breakup* is analogous to the bursting of soap bubbles blown from a soap film attached to a ring. A thin hollow bag is blown downstream while it is still attached to a more massive toroidal rim. The bag eventually burst, forming a large number of small fragments; the rim disintegrates in a short time later, producing a small number of large fragments.

3. In the *bag and stamen breakup* a thin bag is blown downstream while being anchored to a massive toroidal rim. A column of liquid - *stamen* - is formed along the drop axis parallel to the approaching flow. The bag bursts first, rim and stamen follow.

4. In the *sheet stripping* a thin sheet disintegrates a short distance downstream from the drop. A coherent residual drop exists during the entire breakup process.

5. In the *wave crests stripping* small waves are formed on the windward surface of the drop. The wave crests are continuously eroded by the action of the flow field over the surface of the drop.

6. In the *catastrophic breakup* large amplitude, long wavelength waves ultimately penetrate the drop creating several large fragments before wave crest stripping can significantly reduce the drop mass.

The mode of the particle disintegration depends on the initial conditions before the fragmentation. The initial condition is characterized by the *Weber* number. Depending on the initial *Weber* number

$$We_d = \frac{\rho_c (V_c - V_d)^2}{\sigma_d / D_d} \qquad (8.39)$$

the experimentally observed modes of fragmentation differ from each other.

Anderson and *Wolfe* [1] (1965) divided the fragmentation mechanisms into two extreme modes: bag breakup and stripping breakup for following reason. The total drag force exerted on the drop consists of two components: (i) the pressure drag due to the pressure distribution over the surface of the drop and (ii) the friction drag due to viscous shear at the interface. The pressure drag at low relative velocities produces bag breakup, while the friction drag at high velocities produces stripping breakup. There are some other characteristic modes in between the above mentioned two extreme modes. *Pilch* et al. [42] (1981) showed that the critical *Rayleigh-Taylor* (RT) instability wavelength

$$\lambda_d \approx 2\pi \left[(1-\rho_c/\rho_d)\rho_d a/\sigma \right]^{-1/2} \pi/2 = 2\pi D_{d0} \left[(1-\rho_c/\rho_d)\frac{3}{4}c_{cd}^d We_{cd} \right]^{-1/2} \pi/2, \qquad (8.40)$$

where

$$a = \frac{3}{4} c_{cd}^d \frac{\rho_c}{\rho_d} \frac{\Delta V_{cd}^2}{D_d} \qquad (8.41)$$

is the drop acceleration, and $c_{cd}^d \approx 1.7$, determines the range of the bag, bag and stamen breakup and sheet stripping: The *Weber* number ≈ 14.4 was predicted in order to have one unstable *RT* wave to fit on the deformed windward drop surface ($\lambda_d \approx 2.3 D_{d0}$) which is approximately the lower limit below which bag breakup is not observed. The *Weber* number of 57.8 is prescribed in order to have two unstable *RT* waves to fit on the deformed windward drop surface ($2\lambda_d \approx 2.3 D_{d0}$), which is approximately the lower limit below which bag and stamen breakup was not observed. The *Weber* number of ≈ 130 is predicted in order to have three unstable *RT* waves to fit on the deformed windward surface ($3\lambda_d \approx 2.3 D_{d0}$) which is approximately the limit above which bag and stamen breakup was not observed. *The Weber number of 350 is the limit below which only primary breakup of the newly born fragments was experimentally observed.*

Sarjeant [48] (1979), reviewing data of 10 literature sources, estimated the boundaries of the different regimes as follows:

1. Bag mode

$$We_{d\infty} < We_{cd} < 25, \qquad (8.42)$$

confirmed also by *Krzeczkowski* [35] (1980), $10 < We_{cd} < 20$ and by *Magarvey* and *Taylor* [40] (1956) for free falling droplet, *Haas* [21] (1980), $5.6 < We_{cd} < 37.5$. *Arcoumanis* et al. [3] supported recently the upper limit of 25.

2. Umbrella mode

$$25 \leq We_{cd} < 50, \tag{8.43}$$

confirmed also by *Krzeczkowski* [35] (1980), $20 \leq We_{cd} < 65$. *Arcoumanis* et al. [3] supported recently the upper limit of 65.

3. Stripping of the ligaments from the periphery of the deformed drop

$$50 \leq We_{cd} < 100; \tag{8.44}$$

4. Stripping of the ligaments from the periphery of the deformed drop and droplets from the crest of waves

$$100 \leq We_{cd} < 1000; \tag{8.45}$$

5. Stripping of droplets from the crests of waves on the windward face

$$1000 \leq We_{cd} < 10^5; \tag{8.46}$$

6. Initial stripping of droplets from the crests of waves on the windward face followed by perforation due to *Taylor* waves

$$10^5 \leq We_{cd} < 10^6; \tag{8.47}$$

Perforation due to *Taylor* waves

$$10^6 \leq We_{cd}. \tag{8.48}$$

Gelfand et al. [18] (1974), *Pilch* et al. [42] (1981) classified the experimentally observed modes of fragmentation as follows:

1. Vibration breakup

$$We_c < 12; \tag{8.49}$$

2. Bag breakup

$$12 \leq We_c < 50 \tag{8.50}$$

(experimentally confirmed by *Simpkins* and *Bales* [51] (1972), $7 \leq We_c < 50$;

3. Bag - and - stamen breakup

$$50 \le We_c < 100; \qquad (8.51)$$

4. Sheet stripping

$$100 \le We_c < 350; \qquad (8.52)$$

5. Wave crest stripping followed by catastrophic breakup

$$350 \le We_c . \qquad (8.53)$$

Obviously there are not clear boundaries between the different regimes which explains the differences due to the subjective judgment of the authors. *Gelfand* and *Pilch's* proposal is formally supported by experimental data which change the character of break time as a function of *Weber* number at the above mentioned boundaries. That is why, for low viscous liquids, we recommend the method proposed by *Gelfand* et al. [18] (1974) and *Pilch* et al. [42] (1981).

In the region of

$$10^{-7} \le 1/La_d < 5 \qquad (8.54)$$

Krzeckowski provided experimental evidence that the limits of the regimes are functions of $1/La_d$, where the relationship between the *Laplace* number and the already introduced *Ohnesorge* number is

$$1/La_d = On_d^2 . \qquad (8.55)$$

For $1/La_d > 10^{-3}$ the influence of the liquid viscosity is negligible. For $1/La_d > 10^{-3}$ the boundaries of the regimes increase with $1/La_d$. For such cases the classification by *Krzeckowski* is useful.

8.3 Relative velocity after fragmentation

The stability criterion, Eqs. (8.36) and (8.37), $We_{d\infty} = const$ refers to the final stable diameter, $D_{d\infty}$, and the *final fragment velocity*, $V_{d\infty}$, reached by the fragments after the fragmentation. Therefore, to predict $D_{d\infty}$ knowing D_d, V_d, V_c one needs a method to predict $V_{d\infty}$.

Using the momentum equations of the continuum and of the dispersed phase, neglecting the inertial terms, dividing by the corresponding densities and summing up the resulting equations, one obtains the following equation governing the relative velocity, ΔV_{cd},

$$\left(1+bc_{cd}^{vm}\right)\frac{\partial}{\partial \tau}\Delta V_{cd} + \left(\frac{1}{\rho_c}-\frac{1}{\rho_d}\right)\nabla p + b\frac{3}{4}\frac{c_{cd}^d}{D_d}|\Delta V_{cd}|\Delta V_{cd} = 0. \tag{8.56}$$

Here

$$b = \left(\alpha_c \rho_c + \alpha_d \rho_d\right)/\alpha_c \rho_c, \tag{8.57}$$

c_{cd}^{vm} is the virtual mass coefficient which for droplets in gas is negligible but for liquid-liquid or bubble-liquid systems is very important.

For drag forces considerably larger than the pressure gradient forces the acceleration of a single droplet is computed using a simplified form of Eq. (8.56),

$$\frac{\partial}{\partial \tau}\Delta V_{cd} = -\Delta V_{cd}/\Delta \tau_{\Delta V}, \tag{8.58}$$

where

$$\Delta \tau_{\Delta V} = \frac{4}{3}\frac{1+bc_{cd}^{vm}}{b}\frac{D_d}{c_{d\infty}^d}\bigg/|\Delta V_{cd}| \tag{8.59}$$

is used to estimate the linearized velocity relaxation time constant. The analytical solution of the above simplified equation is

$$\frac{\Delta V_{cd\infty}}{\Delta V_{cd}} = \frac{1}{1+\dfrac{\Delta \tau}{\Delta \tau_{\Delta V}}}. \tag{8.60}$$

Eq. (8.56) is an equation of the *Riccati* type with respect to ΔV_{cd} and possesses a hyperbolic solution. Substituting

$$a_1 = \left(b\frac{3}{4}\frac{c_{cd}^d}{D_{d\infty}}\right)\bigg/\left(1+bc_{cd}^{vm}\right), \tag{8.61}$$

$$a_2 = \left(\frac{1}{\rho_c}-\frac{1}{\rho_d}\right)\nabla p\bigg/\left(1+bc_{cd}^{vm}\right) \tag{8.62}$$

and using the initial condition $\Delta V_{cd}(\tau) = \Delta V_{cd,0}$ we obtain the velocity difference ΔV_{cd} after the time interval $\Delta \tau$.

$$\Delta V_{cd} = \frac{\Delta V_{cd,0}(a_1 a_2)^{1/2} + a_1 \tanh\left[(a_1 a_2)^{1/2} \Delta \tau\right]}{(a_1 a_2)^{1/2} + a_1 \Delta V_{cd,0} \tanh\left[(a_1 a_2)^{1/2} \Delta \tau\right]}, \tag{8.63}$$

see *Kamke* [30] (1959). If the time interval is equal to the time needed to complete the fragmentation, $\Delta \tau = \Delta \tau_{br}$, Eq. (8.63) provides the velocity difference between the cloud of fragments and its surroundings at the end of the fragmentation. Note that the drag coefficient of the deforming particles divided by the changing diameter of the fragmented particles during the breakup period $\Delta \tau_{br}$ is a complicated function of time

$$c_{cd}^d / D_d = f(\tau^*) \quad \text{for} \quad \tau < \tau^* < \tau + \Delta \tau_{br} \tag{8.64}$$

so that the averaged quotient is in any case greater than the final one

$$\overline{c_{cd}^d / D_d} > c_{cd}^d / D_d. \tag{8.65}$$

Krzeczkowski [35] (1980) provided very important measurements on the kinematics of the fragmentation. *A common feature of all types of fragmentation is that they start with a basic shape the so called "liquid disk"*. Data for D_d / D_{d0}^v as a function of time are plotted for $13.5 < We_{cd} < 101$.

The *Baines* and *Buttery* [5] (1978) experimental data for deformation preceding the breakup for liquid-gas ($1.3 \times 10^3 \le We_{cd} \le 7.1 \times 10^5$) as well for liquid-liquid ($2.5 \times 10^2 \le We_{cd} \le 3.9 \times 10^3$) are correlated by *Pilch* et al. [42] (1981) with

$$D_d / D_{d0}^v - 1 \approx 1.4 \Delta \tau^* - 0.3 \Delta \tau^{*2}, \quad 0 \le \Delta \tau^* \le 2 \tag{8.66}$$

$$D_d / D_{d0}^v - 1 \approx 0.12 + 1.28 \Delta \tau^* - 0.27 \Delta \tau^{*2}, \quad 2 \le \Delta \tau^* \le 5.5 \tag{8.67}$$

where $\Delta \tau^* = \tau \Delta V_{cd} (\rho_c / \rho_d)^{1/2} / D_{d0}^v$.

Simpkins and *Bales* [51] (1972) found experimentally that the drag coefficient for distorted drops is similar to the rigid sphere value for $Re_d < 10^3$. When compressibility effects become significant, a mean value of $c_{cd}^d = 2.5$ is observed for $10^3 < Re_d < 10^5$. The later result is comparable to measurements taken from rigid disks in similar flow conditions. *Pilch* et al. [42] (1981) reviewing experimental data from several authors found that drag coefficients for fragmenting drops are 2 to 3 times larger than the rigid sphere drag coefficient (≈ 2.5 for compressible flow, ≈ 1.54 for fragmenting drops in incompressible flow). That is the reason leading some authors to look for an empirical description of the final fragment velocity by modifying solutions of the momentum equations using experimental data.

The fragmentation cloud velocity that is reached after completion of the breakup process for regimes (a) through (e) was correlated by *Pilch* et al., p.22, by a modified solution of the simple force balance equation ($\tau = 0$, $V_d = 0$; $\tau = \Delta \tau_{br}$, $V_d = V_{d\infty}$, $V_c = const$)

$$V_{d\infty} = V_c \left(\rho_c / \rho_d \right)^{1/2} \left(\frac{3}{4} c_{cd}^d \Delta \tau_{br} + 3b\Delta \tau_{br}^2 \right) \qquad (8.68)$$

or

$$\left(V_c - V_{d\infty} \right)^2 = V_c^2 \left[1 - \left(\rho_c / \rho_d \right)^{1/2} \left(\frac{3}{4} c_{cd}^d \Delta \tau_{br} + 3b\Delta \tau_{br}^2 \right) \right]^2 . \qquad (8.69)$$

The experimental data are reproduced using the tuning constants $c_{cd}^d = 0.5$, $b = 0.0758$ for incompressible ($Ma < 0.1$), and $c_{cd}^d = 1$, $b = 0.116$ for compressible flow ($Ma > 0.1$). The different form during the different deformation modes explains the differences in c_{cd}^d.

For the liquid - liquid system *Pilch* et al. [42] (1981), p.16, used

$$V_{d\infty} = V_c \left(\rho_c / \rho_d \right)^{1/2} \left(\frac{3}{4} c_{cd}^d \Delta \tau_{br} \right) \bigg/ \left[1 + \frac{3}{4} c_{cd}^d \Delta \tau_{br} \left(\rho_c / \rho_d \right)^{1/2} \right] \qquad (8.70)$$

or

$$\left(V_c - V_{d\infty} \right)^2 = V_c^2 \left\{ 1 - \left(\rho_c / \rho_d \right)^{1/2} \left(\frac{3}{4} c_{cd}^d \Delta \tau_{br} \right) \bigg/ \left[1 + \frac{3}{4} c_{cd}^d \Delta \tau_{br} \left(\rho_c / \rho_d \right)^{1/2} \right] \right\}^2 .$$

$$(8.71)$$

Thus, for computation of the final velocity difference after the fragmentation, either the stepwise analytical solution (8.63) together with the deformation approximation (8.66, 8.67) for computing c_{cd}^d / D_d or the empirical correlation (8.68, 8.69) provided by *Pilch* et al. [42] (1981), which is simpler, should be used.

8.4 Breakup time

Next we concentrate our attention on the modeling of the time scale of the fragmentation process $\Delta \tau_{br}$, see Fig. 8.9.

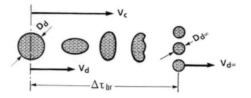

Fig. 8.9. Definition of the parameters characterizing the break-up process

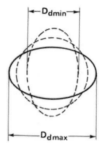

Fig. 8.10. Particle vibration

The vibration breakup mechanism, see Fig. 8.10, is probably the best analyzed fragmentation mechanism up to now. According to *Lamb* (1932), see in [37] (1945), the natural frequency of the *n*-th order mode of a spherical drop or bubble (d) performing *oscillations* in continuum (c) is given by

$$(2\pi f_n)^2 = b \frac{8\sigma_d}{D_d^3} \frac{n(n+1)(n-1)(n+2)}{(n+1)\rho_d + n\rho_c}, \quad b \approx 1, \qquad (8.72)$$

where $n = 2, 3, ..., n$. *Riso* and *Fabre* [58] reported in 1998 that bubbles in turbulent field manifest the strongest oscillation at the first mode ($n = 2$). For droplets the first mode of oscillation ($n = 2$) is the only one observed experimentally. The factor *b* should approach unity for small amplitude oscillations. *Schröder* and *Kitner* [49] (1965) describe 132 data points for 19 disperse - continuum systems with the correlation

$$b = 1 - \frac{D_{d,\max} - D_{d,\min}}{2D_d^v} \approx 0.805 D_d^{v\,0.225} \qquad (8.73)$$

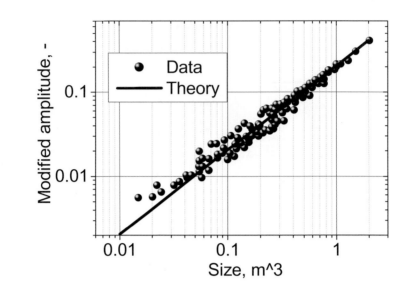

Fig. 8.11. Amplitude corrected data $b(\sigma_d/f_n^2)/(3\rho_d + 2\rho_c)$ vs. drop size D_d^3. Values of b from Eq. (8.79). *Schröder* and *Kitner* [49] (1965)

with an average error of 9.01 % - see Fig. 8.11. Here D_d^v is the volume-averaged diameter. The data cover the region $0.01 < (D_d^v)^3 < 3$ and take values within $0.005 < \dfrac{b\sigma}{f_2^2(3\rho_d + 2\rho_c)} < 0.5$. They observed oscillations only in the presence of a vortex tail which requires a *Reynolds* number of at least 200. *Hsiang* and *Faeth*, see in [14] p. C0-7, summarized a relatively large data base on maximum drop deformation for steady disturbances, considering both drop-gas and drop-immiscible environments for

$$We_d \leq 20, \quad D_{d,\max}/D_{d,\min} = \left(1 + 0.07 We_d^{1/2}\right)^3, \tag{8.74}$$

and

$$D_{d,\max}^2 D_{d,\min} = \left(D_d^v\right)^3. \tag{8.75}$$

The time needed for breakup of an unstable, vibrating particle should therefore be of the order

$$\Delta \tau_{br} \approx 1/f_2 \qquad (8.76)$$

- see Fig. 8.11. This was first proposed by *Laftaye* [36] (1943) as a rough estimate of the time of splitting-up for the vibration breakup. This, however is correct if the relative velocity is just equal to the critical speed

$$\Delta V_{cd}^* = \left(We_{d\infty} \frac{\sigma_d}{D_d \rho_c}\right)^{1/2}. \qquad (8.77)$$

If the relative velocity exceeds the critical speed, ΔV_{cd}^*, the splitting time is shorter than the time determined by the natural vibration period. The *Kelvin-Helmholtz* instability analysis, see [25] (1981) provides the period for the droplet fragmentation

$$\Delta \tau_{br}^* = \frac{\Delta \tau_{br}}{\delta_{dKH} / \left[\Delta V_{cd} \left(\rho_c/\rho_d\right)^{1/2}\right]} \approx \frac{\sqrt{3}}{2\pi}(1+\rho_c/\rho_d) \approx 2.28, \qquad (8.78)$$

that is the lowest limit of the experimentally observed fragmentation times. The expression $\delta_{dKH}/\left[\Delta V_{cd}\left(\rho_c/\rho_d\right)^{1/2}\right]$ has the dimension of a time and is usually used as a time scale for modeling of droplet fragmentation in the form $D_d/\left[\Delta V_{cd}\left(\rho_c/\rho_d\right)^{1/2}\right]$. In this sense the time needed for the first mode of oscillation in accordance with Eq. (8.78) is

$$\frac{\Delta \tau_{br}}{D_d/\left[\Delta V_{cd}\left(\rho_c/\rho_d\right)^{1/2}\right]} \approx \frac{\pi}{4}\left[\frac{1}{b}\left(1+\frac{2}{3}\frac{\rho_c}{\rho_d}\right)We_{cd}\right]^{1/2}. \qquad (8.79)$$

Having in mind that the vibration mode is observed for $\approx 12 < We_{cd} < 18$, the dimensionless breakup time should be expected to be of the order of

$$\Delta \tau_{br}^* = 6.6 \text{ to } 8.2 \qquad (8.80)$$

for $b \approx 1/2$, which is in fact the upper limit for the fragmentation time observed in all sources known to the author.

Pilch et al. [42] (1981) correlated experimental data for the total breakup period for droplets in gas from 8 sources, see Fig. 8.12, for the non-vibration modes (b) through (e) in the form

$$\Delta\tau_{br}^* = c\left(We_{cd} - We_{d\infty}\right)^m, \quad \text{for } We_{cd}^* < We_{cd} < We_{cd}^{**} \tag{8.81}$$

where the dimensionless time $\Delta\tau_{br}^*$ was defined as

$$\Delta\tau_{br}^* = \frac{\Delta\tau_{br}}{D_d / \left[V_c\left(\rho_c/\rho_d\right)^{1/2}\right]}, \tag{8.82}$$

the critical *Weber* number,

$$We_{cd} = 12, \tag{8.83}$$

and the constants are

c	m	We_{cd}^*	We_{cd}^{**}	
7	0	1	12	Vibration mode
6	$-1/4$	12	18	Bag breakup
2.45	$+1/4$	18	45	Bag - and - stamen breakup
14.1	$-1/4$	45	351	Sheet stripping
0.766	$+1/4$	351	2670	Wave crest stripping followed by catastrophic breakup
5.5	0	2670	∞	

The data base of *Sarjeant*, see Fig. 4 in [48] (1979), confirms also Eq. (8.81) with some differences for the boundaries of the regimes. For *Weber* number > 500 c is estimated by *Hsiang* (see in [14] p. C0-7, fig. 8) to be 5.

The accuracy of the data representation is of order of

$$\pm D_d / \left[V_c\left(\rho_c/\rho_d\right)^{1/2}\right]. \tag{8.84}$$

Experimental data for the bag breakup are provided by *Haas* [21] (1964). Using mercury drop in air *Haas* observed bag breakup in the region of $5.6 < We_{cd} < 37.5$ and $200 < Re_d < 17600$. *Haas* provided measurements for the duration of the three stages of breakup (a) time for disk formation, (b) time for the bag formation and (c) time for the global breakup. The constant c in Eq. (8.81) derived from his data is ≈ 2.3 in

$$\Delta\tau_{br}^* \approx 2.3 We_{cd}^{1/4}. \tag{8.85}$$

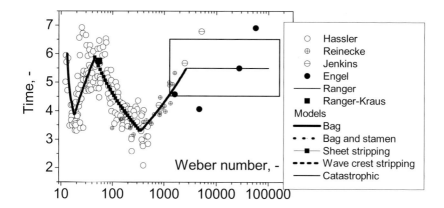

Fig. 8.12. Total break-up time $\Delta \tau_{br} \left[V_c \left(\rho_c / \rho_d \right)^{1/2} \right] / D_d$ as a function of *Weber* number We_{cd} for gas-liquid systems. The summary was prepared by *Pilch* et al. in 1981

The order of magnitude of the constant was confirmed later by the measurements of *Krzezkowski* [35] (1980)

$$\Delta \tau_{br}^* \approx (1.9 \text{ to } 2.7) \, We_{cd}^{1/4}. \tag{8.86}$$

Note that the relationship for the fourth regime is based on the experimental observations made by *Simpkins* and *Bales* [51] (1972) and other authors. The original relationship for the onset of *Taylor* instability of water droplets in gas, i.e. the least squares fit of the minimum value for the breakup time, is

$$\frac{\Delta \tau_{br}}{D_d / \left[\Delta V_{cd} \left(\rho_c / \rho_d \right)^{1/2} \right]} = \frac{1}{2} 22 Bo_d^{-1/4} \approx 13.3 We_{cd}^{-1/4}, \tag{8.87}$$

where

$$Bo_d = \left[\rho_d \frac{dV_d}{d\tau} (D_d / 2)^2 \right] / \sigma_d = \frac{3}{16} c_{cd}^d We_{cd} \tag{8.88}$$

and

$$c_{cd}^d = f(Re_{cd}), \text{ for rigid sphere, } Re_{cd} < 10^3, \tag{8.89}$$

$$c_{cd}^d = 2.5, \text{ for } 10^3 \leq Re_{cd} < 10^5. \tag{8.90}$$

The experimental data show that the breakup time is in any case less than the time prescribed by the equation

$$\frac{\Delta\tau_{br}}{D_d/\left[\Delta V_{cd}\left(\rho_c/\rho_d\right)^{1/2}\right]} \approx \frac{1}{2}65Bo_d^{-1/4} \approx 39.9We_{cd}^{-1/4}. \tag{8.91}$$

The experimental data of *Krzeckowski* [35] (1980) support the order of magnitude of the constant

$$\Delta\tau_{br}^* \approx (12 \text{ to } 22) We_{cd}^{-1/4}. \tag{8.92}$$

The reader should be careful when comparing such correlations because, as already mentioned, the definition of the end of the breakup process may differ depending on the author's subjective view. Next an example will be given.

For liquid-liquid systems, *Patel* and *Theofanous* [41] (1981) modified the correlation by *Simpkins* and *Bales* [51] (1972), Eq. (8.93), as follows

$$\frac{\Delta\tau_{br}}{D_d/\left[\Delta V_{cd}\left(\rho_c/\rho_d\right)^{1/2}\right]} = cWe_{cd}^{-1/4}, \tag{8.93}$$

where c takes values between 0.9 and 3.6, which means that the dimensionless breakup time is an order of magnitude shorter than for the case of water - gas systems for the same *Weber* number. More recent experiments with a mercury-water system reported by *Yuen* et al. [54] motivated a revision the equation (8.91) to

$$\frac{\Delta\tau_{br}}{D_d/\left[\Delta V_{cd}\left(\rho_c/\rho_d\right)^{1/2}\right]} \approx 14.8Bo_d^{-1/4} \approx 17.8We_{cd}^{-1/4}, \tag{8.94}$$

which is essentially the same as *Krzeckowski*'s [35] finding for gas-liquid systems. Recently *Chen* et al. [8] (1995) reported data for $Bo_d = 10^2, 10^3, 10^4$ giving dimensionless fragmentation times 2.8, 1.6, and 0.9, respectively, which means

$$\Delta\tau_{br}^* \approx 9Bo_d^{-1/4} \approx 10.8We_{cd}^{-1/4}. \tag{8.95}$$

Obviously the spreading of the experimental data and the subjective judgment of the end of the fragmentation process explains the variation of the controlling constant.

Gelfand [17] (1975) proposed to take into account the viscosity of the droplets in the following manner

$$\Delta \tau_{br}^* = 4.5\left(1+1.2 On_d^{1.64}\right), \quad \text{for} \quad We_d \leq 228 \tag{8.96}$$

where On_d is the *Ohnesorge* number. *Pilch* et al. [42] envisage the possibility of taking the viscosity of the droplet into account by multiplying their correlation by $1 + 1.2\ On_d$, where D_d is the instantaneous, deformed drop diameter measured normal to the approaching free stream. In a later work *Pilch* and *Erdman* [43] (1987) used $1 + 2.2\ On_d$. *Hsiang* and *Faeth* [27] approximated provisionally their few data for $On_d < 3.7$ and $We_d < 10^3$ with the correction $1/(1 - On_d/7)$ instead.

Let us summarize the results presented in this section: The final particle size for acceleration induced fragmentation is assumed to be the stable particle size governed by the stability criterion. The stability criterion is defined by the ratio of the shear force acting to destroy the particle and the surface tension force acting to retain the particle form. This ratio is found to be constant. The stability criterion contains the difference between the velocities of the continuum and of the particle adjusted after the fragmentation. This velocity difference is predicted either by analytical or by modified analytical solution of simplified momentum equations. The velocity difference computed by means of the stability criterion with the initial particle size is called *critical velocity*. The higher the relative velocity compared with the critical velocity the shorter the time interval in which the fragmentation occurs. There are 5 types of disintegration of particles exposed in continuum. The duration of the vibration type of fragmentation is estimated analytically. The duration of the other kinds of fragmentation is described empirically.

Knowing (i) the breakup time $\Delta \tau_{br}^*$, (ii) the final velocity difference between the stable particles and their environment, $V_c - V_{d\infty}$, and, (iii) using the hydrodynamic stability criterion Eqs. (8.36) and (8.37), we can compute the final maximum of the diameter of the fragments

$$D_{d\infty} \approx We_{d\infty} \sigma_d \Big/ \left[\rho_c \left(V_c - V_{d\infty} \right)^2 \right] \tag{8.97}$$

which is the first step of the modeling of the fragmentation process.

This method known as "maximum stable diameter concept" was introduced by *Pilch* et al. The maximum stable diameter concept is useful for estimating the largest stable fragment sizes resulting from droplet breakup, provided that the initial *Weber* number exceeds about 350. The authors found that in this region the ratio of the maximum to the mass median fragment size is

$$D_{d\infty,\max} / D_{d\infty} = 2.04 \qquad (8.98)$$

independent of the *Weber* number.

Thus, we recommend the *Pilch* et al. method for prediction of the mass median fragment size for *Weber* numbers exceeding 350. The successful comparison of this method given in [6] (1981), p.111 Fig. 3.2, is valid for $\approx 350 < We_{cd} < 10^5$.

If the computed mass median fragment size for $We_{cd} > 5.28$ after the fragmentation is larger than the initial drop size only vibration breakup can be expected.

8.5 Particle production rate correlations

Knowing the characteristics of the different fragmentation modes from experiments we can estimate approximately the particle production rate. Next we give the expressions for the production rates incorporating our knowledge of the different regimes of fragmentation.

8.5.1 Vibration breakup

Each vibration breakup leads to particle splitting. Therefore the particle production rate is

$$\dot{n}_{sp} \approx \frac{2n_d - n_d}{\Delta\tau_{br}} = n_d \frac{1}{\Delta\tau_{br}} = n_d f_{d,sp} \qquad (8.99)$$

and the fragmentation frequency is simply the reciprocal value of the fragmentation period

$$f_{d,sp} = 1/\Delta\tau_{br}, \qquad (8.100)$$

where $\Delta\tau_{br}$ is computed using Eqs. (8.79) and (8.73).

8.5.2 Bag breakup

As shown by *Magarvey* and *Taylor* [40] (1956) for free falling 15 *mm* water drops each bag breakup produces a spectrum of particle sizes that can be roughly divided into two groups:

1. Macroscopic particles produced by the rim breakup N_{dr} (e.g. $N_{dr} \approx 2$ to 8) having 70 to 75% of the drop mass - see *Komabaysi* et al., [34] (1964) and *Gelfand* et al. [20] (1976). For the following analysis we assume $N_{dr} \approx 6$. Assuming 75% the resulting particle size is

$$D_{dr} \approx \left(\frac{0.75}{N_{dr}}\right)^{1/3} D_d. \tag{8.101}$$

2. Microscopic particles produced by bag fragmentation $N_{db} \approx 30$ to 100 as shown by *Fournier* et al. [16] (1955), and *Magarvey* and *Taylor* [40] (1956), having a size governed by the capillary surface waves amplitude. For the following analysis we assume $N_{db} \approx 65$. Assuming that the microscopic droplets have 25% of the initial mass we have for their size

$$D_{db} \approx \left(\frac{0.25}{N_{db}}\right)^{1/3} D_d. \tag{8.102}$$

Therefore the particle production rate is

$$\dot{n}_{sp} \approx \frac{n_d (N_{dr} + N_{db}) - n_d}{\Delta \tau_{br}} = n_d \frac{N_{dr} + N_{db} - 1}{\Delta \tau_{br}} = n_d f_{d,sp}, \tag{8.103}$$

and the fragmentation frequency is

$$f_{d,sp} = \frac{N_{dr} + N_{db} - 1}{\Delta \tau_{br}} \approx (N_{dr} + N_{db}) \frac{1}{\Delta \tau_{br}} \approx 70/\Delta \tau_{br}, \tag{8.104}$$

where $\Delta \tau_{br}$ is computed using Eq. (8.81).

The above information is sufficient to compute the resulting mass mean diameter

$$D_{d\infty} = D_d / (N_{dr} + N_{db})^{1/3} \approx D_d / 4.12, \tag{8.105}$$

the *Souter* mean diameter

$$D_{d\infty}^{sm} = \left(\frac{N_{dr}D_{dr}^2 + N_{db}D_{db}^2}{N_{dr} + N_{db}}\right)^{1/2} = \left[\frac{N_{dr}\left(\frac{0.75}{N_{dr}}\right)^{2/3} + N_{db}\left(\frac{0.25}{N_{db}}\right)^{2/3}}{N_{dr} + N_{db}}\right]^{1/2} D_d,$$

(8.106)

and finally the ratio

$$\left(\frac{D_{d\infty}^{sm}}{D_{d\infty}}\right)^2 = \frac{N_{dr}\left(\frac{0.75}{N_{dr}}\right)^{2/3} + N_{db}\left(\frac{0.25}{N_{db}}\right)^{2/3}}{\left(N_{dr} + N_{db}\right)^{1/3}} = \frac{N_{dr}\left(\frac{0.75}{N_{dr}}\right)^{2/3} + N_{db}\left(\frac{0.25}{N_{db}}\right)^{2/3}}{\left(N_{dr} + N_{db}\right)^{1/3}}$$

$$= \frac{5\left(\frac{0.75}{5}\right)^{2/3} + 65\left(\frac{0.25}{65}\right)^{2/3}}{(5+65)^{1/3}} = 0.73$$

(8.107)

which is important for computation of heat and mass transfer at the surface. Note that *Hsiang* and *Faeth* (1992) [27] found experimentaly $0.83^2 = 0.79$ which is very close to the above estimate for this case.

Sophisticated application of this theory needs at least two velocity fields for the droplets description: microscopic and macroscopic, and more reliable experimental information about N_{db}.

8.5.3 Bag and stamen breakup

A similar approach to that used for the bag breakup is useful also for bag and stamen breakup. For this case N_{dr} should be interpreted as the number of fragments produced by the rim and stamen breakup.

8.5.4 Sheet stripping and wave crest stripping following by catastrophic breakup

The maximum of the fragment size for the sheet stripping and wave crest striping followed by catastrophic breakup is well described by Eqs. (8.48) and (8.49). The ratio of the maximum to mass median fragment size has a constant value of 2.04 regardless of the initial *Weber* number for $We_d > 300$.

8.5.4.1 Model without taking into account the primary breakup

Both mechanisms are associated with a very short breakup period, which is the reason for assuming that the production rate as a time function during $\Delta\tau_{br}$ is equal to the averaged production rate, namely

$$\dot{n}_d \approx \frac{n_{d\infty} - n_d}{\Delta\tau_{br}} = n_d \left[\left(\frac{D_d}{D_{d\infty}}\right)^3 - 1\right] \Big/ \Delta\tau_{br} \approx n_d \left(\frac{D_d}{D_{d\infty}}\right)^3 \frac{1}{\Delta\tau_{br}}. \qquad (8.108)$$

Next it will be shown that this expression is essentially equivalent to the one obtained by the assumption of linear reduction of the mass of the donor droplet with respect to the time.

It is convenient experimentally to measure the size or the mass of the remaining droplet and therefore

$$\frac{M_d}{M_{d0}} = f(\tau). \qquad (8.109)$$

If the entrained fragment size is $D_{d\infty}$, the fragmentation source term can be computed as follows

$$\dot{n}_d = -\frac{n_d}{\rho_d \frac{\pi D_{d\infty}^3}{6}} \frac{dM_d}{d\tau} = -n_d \left(\frac{D_d}{D_{d\infty}}\right)^3 \frac{d}{d\tau} \frac{M_d}{M_{d0}}. \qquad (8.110)$$

For linear mass reduction we have

$$M_d / M_{d0} = 1 - \tau/\Delta\tau_{br}, \qquad (8.111)$$

$$\dot{n}_d = n_d \left(\frac{D_d}{D_{d\infty}}\right)^3 \frac{1}{\Delta\tau_{br}}. \qquad (8.112)$$

If the entrained droplets have just the stable size at the local relative velocity $V_{d\infty}$ which is close to the V_d we obtain

$$\frac{D_d}{D_{d\infty}} = \frac{We_d}{We_{d\infty}} \left(\frac{V_c - V_{d\infty}}{V_c - V_d}\right)^2 \approx \frac{We_d}{We_{d\infty}} \qquad (8.113)$$

and therefore

$$\dot{n}_{d,sp} \approx n_d \left[\left(\frac{We_d}{We_{d\infty}} \right)^3 - 1 \right] / \Delta \tau_{br} . \qquad (8.114)$$

This approach is used by the author of this work in the computer code IVA3 [31, 32, 33] (1991) and is recommended for $We_c > 350$.

For very high *Weber* numbers we have

$$\dot{n}_{d,sp} \approx \frac{n_d}{\Delta \tau_{br}} \left(\frac{We_d}{We_{d\infty}} \right)^3 \qquad (8.115)$$

Note the difference between Eq. (8.115) and the intuitive proposal made by Kalinin [29] (1970)

$$\dot{n}_{d,sp} \approx \frac{n_d}{\Delta \tau_{br}} \left[\exp(-1/We_d) \right], \qquad (8.116)$$

which is similar to Eq. (8.115) in the qualitative dependence on We_d but neglects the influence of the velocity ratio.

Hsiang and *Faeth* (1992) [28] succeeded to correlate measurements for the final mass median diameter for all regimes of fragmentation which is dependent on the initial droplet size and *Reynolds* number

$$\frac{We_{d\infty}}{We_d} = 7.44 \left(\rho_d / \rho_c \right)^{1/4} / Re_d^{1/2} \qquad (8.117)$$

for the region of the $\left(\rho_d / \rho_c \right)^{1/4} We_d / Re_d^{1/2} \approx 0.3$ to 10, $We_d = 0.5$ to 1000, $Oh_d = 0.0006$ to 4, $\rho_d / \rho_c = 580$ to 12 000, and $Re_d = 300$ to 16 000. The data shows that the *Weber* number of the drop after the end of the breakup was generally greater than the critical *Weber* characteristic for the final particle diameter and the new adjusted velocity ($We_{d\infty} > 13$). The *Souter* mean diameter was found to be 0.83 of the mass median diameter. Using this expression we obtain for the linearized fragmentation rate

$$\dot{n}_{d,sp} \approx \frac{n_{d\infty} - n_d}{\Delta \tau_{br}} = n_d 0.00243 \left(\rho_c / \rho_d \right)^{3/4} Re_d^{3/2} / \Delta \tau_{br} . \qquad (8.118)$$

From Fig. (8.13) we see that

$$\Delta \tau_{br} \approx 5.5 D_d / \left[\Delta V_{cd} \left(\rho_c / \rho_d \right)^{1/2} \right]. \qquad (8.119)$$

Substituting into the above equation we finally obtain

$$\dot{n}_{d,sp} \approx n_d 0.00044 \left(\rho_c / \rho_d \right)^{5/4} \frac{\Delta V_{cd}}{D_d} Re_d^{3/2}. \qquad (8.120)$$

8.5.4.2 Models taking into account the primary breakup

Sheet stripping and wave crest stripping followed by catastrophic breakup are associated with a *primary breakup* leading to $N'_d = 3$ to 5 fragments after $\Delta \tau'_{br}$ for initial *Weber* number < 350 defined by

$$\frac{\Delta \tau'_{br}}{D_d / \left[\Delta V_{cd} \left(\rho_c / \rho_d \right)^{1/2} \right]} \approx 1 \text{ to } 1.5 \quad \text{for drop in gas} \qquad (8.121)$$

and

$$\frac{\Delta \tau'_{br}}{D_d / \left[\Delta V_{cd} \left(\rho_c / \rho_d \right)^{1/2} \right]} \approx 2 \quad \text{for drop in liquid.} \qquad (8.122)$$

Therefore for $0 < \Delta \tau \leq \Delta \tau'_{br}$ we have

$$\dot{n}_d \approx \frac{n_{d\infty} - n_d}{\Delta \tau'_{br}} \approx \frac{N'_d n_d - n_d}{\Delta \tau'_{br}} = n_d \frac{N'_d - 1}{\Delta \tau'_{br}} \approx n_d \frac{3}{\Delta \tau'_{br}}, \qquad (8.123)$$

or

$$\dot{n}_d \approx n_d 2 \frac{\Delta V_{cd}}{D_d} \left(\rho_c / \rho_d \right)^{1/2} \qquad (8.124)$$

for drop in gas and

$$\dot{n}_d \approx n_d \frac{3}{2} \frac{\Delta V_{cd}}{D_d} \left(\rho_c / \rho_d \right)^{1/2} \qquad (8.125)$$

for liquid in liquid. The size of the newly produced unstable fragments is

$$D'_d = D_d / N'^{1/3}_d. \tag{8.126}$$

During this time $\Delta\tau > \Delta\tau'_{br}$ fragments with diameter λ_{KH} are produced due to sheet stripping. Thus the maximum number of particles produced from the stripping of a single donor droplet is

$$\frac{D'^3_d - D'^3_{d\infty}}{\lambda^3_{KH}}. \tag{8.127}$$

Here we denote with $D_{d\infty}$ the size of the donor particle remaining stable after the sheet stripping. The correlation of *Hsiang* and *Faeth* (1992) [28] can be used also in this case.

Thus the rate of particle production during the time interval $\Delta\tau_{br} - \Delta\tau'_{br}$ is

$$\dot{n}_d \approx \frac{N'_d n_d \dfrac{D'^3_d - D'^3_{d\infty}}{\lambda^3_{KH}} - N'_d n_d}{\Delta\tau_{br} - \Delta\tau'_{br}} = N'_d n_d \frac{\left(\dfrac{D'_d}{\lambda_{KH}}\right)^3 - \left(\dfrac{D'_{d\infty}}{\lambda_{KH}}\right)^3 - 1}{\Delta\tau_{br} - \Delta\tau'_{br}}$$

$$= N'_d n_d \frac{\dfrac{1}{N'_d}\left(\dfrac{D_d}{\lambda_{KH}}\right)^3 - \left(\dfrac{D'_{d\infty}}{\lambda_{KH}}\right)^3 - 1}{\Delta\tau_{br} - \Delta\tau'_{br}} \tag{8.128}$$

The time-averaged production rate is therefore

$$\dot{n}_d \approx \frac{n_d}{\Delta\tau_{br}}\left[\left(\frac{D_d}{\lambda_{KH}}\right)^3 - N'_d\left(\frac{D'_{d\infty}}{\lambda_{KH}}\right)^3 - 1\right]$$

$$\approx n_d \frac{\Delta V_{cd}}{D_d}(\rho_c/\rho_d)^{1/2}\frac{1}{5.5}\left[\left(\frac{D_d}{\lambda_{KH}}\right)^3 - 4\left(\frac{D'_{d\infty}}{\lambda_{KH}}\right)^3\right]. \tag{8.129}$$

The total number of droplets after the fragmentation of a single droplet is

$$N'_d + N'_d \frac{D'^3_d - D'^3_{d\infty}}{\lambda^3_{KH}} \tag{8.130}$$

and the resulting mass median diameter is

$$D_{d\infty} = D_d / \left(N'_d + N'_d \frac{D'^3_d - D'^3_{d\infty}}{\lambda^3_{KH}} \right)^{1/3} = \frac{D_d}{N'^{1/3}_d} / \left\{ 1 + \left(\frac{D_d}{\lambda_{KH}} \right)^3 \left[\frac{1}{N'_d} - \left(\frac{D'_{d\infty}}{D_d} \right)^3 \right] \right\}^{1/3}.$$
(8.131)

8.5.4.3 Other models

Reinecke and *Waldman* [45, 46] (1970, 75) recommended the following empirical equation to describe the change of the single drop mass

$$M_d / M_{d0} = \frac{1}{2}\left[1 - \cos\left(\pi \frac{\tau}{\Delta \tau_{br}} \right) \right], \quad 0 < \tau \leq \Delta \tau_{br}$$
(8.132)

due to sheet stripping. Here the dimensionless breakup time is assumed to be $\Delta \tau^*_{br} \approx 4$ and primary breakup was not taken into account. For high pressure shock waves (200, 340 and 476 *bar* amplitudes) *Yuen* et al. [53] provided experimental data for mercury-water systems which are described by the linear law

$$M_d / M_{d0} = 1 - \tau/\tau_{br} = 1 - \frac{1}{c}\tau We_{cd}^{1/4}, \quad 0 < \tau \leq \Delta \tau^*_{br},$$
(8.133)

see *Yuen* et al. [55], Fig. 1, where $\Delta \tau_{br}$ = Eq. (8.95). In addition the authors demonstrated that the *Reineke* and *Waldman* correlation does not represent the *Yuen* et al. data, see Fig. 3 in [54], and demonstrated that linear dependence, Eq. (8.133) is the better one, compare with Fig. 1 in [54].

For the region of sheet stripping and wave crest stripping with catastrophic breakup, *Chu* and *Corradini* proposed two models. The first model [10],

$$M_d / M_{d0} \approx \exp\left(-3c_1 \tau^{c_2} We_{cd}^{c_3} \right)$$
(8.134)

where $c_1 = 0.1708 - 0.149 \left(\rho_d / \rho_c \right)^{1/2}$, $c_2 = 0.772$, and $c_3 = 0.246$, describes the droplet diameters D_d as a function of the time since the beginning of the fragmentation, τ. The second model [11],

$$M_{d2} / M_{d1} \approx \left(1 - c_0 \Delta \tau_{21}^{c_2} We_{cd}^{c_3} \right)^3$$
(8.135)

where $c_0 = 0.1093 - 0.078 \left(\rho_d / \rho_c \right)^{1/2}$, gives the relationship between two average particle masses M_{d2} and M_{d1} for two different times τ_2 and τ_1 during the fragmentation process. The database for the second model consists of the *Baines* and *Buttery* experimental data for $We_{cd} \approx 200 - 3860$ [5].

Using Eq. (8.132), and assuming that the produced small particles have the size $D_{d\infty}$, the particle number production rate is

$$\dot{n}_{d,sp} = -\frac{n_d}{\rho_d \frac{\pi D_{d\infty}^3}{6}} \frac{dM_d}{d\tau} = n_d \frac{\pi}{8} \frac{\Delta V_{cd}}{D_d} (\rho_c/\rho_d)^{1/2} \left(\frac{D_d}{D_{d\infty}}\right)^3 \sin\left(\pi \frac{\tau}{\Delta \tau_{br}}\right) = n_d f_{d,sp}.$$

(8.136)

Note that the time used in this correlation is the time elapsed from the beginning of the fragmentation process which has to be kept in storage during computational analysis. This is very inconvenient. Using Eq. (8.133) we obtain

$$\dot{n}_{d,sp} = n_d \frac{\Delta V_{cd}}{D_d} (\rho_c/\rho_d)^{1/2} \left(\frac{D_d}{D_{d\infty}}\right)^3 \frac{1}{\Delta \tau^*} = n_d f_{d,sp},$$

(8.137)

which is time independent and very convenient for use in computer codes.

Keeping in mind that the droplet surface per unit volume of the mixture, (F_d/Vol), is about $n_d \pi D_d^2$ the term

$$\dot{n}''_{d,sp} = \frac{1}{2\Delta \tau_s^*} \frac{\Delta V_{cd}}{D_d^3 (\rho_d/\rho_c)^{1/2}} \sin\left(\pi \frac{\Delta \tau^*}{\Delta \tau_s^*}\right)$$

(8.138)

or

$$\dot{n}''_{d,sp} = \frac{1}{\pi \Delta \tau_{br}^*} \frac{\Delta V_{cd}}{D_d^3 (\rho_d/\rho_c)^{1/2}}$$

(8.138)

for the high pressure liquid-liquid case, can be interpreted as the number of the entrained micro-droplets per unit surface of the macroscopic droplet (particle flow rate). So the idea naturally arises to model the surface particle flow rate not only at droplet surfaces but also at surfaces with arbitrary form, namely

$$\dot{n}_{d,sp} = (F_d/Vol)\dot{n}''_{d,sp}.$$

(8.140)

As already mentioned, some authors prefer to use interfacial area concentration as the convected variable instead of the particle number density. For such an approach the generated interfacial area per unit time and unit mixture volume is easily computed multiplying the above equation with $\pi D_{d\infty}^2$. The result is

$$\dot{F}_d = (F_d / Vol) \frac{1}{2\Delta\tau_s^*} \frac{\Delta V_{cd}}{D_d^3 (\rho_d / \rho_c)^{1/2}} \sin\left(\pi \frac{\Delta\tau^*}{\Delta\tau_s^*}\right) \qquad (8.141)$$

or

$$\dot{F}_d = (F_d / Vol) \frac{1}{\Delta\tau_{br}^*} \frac{\Delta V_{cd}}{D_d^3 (\rho_d / \rho_c)^{1/2}} \qquad (8.142)$$

for high pressure liquid-liquid case. A similar expression,

$$\dot{F}_d = (F_d / Vol) 0.089 c_{cd}^{d\ 3/4} We_{cd}^{1/4} \frac{|\Delta V_{cd}|}{D_d (\rho_d / \rho_c)^{1/2}} \qquad (8.143)$$

was proposed by *Pilch* and *Young* and used by *Young* [56] (1989) in the computer code IFCI, where c_{cd}^d is the drag coefficient between continuum and droplet. This idea was used in terms of diameter change by *Young* [56],

$$\dot{D}_d^{donor} = -\frac{\left[1 - (N_d')^{-1/3}\right] \Delta V_{cd}^{donor}}{\Delta\tau_{br}' (\rho_d / \rho_c)^{1/2}}. \qquad (8.144)$$

Two points should be considered in the application of the detailed sheet stripping models in system computer codes:

1. The detailed description of the sheet stripping needs additional computer storage to store the time when the sheet stripping process starts and the particle size at this moment in each computational cell (see the comments by *Fletcher* and *Anderson* [15] (1990)), because the time in Eq. (8.116) is counted from the beginning of the sheet striping in a *Lagrangian* manner.

2. At least two velocity fields for representation of macro- and microscopic droplets are formally needed in order to distinguish between the donor drop and the produced fragments. Within the donor drop field the primary breakup can be easily modeled if the time from the beginning of the sheet stripping reaches the time for primary breakup. Within the next actual time step $\Delta\tau$ the particle production rate in the donor drop field is

$$\dot{n}_{d,sp}^{donor} = N_d' n_d^{donor} / \Delta\tau . \qquad (8.145)$$

The further sheet stripping is allowed until the donor cell reduces its size to the size of the produced stable fragments. Another modeling approach is the use of the averaged production rate during the period $\Delta\tau_{br}'$

$$\dot{n}_{d,sp}^{donor} = \frac{n_d^{donor} - N_d' n_d^{donor}}{\Delta \tau_{br}'} = n_d^{donor} \frac{1-N_d'}{\Delta \tau_{br}'} = n_d^{donor} \frac{1-N_d'}{\Delta \tau_{br}^*} \frac{\Delta V_{cd}^{donor}}{D_d^{donor} (\rho_d/\rho_c)^{1/2}}. \quad (8.146)$$

This idea was used in terms of diameter change by *Young* [56] (1989)

$$\dot{D}_d^{donor} = -\frac{1-N_d'^{-1/3}}{\Delta \tau_{br}^*} \frac{\Delta V_{cd}^{donor}}{(\rho_d/\rho_c)^{1/2}}. \quad (8.147)$$

Acceleration induced fragmentation seems to be the most frequently investigated fragmentation phenomenon and reliable data are available for practical *approximate* modeling of dynamic fragmentation.

The questions that are not resolved by this approach are the following: Acceleration induced fragmentation strongly depends on the relative velocities between neighboring velocity fields. The relative velocities are governed by the modeling of drag, lift, and virtual mass forces. Drag, lift, and virtual mass coefficients are usually measured and correlated for simple steady state flows. Little is known about the drag coefficients for transient conditions, for molten drops in water with strong evaporation, drag coefficients in mutual inter-penetration of three fluids. This inevitably affects the accuracy of the prediction of the relative velocity and therefore the accuracy of the prediction of the dynamic fragmentation. Obviously improvement of the knowledge in these fields is necessary in order to increase the accuracy of the theoretical predictions.

8.6 Droplets production due to highly energetic collisions

Consider collision of two droplets with relative velocity of $V^{rel} = \Delta V_{dd}^{no}$ being very high. The collision will inevitably lead to a strong instability of the resulting droplet and eventually to fragmentation of the resulting droplet. If after the collision of two droplets two new droplets are produced the coalescence frequency of a single droplet is equal to zero

$$\left(f_{coal}^p\right)^{no} = 0, \quad (8.148)$$

and there is no effective droplet production due to collision. If more than two droplets are produced the coalescence frequency of a single droplet due to collision is zero and there is a production of particles due to collision which may be expressed with negative coalescence frequency of a single droplet

$$f_{coal}^p < 0. \tag{8.150}$$

The interactions observed for colliding water droplets by *Ashgriz* and *Givi* [4] (1987) may involve:

1) a bouncing collision;
2) a grazing collision, in which the droplets just touch each other slightly without coalescence;
3) a permanent coalescence;
4) temporally coalescence followed by separation in which satellite droplets are generated;
5) a shattering collision, occurring at high energy collisions, in which numerous tiny droplets are expelled radially from the periphery of the interacting drops.

An empirical description of the dynamics of such collisions for gas droplet suspensions was made by *Podovisotsky* and *Schreiber* [44] (1984). According to the authors, a target droplet *d1* colliding with the projectile *d2* (smaller than the target droplet) undergoes a main change in mass, given as

$$\Delta m_{d1} / m_{d2} = 1 - 0.246 Re_{d12}^{0.407} Lp_{d1}^{-0.096} \left(D_{d1} / D_{d2} \right)^{-0.278} - \Phi_{d12} \tag{8.151}$$

(for $30 < Re_{d12} < 6000$; $5 < Lp_{d1} < 3 \times 10^5$; $1.9 < D_{d1} / D_{d2} \leq 12$), where m is the droplet mass; Re_{d12} is the *Reynolds* number for a small droplet penetrating into the larger droplet,

$$Re_{d12} = D_{d2} |V_{d2} - V_{d1}| \rho_d / \eta_d, \tag{8.152}$$

where D_{d2} is the droplet diameter, V_d is the droplet velocity, ρ_d is its density and η_d is the liquid particle viscosity; Lp_{d1} is the *Laplace* number, indicating a ratio between surface tension forces and viscous drag,

$$Lp_{d1} = D_{d1} \rho_d \sigma / \eta_d^2, \tag{8.153}$$

where σ is the surface tension between droplet and gas; Φ_{d12} is a correction term accounting for the gas flow, given as

$$\Phi_{d12} = 0.00446 A \quad \text{for} \quad A \leq 40.6, \tag{8.154}$$

$$\Phi_{d12} = 11.85 (0.01A)^{4.64} \quad \text{for} \quad 40.6 \leq A \leq 120, \tag{8.155}$$

$$A = Re_{d12}^{0.285} Lp_{d1}^{0.2} \left(D_{d1}/D_{d2}\right)^{0.4} We_{cd1}^{0.442}, \tag{8.156}$$

and We_{cd1} is the *Weber* number, indicating a ratio between the inertia force and the surface tension force,

$$We_{cd1} = \rho_c \left(V_c - V_{d1}\right)^2 D_{d1}/\sigma, \tag{8.157}$$

where V_c is the local gas velocity. When

$$\Delta m_{d1}/m_{d2} = 1, \tag{8.158}$$

the droplets coalesce. When

$$0 < \Delta m_{d1}/m_{d2} < 1, \tag{8.159}$$

some of the projectile droplets are fragmented. For

$$-\Delta m_{d1}/m_{d2} < \Delta m_{d1}/m_{d2} < 0 \tag{8.160}$$

the projectile droplet is fully fragmented while the target droplet is fragmented partially or fully. Note that the condition

$$-\Delta m_{d1}/m_{d2} < \Delta m_{d1}/m_{d2} \tag{8.161}$$

is imposed by conservation of mass. We see that this interpretation of $\Delta m_{d1}/m_{d2}$ resembles the coalescence probability

$$\left(f_{dcoal}^p\right)^{no} \approx \Delta m_{d1}/m_{d2}. \tag{8.162}$$

8.7 Acceleration induced bubble fragmentation

Bubbles can disintegrate due to different kinds of instabilities, e.g. acceleration or turbulence induced fragmentation. To model the dynamic fragmentation means to compute the bubble number production rate in Eq. (7.2), i. e. the estimation of the bubble diameter at the end of the fragmentation process, $D_{d\infty}$, and the estimation of the duration of the fragmentation, $\Delta \tau_{br}$. The production rate due to bubble splitting is computed by means of Eq. (7.11). The hydrodynamic stability criterion already introduced for droplets

$$We_{d\infty} = \rho_c (V_c - V_{d\infty})^2 D_{d\infty} / \sigma \approx const \qquad (8.163)$$

for acceleration induced fragmentation is also valid for bubble fragmentation. The most frequently used value for the hydrodynamic stability criterion in the two-phase literature is

$$We_{d\infty} = \rho_c (V_c - V_{d\infty})^2 D_{d\infty} / \sigma \approx 12 \qquad (8.164)$$

Next we discuss first the upper and then the lower values of $D_{d\infty}$.

There is a bubble flow regime (bubble diameter approximately 0.001 to 0.02 m) in which the terminal velocity

$$\Delta V_{cd\infty} = K u V_K, \qquad (8.165)$$

see *Wallis* [57] (1969), is independent of the size and is proportional to the so called *Kutateladze* terminal velocity

$$V_K = \left[\sigma g (\rho_c - \rho_d) / \rho_c^2 \right]^{1/4}, \quad Ku = \sqrt{2}. \qquad (8.166)$$

Substituting Eq. (8.165) into Eq. (8.164) and solving for $D_{d\infty}$ we obtain the upper limit of the bubble size in the flow

$$D_{d\infty,max} \approx We_{d\infty} \frac{1}{2} \lambda_{RT}. \qquad (8.167)$$

Brodkey [7] (1967) found that bubbles having a diameter less than

$$D_{d\infty,min} \approx 2.53 \frac{1}{2} \lambda_{RT} \qquad (8.168)$$

behave as solid spheres and are not subject to further splitting. *Berenson* observed experimentally that during film boiling on horizontal surfaces the bubbles departing the film have a size

$$D_{d\infty,min} \approx 4.7 \lambda_{RT}. \qquad (8.169)$$

Thus, the stable bubble size after the fragmentation should be

$$D_{d\infty} \approx We_{d\infty} \frac{\sigma}{\rho_c (V_c - V_{d\infty})^2}, \qquad (8.170)$$

within the bounds

$$D_{d\infty,\min} < D_{d\infty} < D_{d\infty,\max}. \qquad (8.171)$$

Some authors have correlated the observed bubble diameters for steady state flows as a function of the local conditions. *Note that these correlations contain integral results of simultaneously happening fragmentation and coalescence.* Two examples are given below.

Ahmad [2] (1979) correlated data modifying the maximum possible bubble diameter observed in the flow by using instead of $We_{d\infty} = 12$, the following correlation

$$We_{d\infty} = \frac{1.8}{1 + 1.34(\alpha_c w_c)^{1/3}} = We_{d\infty}^* \quad \text{for} \quad \alpha_d < 0.1 \qquad (8.172)$$

$$We_{d\infty} = We_{d\infty}^* \left(\frac{9\alpha_d}{1-\alpha_d}\right)^{1/3} \quad \text{for} \quad 0.1 < \alpha_d < 0.99. \qquad (8.173)$$

Serizawa and *Kataoka* [50] (1987) correlated their data for the interfacial area concentration for vertical bubble flow in a pipe with 0.3 *m* diameter in the region $\alpha_c w_c$ =1 to 5 *m/s* by the following correlation

$$\frac{6\alpha_d}{D_{d\infty}} = 1030 \alpha_d^{0.87} (\alpha_c w_c)^{0.2} \quad \text{or} \qquad (8.174)$$

$$D_{d\infty} = 0.00582 \alpha_d^{0.13} / (\alpha_c w_c)^{0.2}. \qquad (8.175)$$

Both correlations already contain the effect of the bubble splitting with increasing liquid velocity and of the coalescence with increasing gas volume fraction.

In what follows we estimate the period of the bubble splitting. By relative displacement of a body in an infinite continuum, compared with the size of the body, the continuum is deformed in front of, and around the body, and fills the region behind it. Similarly, bubbles displace liquid due to their relative motion with respect to the continuum. Bubble flows in which the bubble velocity differs from the liquid velocity possess natural fluctuations. We estimate the period of these fluctuations by dividing the average distance between two adjacent bubbles by the time needed for one bubble to sweep out this distance. The average distance between two adjacent bubbles, $\Delta \ell_d$, in a flow with bubble volume fraction α_d is

$$\Delta \ell_d \approx D_d \left(\frac{\pi \sqrt{2}}{6 \alpha_d} \right)^{1/3} \approx 0.9047 D_d / \alpha_d^{1/3}. \qquad (8.176)$$

The time needed for one bubble to sweep out this distance is

$$\Delta \tau_d^{nat} \approx \Delta \ell_d / |\Delta V_{cd}| \approx \alpha_d^{1/3} \left(\frac{6}{\pi \sqrt{2}} \right)^{1/3} \bigg/ |\Delta V_{cd}|. \qquad (8.177)$$

In fact the real picture is more complicated. The fluctuations are dumped due to the bubble compressibility, and the bubbles are interacting with each other and with the eddies behind the bubbles.

Thus the natural fluctuation period in the bubble flow is of the order of

$$\Delta \tau_d^{nat} \approx const \; \alpha_d^{1/3} \left(\frac{6}{\pi \sqrt{2}} \right)^{1/3} \bigg/ |\Delta V_{cd}|. \qquad (8.178)$$

If the bubble is unstable,

$$We_d > We_{d\infty}, \qquad (8.179)$$

the fragmentation is likely to happen within $\Delta \tau_d^{nat}$ i.e.

$$\Delta \tau_{br} \approx const \; \alpha_d^{1/3} \left(\frac{6}{\pi \sqrt{2}} \right)^{1/3} \bigg/ |\Delta V_{cd}|. \qquad (8.180)$$

Thus, bubbles having relative velocity with respect to the surrounding liquid are unstable if $We_d > We_{d\infty}$, where $We_{d\infty}$ is found to be equal 12 within the time interval $\Delta \tau_{br}$ defined by Eq. (8.178). The final diameter after the fragmentation can be assumed equal to that defined by Eq. (8.170), within the limits given with the inequality (8.171).

Gelfand et al. [18] (1975) found experimentally that bubbles behind the front of the waves are stable for $We_{d\infty} < 5$. *Gelfand* et al. [19] (1974) have successfully used the time scale

$$\Delta\tau_{br}^* = \frac{\Delta\tau_{br}}{\delta_{dKH} / \left[\Delta V_{cd} \left(\rho_d / \rho_c\right)^{1/2}\right]} \approx 1 \qquad (8.181)$$

for bubble fragmentation behind the front of the pressure wave.

There are some unresolved problems in this approach:

- Why should the swarm of bubbles with an unstable average diameter D_d after the fragmentation take just the stable averaged diameter $D_{d\infty}$?

- How large is the constant in Eq. (8.180), or is it a constant at all?

Further experiments are needed to resolve this particular problem.

Nomenclature

Latin

Ar	Archimedes number, *dimensionless*
Bo_d	$= \left[\rho_d \dfrac{dV_d}{d\tau}(D_d/2)^2\right] / \sigma_d = \dfrac{3}{16} c_{cd}^d We_{cd}$, Bond number, *dimensionless*
c_{cd}^d	drag coefficient, *dimensionless*
c_{cd}^{vm}	virtual mass coefficient, *dimensionless*
D	particle diameter, *m*
D_d^v	volume equivalent diameter for non-spherical particle, *m*
$D_{d,\max}$	maximum diameter during the oscillation, *m*
$D_{d,\min}$	minimum diameter during the oscillation, *m*
D_{dr}	particle diameter produced by the rim breakup, *m*
D_{db}	particle diameter produced by the bag breakup, *m*
$D_{d\infty}^{sm}$	Souter mean diameter, *m*
$\dot{D}_d^{donor}$	change of the donor particle diameter per unit time, *m/s*
$D_{d\infty}$	stable particle diameter after the fragmentation is finished, *m*
D_{d1}	target droplet diameter, *m*
D_{d2}	projectile droplet diameter (smaller than the target droplet diameter), *m*
d	differential

$\dot{F}_d$ generated interfacial area per unit time and unit mixture volume, $m^2/(sm^3)$
f_n natural frequency of the n-th order mode of a spherical drop or bubble performing oscillations in continuum, where $n = 2, 3, ..., n, 1/s$
$f_{d,sp}$ $= 1/\Delta\tau_{br}$, fragmentation frequency, $1/s$
g gravitational acceleration, m/s^2
La_d $= 1/On_d^2$, Laplace number, *dimensionless*
Lp_{d1} $= D_{d1}\rho_d\sigma/\eta_d^2$, Laplace number, ratio between surface tension forces and viscous drag, *dimensionless*
M_d mass of a single droplet, kg
M_{d0} initial mass of a single droplet before fragmentation, kg
Ma Mach number, *dimensionless*
$N_{\eta d}$ viscosity number, *dimensionless*
N_{dr} macroscopic particles produced by the rim breakup, -
N_{db} microscopic particles produced by bag fragmentation, -
N'_d microscopic particles produced by primary breakup, -
$\dot{n}_{sp}$ particle production rate, $1/(m^3s)$
$\dot{n}''_{d,sp}$ number of the entrained micro-droplets per unit surface of the macroscopic droplet (particle flow rate), $1/(m^2s)$
$\dot{n}_{d,sp}^{donor}$ particle production rate in the donor drop field, $1/(m^3s)$
On_d $= \eta_d/\sqrt{\rho_d D_d \sigma_d} = We_d^{1/2}/Re_d$, Ohnesorge number, dimensionless
Re Reynolds number, *dimensionless*
Re_{d12} $= D_{d2}|V_{d2}-V_{d1}|\rho_d/\eta_d$, Reynolds number for a small droplet penetrating into the larger droplet, *dimensionless*
t tangential force per unit surface, N/m^2
V velocity, m/s
$V_{d\infty}$ final fragment velocity, m/s
We Weber number, *dimensionless*

Greek

Δm_{d1} mass change of the target droplet *d1* after colliding with the projectile droplet *d2* having mass m_{d2}, kg
m_{d2} mass of the projectile droplet, kg
$\Delta\rho_{dc}$ density difference, kg/m^3
ΔV_{13} velocity difference, m/s
$\Delta\tau_{\Delta V}$ linearized velocity relaxation time constant, s

$\Delta \tau_{br}$ fragmentation period, s
$\Delta \tau'_{br}$ duration of the primary breakup, s
$\Delta \tau *$ time, *dimensionless*
$\delta_{d,KH}$ most unstable wavelength, *Kelvin-Helmholtz* wavelength, m
Φ_{d12} correction term accounting for the gas flow, *dimensionless*
σ surface tension, N/m
τ time, s
λ_{bk} liquid boundary layer thickness inside the droplet, m
λ_{RT} $= \left[\sigma_d / (g \Delta \rho_{dc}) \right]^{1/2}$, scale of the *Rayleigh-Taylor* instability wavelength, m
ρ density, kg/m^3
η dynamic viscosity, $kg/(ms)$

Subscripts

0   initial, before fragmentation
1   gas
2   liquid
3   droplet, particles
c   continuous
d   disperse
∞   at the end of the fragmentation
d1   target droplet
d2   projectile droplet (smaller than the target droplet)

References

1. Anderson WH and Wolfe HE (1965) Aerodynamic breakup liquid drops - I. Theoretical, Proc. Int. Shock Tube Symposium, Naval Ordinance Lab. White Oak, Maryland, USA, pp 1145-1152
2. Ahmad SY (1970) Axial distribution of bulk temperature and void fraction in a heated channel with inlet subcooling, J. Heat Transfer, vol 92 pp 595
3. Arcoumanis C, Khezzar L, Whitelaw DS and Warren BCH (1994) Breakup of Newton and non-Newton fluids in air jets, Experiments in Fluids, vol 17 pp 405-414
4. Ashgriz N and Givi P (1987) Binary collision dynamics of fuel droplets, Int. J. Heat Fluid Flow, vol 8 pp 205-210
5. Baines M and Buttery NE (Sept. 1978) Differential velocity fragmentation in liquid - liquid systems, RD/B/N4643, Berkley Nuclear Laboratories

6. Brauer H (1992) Umströmung beschleunigter und verzögerter Partikel, Wärme und Stoffübertragung, vol 27 pp 93-101
7. Brodkey RS (1967) The phenomena of fluid motions, Addison-Wesley Press
8. Chen X, Yuen WW and Theofanous TG (1995) On the constitutive description of microinteractions concept in steam explosions, Proceedings of the Seventh International Topical Meeting on Nuclear Reactor Thermal Hydraulics NURETH-7, New York, USA, NUREG/CP-0142
9. Chandrasekhar S (1981) Hydrodynamic and hydromagnetic stability, Dover Publ. Inc., New York
10. Chu CC and Corrardini MC (Nov.1984) Hydrodynamics fragmentation of liquid droplets, ANS Transaction, Wash. DC, vol 47
11. Chu CC and Corrardini MC (Febr. 2-6, 1986) One-dimensional model for fuel coolant fragmentation and mixing analysis, Proceedings of the International ANS/ENS Topical Meeting on Thermal Reactor Safety, San Diego, California, U.S.A., vol 1 pp II.2-1-II.2-10
12. De Jarlais G, Ishii M and Linehan J (Febr. 1986) Hydrodynamic stability of inverted annular flow in an adiabatic simulation, Transactions of ASME, Journal of Heat Transfer, vol 108 pp 85-92
13. Dinh AT, Dinh TN, Nourgaliev RR and Sehgal BR (19th-21th May 1997) Hydrodynamic fragmentation of melt drop in water, OECD/CSNI Specialist Meeting on Fuel Coolant Interactions, JAERI-Tokai Research Establishment, Japan
14. Faeth GM (April 3-7, 1995) Spray combustion: a review, Proc. of The 2nd International Conference on Multiphase Flow '95 Kyoto, Japan
15. Fletcher DF and Anderson RP (1990) A review of pressure-induced propagation models of the vapor explosion process, Progress in Nuclear Energy, vol 23 no 2 pp 137-179
16. Fournier D'Albe EM and Hidayetulla MS (1955) Quartery Journal Royal Meterological Society, Kondon, vol 81 pp 610-613
17. Gelfand BE, Gubin SA, Kogarko SM and Komar SP (1973) Osobennosti Razrushenija Kapel Vijaskoi Zhidkosti v Udarnich Volnach (Feature of Voscous Liquid Drop Breakup Behind Shock Wave) Inzenerno Physicheski Journal, vol 25 no 3 pp 467-470
18. Gelfand BE et al (1974) The main models of the drops breakup; Inzenerno Physicheski Journal, vol 27 no 1 pp 120-126
19. Gelfand BE, Gubin SA et al (1975) Breakup of air bubbles in liquid, Dokl. USSR Ac. Sci., vol 220 no 4 pp 802-804
20. Gelfand BE, Gubin SA and Kogarko SM (1976) Various forms of drop fragmentation in shock waves and their spatial characteristics, J. Eng. Phys., vol 27
21. Haas F (November 1964) Stability of droplets suddenly exposed to a high velocity gas steam, A.I.Ch.E. Journal, vol 10 no 6 pp 920-924
22. Hanson AR, Domich EG and Adams HS (August 1963) Shock tube investigation of the breakup of drop by air blasts, Phys. Fluids, vol 6 no 8 pp 1070 - 1080
23. Hassler G (1971) Untersuchungen zur Verformung und Auflösung von Wassertropfen durch aerodynamische Kräfte im stationären Luft - und Wasserstrom für Unterschallgeschwindigkeit, Dissertation, Universität Karlsruhe
24. Hinze JO (1949) Critical speed and sizes of liquids globules, Appl. Sci. Res., vol A1 pp 273-288
25. Hinze JO (1949) Forced deformation of viscous liquid globules, Appl. sci. Res., vol A1 pp 263-272

26. Hinze JO (1955) Fundamentals of hydrodynamics of splitting in dispersion processes, AIChE Journal, vol 1 pp 284-295
27. Hsiang LP and Faeth GM (1992) Near-limit drop deformation and secondary breakup, Int. J. Multiphase Flow, vol 18 no 5 pp 635-652
28. Ishii M (1977) One-dimensional drift-flux model and constitutive equations for relative motion between phases in various two-phase flow regimes, ANL-77-47
29. Kalinin AV (May 1970) Derivation of fluid mechanics equations for two phase medium with phase change, Heat Transfer - Sov. Res. Vol 2 no 3
30. Kamke E (1959) Differentialgleichungen, Lösungsmethoden und Lösungen, Bd.I Gewöhnliche Differentialgleichungen. Leipzig: Gees & Portig
31. Kolev NI (Sept. (1991) A three-field model of transient 3D multi-phase three-component flow for the computer code IVA3, Part 1: Theoretical basics: Conservation and state equations, numerics. Kernforschungszentrum Karlsruhe, KfK 4948
32. Kolev NI (Sept. 1991) A three-field model of transient 3D multi-phase three- component flow for the computer code IVA3, Part 2: Models for interfacial transport phenomena. Code Validation. Kernforschungszentrum Karlsruhe, KfK 4949
33. Kolev NI (Sept. 1991) IVA3: Computer code for modeling of transient three dimensional three phase flow in complicated geometry, Program documentation: Input description. Kernforschungszentrum Karlsruhe, KfK 4950
34. Komabayasi MT, Gonda T and Isono K (1964) Life time of water drops before breaking in size distribution fragment droplet, J. Met. Soc. Japan, vol 42 no 5 pp 330-340
35. Krzeczkowski S (1980) Measurement of liquid droplet disintegration mechanisms, Int. J. Multiphase Flow, vol 6 pp 227-239
36. Laftaye G (1943) Sur l'atomatisation d'un jet liquide, C. R. Acad. Sci., Paris 217 p 340
37. Lamb MA (1945) Hydrodynamics. Cambridge, At the University Press
38. Lane WR (June 1951) Shatter of drops in streams of air, Ind. Eng. Chem., vol 43 pp 1312-1317
39. Li MK and Folger HS (1978) Acoustic emulsification, Part 2. Breakup of large primary oil droplets in water medium, J. Fluid Mech., vol 88 no 3 pp 513-528
40. Magarvey RH and Taylor BW (Oct.1956) Free fall breakup of large drops, Journal of Applied Physics, vol 27 no 10 pp 1129 - 1135
41. Patel PD and Theofanous TG (1981) Hydrodynamic fragmentation of drops, J. Fluid Mech., vol 103 pp 307-323
42. Pilch M, Erdman CA and Reynolds AB (Aug.1981) Acceleration induced fragmentation of liquid drops, Charlottesville, VA: Department of Nucl. Eng., University of Virginia, NUREG/CR-2247
43. Pilch MM and Erdman CA (1987) Use of the breakup time data and velocity history data to predict the maximum size of stable fragments for acceleraton-induced breakup of a liquid drops, Int. J. Multiphase Flow, vol 13 no 6 pp 741-757
44. Podovisotsky AM and Schreiber AA (1984) Coalescence and break-up of drops in two-phase flows, Int. J. Multiphase Flow, vol 10 pp 195-209
45. Reineke WG and Waldman GD (Aug. 11-13, 1970) An investigation of water drop disintegration in region behind strong shock waves, Third International Conference on Rain Erosion and Related Phenomena, Hampshire, England
46. Reineke WG and Waldman GD (Jan. 20-22, 1975) Shock layer shattering of cloud drops in reentry flight, AIAA Paper 75-152, AIAA 13th Aerospace Sciences Meeting, Pasadena California

47. Ruft K (1977) Maximale Einzeltropfen bei stationärer Bewegung in einer niedrigviscosen kontinuierlichen Phase, Chemie Ingenieur Technik, vol 49 no 5 pp 418-419
48. Sarjeant M (4th-6th April 1979) Drop breakup by gas streams, Third European Conference on MIXING, Held at the University of York, England, pp 225-267
49. Schröder RR and Kitner RC (Jan. 1965) Oscillation of drops falling in liquid field, A.I.Ch.E. Journal, vol 11 no 1 pp 5-8
50. Serizawa A and Kataoka I (May 24-30, 1987) Phase distribution in two-phase flow, Proc. ICHMT Int. Sem. Transient Two-Phase Flow, Dubrovnik, Yugoslavia, Invited Lecture
51. Simpkins PG and Bales EL (1972) Water-drop response to sudden accelerations, J. Fluid Mech., vol 55 pp 629-639
52. Taylor GI (1949) The shape and acceleration of a drop in a high - speed air stream, Min. of Supply Paper AC10647/Phys. C69. See also (1963) "Scientific Papers", Cambridge University Press, vol 3 pp 457-464
53. Tricbnigg H (1929) Der Einblase- und Einspritzvorgang bei Dieselmotoren, Vienna
54. Yuen WW, Chen X and Theofanous TG (Sept.21-24, 1992) On the fundamental micro-interactions that support the propagation of steam explosions, Proc. at the Fifth Int. Top. Meeting on Reactor Thermal Hydraulics, Salt Lake City, UT. NURETH-5, vol 11 pp 627-636
55. Yuen WW, Chen X and Theofanous TG (1994) On the fundamental micro-interactions that support the propagation of steam explosions, NED, vol 146 pp 133-146
56. Young MF (1989) Application of the IFCI integrated fuel-coolant interaction code to FITS-type pouring mode experiments, SAND 89 -1962C, Sandia National Laboratories
57. Wallis GB (1969) One-dimensional two-phase flow, New York: McGraw Hill
58. Riso F and Fabre J (1998) Oscillation and breakup of bubbles immersed in turbulent field, J. Fluid Mech. vol 372 pp 323-355

9. Turbulence induced particle fragmentation and coalescence

Particles in continuum react differently on the fluctuation of the mean continuum velocity. While bubbles follow quickly the continuum, heavy droplet may considerably delay following the continuum fluctuation. This makes the main difference in the criteria for fragmentation of bubbles and particles. In this Chapter we will give a brief characterization of the homogeneous isotropic turbulence, and then we will analyze the reaction capability of a particle to follow the changes in velocity of the surrounding continuum. We will derive expressions for the maximum relative velocity, which creates distortion and possible fragmentation. Finally we will look for quantitative information describing, as in the previous section, (a) the final diameter after the fragmentation, $D_{d\infty}$, and (b) the time interval in which the fragmentation occurs, $\Delta\tau_{br}$. The different components of the turbulent energy dissipation are analyzed. For channel flows expression are given for approximate estimation of the dissipation rate of the turbulent kinetic energy as a function of the frictional pressure drop. The dissipation rate of the turbulent kinetic energy due to relative phase motion is also analyzed. The probability of the bubble and droplet coalescence is then estimated.

9.1. Homogeneous turbulence characteristics

Large scale motion: Observing turbulent motion of continuum one realizes that random eddies overlay the mean flow. The eddies have different size ℓ at this length scale. The largest scale of the eddies is limited by the geometrical boundary of the systems. In a pipe flow the natural limit is the pipe diameter

$$\ell \le D_h.\qquad(9.1)$$

Large scale eddies contain the main part of the turbulent kinetic energy. Viscous forces have no effect on the large scale motion. There is no energy dissipation. The large scale motion is characterized by the velocity difference over the scale of the eddies ℓ, V'_ℓ. Usually the turbulence *Reynolds* number

$$\mathrm{Re}'_\ell = \frac{V'_\ell \ell}{\nu} \qquad (9.2)$$

is used for characterizing the turbulence.

Small scale motion: Small scale eddies with a size

$$\ell_e < \ell \qquad (9.3)$$

contains only a small part of the kinetic energy of turbulent motion. There are characterized by

$$\mathrm{Re}'_{\ell_e} = \frac{V'_\ell \ell_e}{\nu}. \qquad (9.4)$$

Viscous limit: The size of a eddy characterized by

$$\mathrm{Re}'_{\ell_0} = \frac{V'_{\ell 0} \ell_0}{\nu} \approx 1 \qquad (9.5)$$

or

$$\ell_0 \approx \frac{\nu}{V'_{\ell 0}} \qquad (9.6)$$

is called *inner scale* or *micro scale* of turbulence. Eddies with

$$\ell < \ell_0 \qquad (9.7)$$

dissipate mechanical energy by viscous friction in internal energy (heat). Fluctuations with such sizes are gradually damped. The very nature of the turbulent motion is the continuous transfer of mechanical energy from larger to smaller eddies.

Dimensional analysis for small scale motion: We already stated that the turbulent motion of scale larger then the viscous limit does not depend on viscosity and is characterized by V'_ℓ, ℓ_e, ρ. Starting with this observation *Kolmogoroff* [13, 14] (1941, 1949) found that the only combination of these flow parameters having dimension of energy dissipated per unit time and per unit volume is $\rho V'^3_{l_e} / \ell_e$, or per unit time and unit mass

$$\varepsilon \approx \frac{V'^3_{l_e}}{\ell_e}. \qquad (9.8)$$

9.1. Homogeneous turbulence characteristics

The mechanical energy of turbulent dissipation per unit time and unit mass of the continuum ε is called *turbulence dissipation rate* in the following. Thus the velocity change over the distance ℓ_e

$$V'_{l_e} = (\varepsilon \ell_e)^{1/3} \tag{9.9}$$

increases with increasing energy dissipation. The characteristic time period of the fluctuation with given size ℓ_e is then

$$\tau'_{l_e} = \frac{\ell_e}{V'_{l_e}} = \left(\frac{\ell_e^2}{\varepsilon}\right)^{1/3}. \tag{9.10}$$

The acceleration of the fluctuation with given size ℓ_e is then

$$\frac{dV'_{l_e}}{d\tau} \approx \frac{V'_{l_e}}{\tau'_{l_e}} = \frac{\varepsilon^{2/3}}{\ell_e^{1/3}}. \tag{9.11}$$

Substituting the fluctuation velocity defined by Eq. (9.9) into the definition equations for the turbulent *Reynolds* number results in

$$Re'_{l_e} = \frac{V'_\ell \ell_e}{\nu} = \varepsilon^{1/3} \frac{\ell_e^{4/3}}{\nu}. \tag{9.12}$$

Viscous limit as a function of the dissipation of the specific kinetic energy of turbulence: Setting the turbulent *Reynolds* number equal to one defines the viscosity limit as follows

$$\ell_0 = \left(\frac{\nu^3}{\varepsilon}\right)^{1/4}, \tag{9.13}$$

$$V'_{\ell 0} \approx \frac{\nu}{\ell_0} = (\varepsilon \nu)^{1/4}. \tag{9.14}$$

The characteristic time period of the fluctuation with given size ℓ_0 is then

$$\tau'_{l_0} = \frac{\ell_0}{V'_{l0}} = \left(\frac{\nu}{\varepsilon}\right)^{1/2}. \tag{9.15}$$

The acceleration of the fluctuation with given size ℓ_0 is then

$$\frac{dV'_{l_e}}{d\tau} \approx \frac{V'_{l_e}}{\tau'_{l_e}} = \frac{\varepsilon^{3/4}}{\nu^{1/4}}. \qquad (9.16)$$

Yaglom [30] found in 1949 the exact expression

$$\frac{dV'_{l_e}}{d\tau} = \sqrt{3}\frac{\varepsilon^{3/4}}{\nu^{1/4}}. \qquad (9.17)$$

Turbulent viscosity: In analogy to the laminar flow

$$\varepsilon \approx \nu'\left(\frac{V'_l}{\ell}\right)^2, \qquad (9.18)$$

a effective dynamic viscosity of turbulence

$$\eta' = \rho V'_l \ell \qquad (9.19)$$

can be defined. The effective kinematic viscosity of turbulence is then

$$\nu' = V'_l \ell. \qquad (9.20)$$

The analogy to the molecular kinetics is obvious: V'_l corresponds to the molecule velocity and ℓ to the main free path. The ratio

$$\frac{\nu}{\nu'} = \frac{\nu}{V'_l \ell} = \frac{1}{Re'_\ell} \qquad (9.21)$$

contains the information that the larger the turbulent *Reynolds* number the more negligible is the molecular viscosity compared to the turbulent viscosity.

There is also a deterministic way to Eq. (9.15). If we assume isotropic turbulence the definition of the turbulence dissipation rate, Eq. (5.156) in Chapter 5 of Volume 1 results in

$$\varepsilon = \frac{15}{2}\nu\left(\frac{\partial V'}{\partial r}\right)^2. \qquad (9.22)$$

Taking $\partial V' \approx V'_l$ and $\partial r \approx \ell$ and rearranging we obtain

$$\frac{V'_l}{\ell} = \left(\frac{2}{15}\frac{\varepsilon}{\nu}\right)^{1/2}, \qquad (9.23)$$

which is very similar to Eq. (9.15). This equation is frequently used in the literature for estimation the order of magnitude of the velocity fluctuation in the viscous turbulence regime

$$V'_\ell = \ell \left(\frac{2\,\varepsilon}{15\,\nu} \right)^{1/2} \quad \text{for} \quad \ell < \ell_0, \tag{9.24}$$

e.g. *Smoluchowski* [27] (1918). This result is used in the literature for instance for modeling of rain droplet agglomeration. If we take as a characteristic distance the size of a particle inside a continuum then we have for $D_d < \ell_0$

$$V'_{D_d} = D_d \left(\frac{2\,\varepsilon_c}{15\,\nu_c} \right)^{1/2}, \tag{9.25}$$

$$\tau' = \frac{D_d}{V'_{D_d}} = \left(\frac{15\,\nu_c}{2\,\varepsilon_c} \right)^{1/2}, \tag{9.26}$$

$$\frac{dV'_{D_d}}{d\tau} \approx \frac{V'_{D_d}}{\tau'} = \frac{2}{15} D_d \frac{\varepsilon_c}{\nu_c}. \tag{9.27}$$

9.2 Reaction of a particle to the acceleration of the surrounding continuum

We start with the simplified momentum equations for the continuous and the disperse phase, denoted with c and d, respectively, neglecting compressibility, interfacial mass transfer, the spatial acceleration and the viscous terms

$$\alpha_c \rho_c \frac{\partial w_c}{\partial \tau} + \alpha_c \nabla p + \alpha_c \rho_c g - f_d^d - f_d^{vm} = 0, \tag{9.28}$$

$$\alpha_d \rho_d \frac{\partial w_d}{\partial \tau} + \alpha_d \nabla p + \alpha_d \rho_d g + f_d^d + f_d^{vm} = 0, \tag{9.29}$$

where the drag force per unit mixture volume

$$f_d^d = -\alpha_d \rho_c \frac{3}{4} \frac{c_d^d}{D_d} |\Delta w_{cd}| (w_c - w_d), \tag{9.30}$$

and the virtual mass force per unit mixture volume

$$f_d^{vm} = -\alpha_d \rho_c c_d^{vm} \frac{\partial}{\partial \tau}(w_c - w_d) \qquad (9.31)$$

are functions of the relative velocities. Dividing by the corresponding volume fraction and subtracting the equations we eliminate the pressure gradient

$$\rho_c \frac{\partial w_c}{\partial \tau} - \rho_d \frac{\partial w_d}{\partial \tau} + (\rho_c - \rho_d)g - \frac{1}{\alpha_c \alpha_d} f_d^d - \frac{1}{\alpha_c \alpha_d} f_d^{vm} = 0. \qquad (9.32)$$

After some rearrangements we have

$$\alpha_c (\rho_c - \rho_d) \frac{\partial w_c}{\partial \tau} + (\alpha_c \rho_d + \rho_c c_d^{vm}) \frac{\partial \Delta w_{cd}}{\partial \tau} + \alpha_c (\rho_c - \rho_d)g$$

$$+ \rho_c \frac{3}{4} \frac{c_d^d}{D_d} |\Delta w_{cd}| \Delta w_{cd} = 0. \qquad (9.33)$$

For a single particle in infinite continuum $\alpha_c \to 1$ we have

$$(\rho_c - \rho_d) \frac{\partial w_c}{\partial \tau} + (\rho_d + \rho_c c_d^{vm}) \frac{\partial \Delta w_{cd}}{\partial \tau} + (\rho_c - \rho_d)g + \rho_c \frac{3}{4} \frac{c_d^d}{D_d} |\Delta w_{cd}| \Delta w_{cd} = 0.$$

$$(9.34)$$

If the acceleration of the continuum $\partial w_c / \partial \tau$ is known the reaction of the particle is governed by the above equation. In a turbulent field the local acceleration of the continuum depends on the local turbulent characteristics and the size of the vortex. Neglecting the gravity and the virtual mass force Eq. (9.34) reduces to Eq. (32.2) by *Levich* [16].

Equation (9.34) contains surprising information. Consider the case of no gravity and no friction

$$\frac{\partial \Delta w_{cd}}{\partial \tau} = \frac{\rho_d - \rho_c}{\rho_d + \rho_c c_d^{vm}} \frac{\partial w_c}{\partial \tau}. \qquad (9.35)$$

For positive acceleration the velocity-difference my increase if the dispersed phase has larger density or decrease if the denser phase has lower density then the continuum density. This means that a bubble in water will rush into the direction of the acceleration of the water.

9.3 Reaction of particle entrained inside the turbulent vortex – inertial range

The relative velocity within the turbulent eddy: Assuming that the particles are large enough that the quadratic drag law holds we replace in Eq. (9.34)

$$\frac{\partial w_c}{\partial \tau} = \frac{\varepsilon_c^{2/3}}{\ell_{ec}^{1/3}} \qquad (9.36)$$

$$\frac{\partial \Delta w_{cd}}{\partial \tau} = \frac{\Delta w_{cd}}{\tau'_{cd}} = \frac{\Delta w_{cd}^2}{\ell_{ec}} \qquad (9.37)$$

$$\tau'_{cd} \approx \frac{\ell_{ec}}{\Delta w_{cd}} \qquad (9.38)$$

as proposed by *Levich* [16] and obtain

$$(\rho_c - \rho_d)\frac{\varepsilon_c^{2/3}}{\ell_{ec}^{1/3}} + (\rho_d + \rho_c c_d^{vm})\frac{\Delta w_{cd}^2}{\ell_{ec}} + (\rho_c - \rho_d)g + \rho_c \frac{3}{4}\frac{c_d^d}{D_d}|\Delta w_{cd}|\Delta w_{cd} = 0 \qquad (9.39)$$

or solved with respect to the relative velocity

$$\Delta w_{cd}^2 = (\rho_d - \rho_c)\frac{\varepsilon_c^{2/3}\ell_{ec}^{2/3} + g\ell_{ec}}{\rho_d + \rho_c c_d^{vm} + \rho_c \frac{3}{4}\frac{c_d^d}{D_d}\ell_{ec}} . \qquad (9.40)$$

The velocity difference is obviously a function of the size of the eddy. The maximum of the relative velocity is reached if

$$\frac{d\Delta w_{cd}^2}{d\ell_{ec}} = 0 , \qquad (9.41)$$

that is if

$$\ell_{ec,max} - \frac{g}{\varepsilon_c^{2/3}} a\ell_{ec,max}^{1/3} - \frac{2}{3}a = 0 , \qquad (9.42)$$

where

$$a = \frac{4}{c_d^d}D_d\left(\frac{\rho_d}{\rho_c} + c_d^{vm}\right). \qquad (9.43)$$

The maximum relative velocity is then

$$\Delta w_{cd,\max}^2 = (\rho_d - \rho_c) \frac{\varepsilon_c^{2/3} \ell_{ec,\max}^{2/3} + g\ell_{ec,\max}}{\rho_d + \rho_c c_d^{vm} + \rho_c \frac{3}{4} \frac{c_d^d}{D_d} \ell_{ec,\max}}. \quad (9.44)$$

Neglecting the virtual mass force and the gravitation influence we obtain the *Levich* equation (33.9) in [16]. Note also by comparing with the *Levich* expression in [16] that he uses turbulence energy dissipation per unit volume and we use turbulence energy dissipation per unit mass. In our notation the *Levich* expression reads

$$\Delta w_{cd,\max}^2 = \frac{\rho_d - \rho_c}{\rho_d} \frac{1}{3} \left(\frac{2^3}{3} \frac{1}{c_d^d} \frac{\rho_d}{\rho_c} \varepsilon_c D_d \right)^{2/3} \approx \frac{\rho_d - \rho_c}{\rho_d} \left(\frac{\rho_d}{\rho_c} \right)^{2/3} (\varepsilon_c D_d)^{2/3}. \quad (9.45)$$

Note that for particles in gas the neglecting of the virtual mass force is a good approximation but the neglecting of the gravitational influence is not always a good one.

Now consider again Eq. (9.9) for a velocity difference inside the continuum between two points having D_d distance between each other

$$V_{l_{ec}}'^2 = (\varepsilon_c D_d)^{2/3} \quad (9.46)$$

It is obvious that for $\rho_d \gg \rho_c$ Eq. (9.44) gives much higher velocities than Eq. (9.46), see the discussion by *Kocamustafaogullari* and *Ishii* [12] on p. 436. Therefore droplets in gas will still obey the mechanical stability criterion for inertial droplet fragmentation as discussed before but with velocity computed by Eq. (9.44). For the case of bubbles in continuum the stability criterion will be modified and additional information is required as discussed below.

9.4 Stability criterion for bubbles in continuum

Bubbles in a turbulent velocity field having sizes, D_d, larger than some characteristic size of the turbulent eddies, ℓ_{ec}, are unstable and disintegrate. The resulting bubble size distribution is likely to be similar to the distribution of the sizes of the turbulent eddies. The characteristic time period in which the disintegration occurs is likely to be equal to the characteristic time scale of the turbulent fluctuations. These are the main ideas leading to a successful quantitative description of turbulence induced bubble fragmentation. In what follows we look for quantitative information describing, as in the previous section, (a) the final diameter after the fragmentation, $D_{d\infty}$, and (b) the time interval in which the fragmentation occurs, $\Delta\tau_{br}$.

The introduction of the hydrodynamic stability criterion in this case, as for the acceleration induced fragmentation, is very useful. However, the nature of the force causing the bubble distortion is different. If the continuous velocity field, c, is highly turbulent, the force acting towards the bubble distortion is the turbulent shear force instead of the drag force - see Fig. 9.1. So the critical *Weber* number is defined as

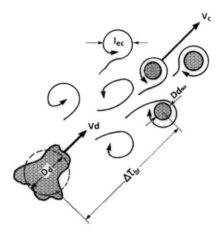

Fig. 9.1. Characteristics of turbulence induced particle fragmentation

$We'_{d\infty}$ = turbulent shear force / surface tension force

$$= \mathbf{t}\pi D_{d\infty}^2 / (\sigma_d \pi D_{d\infty}) \approx const, \qquad (9.47)$$

where

$$\mathbf{t} = \rho_c V_c'^2 \qquad (9.48)$$

is the turbulent shear stress. It is customary to assume that the turbulence is locally isotropic and that the particle size lies in the inertial sub range. Using the *Kolmogoroff* [13, 14] (1941, 1949) hypothesis for an isotropic turbulent structure to compute V_c' for the lower end of the equilibrium range where the energy transfer through the spectrum is independent of the viscosity

$$V_c'^2 = \left(1/c_1\right)^{2/3} \left(\ell_{ec}\varepsilon_c\right)^{2/3}, \qquad (9.49)$$

where

$$(1/c_1)^{2/3} = 2 \quad \text{or} \quad c_1 = 0.3536, \tag{9.50}$$

and assuming that the maximum particle diameter, $D_{d\infty}$, is equal to the characteristic size of the turbulent eddies, ℓ_{ec}, the *Weber* number can be expressed by

$$We'_{d\infty} = D_{d\infty}^{5/3} \left(\varepsilon_c / c_1\right)^{2/3} \rho_c / \sigma_d . \tag{9.51}$$

Hinze [7] (1955) used *Clay*'s data [4] (1940) for a *liquid-liquid* system and found

$$We'_{d\infty} = 1.17 . \tag{9.52}$$

Thomas [29] reviewed several literature sources and after comparing with experimental data confirms the value 1.17.

Another argument supplying the order of magnitude of this constant comes from *Sevik* and *Park* [26] (1973). They analyzed theoretically the splitting of dispersed particles in continuum using the following argument. According to *Lamb*, see in [15] (1945), the natural frequency of the *n*-th order mode of a spherical drop or bubble (d) performing small amplitude oscillations in continuum (c) is given by Eq. (8.72). Due to the small amount of damping the amplitude of the response of a bubble to a steady-state sinusoidal external pressure increases enormously if one of the natural frequencies is reached. The characteristic frequency of the turbulent flow is given by

$$f = \sqrt{V'^2}/\ell_{ec} . \tag{9.53}$$

To obtain the *Weber* number at which particle breakup will occur, *Sevik* and *Park* set f equal to the natural frequency at the lowest fixed volume oscillation of the particle and ℓ_{ec} equal to the particle diameter. The resulting critical *Weber* number for *liquid-liquid* systems with density ratio equal to one is 1.04, which is very close to the *Hinze*'s result, and for *bubbles in water* is

$$We'_{d\infty} \cong 2.48 . \tag{9.54}$$

This value describes very well the breakup of a *gas jet in a turbulent liquid stream* observed experimentally by *Sevik* and *Park*.

Thus the values for the critical *Weber* number

$$We'_{d\infty} = 1.17 \tag{9.55}$$

for *droplets in turbulent liquid* and

$$We'_{d\infty} \cong 2.48 \tag{9.56}$$

for *bubbles in turbulent liquid* can be considered reliable. Modification of Eq. (9.49) taking into account the viscosity of the dispersed system

$$We'_{d\infty} = 1.17\left[1+1.077 On_d^{1.64}\right] \tag{9.56}$$

is also recommended.

For a particle concentration not affecting the main turbulent characteristics of the continuum i.e. $\alpha_d \to 0$, Eq. (9.51) solved for $D_{d\infty}$

$$D_{d\infty} \approx \left(c_1/\varepsilon_c\right)^{2/5}\left(We'_{d\infty}\sigma_d/\rho_c\right)^{3/5} \tag{9.57}$$

is all what we need to compute the particle size after the fragmentation as a function of the turbulence dissipation rate. Obviously increasing the turbulence dissipation rate leads to a decrease of the particle size which is the intuitively expected relationship.

The influence of the volumetric particle concentration and of the viscosity [28, 7] (1935, 1955) of both fluids on the turbulent characteristics and therefore on the stable particle size is taken into account by *Calderbank* [3] (1967) by introducing the correction factor, $\alpha_c^n \left(\eta_d/\eta_c\right)^{0.25}$, for liquid-liquid and bubble-liquid systems

$$D_{d\infty} \approx \left(c_1/\varepsilon_c\right)^{2/5}\left(We'_{d\infty}\sigma_d/\rho_c\right)^{3/5} \alpha_c^n \left(\eta_d/\eta_c\right)^{0.25}, \tag{9.58}$$

where $n \approx 0.45$ to 0.65. A similar correction factor $\left(100\alpha_c\right)^{0.15}\left(\eta_d/\eta_c\right)^{0.25}$ was introduced by *Meusel* [18] (1989) to take into account the interference of the particles in the flow. Note that Eq. (9.58) generalizes integral results of *simultaneously happening fragmentation and coalescence*.

The characteristic time constant for the fragmentation should be comparable with the period of oscillation of a single turbulent eddy

$$\Delta\tau_{br} \approx \Delta\tau' \approx \ell_{ec}/\sqrt{V_c'^2} \approx D_{d\infty}/\sqrt{V_c'^2} \approx \left(D_{d\infty}^2/\varepsilon_c\right)^{1/3}. \tag{9.59}$$

We see that the key parameter for an estimation of the local particle size by turbulence induced fragmentation is the rate of the local dissipation of turbulent kinetic energy, ε_c, of the continuous velocity field.

To decide whether acceleration or turbulence induced fragmentation is the governing mechanism we compare the stable particle size computed using the critical *Weber* number for acceleration induced fragmentation $We_{d\infty}$ as discussed in the previous Section with Eq. (9.58). If

$$\frac{We_{d\infty}\sigma_d}{\rho_c(V_c-V_d)^2} > (c_1/\varepsilon_c)^{2/5}(We'_{d\infty}\sigma_d/\rho_c)^{3/5}\alpha_c^n(\eta_d/\eta_c)^{0.25} \qquad (9.60)$$

turbulence induced droplet fragmentation is probably the predominant fragmentation mechanism, otherwise the acceleration induced droplet fragmentation is probably predominant. Eq. (9.60) needs experimental confirmation.

Paranjape et all [31] developed in 2003 a correlation for bubble breakup by impact of turbulent eddies in the form

$$\dot{n}_{d,sp} = C_{TI}\frac{1}{3}\frac{n_d}{D_d}V_c\sqrt{1-6/We}\exp(-6/We)$$

based on data for upwards and downwards air water flow in vertical pipes with 25.5 and 50.8mm ID for j_c = 0.62 to 5m/s and j_d =0.004 to 1.977m/s at 6.9bar, where $We = \rho_c w_c^2 D_d/\sigma > 6$, C_{TI} = 0.085 for upward flow and 0.034 for downward flow. *Morel, Yao* and *Bestion* [32] developed in the same time the following correlation for the above discussed case

$$\dot{n}_{d,sp} = 1.6\frac{\varepsilon_c^{1/3}\alpha_d(1-\alpha_d)}{D_d^{11/3}}\cdot\frac{\exp(-\sqrt{We'_d/We'_{d\infty}})}{1+0.42(1-\alpha_d)\sqrt{We'_d/We'_{d\infty}}},$$

where $We'_{d\infty} = 1.24$, $\alpha_{d,\max}$ =0.52, $We'_d = 2D_d^{5/3}\varepsilon_c^{2/3}\rho_c/\sigma_d$.

9.5 Turbulence energy dissipation due to the wall friction

Now let us consider the turbulence induced droplet fragmentation in channels. As for the turbulence induced droplet fragmentation in pool flow the stable droplet diameter can be computed by means of Eq. (9.57) or one of its modified forms. The key parameter here is the estimation of the rate of dissipation of the turbulent kinetic energy. For channel flow, the knowledge of the frictional pressure drop allows one to compute the dissipation rate of the turbulent kinetic energy and therefore to estimate the dynamic fragmentation source. For 3D flows in structures resembling a porous body, the frictional pressure drop is strongly related to the dissipation rate of the turbulent kinetic energy of the continuum, namely

$$\alpha_c \rho_c \varepsilon_c = u\overline{R}_u + v\overline{R}_v + w\overline{R}_w,$$ (9.61)

where u, v and w are the mixture velocity components in r, θ and z directions and $\overline{R}_u$, $\overline{R}_v$ and $\overline{R}_w$ are the pressure drops in each particular direction per unit length. In accordance with *Lockhart* and *Martinelli*

$$\overline{R}_u = \frac{1}{2}\frac{1}{\rho_c}(\rho u)^2 \Phi_{c0u}^2 \left(\lambda_{Ru}/D_{hu} + \xi_u/\Delta r \right),$$ (9.62)

$$\overline{R}_v = \frac{1}{2}\frac{1}{\rho_c}(\rho v)^2 \Phi_{c0v}^2 \left[\lambda_{Rv}/D_{hv} + \xi_v/(r^\kappa \Delta\theta) \right],$$ (9.63)

$$\overline{R}_w = \frac{1}{2}\frac{1}{\rho_c}(\rho w)^2 \Phi_{c0w}^2 \left(\lambda_{Rw}/D_{hw} + \xi_w/\Delta z \right),$$ (9.64)

ρ is the mixture density and Φ_{c0}^2 are the two-phase multipliers for each particular direction and the friction factors λ_{Ru}, λ_{Rv}, λ_{Rw} are computed assuming that the total mass flow rate ρV has the properties of the continuum,

$$\lambda_R = \lambda_R \left(\rho V D_h / \eta_c, \delta_k / D_h \right).$$ (9.65)

A simplified version of this approach for a pipe flow was used by *Rozenzweig* et al. [24] (1980). These authors computed the dissipation rate caused by wall friction as follows: $\varepsilon_c = \frac{1}{2}\frac{\lambda_R}{D_h} w_c^3$, where $\lambda_R \approx 0.3164/Re_c^{1/4}$, $Re_c = D_h w_c / \nu_c$.

Bello [1] (1968) computed successfully the stable bubble diameter in one-dimensional pipe flow using Eqs. (9.57) and (9.61) and

$$We'_{d\infty} \approx 0.437.$$ (9.66)

A slightly modified approach for pipes was used later by *Kocamustafaogullari* et al [11] in 1994. The particle size thus obtained is some averaged size for the cross section of the channel. In fact in channel flow the turbulence is generated mainly by the wall and the length scale of turbulence outside the laminar sublayer

$$\ell_{ec} \approx \text{const} \frac{r}{D_h} \left(\frac{v_c^3 V_c^{*3}}{D_d} \right)^{1/4}, \qquad \text{Kolmogoroff [14] (1949)}, \qquad (9.67)$$

depends on the distance from the tube axis, r. Here V_c^* is the friction velocity. Therefore in tubes one should expect a spectrum of the particle sizes depending on the distance from the wall.

Modeling of turbulence induced fragmentation (and coalescence) needs modeling of turbulence in multi-phase flows. For channel flow the knowledge of the frictional pressure drop allows to compute the dissipation rate of the turbulent kinetic energy and therefore to estimate the dynamic fragmentation sources. For pool flows or 3D flows in complicated geometries multi-phase turbulence modeling needs much more attention because a reliable model of turbulence does not yet exist. That is why the theoretical estimation of the dynamic fragmentation (and coalescence) characteristics is not possible today. Exceptions are some simple geometries or flows with small concentrations of the dispersed phase where the turbulence modeling consists mainly of one phase turbulence models or cases where the input of the mechanical dissipation energy per unit time from external sources is known (e.g. some chemical equipment).

9.6 Turbulence energy dissipation due to the relative motion

The wall friction is one cause for turbulence generation and dissipation. Another cause is the relative motion between particles and continuum. The dissipation rate of the turbulent kinetic energy in the wakes behind particles, d, can be estimated as the β_c part

$$\alpha_c \rho_c \varepsilon_c = \beta_c \alpha_d \rho_c \frac{3}{4} \frac{1}{D_d} \left(c_{cd,r}^d \left| u_c - u_d \right|^3 + c_{cd,\theta}^d \left| v_c - v_d \right|^3 + c_{cd,z}^d \left| w_c - w_d \right|^3 \right)$$

(9.68)

of the specific power needed to move the particles with relative velocity

$$\left| \Delta V_{cd} \right| = \left(\Delta u_{cd}^2 + \Delta v_{cd}^2 + \Delta w_{cd}^2 \right)^{1/2}. \qquad (9.69)$$

Here Δu, Δv, and Δw are the components of the vector $\Delta \mathbf{V}$ in each particulate coordinate r, θ and z.

For low *Reynolds* number (laminar flow), $0 < Re_{cd} < Re_{cd}^*$, there is no energy dissipation due to particle produced turbulence

$$\beta_c \approx 0. \tag{9.70}$$

In the transition regime $Re_{cd}^* < Re_{cd} < Re_{cd}^{**}$ of the periodic particle deformation

$$\beta_c \approx \left(Re_{cd} - Re_{cd}^*\right) / \left(Re_{cd}^{**} - Re_{cd}^*\right) \tag{9.71}$$

only a part of the kinetic energy is dissipated. For high *Reynolds* numbers $Re_{cd} > Re_{cd}^{**}$, the particles are stochastically deformed and the flow is turbulent wake flow in which the kinetic energy need to move the particles with respect to the continuum is totally dispersed

$$\beta_c \approx 1. \tag{9.72}$$

Equation (9.68) contains in its RHS the density of the continuum as a multiplier. For gas as a continuum phase the resulting RHS is much smaller than for liquid as a continuum phase, i.e. the bubbles themselves produce considerable turbulence in a liquid.

Reichardt [23] (1942) derived theoretical equations for both limiting *Reynolds* numbers

$$Re_{cd}^* = 3.73 \left(\frac{\rho_c \sigma_c^3}{g \eta_c^4}\right)^{0.209} = 3.73 \left(\frac{\Delta \rho_{cd}}{\rho_c} Ar^2\right)^{0.209} \tag{9.73}$$

and

$$Re_{cd}^{**} = 3.1 \left(\frac{\rho_c \sigma_c^3}{g \eta_c^4}\right)^{0.25} = 3.1 \left(\frac{\Delta \rho_{cd}}{\rho_c} Ar^2\right)^{0.25}. \tag{9.74}$$

There is an alternative approach to compute the dissipation rate of the turbulent kinetic energy of turbulence in the liquid. *Bhavaraju* et al. [2] (1978) developed an expression for the power input to gas sparked vessels, based on the expansion of bubbles as they ascend up a column from pressure p_2 to p_1. The resulting expression can be generally written as

$$\alpha_c \rho_c \varepsilon_c = \alpha_d w_d g \frac{p_2}{p_1 - p_2} \ln \frac{p_1}{p_2}. \tag{9.75}$$

Let us summarize the results of this section. In a pool flow the turbulence is produced mainly due to (1) velocity gradients in space and (2) wakes behind the bubbles. For the estimation of the first component models are needed to describe the turbulence evolution of bubble flow in time and space. Such models are at the beginning of their development and no reliable method is known up to now to the author. For the estimation of the second component, Eq. (9.68) is recommended. With an approximate estimation of the dissipation rate of the turbulent kinetic energy, the stable diameter after the fragmentation can be estimated using Eq. (9.57) or (9.58), and for the breakup time Eq. (9.59) can be used.

For channel flow or flow in porous structures with significant wall friction the dissipation rate of the turbulent kinetic energy of the liquid is

$$\alpha_c \rho_c \varepsilon_c = \text{RHS of Eq. (9.61)} + \text{RHS of Eq. (9.68)}. \qquad (9.76)$$

Here RHS stands for right hand side. After estimating ε_c the procedure to compute the production rate using Eq. (9.5) is the same as for pool flow.

9.7 Bubble coalescence probability

The probability of coalescence, sometimes called efficiency of coalescence of bubbles, is defined as a function of the ratio of the time interval within which the eddies are touching each other, called *collision time* interval $\Delta\tau_{col}$ and the time interval required to push out the surrounding liquid and to overcome the strength of the capillary micro-layer between the two bubbles, called *coalescence time* interval $\Delta\tau_{coal}$

$$\left(f_{d,coal}^p\right)^o \approx f\left(\Delta\tau_{col}/\Delta\tau_{coal}\right). \qquad (9.77)$$

An example of such functional relationship given by *Coulaloglou* and *Tavlarides* [5] (1977)

$$\left(f_{d,coal}^p\right)^o \approx \exp\left(-\Delta\tau_{coal}/\Delta\tau_{col}\right), \qquad (9.78)$$

will be used here. Note that there are also other proposals for the functional dependence in the literature. Obviously coalescence is possible if there is enough contact time for completing the coalescence

$$\Delta\tau_{col} > \Delta\tau_{coal}. \qquad (9.79)$$

If the bubbles are jumping apart before the liquid is pushed out completely between them and before the micro skin between them is broken there is effectively no coalescence. Usually the change of the velocity along the distance equal to the bubble size D_d is estimated by the *Kolmogoroff* velocity scale, Eq. (9.9),

9.7 Bubble coalescence probability

$$V^{rel} \approx (D_d \varepsilon_c)^{1/3}. \qquad (9.80)$$

Assuming that the bubble is following completely the turbulent fluctuation the collision time interval is then

$$\Delta \tau_{col} \approx D_d / V_d' = \frac{D_d^{2/3}}{\varepsilon_c^{1/3}}. \qquad (9.81)$$

Thus the collision is completely controlled by the turbulence of the continuum.

What remains to be estimated is the coalescence time. The time needed to push out the liquid between the bubbles is approximated by *Thomas* [29] (1981) by the time required to push out the liquid between two discs each of them of radius r (a classical lubrication problem)

$$\Delta \tau_{coal} = \frac{3\pi \eta_c r^4}{2F\delta^2}, \qquad (9.82)$$

r is the radius of common surface of the both bubbles at the moment of coalescence,

$$\delta \approx 10^{-7} m \qquad (9.83)$$

is the radial film thickness between them and

$$F = 4\pi r^2 \sigma / D_d \qquad (9.84)$$

is the force acting from inside of the bubble on the common surface that is equal to the pulsation pressure force

$$F \approx tD_d^2 = \rho_c V_d'^2 D_d^2. \qquad (9.85)$$

Solving the last two equations for D_d and substituting into Eq. (9.80) we obtain

$$\Delta \tau_{coal} = \frac{3}{32\pi} \eta_c \rho_c V_d'^2 D_d^2 \left(\frac{D_d}{\sigma \delta} \right)^2. \qquad (9.86)$$

Having in mind Eq. (9.81) we have for the time ratio

$$\frac{\Delta \tau_{col}}{\Delta \tau_{coal}} = \frac{32\pi}{3} \frac{(\sigma\delta)^2}{\eta_c \rho_c (V_d' D_d)^3} = \frac{32\pi}{3} \frac{(\sigma\delta)^2}{\eta_c \rho_c D_d^4 \varepsilon_c} \qquad (9.87)$$

or for the coalescence probability

$$\left(f_{d,coal}^p\right)^o \approx \exp\left(-\Delta\tau_{coal}/\Delta\tau_{col}\right) = \exp\left(-\frac{3}{32\pi\delta^2}\frac{\eta_c\rho_c D_d^4 \varepsilon_c}{\sigma^2}\right) \quad (9.88)$$

for $\Delta\tau_{col}/\Delta\tau_{coal} \geq 1$.

$$\left(f_{d,coal}^p\right)^o \approx 0, \quad (9.89)$$

for $\Delta\tau_{col}/\Delta\tau_{coal} < 1$. We see from this expression that the coalescence probability decreases with in-creasing the bubble diameter and with increasing the fluctuation velocity. There should be a limiting diameter called coalescence escape limit dividing the regimes of coalescence and no coalescence. It is computed from the condition

$$\Delta\tau_{col}/\Delta\tau_{coal} = 1, \quad (9.90)$$

namely

$$D_{d,\lim} = \left[\frac{32\pi}{3}\frac{(\sigma\delta)^2}{\eta_c\rho_c}\right]^{1/3} / V_d', \quad (9.91)$$

or replacing by using Eq. (9.80) and solving with respect to the diameter

$$D_{d,\lim} \approx \left(\frac{32\pi}{4}\right)^{1/4} (\sigma\delta)^{1/2} / (\eta_c\rho_c\varepsilon_c)^{1/4}. \quad (9.92)$$

Thus for

$$D_d > D_{d,\min} \quad (9.93)$$

no coalescence is expected. This regime is called *coalescence escaping* regime. It is important to note that if the bubble diameters are larger than the coalescence escaping diameter, but lower than the critical bubble diameter

$$D_{d,\min} < D_d < D_{d\infty} \quad (9.94)$$

there is neither fragmentation nor coalescence. Obviously, for small relative velocities causing bubble coalescence, $D_{d,\lim}$ can increase and take values larger than the stable bubble diameter $D_{d\infty}$. In this case coalescence and fragmentation simultaneously take place. The values of the limiting relative bubble velocity, $V_{\lim}^{rel}$, is easily obtained by equating $D_{d,\lim}$ and $D_{d\infty}$

$$\left[\frac{32\pi}{3}\frac{(\sigma\delta)^2}{\eta_c\rho_c}\right]^{1/3}/V'_{d,\lim} = \frac{We_{d\infty}\sigma}{\rho_c(V_c-V_{d\infty})^2}, \qquad (9.95)$$

or

$$V'_{d,\lim} = \frac{\rho_c(V_c-V_{d\infty})^2}{We_{d\infty}\sigma}\left[\frac{32\pi}{3}\frac{(\sigma\delta)^2}{\eta_c\rho_c}\right]^{1/3}. \qquad (9.96)$$

This theory is already applied by *Thomas* [29] (1981) for turbulence induced coalescence.

In reality *two regimes* of bubble formation with stable diameters are observed for turbulent two-phase flow. They depend on the amount of dissipated mechanical energy of turbulence. For low values, the controlling mechanism is the escaping from coalescence, and for greater values - the destroying of unstable bubbles.

Recently *Hibiki* and *Ishii* [6] (1999) recommended to use the *Oolman* and *Blanch* [21, 22] results for thinning of the liquid film between bubbles of equal size

$$\Delta\tau_{coal} = \frac{1}{8}\left(\frac{\rho_c D_d^3}{2\sigma}\right)^{1/2}\ln\frac{\delta_{init}}{\delta_{crit}} = 0.814\left(\frac{\rho_c D_d^3}{\sigma}\right)^{1/2}, \qquad (9.97)$$

where the initial film thickness is

$$\delta_{init} = 10^{-4}\,m, \qquad (9.98)$$

Kirkpartick and *Lockett* [10] (1974), and the final thickness is

$$\delta_{crit} = 10^{-8}\,m, \qquad (9.99)$$

Kim and *Lee* [9] (1987). Using Eqs. (9.81) and (9.80) *Hibiki* and *Ishii* [6] (1999) corrected the constant in the ratio

$$\frac{\Delta\tau_{coal}}{\Delta\tau_{col}} = 0.814\left(\frac{\rho_c D_d}{\sigma}\right)^{1/2}V^{rel} = 0.814\left(\frac{\rho_c^3 D_d^5 \varepsilon_c^2}{\sigma^3}\right)^{1/6} \qquad (9.100)$$

to 1.29 by comparing with experiments and used finally the *Coulaloglou* and *Tavlarides* [5] (1977) expression for the coalescence probability

$$f_{d,coal}^p \approx \exp(-\Delta\tau_{coal}/\Delta\tau_{col}) = \exp\left[-1.29\left(\frac{\rho_c^3 D_d^5 \varepsilon_c^2}{\sigma^3}\right)^{1/6}\right]. \tag{9.101}$$

which differs considerably from the *Thomas* [29] (1981) expression

$$\left(f_{d,coal}^p\right)^o \approx \exp(-\Delta\tau_{coal}/\Delta\tau_{col}) = \exp\left(-\frac{3}{32\pi\delta^2}\frac{\eta_c\rho_c D_d^4 \varepsilon_c}{\sigma^2}\right). \tag{9.102}$$

Using Eq. (7.33) the *Hibiki* and *Ishii* final expression for the coalescence frequency due to turbulence in bubbly flow is given as

$$\left(f_{d,col}^p f_{d,coal}^p\right)^o = const \frac{\alpha_d}{0.52 - \alpha_d}\frac{6}{D_d}V^{rel}\exp\left[-1.29\left(\frac{\rho_c^3 D_d^5 \varepsilon_c^2}{\sigma^3}\right)^{1/6}\right]$$

for $\Delta\tau_{col}/\Delta\tau_{coal} \geq 1$, (9.103)

in which the empirical constant is introduced by comparison with experimental data.

Paranjape et all [31] developed in 2003 a correlation in the form

$$\left(f_{d,col}^p f_{d,coal}^p\right)^o = \frac{1}{3}\left(\frac{6}{\pi}\right)^{2/3} n_d^{4/3} \alpha_d^{2/3} \begin{bmatrix} \frac{0.00141V_c}{\alpha_{d,max}^{1/3}\left(\alpha_{d,max}^{1/3} - \alpha_d^{1/3}\right)}\exp\left(1 - 0.3\frac{\alpha_{d,max}^{1/3}\alpha_{d,c}^{1/3}}{\alpha_{d,max}^{1/3} - \alpha_d^{1/3}}\right) \\ + 0.002c_d^{drag}\left(V_d - V_c\right) \end{bmatrix}$$

based on data for upwards and downwards air water flow in vertical pipes with 25.5 and 50.8mm ID for j_c = 0.62 to 5m/s and j_d =0.004 to 1.977m/s at 6.9bar, where $\alpha_{d,max}$ =0.75. The second term in the brackets takes into account the bubble coalescence due to wake entrainment. *Morel, Yao* and *Bestion* [32] developed in the same time the following correlation for the above discussed case

$$\left(f_{d,col}^{p} f_{d,coal}^{p}\right)^{o} = 2.86 \frac{\varepsilon_{c}^{1/3} \alpha_{d}^{2}}{D_{d}^{11/3}} \frac{\exp\left(-1.017\sqrt{We_{d}'/We_{d\infty}'}\right)}{\frac{\alpha_{d,max}^{1/3} - \alpha_{d}^{1/3}}{\alpha_{d,max}^{1/3}} + 1.992 \alpha_{d} \sqrt{We_{d}'/We_{d\infty}'}},$$

where $We_{d\infty}' = 1.24$, $\alpha_{d,max} = 0.52$, $We_{d}' = 2 D_{d}^{5/3} \varepsilon_{c}^{2/3} \rho_{c} / \sigma_{d}$.

9.8 Coalescence probability of small droplets

Consider liquid droplets that follow the turbulent motion exactly. In this case the turbulent pulsation causes particle collisions and under some circumstances coalescence. As already discussed before, large particles delay in following the pulsation in the gas continuum. Larger than a given size they do no react any more to the gas pulsations.

Again as for the bubbles coalescence, the assumption of small particles results in the use of Eqs. (9.80) and (9.81) for computation of the pulsation velocity and collision time. *Meusel* [17, 19] (1980, 1989) proposed the following functional relationship for the coalescence probability

$$\left(f_{d,coal}^{p}\right)^{o} \approx const \left(\Delta \tau_{col} / \Delta \tau_{coal}\right)^{1/3}, \quad \text{for} \quad \Delta \tau_{col} / \Delta \tau_{coal} \geq 1, \qquad (9.97)$$

where

$$\Delta \tau_{col} \approx D_{d} V^{rel} \qquad (9.104)$$

is the time available for a possible collision, and $\Delta \tau_{coal}$ is the time needed to complete the coalescence. *Meusel* found by comparison with experiments with bubble flows for the constant the following value

$$const = 0.032. \qquad (9.105)$$

Next we estimate the coalescence time. The coalescence time for droplets should be comparable with the period of oscillation, Eqs. (8.72) and (8.73). We assume that after two droplets are touching each other the maximum unstable diameter is $D_{d,max} \approx 2 D_{d}$ - see the photographs of droplet collisions presented in Figs. 3a,b given by *Schelle* and *Leng* in [25] (1971). The lowest limit for the diameter of the newly produced oscillating droplet is assumed to be $D_{d,min} \approx D_{d}$. For $n = 2$ Eq. (8.72) results in

$$\Delta \tau_{coal} \approx 0.64 \left[\frac{D_d^3 (3\rho_d + 2\rho_c)}{\sigma_d} \right]^{1/2}. \qquad (9.106)$$

Therefore we have

$$\Delta \tau_{col} / \Delta \tau_{coal} \approx 1.56 \left[\frac{\sigma_d}{D_d^3 (3\rho_d + 2\rho_c)} \right]^{1/2} / V^{rel}. \qquad (9.107)$$

For comparison we give here the coalescence time for water droplets as correlated by *Jeffreys* and *Hawksley* [9] (1965)

$$\Delta \tau_{coal} = 1.96 \eta_2^{1/2} \Delta \tau_{1/2} \quad (\approx 1.01 \Delta \tau_{1/2}), \qquad (9.108)$$

where

$$\Delta \tau_{1/2} = 4.53 \times 10^5 \left[\frac{\eta_d^{1/2} (\rho_d - \rho_c)^{1.2}}{\sigma^2} \left(\frac{T_d - 273.15}{25} \right)^{-0.7 \eta_d^{1/2}} D_d^{0.021} \left(\frac{\sigma^2}{\eta_d^{1/5}} \right)^{0.55} \right]^{0.91}$$

$$\qquad (9.109)$$

is the half time period $(\Delta \tau_{1/2} \to 10s)$ (see *Pfeifer* and *Schmidt* [20] p. 29). Obviously the experiment confirms the trend of the dependence of $\Delta \tau_{coal}$ on D_d, ρ_d and σ_d.

Again the key parameter controlling agglomeration is the dissipation rate of the turbulent kinetic energy, ε_c, of the continuous field.

Nomenclature

Latin

c_{cd}^d	drag coefficient, *dimensionless*
c_{cd}^{vm}	virtual mass coefficient, *dimensionless*
D_h	pipe diameter, *m*
$D_{d\infty}$	final diameter after the fragmentation, *m*
D_d	size of a particle inside a continuum, *m*
$D_{d,\lim}$	coalescence escape limit diameter dividing the regimes of coalescence and no coalescence, *m*

F	force acting from inside of the bubble on the common surface
f	$=\sqrt{V'^2}/\ell_{ec}$, characteristic frequency of the turbulent flow, 1/s
$\left(f_{d,coal}^p\right)^o$	probability of coalescence, efficiency of coalescence, *dimensionless*
ℓ	eddies size, *m*
ℓ_e	small scale eddies (contains only a small part of the kinetic energy of turbulent motion), *m*
ℓ_0	inner scale or micro scale of turbulence, *m*
p	pressure, *Pa*
r	radius of the common surface of two bubbles at the moment of coalescence, *m*
Re'_ℓ	$=\dfrac{V'_\ell \ell}{v}$, turbulence *Reynolds* number, *dimensionless*
Re'_{ℓ_e}	$=\dfrac{V'_\ell \ell_e}{v}$, turbulent *Reynolds* number, *dimensionless*
$\bar{R}_u, \bar{R}_v, \bar{R}_w$	pressure drops in each particular direction per unit length, *Pa/m*
t	$=\rho_c V_c^2$, turbulent shear stress, *N/m²*
u, v, w	mixture velocity components in r, θ and z directions, *m/s*
V'_{l_e}	velocity change over the distance ℓ_e, *m/s*
V'_c	spatial velocity fluctuation at the lower end of the equilibrium range, *m/s*
V'_ℓ	velocity difference over the scale of the eddies ℓ for large scale motion, *m/s*
V	velocity, *m/s*
$We'_{d\infty}$	=turbulent shear force / surface tension force, critical *Weber* number, *dimensionless*

Greek

α_d	particle volume fraction, *dimensionless*
$\Delta\tau_{col}$	collision time, *s*
$\Delta\tau_{coal}$	coalescence time interval, *s*
$\Delta\tau_{br}$	time interval in which the fragmentation occurs, *s*
$\Delta w_{cd,\max}$	maximum relative velocity, *m/s*
δ	radial film thickness between two colliding bubbles, *m*
δ_{init}	initial film thickness, *m*
δ_{crit}	final thickness, *m*
ε	turbulence dissipation rate, mechanical energy of turbulence dissipated per unit time and unit mass of the continuum, *W/kg*
Φ_{c0}^2	two-phase multipliers for each particular direction, *dimensionless*

ρ density, kg/m^3
ρV total mass flow rate, $kg/(m^2 s)$
λ_{Ru}, λ_{Rv}, λ_{Rw} friction factors for each particular direction, *dimensionless*
η dynamic viscosity, $kg/(sm)$
η^t $= \rho V_l' \ell$, effective dynamic viscosity of turbulence, $kg/(sm)$
v^t $= V_l' \ell$, effective kinematic viscosity of turbulence, m^2/s
τ'_{l_e} $= \dfrac{\ell_e}{V'_{l_e}}$, characteristic time period of the fluctuation with given size ℓ_e, s

Subscripts

l = 1 first fluid – gas
       = 2 second fluid – liquid
       = 3 third fluid - another liquid
d disperse
c continuous

References

1. Bello JK (1968) Turbulent flow in channel with parallel walls, Moskva, Mir, in Russian
2. Bhavaraju SM, Russel TWF, Blanch HW (1978) The design of gas sparged devices for viscous liquids systems, AIChE J., vol 24 p 3
3. Calderbank PH (October 1967) Gas absorption from bubbles, The Chemical Engineer CE209-CE233
4. Clay PH (1940) Proceedings of the Royal Academy of Science (Amsterdam), Vol. 43 p 852
5. Coulaloglu CA, Tavlarides LL (1977) Description of Interaction Processes in Agitated Liquid - Liquid Dispersion, Chem. Eng. Sci., vol 32 p 1289
6. Hibiki T, Ishii M (August 15-17, 1999) Interfacial area transport of air-water bubbly flow in vertical round tubes, CD proc. of the 33rd Nat. Heat Transfer Conf., Albuquerque, New Mexico
7. Hinze JO (1955) Fundamentals of hydrodynamics of splitting in dispersion processes, AIChE Journal, vol 1 pp 284-295
8. Jeffreys GV, Hawksley L (1965) Coalescence of liquid droplets in two-component - two-phase systems, Amer. Inst. Chem. Engs. J., vol 11 p 413
9. Kim WK, Lee KL (1987) Coalescence behavior of two bubbles in stagnant liquid, J. Chem. Eng. Japan, vol 20 pp 449-453
10. Kirkpartick RD, Lockett MJ (1974) The influence of the aproach velocity on bubble coalescence, Chem. Eng. Sci., vol 29 pp 2363-2373
11. Kocamustafaogullari G, Huang W D, Razi J (1994) Measurements and modelling of average void fraction, bubble size and interfacial area, Nucl. Eng. Des., vol 148 p 437
12. Kocamustafaogullari G, Ishii M (1995) Foundation of the interfacial area transport equation and ist closure relations, Int. J. Heat Mass Transfer, vol 38 no 3 pp 481-493

13. Kolmogoroff AN (1941) The local structure of turbulence in incompressible viscous fluid for very large Reynolds numbers, C. R. Acad. Sci. U.S.S.R., vol 30 pp 825-828
14. Kolmogoroff AN (1949) On the disintegration of drops in turbulent flow, Doklady Akad. Nauk. U.S.S.R., vol 66 p 825
15. Lamb MA (1945) Hydrodynamics. Cambridge, At the University Press
16. Levich VG (1962) Physicochemical hydrodynamics, Prentice-Hall, Inc. Englewood Cliffs, N.J.
17. Meusel W (1980) Einfluß der Partikelkoaleszenz auf den Stoffübergang in turbulenten Gas-flüssig-Systemen, Ph. D. Thesis, Ingenieurhochschule Köthen
18. Meusel W (1989) Beitrag zur Modellierung von Gas - Flüssigkeits - Reaktoren auf der Basis relevanter Mikroprozesse, Doctoral Thesis, Ingenieurhochschule Köthen
19. Meusel W (1989) Beitrag zur Modellierung von Gas - Flüssigkeits - Reaktoren auf der Basis relevanter Mikroprozesse, Doctoral Thesis, Ingenieurhochschule Köthen
20. Pfeifer W, Schmidt H (April 1978) Literaturübersicht zu den fluiddynamischen Problemen bei der Auslegung gepulster Siebböden–Kolonnen, Kernforschungszentrum Karlsruhe, KfK 2560
21. Oolman T, Blanch HW (1986) Bubble coalescence in air-sparged bio-reactors, Biotech. Bioeng., vol 28 pp 578-584
22. Oolman T, Blanch HW (1986) Bubble coalescence in stagnant liquids, Chem. Eng. Commun., vol 43 pp 237-261
23. Reichardt H (Mai/Juni 1942) Gesetzmäßigkeiten der freien Turbulenz, VDI-Forschungsh. Nr. 414, Beilage zu "Forschung auf dem Gebiet des Ingenieurwesens", Ausgabe B, Band 13
24. Rosenzweig AK, Tronov VP, Perguschev LP (1980) Coaliszencija Kapel Vody v Melkodispersnyh Emulsijach Tipa Voda v Nevti, Journal Pricladnoj Chimii, no 8 pp 1776-1780
25. Schelle GF, Leng DE (1971) An experimental study of factors which promote coalescence of two colliding drops suspended in Water-I, Chemical Engineering Science, vol 26 pp 1867-1879
26. Sevik M, Park SH (1973) The splitting of drops and bubbles by turbulent fluid flow, Trans. ASME J. Fluid Engs., vol 95 p 53
27. Smoluchowski M (1918) Versuch einer mathematischen Theorie der Koagulationskinetik kolloider Lösungen, Zeitschrift für Physikalische Chemie, Leipzig Band XCII pp 129-168
28. Taylor GI (1935) Proc. Roy. Soc. A, vol 151 p 429. See also (1950) The instability of liquid surface when accelerated in a direction perpendicular to their plane, Proc. Roy. Soc. A, vol 201 pp 192-196
29. Thomas RM (1981) Bubble coalescence in turbulent flows, Int. J. Multiphase Flow, vol 6 no 6 pp 709-717
30. Yaglom AM (1949) Doclady AN SSSR, vol 67 no 5
31. Paranjape SS et al (5-9 October 2003) Interfacial structure and area transport in upward and downward two-phase flow, 10th Int. Top. Meeting on Nuclear Reactor Thermal Hydraulic (NURETH-10) Seul, Korea
32. Morel C, Yao W and Bestion D (5-9 October 2003) Three dimensional modeling of boiling flow for NEPTUNE code, 10th Int. Top. Meeting on Nuclear Reactor Thermal Hydraulic (NURETH-10) Seul, Korea

10. Liquid and gas jet disintegration

10.1 Liquid jet disintegration in pools

The mathematical description of jet breakup has attracted the attention of outstanding scientists in the past. Lord *Rayleigh* analyzed for the first time in 1878 [16] the instability of jets. *Nils Bohr* extended *Rayleigh*'s analysis to include viscous effects, in a prize winning paper on the evaluation of the surface tension in 1909 [2]. *Constantin Weber* went on to obtain the breakup length for a viscous jet in 1936 [35]. *Wolfgang von Ohnesorge* classified in 1936 [24] his experimental observation with a high speed camera (200 to 12 000 frames per second) of four different regimes of jet breakup, introduced a dimensionless number quantifying the properties of the jet and described successfully two jet transition boundaries. *Weber* in 1936 [35] and later *Taylor* in [1] introduced in addition to the previous analyses the influence of the environment on the jet breakup. These are the fundamental works in jet fragmentation theory used in almost all later works on this topic. In recent time the nozzle geometry is found to influence strongly the jet dynamics. Geometries not allowing the establishment of a turbulent boundary layer produce more stable jets than those which promote turbulent boundary layers at the jet interface - see *Iciek* [17].

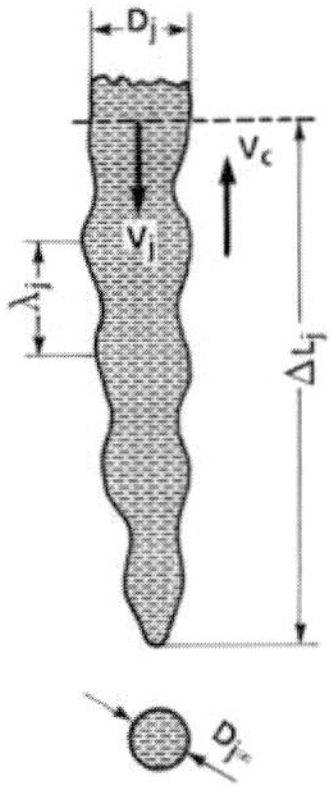

Fig. 10.1. Characteristics of jet disintegration in a pool

A jet disintegrates due to internally or externally excited and unboundedly growing instabilities, due to surface entrainment or due to all mechanisms working together with different strength. The products of the fragmentations are primary ligaments which may continue to fragment to secondary droplets.

We use the *Ohnesorge* classification of the jet fragmentation mechanism (see Fig. 10.1).

0) Slow dropping from the nozzle under the gravity influence without jet formation (depends on the density ratio);

I) Cylindrical jet disintegration due to symmetric interface oscillations as discussed by *Rayleigh* (later called varicose regime). In this regime the relative velocity has no influence and increasing jet velocity causes increasing penetration length.

II) Jet disintegration due to asymmetric waves which deform the core to the snake-like shape as discussed by *Weber* and *Haenlein* (called sinuous jet breakup);

III) Jet atomization due to internal turbulization and surface entrainment.

The dimensionless numbers required to describe jet dynamics are given below.

$\Delta L_j / D_j$ relative jet penetration length

$We_j = \rho_j V_j^2 D_j / \sigma$, jet *Weber* number

$We_{cj} = \rho_j \Delta V_{cj}^2 D_j / \sigma$, continuum *Weber* number based on the relative velocity

$Fr_{cj} = \Delta V_{cj}^2 / (g D_j)$, *Froud* number based on the relative velocity, appropriate for jets causing boiling at the interface so that the environment velocity is governed by buoyancy driven convection

$Fr_j = V_j^2 / (g D_j)$, *Froud* number based on the jet velocity, appropriate for gravitationally driven jets

$Ta_{cj} = (\rho_j / \rho_c) [\sigma / (\eta_j \Delta V_{cj})]^2$, *Taylor* number based on relative velocity

$Ta_c = (\rho_j / \rho_c) [\sigma / (\eta_j V_c)]^2$, *Taylor* number based on the continuum velocity

$Lp_j = \rho_j \sigma_j D_j / \eta_j^2$, *Laplace* number

$Re_j = \rho_j V_j D_j / \eta_j$, liquid jet *Reynolds* number

$On_j = We_j^{1/2} / Re_j = 1 / Lp_j^{1/2} = \eta_j / (D_j \sigma_j \rho_j)^{1/2}$, jet *Ohnesorge* number [25]

$On_{j\lambda} = \eta_j / (\lambda_{RT} \sigma_j \rho_j)^{1/2}$, jet viscosity number based on *Rayleigh-Taylor* instability wavelength

$On_j^* = (3\eta_j + \eta_c) / (D_j \sigma_j \rho_j)^{1/2}$, modified *Ohnesorge* number

$On_c = \eta_c / (D_j \sigma_j \rho_c)^{1/2}$, ambient *Ohnesorge* number

Here we denote with j the first continuum and with c the second continuum. The jet penetration length, ΔL_j, is defined as the length for complete disintegration measured from the nozzle outlet. Note that the jet velocity, V_j, is not necessarily equal to the relative velocity between the jet and the environment, ΔV_{cj}. Both velocities are equal for relatively slow jet penetration in a stagnant continuum.

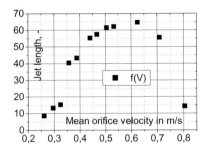
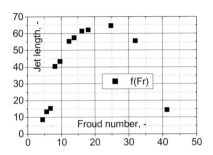
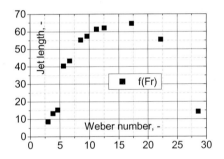

Fig. 10.2. Jet penetration length $\Delta L_j / D_j$ as a function of jet velocity in three different presentations. Liquid-liquid jet fragmentation experiment by Meister and Scheele. Heptane in water: $D_j = 0.0016m$, $\eta_j = 0.000393\ Pa\ s$, $\rho_j = 683 kg/m^3$, $\sigma_j = 0.0362 N/m$

Figure 10.2 digitized from [21] gives possible regimes of jet breakup. The penetration length $\Delta L_j / D_j$ is presented as a function of the nozzle velocity (alternatively as a function Fr_j, or $We_{cj} = We_j$ for liquid jet penetration from the bottom of a reservoir filled with stagnant liquid.

To model jet fragmentation for engineering applications we need to know

(a) the boundaries of the different fragmentation mechanisms,
(b) the size of the generated particles, and

(c) the dependence of the jet penetration length on the jet velocity and other local parameters.

10.2 Boundary of different fragmentation mechanisms

a) The regime (0) is seldom interesting for practical application.

b) The transition from regime (I) to regime (II) can be deduced from the *Ohnesorge* diagram [24]

$$Re_j = 46/On_j^{4/5}. \tag{10.1}$$

c) In the same work *Ohnesorge* found for the transition between regime (II) and regime (III)

$$Re_j = 270/On_j^{4/5}. \tag{10.2}$$

This transition criterion was modified in 1947 by *Merrington* and *Richardson* [22], $Re_j = 300/On_j^{4/5}$, in 1954 by *Tanazawa* and *Toyoda* [32], $Re_j = 370/On_j^{0.318}$, and later by *Grant* and *Middleman* [15] based on their own experiments, $Re_j = 325/On_j^{0.28}$. The transition boundary from sinuous to turbulent jets was demonstrated to happen for $Re_j \approx (2 \text{ to } 5) \times 10^4$, for water which corresponds to the *Ohnesorge* experiment performed in 1931 and the result obtained by *Lienhard* and *Day* compare with Fig.5 in [19] where $Re_j \approx (4 \text{ to } 6) \times 10^4$. For liquid-liquid systems additional dependence of the ambient condition was reported by *Takahashi* and *Kitamura* [30, 31],

$$Re_j = \left(98/On_c^{0.11}\right)/On_j^{0.30}. \tag{10.3}$$

Grant and *Middleman* [15] indicated that it may depend on the system geometry which can promote or dampen turbulence in the jet surface. *Iciek* [17] reported a data set for nozzles with different geometries that strongly supports this conclusion. *Faeth* [11, p. C0-11, fig.15] demonstrated experimentally that the liquid jet *Reynolds* number, and the nozzle length to diameter ratio, govern the internal structure of the jet turbulence. For

$$L_{nozzle}/D_j < 5, \tag{10.4}$$

the jet remain laminar. The higher the nozzle length the better the turbulence develops.

In regime (II) the influence of the ambient conditions starts to play an important role. It is known that water jets with coaxial gas injection producing gas velocity equal to the jet velocity are more stable than jets with the same velocity in stagnant gas. The transition to regime (III) can be accelerated if the ambient relative velocity increases. It is associated by some authors with some critical *Weber* number. *Fenn* and *Middleman* [12] found experimentally that for ambient *Weber* number

$$We_{cj} < 5.3, \qquad (10.5)$$

there is no influence of the ambient aerodynamic pressure forces on the fragmentation. For larger We_{cj} the surrounding continuum conditions begin to reduce the core breakup length.

Smith and *Moss* [28, see Fig. 8] found in 1917 experimenting with mercury in aqueous solutions that for

$$We_{cj} > 100 \qquad (10.6)$$

atomization starts. In the atomization regime increasing relative velocity decreases the penetration length. The atomization is associated with dramatic surface droplet entrainment. *De Jarlais* et al. [9] (1986) found that at

$$On_{j\lambda}^{0.8} Ta_{cj}^{1/2} \leq 1 \quad \text{for} \quad On_{j\lambda} < 1/15, \qquad (10.7)$$

and

$$0.1146 Ta_{cj}^{1/2} \leq 1 \quad \text{for} \quad On_{j\lambda} \geq 1/15, \qquad (10.8)$$

roll wave entrainment starts [9]. Higher relative velocities cause droplet entrainment which thin the jet. Breakup may occur in such cases before the core exhibits any sinuous jet behavior.

Conclusions:

1. Use Eq.(10.1) to recognize the transition varicose/sinuous.
2. Use Eq.(10.2) or its recent modifications to recognize the transition sinuous/atomization.
3. Check by Eq.(10.5) whether the ambient conditions are important.
4. Check by Eqs.(10.7) and (10.8) whether surface entrainment starts.
5. Decide in favor of the regime that is most stable at the given local conditions.

10.3 Size of the ligaments

The size of the dropping liquid is observed by *Ohnesorge* [24] to be in order of

$$Lp_j \approx 0.01 \text{ to } 1. \tag{10.9}$$

The size of the ligaments resulting from primary breakup for inviscid fluid is estimated by *Rayleigh*

$$D_{d\infty}/D_j = 9.02. \tag{10.10}$$

Timotika [34] found in 1935 theoretically that the jet properties influence the size of the ligaments for the varicose and sinuous breakup as follows

$$D_{d\infty}/D_j = 13 On_j^{1/2}. \tag{10.11}$$

Recently the size of the gross ligaments resulting from the primary varicose or sinuous breakup has been correlated by *Teng* et al. [33] by considering the influence of the ambient continuum properties in addition to the jet properties as follows

$$D_{d\infty}/D_j = 1.88\left(1.1303 + 0.0236 \ln On_j^*\right)\left(1 + On_j^*\right)^{1/6}. \tag{10.12}$$

This correlation was compared with data for seventeen *Newtonian* and power - law/*Newtonian* liquid-in-liquid systems. The averaged deviation was 8.1 and 5.1%, respectively. For water jets in air, the averaged deviation was 1.6%. The secondary breakup stability criterion corresponds to that already discussed assuming the new born primary drop has 0.9 from the jet velocity as experimentally measured by *Faeth* in [11].

Turbulence inside the jet can produce surface instabilities and therefore surface entrainment. This fact is widely used in different technologies to fragment jets in smaller particles. Simple approach to the size of the particles jumping apart the surface, $D_{d\infty}$, is to equate the pulsation pressure at the interface, $\frac{1}{2}\rho_j V_j^{\prime 2} \pi D_{d\infty}^2/4$, to the surface tension force, $\pi D_{d\infty} \sigma$, which results in the *Weber* number stability criterion based on the pulsation velocity equal to 8. In fact, as already discussed in Section 9.4, this is of the order of 1.17 to 2.48, and therefore, $\rho_j V_j^{\prime 2} D_{d\infty}/\sigma \approx (1.17$ to $2.48)$. Having in mind that the pulsation velocity is only a part of the jet velocity, $V_j' \approx 0.07 V_j$, see *Faeth* [11], we obtain simply

$$D_{d\infty}/D_j \approx (240 \text{ to } 500)/We_j, \qquad (10.13)$$

which is not a function of the jet diameter. This simple estimation is in quite good agreement with the experimental observation reported by *Faeth* [11]

$$D_{d\infty}/D_j \approx 133/We_j^{0.74}, \qquad (10.14)$$

which shows some dependence of the jet diameter.

Thus, I recommend to take as a final size after the primary and secondary fragmentation the minimum of all relevant mechanisms. The justification is: If the inertia fragmentation produces the smaller particles they will be stable in the environment. If the relative velocity method predicts the smaller particle size the primary entrained droplet will further fragment to reach a stable state following relative velocity fragmentation.

10.4 Unbounded instability controlling jet fragmentation

10.4.1 No ambient influence

Lord Rayleigh [16] found in 1878 by perturbation analysis for the varicose mode of water jet breakup

$$\Delta L_j/D_j = const\ We_j^{1/2}, \qquad (10.15)$$

where $const = 8.46$ for $Lp_j^{1/2} \gg 1$. The experiments performed by *Smith* and *Moss* [28, p 388] in 1917 shows that the constant is

$$const = 13. \qquad (10.16)$$

Weber [35] analyzed in 1931 the stability in the varicose mode of vibrations. His remarkable result for the fragmentation time is

$$\Delta \tau_j = \ln\left[D_j/(2\lambda_0)\right]\left(\rho_j D_j/\sigma_j\right)^{1/2} D_j \left(1+3On_j\right) \qquad (10.17)$$

or having in mind that $\Delta \tau_j = \Delta L_j/V_j$

$$\Delta L_j/D_j = \ln\left[D_j/(2\lambda_0)\right]\left(We_j^{1/2} + 3We_j/Re_j\right)$$

$$= \ln\left[D_j/(2\lambda_0)\right]\left(1+3On_j\right)We_j^{1/2}, \qquad (10.18)$$

where the symbol λ_0 denotes the initial deviation from the non-disturbed jet radius and $\ln\left[D_j/(2\lambda_0)\right]$ is assumed to be equal 12. This result compares well with data for $10^{-4} \le \frac{1}{9}/On_j^2 < 10^4$. Remarkably, the result reduces to the *Rayleigh* equation (10.15) with a constant very similar to those previously estimated by *Smith* and *Moss* [28]. *Grant* and *Middleman* [15] found experimentally that the excitation of interface instability is also a function of the jet properties and fitted their data with the correlation

$$\ln\left[D_j/(2\lambda_0)\right] = 7.68 - 2.66\ln On_j. \qquad (10.19)$$

The authors correlated their own experimental data in the region of $2 < We_j^{1/2} + 3We_j/\mathrm{Re}_j < 200$ with the slightly modified form of the *Weber* result, namely

$$\Delta L_j/D_j = 19.5\left[\left(1+3On_j\right)We_j^{1/2}\right]^{0.85} \qquad (10.20)$$

for $\mathrm{Re}_j < 325/On_j^{0.28}$, and

$$\Delta L_j/D_j = 8.51\left(We_j^{1/2}\right)^{0.64} \qquad (10.21)$$

for larger jet *Reynolds* numbers - turbulent jets. For the last regime *Iciek* [17] obtained a slightly modified form

$$\Delta L_j/D_j = 11.5\left(We_j^{1/2}\right)^{0.62} \qquad (10.22)$$

on the base of an additional 56 data points.

10.4.2 Ambient influence

Taylor, see in [1], starts with the continuity equation for a steady state jet

$$d\left(V_j D_j^{*2}\right)/dz = -4D_j^* u_{j\sigma} \quad \text{or} \quad dD_j^*/dz = -2u_{j\sigma}/V_j \qquad (10.23)$$

Here D_j^* is a function of z, V_j is the jet velocity, assumed to be constant, and $u_{j\sigma}$ is the interface averaged entrainment velocity. The continuity equation, integrated for $z = 0$, $D_j^* = D_j$ and $z = \Delta L_j$, $D_j^* = 0$ and constant V_j and $u_{j\sigma}$, gives

$$\Delta L_j / D_j = \frac{1}{2} V_j / u_{j\sigma}. \qquad (10.24)$$

The interface averaged entrainment velocity is computed as

$$u_{j\sigma} = const \left(\rho_c / \rho_j \right)^{1/2} \Delta V_{cj} \lambda_m^* f_m^* \qquad (10.25)$$

where λ_m^*, defined by

$$2\pi \lambda_m^* = \lambda_m \rho_c \Delta V_{cj}^2 / \sigma, \qquad (10.26)$$

and f_m^*, defined by

$$2 f_m^* = f_m / \left[\left(\rho_c / \rho_c \right)^{1/2} \rho_c \Delta V_{cj}^3 / \sigma \right], \qquad (10.27)$$

are the dimensionless wavelength and the frequency of the fastest growth of the unstable surface perturbation waves which are complicated functions of Ta_{cj}, being $f_m^* \le \sqrt{3}/9$ and $0.04 \le \lambda_m^* f_m^* \le \sqrt{3}/6$ for $10^{-5} \le Ta_{cj} \le 10^4$ and $\lambda_m^* f_m^* = \sqrt{3}/6$ for $Ta_{cj} > 10^4$. Substituting Eq. (10.25) in Eq. (10.24) results in

$$\Delta L_j / D_j = \frac{1}{2} V_j / \left[const\, \lambda_m^* f_m^* \left(\rho_c / \rho_j \right)^{1/2} \Delta V_{cj} \right], \qquad (10.28)$$

which is a remarkable result. Comparison with experimental data for engine spray where $\Delta V_{cj} = V_j$, indicates that $\frac{1}{2} / \left(const\, \lambda_m^* f_m^* \right)$ is of order of 7, see *Bracco* [3]. Rewritten in terms of *Weber*'s numbers the *Taylor* equation takes the form

$$\Delta L_j / D_j = 7 We_j^{1/2} / We_{cj}^{1/2}. \qquad (10.29)$$

Epstein and *Fauske* [10] (1985) investigated theoretically the stability of a liquid jet in a continuum, making the following assumptions: The wavelength of the surface disturbance is small compared with the diameter of the jet. Considering linear *Kelvin - Helmholtz* stability of three parallel inviscid streams of jet, steam film and liquid, the authors obtained the following expression for the jet penetration distance:

$$\Delta L_j / D_j = \frac{\sqrt{3}}{2}\left(1+\rho_c/\rho_j\right)/\left(\rho_c/\rho_j\right)^{1/2}. \tag{10.30}$$

which for $\rho_c \ll \rho_j$ reduces to the *Taylor* result for $\Delta V_{cj} = V_j$ giving the lowest possible estimate for the constant $\sqrt{3}/2$. *Nigmatulin* et al. [23] reported similar analysis for cylinder geometry coming to an expression very similar to the above mentioned *Rayleigh* result. The authors showed that the *Epstein* and *Fauske* analysis holds for jet diameters larger the 1 *cm*, see [10, p 94] and pointed out the importance of the accurate estimation of the surrounding continuous medium in case of film boiling. *De Jarlais* et al. [9] (1986) investigated jet disintegration in confined pipe geometry. The authors find that the confinement influences the disintegration process and corrected the existing correlation for jet fragmentation by introducing the factor $(1-\alpha_j)$, which reduces to unity for jets in an infinite environment. The obtained correlations are: For $We_{cj}/(1-\alpha_j)^2 < 5.3$, the modified *Weber* result was recommended

$$\Delta L_j / D_j = 12\left(1+3On_{j\lambda}\right)We_j^{1/2}, \tag{10.31}$$

and for $We_{cj}/(1-\alpha_j)^2 \geq 5.3$ the *Grant* and *Middleman* [15] result was recommended

$$\Delta L_j / D_j = 35\left(1+3On_{j\lambda}\right)We_j^{1/2} / \left[We_{cj}/(1-\alpha_j)^2\right]^{0.645}. \tag{10.32}$$

The last correlation contains elements from *Weber*'s and from *Taylor*'s theory.

There are similar correlations to *Taylor*'s, giving decreasing $\Delta L_j / D_j$ with increasing We_{cj} and taking into account the viscosity, e.g.

$$\Delta L_j / D_j = 50.3 On_j^{0.192} / \left[\left(\rho_c/\rho_j\right)^{1/2} We_{cd}^{0.83}\right]. \tag{10.33}$$

Conclusion: For the time being the following approach seems to be meaningful.

(1) Compute the ambient *Weber* number defining the transition into atomization as follows

$$We_{cj}^* = 49 / \left[\left(7.68 - 2.66 \ln On_j \right) \left(1 + 3On_j \right) \right]^2. \quad (10.34)$$

This is a very interesting result. It simply says that for non-viscous jets the transition's ambient *Weber* number is of the order of $We_{cj}^* \approx \rho_j / \rho_c$, which means that jets in gas environment experience transitions into the atomization regime for much higher relative velocities than jets in liquids.

(2) For $We_{cj} \leq We_{cj}^*$ use the *Weber* solution with the *Smith* and *Moss* coefficient for non-atomization regimes,

$$We_{cj}^* = \left(7.68 - 2.66 \ln On_j \right) \left(1 + 3On_j \right) We_j^{1/2}. \quad (10.35)$$

(3) For $We_{cj} > We_{cj}^*$ use the *Taylor* solution with the *Braco* constant, that is Eq. (10.29) to estimate the coherent jet length during atomization.

10.4.3 Jets producing film boiling in the ambient liquid

Now we consider a situation which differs from the above discussed. Consider a jet of molten metal, e.g. iron penetrating water and causing considerable evaporation. In this case the jet surrounding continuum is vapor. Following *Schneider* et. al. [27] (1992) and assuming that (a) the surface entrainment velocity is equal to the jet side surface friction velocity,

$$u_\sigma^* = u_{j\sigma} = \left(\tau_{j\sigma} / \rho_j \right)^{1/2}, \quad (10.36)$$

and (b) the jet side surface shear stress is equal to the vapor side shear stress due to buoyancy driven upwards gas flow

$$\tau_{j\sigma} = c_{1d}^d \frac{1}{2} \rho_1 \Delta V_{1j}^2 \approx \left(D_j^* / 4 \right) \left(\rho_2 - \rho_1 \right) g, \quad (10.37)$$

the jet mass conservation equation reduces to

$$dD_j^* / dz = -2u_{j\sigma} / V_j = -2 \left[\left(D_j^* / 4 \right) \left(\rho_2 - \rho_1 \right) g / \rho_j \right]^{1/2} / V_j, \quad (10.38)$$

or after integrating for $z = 0$, $D_j^* = D_j$ and $z = \Delta L_j$, $D_j^* = 0$ and constant V_j

$$\Delta L_j / D_j = 2 Fr_j / \left[\left(\rho_2 - \rho_1 \right) / \rho_j \right]^{1/2}. \quad (10.39)$$

Low pressure experiments performed by *Saito* et al. [26] (1988) show that the penetration length of the very hot jet into coolant with strong evaporation tends to increase with the jet velocity. These authors correlated their data with the expression

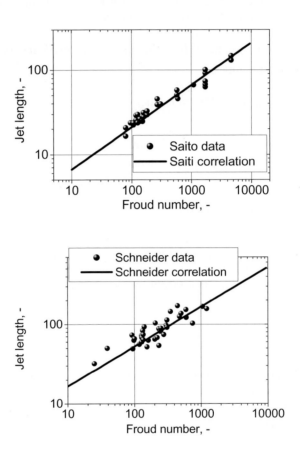

Fig. 10.3. (a) Film boiling jet penetration length $\left(\Delta L_j / D_j\right)\left(\rho_c / \rho_j\right)^{1/2}$ as a function of the Froud number Fr_j. (b) Film boiling jet penetration length $\Delta L_j / D_j$ as a function of the Froud number Fr_j. Comparison between the prediction of the Saito correlation with the experimental data of Saito and Schneider.

$$\Delta L_j / D_j = 2.1 Fr_j^{1/2} / \left(\rho_2 / \rho_j\right)^{1/2} \tag{10.40}$$

for $Fr_j < 5 \times 10^3$, which is surprisingly close to the analytical solution. For application we recommend not to neglect the vapor density because it will restrict the validity of *Saito*'s finding only to low pressure. The correlation with film boiling (10.40) predicts larger penetration lengths than experimentally observed for adiabatic liquid-liquid penetration - see *Sun* [29] 1998

$$\Delta L_j / D_j = 1.45\, Fr_j^{1/2} / (\rho_2 / \rho_j)^{1/2}$$

valid for $90 < Fr_j < 2 \times 10^3$. *Schneider* et al. [27] (1992) provided a theoretical explanation of the observations made by *Saito* et al. Interface jet entrainment caused by the surrounding vapor phase was made responsible for this process as discussed above. In a similar way as the above the authors obtained the analytical equation

$$\Delta L_j / D_j = \frac{1}{2} Fr_j^{1/2} \left\{ \left[1 + 1 / \left(\frac{2}{5} Fr_j^{1/2} (\rho_2 / \rho_j)^{1/2} \right) \right]^{8/5} - 1 \right\} \qquad (10.41)$$

which can successfully be approximated by *Saito*'s empirical equation. Equation (10.41) is experimentally verified to $Fr_j < 1400$ or $We_j < 67$ - see Fig. 10.3. In accordance with *Saito* and *Schneider*'s experiments for $\rho_2 < \rho_j$, the denser the jet surrounding continuum, 2, the smaller the jet penetration distance ΔL_j.

Note: *Saito*'s type of correlations has to be used only for jets causing film boiling. Do not use them in other cases.

10.4.4 An alternative approach

The success of the simple *Taylor-Scheider* approach is very promising also for non boiling environment. The internal turbulent pulsation can cause also surface entrainment not influenced by the environment directly. That is why I recommend to use in the mass conservation equation the interface averaged entrainment velocity

$$u_{j\sigma} = u_\sigma^* + u_j' = \left(c_{cd}^d \frac{1}{2} \rho_c \Delta V_{cj}^2 / \rho_j \right)^{1/2} + cV_j. \qquad (10.42)$$

Here the interface averaged pulsation velocity $u_j' = cV_j$ is proportional to jet velocity. The proportionality factor can be of the order of 0.1, see *Faeth* [11]. The approximate solution of the mass conservation equations looks like

$$\Delta L_j / D_j = \frac{1}{2} V_j / u_\sigma^* = \frac{1}{2} V_j / \left\{ \left(c_{cd}^d / 2 \right)^{1/2} \left(\rho_c / \rho_j \right)^{1/2} \Delta V_{cj} + c V_j \right\}. \quad (10.43)$$

Equation (10.43) can be expressed in terms of *Weber* numbers

$$\Delta L_j / D_j = \frac{1}{2} We_j^{1/2} / \left[\left(c_{cd}^d / 2 \right)^{1/2} We_{cj}^{1/2} + c We_j^{1/2} \right]. \quad (10.44)$$

The above simple equations have some surprising properties. Neglecting the influence of the internal jet turbulence, $c = 0$, we obtain an equation very similar to the *Taylor* result, compare also with Eq. (2) published by *Buerger* et al. [4], for parallel flow surface instability. The friction drag coefficient depends on the interface waviness, which itself depends on the ambient relative velocity, and is expected to be less than 0.33. The constant 7, experimentally obtained by *Braco* in the *Taylor* equation, corresponds to the friction coefficient $c_{cd}^d \approx 0.01$. For jets in stagnant environment, $\Delta V_{cj} \approx V_j$, the above equation resembles the *Epstein* and *Fauske* [10] equation. For strong turbulized jets, $\left(c_{cd}^d / 2 \right)^{1/2} \left(\rho_c / \rho_j \right)^{1/2} \Delta V_{cj} \ll c V_j$, and where c is order of unity, $\Delta L_j / D_j = 1/(2c)$ is as expected, very short. Equating the predictions of Eqs. (10.35) and (10.44)

$$We_{cj}^* = \left(c_{cd}^d / 2 \right) \left\{ 1 / \left[2 \left(7.68 - 2.66 \ln On_j \right) \left(1 + 3 On_j \right) \right] - c We_j^{1/2} \right\}^2 \quad (10.45)$$

we find the transition to atomization boundary.

10.4.5 Jets penetrating two-phase mixtures

If a jet is penetrating a two-phase mixture in a sandwich structure, consisting of gas layers with overall thickness $\alpha_1^* \Delta L_j$, and liquid layers with overall thickness $(1 - \alpha_1^*) \Delta L_j$, where $\alpha_1^* = \alpha_1 / (1 - \alpha_1)$, the penetration length is

$$\Delta L_j / D_j = \frac{1}{2} V_j / \left\{ \begin{array}{l} \alpha_1^* \left(c_{1d}^d / 2 \right)^{1/2} \left(\rho_1 / \rho_j \right)^{1/2} \Delta V_{1j} \\ + \left(1 - \alpha_1^* \right) \left(c_{2d}^d / 2 \right)^{1/2} \left(\rho_2 / \rho_j \right)^{1/2} \Delta V_{2j} + c V_j \end{array} \right\}. \quad (10.46)$$

10.4.6 Particle production rate

The application of these correlations in the form of rate correlation in a space resembling a pool can be performed using Eq. (7.2) as follows. As already mentioned if the number of particles in the volume under consideration is less than one, the fluid should be considered as continuous, c. After a distance ΔL_j, the jets disintegrate in particles having size $D_{d\infty}$. Obviously a "marked control volume", being in the continuum jet and traveling the distance ΔL_j with the velocity V_j, should experience fragmentation. It means that the time interval needed by this "marked control volume" to disintegrate is of the order of $\Delta L_j / V_j$. This is approximately the characteristic time interval in which the jet fragmentation is completed $\Delta \tau_{br} \approx \Delta L_j / V_j$. Using this time and $D_{d\infty}$ we can estimate the particle production rate by means of Eq. (7.11).

10.5. Jet erosion by high velocity gas environment

There are some attempts in the literature to describe entrainment from liquid film for very high gas velocity. Next, we give two possible approaches proposed by *Mayer* [20] (1961) and *Chawla* [5, 6] (1975, 1976).

Mayer [20] (1961) investigated the interaction of a high velocity gas stream with a large liquid surface. For capillary surface waves (ripples), the change of the amplitude, A, with the time was described by

$$\frac{dA}{d\tau} = \frac{A}{\Delta \tau}, \tag{10.47}$$

where the frequency of the waves is a function of their wavelength besides other parameters

$$\frac{1}{\Delta \tau} = c_1 \sqrt{\pi/2} \frac{\rho_1 \Delta V_{12}^2}{\sqrt{\sigma/\rho_2}} \frac{1}{\delta^{1/2}} - 8\pi^2 \frac{\eta_2}{\rho_2} \frac{1}{\delta^2}, \tag{10.48}$$

using for the *sheltering* parameter $c_1 = 0.3$. Here $\Delta \tau$ is the time constant of the process and δ is the wavelength. *Mayer* [20] (1961) found that if the generated wavelengths are greater than a minimum value obtained by the condition $1/\Delta \tau = 0$,

$$\delta > \delta_{\min} = 2\pi \sqrt{16} \left(\frac{\eta_2 \sqrt{\sigma/\rho_2}}{c_1 \rho_1 \Delta V_{12}^2} \right)^{2/3}, \tag{10.49}$$

the capillary wave will grow unboundedly and the wave is eroded as a ligament from which droplets of size proportional δ,

$$D_{3E} \approx c_2 \delta, \qquad (10.50)$$

with $c_2 \approx 0.14$ are entrained. The average size of the entrained particles was found to be

$$\overline{D_{3E}} = 9\pi\sqrt[3]{16 c_1} \left(\frac{\eta_2 \sqrt{\sigma/\rho_2}}{c_1 \rho_1 \Delta V_{12}^2} \right)^{2/3}. \qquad (10.51)$$

The constants c_1 and c_2 are derived from a single experimental point. Later, *Wolfe* and *Andersen* [36] (1965) compared *Mayer*'s theory with their own data in the region

$$9 \times 10^{-5} < 9\pi\sqrt[3]{16 c_1} \left(\frac{\eta_2 \sqrt{\sigma/\rho_2}}{c_1 \rho_1 \Delta V_{12}^2} \right)^{2/3} < 0.0011, \text{ (see also Fig. 12 in [36])} \quad (10.52)$$

and found for the sheltering parameter $c_1 = 0.18$.

In the sense of *Mayer*'s theory the average number of the entrained droplets per unit surface is

$$\dot{n}_{23}'' = \frac{1}{\delta^2 \Delta \tau(\delta)}, \qquad (10.53)$$

where $\overline{\delta} = \overline{D_{3E}}$ from Eq. (10.51), $\Delta\tau$ is computed using Eq. (10.48). The erosion mass flow rate is then

$$(\rho w)_{23} = \rho_2 \frac{\pi}{6} \delta^3 \dot{n}_{23}'' = \rho_2 \frac{\pi}{6} \frac{\delta}{\Delta \tau(\delta)}. \qquad (10.54)$$

Chawla [5, 6] (1975, 1976) analyzed the instabilities of the interface of a gas jet entering a liquid and the resulting droplets entrainment. He found that when the amplitude of the disturbance becomes large enough, the liquid at the wave crests (protruded into the gas jets) is torn off by the gas jet. The size of the resulting droplets is governed by *Kelvin-Helmholz* instabilities

$$D_{3E} \approx c_0 \delta_{3E} + c_1 \delta_{3E}^*, \qquad (10.55)$$

where

$$We_{12E} = \delta_{3E}\rho_1 \Delta V_{12}^2 / \sigma_2 = 3\pi \left(1 - Ma^2\right)^{1/2}, \quad Ma < 1, \quad (10.56)$$

$$We_{12E}^* = \delta_{3E}^* \rho_1 \Delta V_{12}^2 / \sigma_2 = \frac{2\pi}{0.803}\left(\rho_1^* / \rho_2\right)^{1/5}, \quad Ma = 1. \quad (10.57)$$

Here, $Ma = \Delta V_{12} / V_1^*$ is the local *Mach* number and ρ_1^*, V_1^* are critical density and sonic gas velocity at the throat conditions for sonic gas jets, respectively. The constants $c_0 \approx 1.5$, and $c_1 \approx 27.14$ are estimated by comparison with experimental data. Thus for low *Mach* numbers the stability criterion is

$$We_{12E} \approx 14. \quad (10.58)$$

It seems that there is a lack of reliable data for the development of a more reliable correlation for high gas velocity entrainment.

10.6. Jet fragmentation in pipes

The liquid jet disintegration is also a kind of transition of flow with continuous liquid into flow with dispersed liquid. *De Jarlais* et al. [9] (1986) considered the hydrodynamic stability of inverted annular flow in an adiabatic situation as a jet disintegration problem in channels - see Fig. 10.4.

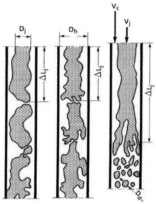

Fig. 10.4. Jet fragmentation in channels

After reviewing the jet stability literature and comparing with their own experimental data, the authors proposed the following correlation which represents their data for the jet core breakup length. For

$$We_{cj}/\alpha_c^2 < 1.73 \tag{10.59}$$

$$\Delta L_j/D_j = 480\, We_j^{1/2}/\mathrm{Re}_j^{0.53}, \tag{10.60}$$

where $\mathrm{Re}_j = \rho_j V_j D_j/\eta_j$ is the liquid jet *Reynolds* number and $We_j = \rho_j V_j^2 D_j/\sigma$ is the liquid jet *Weber* number. The averaged wavelength was found to be

$$\lambda_j \approx 5.8 D_j. \tag{10.61}$$

For

$$We_{cj}/\alpha_c^2 \geq 1.73 \tag{10.62}$$

$$\Delta L_j/D_j = 685\left(We_j^{1/2}/\mathrm{Re}_j^{0.53}\right)/\left(We_{cj}/\alpha_c^2\right)^{0.645}, \tag{10.63}$$

where $We_{cj} = \rho_c \Delta V_{cj}^2 D_h/\eta_c$ is the gas *Weber* number. The average wavelength, λ_j, was found to be

$$\lambda_j \approx 7.6\left(We_{cj}/\alpha_c^2\right)^{-1/2}. \tag{10.64}$$

Again the time needed for the fragmentation is

$$\Delta \tau_{br} \approx \Delta L_j/\Delta V_{cj}. \tag{10.65}$$

Thus the particle production rate in this case is

$$\dot{n}_3 \approx \alpha_j/\left(\frac{\pi}{4}D_j^2 \lambda_j \Delta \tau\right). \tag{10.66}$$

The entrainment of micro-droplets from the surface of the jet was also a subject of investigation by the above mentioned authors but the obtained information is not sufficient for reliable conclusions to be drown.

10.7. Gas jet disintegration in pools

There are many industrial processes in which gas is injected into a liquid. For mathematical modeling of such processes one needs quantitative information on

10.7. Gas jet disintegration in pools

how gas injected into the liquid through an orifice behaves. This information can be used as a boundary condition for modeling the processes in the volume of interest. That is why in this section we concentrate our attention on gas jet disintegration produced by orifices.

Consider vertical gas injection into a stagnant liquid pool. The gas velocity at the orifice is V_{10} and the orifice diameter D_{10}. The volumetric gas flow is

$$\dot{V}_{10} = F_0 V_{10}, \qquad (10.67)$$

where $F_0 = \pi D_0^2 / 4$ is the cross section of the orifice. The cross section of the pool above the orifice is F, the gas volume fraction immediately after the orifice is α_1 and at the bubble velocity $V_{1\infty}$. For approximately constant pressure we have

$$\alpha_1 V_{1\infty} = \frac{F_0}{F} V_{10}. \qquad (10.68)$$

In the following we are interested in the diameter of the *largest* bubble, $D_{1\infty}$ which can be in static equilibrium. For this purpose we write a simple momentum balance: buoyancy force = surface force i.e.

$$\frac{\pi D_{1\infty}^3}{6} g(\rho_2 - \rho_1) = \pi D_{10} \sigma \qquad (10.69)$$

or

$$D_{1\infty} = \left(\frac{6\sigma D_{10}}{g \Delta \rho_{21}}\right)^{1/3}. \qquad (10.70)$$

see *Fritz* and *Ende* [13].

Kutateladze and *Styrikovich* [18] (1958) observed experimentally

$$D_{1\infty} \approx \left(\frac{4\sigma D_{10}}{g \Delta \rho_{21}}\right)^{1/3}. \qquad (10.71)$$

Obviously $D_{1\infty}$ should be at least of the order of D_{10}, i.e.

$$D_{1\infty} < \left(\frac{\sigma}{g \Delta \rho_{21}}\right)^{1/2} < \lambda_{RT}. \qquad (10.72)$$

The simple momentum balance, Eq. (10.69), is not sufficient to describe the real bubble size after the jet fragmentation which is smaller than the largest bubble diameter defined by Eq. (10.71). The effect of the gas momentum is not taken into account. *Davidson* and *Harrison*, see in [8] (1963), obtained an empirical correlation which takes into account the influence of the volumetric gas flow on $D_{1\infty}$

$$\pi D_{1\infty}^3 / 6 \approx 1.138 \dot{V}_{10}^{6/5} / g^{3/5}. \tag{10.73}$$

Later work by *Darton* et al. [7] (1977) supported the order of magnitude of the constant but changed the value from 1.138 to 2.27 in the above equation (10.73). Similar conclusions are reached by *Davidson* and *Schüler*, see in [8] (1963)

$$\pi D_{1\infty}^3 / 6 = \left(\frac{4\pi}{3}\right)^{1/4} \left[\frac{15\eta_2 \dot{V}_{10}}{2g(\rho_2 - \rho_1)}\right]^{3/4}. \tag{10.74}$$

The next step of sophistication of the theory is to use the steady state momentum balance in the form: buoyancy + momentum = drag + virtual mass force + surface force, i.e.

$$\frac{\pi D_{1\infty}^3}{6}\left[g(\rho_2 - \rho_1) + \rho_1 V_{10}^2\right]$$

$$= 3\pi\eta_2 D_{10} V_{10}\left[1 + 0.1\left(D_{10}\rho_2 V_{10}/\eta_2\right)^{3/4}\right] + \frac{1}{2}\frac{\pi D_{1\infty}^3}{6}\rho_2 V_{10}\frac{V_{10}}{D_{1\infty}} + \pi D_{10}\sigma \tag{10.75}$$

and to solve the resulting transcendental equation with respect to $D_{1\infty}$. Further improvement of the theory is reported by *Ruft* [25] (1972) who developed a semi-empirical model modifying the above equation and by *Geary* and *Rice* [14] (1991). *Geary* and *Rice* proposed a theory of transient departure which describes the available experimental data without any adjustable parameter. The final analytical solution is voluminous and will not be repeated here. Nevertheless we recommend this solution as the most accurate to describe jet fragmentation by an orifice.

The number of bubbles generated at the orifice per unit time is easily computed by dividing the volumetric gas flow rate $\alpha_1 V_{1\infty}$ by the volume to the single bubble

$$\dot{n}_{1,sp} = \alpha_1 V_{1\infty} / \frac{\pi D_{1\infty}^3}{6} = \frac{F_0}{F} V_{10} / \frac{\pi D_{1\infty}^3}{6}. \tag{10.76}$$

The time in which the bubble remains attached at the orifice is

$$\Delta \tau_{br} = \frac{\pi D_{1\infty}^3}{6} / \dot{V}_{10} .\qquad(10.77)$$

If the raising velocity of the bubbles after the fragmentation is equal to the *Kutateladze* velocity

$$\Delta V_{12\infty} = \sqrt{2}\left[\frac{\sigma g(\rho_2 - \rho_1)}{\rho_2^2}\right]^{1/2}\qquad(10.78)$$

the bubble leaving the orifice should travel the distance $\Delta V_{12\infty}\Delta \tau_{br}$ during $\Delta \tau_{br}$. If this distance is smaller than the bubble diameter after the fragmentation, i.e.

$$\Delta V_{12\infty}\Delta \tau_{br} < D_{1\infty}\qquad(10.79)$$

or

$$\frac{V_{10}}{\Delta V_{12\infty}} > \frac{2}{3}\left(\frac{D_{1\infty}}{D_0}\right)^2,\qquad(10.80)$$

bubbles are no longer formed individually, but the gas leaves the orifice in the form of a jet which eventually breaks into individual bubbles. This line of argument was first proposed by *Kutateladze* and *Styrikovich* [18] (1958).

Summarizing the results discussed in this section we can say that:

1. To have a jet behind the orifice a critical gas velocity should be exceeded in order that the orifice provides more volumetric flow than that which can be transferred away by buoyancy driven bubble raise;

2. The generated number of bubbles per unit mixture volume is uniquely defined if the stable bubble size after the fragmentation, the dimensions of the orifice and of the pool, and the volumetric gas flow through the orifice are known.

Nomenclature

Latin

A amplitude, *m*
D diameter, *m*
$\overline{D_{3E}}$ average size of the entrained particles, *m*

$D_{d\infty}$ size of the ligaments resulting from primary breachup, m

D_{10} orifice diameter, m

d differential

F_0 $= \pi D_0^2 / 4$, cross section of the orifice, m^2

f_m^* dimensionless frequency of the fastest growing of the unstable surface perturbation waves

Fr_{cj} $= \Delta V_{cj}^2 / (gD_j)$, *Froud* number based on the relative velocity, appropriate for jets causing boiling at the interface so that the environment velocity is governed by buoyancy driven convection, *dimensionless*

Fr_j $= V_j^2 / (gD_j)$, *Froud* number based on the jet velocity, appropriate for gravitationally driven jets gravitational acceleration, *dimensionless*

Lp_j $= \rho_j \sigma_j D_j / \eta_j^2$, *Laplace* number, *dimensionless*

Ma local *Mach* number, *dimensionless*

$\dot{n}_{23}''$ average number of the entrained droplets per unit surface, $1/m^2$

$\dot{n}_{1,sp}$ number of bubbles generated at the orifice per unit time, $1/s$

$\dot{n}_3$ particle production rate, $1/(sm^3)$

On_j $= We_j^{1/2} / \mathrm{Re}_j = 1/ Lp_j^{1/2} = \eta_j / (D_j \sigma_j \rho_j)^{1/2}$, jet *Ohnesorge* number, *dimensionless*

$On_{j\lambda}$ $= \eta_j / (\lambda_{RT} \sigma_j \rho_j)^{1/2}$, jet viscosity number based on *Rayleigh-Taylor* instability wave length, *dimensionless*

On_j^* $= (3\eta_j + \eta_c)/(D_j \sigma_j \rho_j)^{1/2}$, modified *Ohnesorge* number, *dimensionless*

On_c $= \eta_c / (D_j \sigma_j \rho_c)^{1/2}$, ambient *Ohnesorge* number, *dimensionless*

Re_j $= \rho_j V_j D_j / \eta_j$, liquid jet *Reynolds* number, *dimensionless*

Ta_{cj} $= (\rho_j / \rho_c) [\sigma / (\eta_j \Delta V_{cj})]^2$, *Taylor* number based on relative velocity, *dimensionless*

Ta_c $= (\rho_j / \rho_c) [\sigma / (\eta_j V_c)]^2$, *Taylor* number based on the continuum velocity, *dimensionless*

$u_{j\sigma}$ interface averaged entrainment velocity, m/s

u_σ^* jet side surface friction velocity, m/s

u_j' interface averaged pulsation velocity, m/s

V velocity, m/s

V_j jet velocity, m/s

V_{10}	gas velocity at the orifice, *m/s*
$\dot{V}_{10}$	$= F_0 V_{10}$, volumetric gas flow, *m³/s*
V_1^*	critical sonic gas velocity at the throat conditions for sonic gas jets, *m/s*
$V_{1\infty}$	bubble velocity, *m/s*
We_j	$= \rho_j V_j^2 D_j / \sigma$, jet *Weber* number, *dimensionless*
We_{cj}	$= \rho_j \Delta V_{cj}^2 D_j / \sigma$, continuum *Weber* number based on the relative velocity, *dimensionless*

Greek

α	volume fraction, *dimensionless*
α_1	gas volume fraction, *dimensionless*
ΔL_j	length for complete disintegration measured from the nozzle outlet, *m*
ΔV_{cj}	relative velocity between the jet and the environment, *m/s*
$\Delta \tau_{br}$	$\approx \Delta L_j / V_j$, characteristic time interval in which the jet fragmentation is completed, *s*
$\Delta \tau_{br}$	$= \dfrac{\pi D_{1\infty}^3}{6} / \dot{V}_{10}$, time in which the bubble remains attached at the orifice, *s*
$\Delta L_j / D_j$	relative jet penetration length, *dimensionless*
δ	wavelength, *m*
λ_{RT}	$= \left[\sigma_d / (g \Delta \rho_{dc}) \right]^{1/2}$, scale of the *Rayleigh - Taylor* instability wavelength, *m*
λ_0	initial deviation from the non-disturbed jet radius, *m*
λ_m^*	dimensionless wavelength frequency of the fastest growing of the unstable surface perturbation waves
λ_j	average wave length, *m*
η	dynamic viscosity, $kg/(ms)$
ρ	density, *kg/m³*
$(\rho w)_{23}$	erosion mass flow rate, *kg/(sm²)*
ρ_1^*	critical density at the throat conditions for sonic gas jets, *kg/m³*
σ	surface tension, *N/m*
τ	time, *s*
$\tau_{j\sigma}$	vapor side shear stress, *N/m²*

Subscripts

Δ   difference
j jet
c continuum
1   gas
2   liquid
3   droplet

References

1. Batchelor GK (ed) (1958) Collected works of Taylor GI, Cambridge Univ. Press, Cambridge, MA
2. Bohr N (1909) Determination of the surface-tension of water by method of jet vibration, Phil. Trans. Roy. Soc. London, Series A., vol 209 p 281
3. Bracco FV (Feb. 25 - March 1, 1985) Modeling of engine sprays, Proc. International Congress & Exposition Detroit, Michigan, pp 113-136
4. Buerger M, von Berg E, Cho SH, Schatz A, (October 25-29, 1993) Modeling of jet breakup as a key process in premixing, Proc. of the Int. Seminar on The Physics of Vapor Explosions, Tomakomai, pp 79-89
5. Chawla TC (1975) The Kelvin - Helmholtz instability of gas - liquid interface, J. Fluid Mechanics, vol 67 no 3 pp 513-537
6. Chawla TC (1976) Drop size resulting from breakup of liquid-gas interfaces of liquid submerged subsonic and sonic gas jets, Int. J. Multiphase Flow, pp 471-475
7. Darton RC, La Nauze RD, Davidson JF and Harrison D (1977) Bubble growth due to coalescence in fluidized beds, Trans. I. Chem. E., vol 55 pp 274-280
8. Davidson JF and Harrison D (1963) Fluidized particles, Cambridge University Press, London
9. De Jarlais G, Ishii M and Linehan J (Febr. 1986) Hydrodynamic stability of inverted annular flow in an adiabatic simulation, Transactions of ASME, Journal of Heat Transfer, vol 108 pp 85-92
10. Epstein M and Fauske K (August 4-7, 1985) Steam film instability and the mixing of core-melt jets and water, ANS Proceedings, National Heat Transfer Conference, Denver, Colorado, pp 277-284
11. Faeth GM (April 3-7, 1995) Spray combustion: a review, Proc. of The 2nd International Conference on Multiphase Flow, Kyoto, Japan
12. Fenn III RW and Middleman S, Newtonian jet stability: The role of air resistance, AIChE Journal, vol 15 no 3 pp 379-383
13. Fritz W and Ende W (1966) Über den Verdampfungsvorgang nach kinematographischen Aufnahmen an Dampfblasen. Phys. Z., vol 37 pp 391-401
14. Geary NW and Rice RG (Feb. 1991) Bubble size prediction for rigid and flexible sparkers, AIChE Journal, vol 37 no 2 pp 161-168
15. Grant RP and Middleman S (July 1966) Newton Jet Stability, AIChEJ, vol 12 no 4 p 669
16. Rayleigh L (1878) On the instability of jets, Proc. London Math. Soc., vol 10 p 7

17. Iciek J (1982) The hydrodynamics of free, liquid jet and their influence on direct contact heat transfer - I, II, Int. J. Multiphase Flow, vol 8 no 3 pp 239-260
18. Kutateladze SS and Styrikovich MA (1958) Hydraulics of gas-liquid systems, Moscow, Wright Field transl. F-TS-9814/V
19. Lienhard JH and Day JB (1970) The breakup of superheated liquid Jets, ASME J. Basic Eng., vol 92 pp 511-522
20. Mayer E (Dec. 1961) Theory of liquid atomization in high velocity gas streams, ARS Journal, vol 31 pp 1787 - 1785
21. Meister BJ and Scheele GF (1969) AIChEJ vol 15 pp 689 - 699
22. Merrington AC and Richardson EG (1947) The breakup of liquids jets, Proc. Phys. Soc., vol 59 pp 1-13
23. Nigmatulin BI, Melikhov OI, and Melikhov VI (Oct.25-29, 1993) Breakup of liquid jets in film boiling, Proc. of Int. Sem. on The Physics of Vapor Explosions, Tomakomi, pp 90-95
24. Ohnesorge W (1936) Die Bildung von Tröpfen an Düsen und die Auflüssiger Strahlen, Z. Angew. Math. Mech., vol 16 pp 335-359
25. Ruft K (1972) Bildung von Gasblasen an Düsen bei konstantem Volumendurchsatz, Chemie-Ing. Techn. vol 44 no 24 pp 1360-1366
26. Saito M, Sato K and Imahori S (July 24-July 27, 1988) Experimental studi on penetration of water jet into Freon-11 and Liquid Nitrogen, ANS Proc. 1988 Nat. Heat Transfer Conference., HTS-Vol.3, Houston, Texas, pp 173-183
27. Schneider JP, Marchiniak MJ, Jones BG (Sept. 21-24, 1992) Breakup of metal jets penetrating a volatile liquid, Proc. of the Fifth Int. Top. Meeting On Reactor Thermal Hydraulics NURETH-5, vol 2 pp 437-449
28. Smith SWJ and Moss H (1917) Experiments with mercury jets, Proc. Roz. Soc., vol A/93 pp 373-393
29. Sun PH (November 4-6, 1998) Molten fuel-coolant interactions induced by coolant injection into molten fuel, SARJ98: The Workshop on Severe Accident Research held in Japan Tokio, Japan
30. Takahashi T and Kitamura Y (1971) Kogaku Kogaku, vol 35 p 637
31. Takahashi T and Kitamura Y (1972) Kogaku Kogaku, vol 36 p 912
32. Tanzawa Y and Toyoda S (1954) Trans. J.S.M.E. vol 20 p 306
33. Teng H, Kinoshita CM and Masutani SM (1995) Prediction of droplet size from the breakup of cylindrical liquid jet, Int. J. Multiphase Flows, vol 21 no 1 pp 129-136
34. Thimotika H (1935) Proc. Roy. Soc., vol.150 p 322: (1936) vol 153 p 302
35. Weber C (1936) Zum Zerfall eines Flüssigketsstrahles, Z. Angew. Math. Mech., vol 11 pp 136-154
36. Wolfe HE and Anderson WH (1965) Aerodynamic breakup of liquid drops, II. Experimental, Proc. Int. Shock Tube Symposium, Naval Ordinance Lab. White Oak, Maryland, USA

11. Fragmentation of melt in coolant

11.1 Introduction

Let us start with some definitions necessary to understand the content of this Chapter:

a) The term *melt* is used here for liquid having solidification temperature higher then the film boiling temperature of the surrounding coolant.

b) *Coolant* is the liquid surrounding the melt.

c) *Mechanical fragmentation* is fragmentation not influenced by local heat and mass transfer processes.

d) *Thermo-mechanical fragmentation* is mechanical fragmentation additionally amplified by the local heat and mass transfer.

e) *Events distorting the film boiling process* are interface instabilities caused by

- inherent vapor-coolant instability and
- externally introduced pressure pulses.

f) *Inherent vapor-coolant instability* is an interface instability caused by

- mechanical fragmentation of the initially unstable particle leading to intimate melt/coolant contact during the fragmentation,
- transition from film boiling to transition boiling, and
- cavitation of vapor bubbles in subcooled liquid in the immediate neighborhood to the particle being in film boiling.

g) *Contact heat transfer* is a local contact between melt and liquid coolant.

If once established stationary, film boiling at the surface of a liquid sphere is a very stable process. Events distorting the film boiling process may lead to intimate melt-coolant contact resulting in effective energy transfer between the hot droplet and the surrounding coolant. The mechanical feed-back to the droplet leads to additional surface fragmentation which is called thermo-mechanical fragmentation. The result is generation of local pressure pulses. If the same happens not only with a single drop but with a family of melt drops, the resulting pressure pulse may contain considerably more energy than the single drop event.

In order to estimate the risk of the steam explosion for nuclear reactors by postulated severe accidents leading to melt-coolant interaction, to know how the thermo-mechanical fragmentation works is very important. Because of the complexity of the mechanisms involved considerable amount of literature appeared during the last 40 years. The purpose of this Chapter is to review the experimental observation and to derive practically useful estimation methods for quantitative analysis of the steam explosion risk based on the experimental observation. In what follows we will discuss separate effects of these phenomena and the way they can be quantitatively described.

Whether non-stable melt being in film boiling will behave as mechanically fragmenting in vapor or in liquid depends on the thickness of the vapor film. *Epstein* and *Fauske* [33] provided an expression derived from instability analysis of both surfaces which can be used to distinguish between thick and thin vapor films

$$\delta_{1F,critc} = 3 \frac{\rho_3 - \rho_1}{\rho_3} \frac{\sigma_3}{\rho_1 \Delta V_{31}^2} . \qquad (11.1)$$

For $\delta_{1F,fc} < \delta_{1F,critc}$ the influence of the vapor film on the fragmentation can be neglected and liquid metal/coolant is the appropriate mode of mechanical interaction. As already mentioned primary mechanical fragmentation of a non-stable particle in film boiling may lead to intimate melt/water contact and therefore may introduce some thermal fragmentation. *Knowles* [54] found theoretically that the pressure impulse required for vapor collapse is a function of the film thickness. This was experimentally confirmed by *Naylor* [74].

The film boiling at spheres is well understood as demonstrated in Chapter 21 by comparison with about 2000 experimental data points in different two-phase flow environments - continuous or dispersed liquids. In the next section we are rather interested on the approximated estimation of the average film thickness. The reasons why the estimation of the average film thickness is important are:

For given void fraction in three-phase flow, part of the void is attached to the droplets surfaces and other part is in form of bubbles between the molten droplets. The local volume concentration in the space continuously occupied by water naturally provides a criterion whether the water will remain continuous or get dispersed. Some important consequences result from this observation for the practical analysis:

a) The pressure wave propagation is known to be much less damped in systems in which the water forms a continuum. Acoustically water/gas discontinuity acts as reflectors to pressure waves. Thus, events initiating external pressure pulses in continuous liquid may lead to explosions, which is not the case when such events happens in a gas environment. We call mixtures having molten particles and being surrounded by continuous water-*potentially explosive mixtures*. Whether they are really explosive depends on many other parameters beside the local geometry flow pattern.

b) The heat and mass transfer processes at the molten particle interface in case of continuous liquid is completely described by the film boiling description methods. The part of the void drifting as bubbles between the films is subject to additional water/vapor heat and mass transfer. This subject is discussed in details in Chapters 13 and 14.

c) The mechanical fragmentation mechanism is much stronger for melt/liquid systems than for melt/gas systems.

11.2 Vapor thickness in film boiling

Consider clouds of hot spheres characterized by volume fraction, diameter, and particle number density designated with α_3, D_3, n_3, respectively. The mass generated due to film boiling per unit mixture volume and unit time is designated with μ_{21}. The film thickness at the equator δ_{1F} has to satisfy the continuity condition

$$\pi D_3 \delta_{1F} \rho_1 \langle w_1 \rangle_1 = \frac{\mu_{21}}{n_3} = \frac{\mu_{21}}{\alpha_3} \frac{\pi D_3^3}{6}. \tag{11.2}$$

Here

$$\langle w_1 \rangle_1 \approx \frac{\Delta \rho_{21} g \cos(\mathbf{g}, \Delta \mathbf{V}_{23}) D_3^2}{12 \eta_1} \left(\frac{\delta_{1F}}{D_3} \right)^2 + \frac{3}{4} \Delta V_{23} \tag{11.3}$$

is the vapor velocity averaged over the film thickness computed for particle having large size compared to the film thickness. $\cos(\mathbf{g}, \Delta \mathbf{V}_{23})$ is the cosine of the angle between the gravitational acceleration and the relative velocity vector particle-water and

$$\Delta V_{32} = \sqrt{\Delta u_{32}^2 + \Delta v_{32}^2 + \Delta w_{32}^2} \tag{11.4}$$

is the magnitude of the velocity difference. The resulting cubic equation

$$\frac{\rho_1 \Delta \rho g \cos(\mathbf{g}, \Delta \mathbf{V}_{23})}{12\eta_1} \delta_{1F}^3 + \frac{3}{4} \Delta V_{23} \rho_1 \delta_{1F} - \frac{\mu_{21}}{\alpha_3} \frac{D_3^2}{6} = 0, \qquad (11.5)$$

rewritten in the compact form

$$\left(\frac{\delta_{1F}}{\delta_{1F,nc}}\right)^3 + r \frac{\delta_{1F}}{\delta_{1F,nc}} - 1 = 0, \qquad (11.6)$$

has the following real solution

$$\frac{\delta_{1F}}{\delta_{1F,nc}} = a^{1/3} - \frac{1}{3} r a^{-1/3}. \qquad (11.7)$$

Here

$$r = \frac{\delta_{1F,nc}}{\delta_{1F,fc}} = \frac{9}{2} \left(\frac{\alpha_3 \rho_1}{\mu_{21} D_3^2}\right)^{2/3} \Delta V_{23} \left[\frac{2\eta_1}{g \cos(\mathbf{g}, \Delta \mathbf{V}_{23})(\rho_2 - \rho_1)}\right]^{1/3} \qquad (11.8)$$

is the ratio of the film thickness for natural convection only, e.g. for $\Delta V_{23} < 0.001$

$$\delta_{1F,nc} = \left[\frac{2 D_3^2 \mu_{21} \eta_1}{\alpha_3 g \cos(\mathbf{g}, \Delta \mathbf{V}_{23}) \rho_1 (\rho_2 - \rho_1)}\right]^{1/3}, \qquad (11.9)$$

and for predominant forced convection

$$\delta_{1F,fc} = \frac{2}{9} \frac{D_3^2 \mu_{21}}{\alpha_3 \rho_1 \Delta V_{23}}, \qquad (11.10)$$

and

$$a = \frac{1}{2} + \frac{1}{18}\sqrt{3(4r^3 + 27)}. \qquad (11.11)$$

The stronger the film boiling, the thicker the vapor film surrounding the drop, the more stable the droplet.

11.3 Amount of melt surrounded by continuous water

The particles in film boiling are surrounded by a film with dimensionless thickness

$$\delta_{1F}^* = \frac{\delta_{1F}}{D_3}, \qquad (11.12)$$

where $\delta_{1F}^* > 0$ as already discussed in the previous section. The ratio of the volume of the sphere consisting of one particle and the surrounding film to the volume of the particle itself is

$$(\alpha_3 + \alpha_{1F})/\alpha_3 = (D_3 + 2\delta_{1F})^3 / D_3^3 = (1 + 2\delta_{1F}^*)^3. \qquad (11.13)$$

Therefore the gas volume fraction of the film is

$$\alpha_{1F} = \alpha_3 \left[(1 + 2\delta_{1F}^*)^3 - 1 \right]. \qquad (11.14)$$

The condition to have three-phase flow with continuous liquid and particles being in film boiling is

$$(\alpha_1 - \alpha_{1F})/(\alpha_1 - \alpha_{1F} + \alpha_2) < 0.52. \qquad (11.15)$$

This is a very important result. It demonstrates simply that particles in film boiling can be surrounded by much less continuous liquid mass than required in case of no film boiling.

As already mentioned, *this conclusion has strong implication on the mechanical fragmentation condition* simply allowing the molten particles to experience fragmentation in continuous liquid by far less liquid required. Thus, only the f_3^+ part of the particles n_3 per unit mixture volume experiences melt/water fragmentation while the $1 - f_3^+$ part the particles experiences melt gas fragmentation. A practicable approach is to define inside a computational cell

$$f_3^+ = 1 \quad \text{for} \quad (\alpha_1 - \alpha_{1F})/(\alpha_1 - \alpha_{1F} + \alpha_2) < 0.52, \qquad (11.16)$$

$$f_3^+ = 0 \quad \text{for} \quad (\alpha_1 - \alpha_{1F})/(\alpha_1 - \alpha_{1F} + \alpha_2) \geq 0.52. \qquad (11.17)$$

11.4 Thermo-mechanical fragmentation of liquid metal in water

The scale of the thermo-mechanical fragmentation is much smaller than those for which averaged conservation equations are usually applied. Therefore the description of the thermo-mechanical fragmentation is in fact associated with providing constitutive physics for source terms for the averaged equations. Thermo-mechanical fragmentation is experimentally observed within systems in which the microscopic relative phase velocity is so small that pure mechanical fragmentation is impossible. In pressure wave systems with melt and water as a constituents, in which mechanical fragmentation is possible pure mechanical fragmentation mechanism does not provide the complete description of the thermal energy release. The reader can find interesting discussions on this topic in [2,3, 4, 7, 12, 13, 18, 24, 31, 37, 44, 45, 53, 68, 84, 86, 97, 98, 101, 102] (1974-1991), among many others, and in the references given therein.

For the discussion in this Section we consider the velocity field 3 to be a liquid metal.

The initial conditions under which thermo-mechanical liquid metal fragmentation were experimentally observed are summarized below.

1. Coexistence of melt and coolant

$$\alpha_2 > 0, \; \alpha_3 > 0. \tag{11.18}$$

2. Continuous liquid and particles in film boiling

$$(\alpha_1 - \alpha_{1F})/(\alpha_1 - \alpha_{1F} + \alpha_2) < 0.52. \tag{11.19}$$

Experiments show that the lower the particle and vapor volume fraction the higher the probability of thermo-mechanical fragmentation. Such systems are called *lean systems*.

3. Surface of the melt is in liquid state. The fragments after thermo-mechanical fragmentation were usually smooth, indicating that fragmentation occurred while the particles were still molten, *Kim* et al. [53] (1989).

4. The droplet surface temperature is higher than the minimum film boiling temperature $T_{3i} > T_{FB,\min}$.

5. The initial droplet size is considerably greater than the size of the final fragments.

All of these conditions must be satisfied in order to have thermo-mechanical fragmentation. Under the above conditions and if physical mechanisms are acting to establish contact between melt and liquid coolant, the thermal fragmentation occurs. The mechanisms can be classified as

a) due to imposed surface instability, or as

b) due to inherent surface instability, e.g. mechanical fragmentation followed by melt/liquid coolant contact immediately after the fragmentation event, transition from film to nucleate boiling.

Next we discuss some of these conditions in more detail.

11.4.1 External triggers

11.4.1.1 Experimental observations

As already mentioned, if the droplet surface temperature is higher than the minimum film boiling temperature $T_3^\sigma > T_{FB,\min}$, mechanical fragmentation was observed only with additional pressure pulses disturbing the film called trigger pressure pulse

$$\frac{dp}{d\tau} > \left|\frac{dp}{d\tau}\right|_{trigger}. \tag{11.20}$$

The lowest limit of the pressure gradient causing film distortion has not been systematically investigated. Experimental evidence shows that it is a function of the melt surface temperature - the higher the surface temperature of the melt, the higher the pressure pulse needed to destroy the film.

Ando and *Caldarola* [2] produced thermal fragmentation on molten copper droplets in water at atmospheric pressure by using pressure pulses of 4.5-9.5 *bar* with approximately triangle form and impulse 50 to 160 *Pa s*. The authors observed two types of fragmentation, *delayed*, caused by impulses of 10 to 70 *Pa s*, and *prompt*, caused by impulses between 70 and 160 *Pa s*. The delay time varies between 2900 and 700 μs for the delayed fragmentation and between 300 and 100 μs for the prompt fragmentation, respectively.

Knowles [54] suggested the following criterion for vapor collapse:

$$\Delta\tau_{pulse}\Delta p_{pulse} \approx \delta_F \frac{1}{2}\rho_2 a_2. \tag{11.21}$$

where a_2 is the coolant sound velocity.

Table 11.1. Experimentally observed trigger pressure change

$\|dp/d\tau\|_{trigger}$ Pa/s	Δp_{pulse} MPa	$\Delta \tau_{pulse}$ μs	Authors, ref.	Rem.
$\approx 10^{12} \pm 20\%$	≈ 1	1	Nelson and Buxton [77] (1980)	4 cm bridge wire $\Delta\tau_{pulse}\Delta p_{pulse} \approx 1\ Pa\ s$
$\approx 5\times 10^{10} \pm 20\%$	≈ 10	20		detonator $\Delta\tau_{pulse}\Delta p_{pulse} \approx 200\ Pa\ s$
$(4\ \text{to}\ 8)\times 10^9$	0.2 to 0.4	50	Kim et al. [53] (1989)	$\Delta\tau_{pulse}\Delta p_{pulse} \approx 15\ Pa\ s$
$\approx 5\times 10^9$			Peppler et al. [81] (1991)	depends on the form of the produced pulse
$(4\ \text{to}\ 7.6)\times 10^9$	0.45 to 0.95	62	Ando and Caldarola [2]	cooper droplets, triangle pulse form $\Delta\tau_{pulse}\Delta p_{pulse} \approx 43.4\ Pa\ s$
$\approx 5\times 10^9$	4	0.8	Huhtiniemi et al. [49]	gas chamber $\Delta\tau_{pulse}\Delta p_{pulse} \approx 3.2\ Pa\ s$
$\approx 6\times 10^7$	3	0.05	Chapman et al. [23]	electromagnetic movable piston $\Delta\tau_{pulse}\Delta p_{pulse} \approx 0.15\ Pa\ s$
$(1\ \text{to}\ 4)\times 10^{12}$	2.5 to 8	1 to 7	Buetner and Zimanowski [20]	triangle pulse form $\Delta\tau_{pulse}\Delta p_{pulse} \approx 8\ \text{to}\ 15 Pa\ s$

Naylor [74] reported experimental data supporting this criterion. As an example for the order of magnitude estimate consider film boiling of water at atmospheric pressure for film thickness of 0.0001 m. The required pressure impulse is then $\Delta\tau_{pulse}\Delta p_{pulse} \approx 130 Pa\ s$.

Thus the data in the table can be interpreted only by knowing the local conditions and the film thickness which have to be destabilized in order to cause thermal fragmentation.

11.4.1.2 Theory

Because the work by *Naylor* [74] will play an important role in the following discussion we summarize the most important result of this work. If two continua having common interface are accelerated perpendicular to the interface and if the viscous effects are negligible, the initial disturbance with wavelength λ_0 will grow as follows

$$\frac{\lambda}{\lambda_0} = \cosh\frac{\tau}{\Delta\tau_{RT}}, \qquad (11.22)$$

where the growing time constant is

$$\Delta\tau_{RT} \approx \left[a \frac{2\pi}{\lambda_{RT}} \frac{\rho_2 - \rho_1}{\rho_2 + \rho_1} \right]^{-1/2}. \qquad (11.23)$$

a is the normal interface acceleration pointing from the heavier to the lighter continuum, and

$$\lambda_{RT} \sim 2\sqrt{3}\pi \left[\frac{\sigma_2}{a(\rho_2 - \rho_1)} \right]^{1/2} \qquad (11.24)$$

is the wavelength of the fastest growing oscillation. *Belman* and *Pennington* obtained in [9] an additional term in the time constant as given below

$$\Delta\tau_{RT} \approx \lambda_{RT}/V_{2,jets} \approx \left[a \frac{2\pi}{\lambda_{RT}} \frac{\rho_2 - \rho_1}{\rho_2 + \rho_1} - \left(\frac{2\pi}{\lambda_{RT}}\right)^3 \frac{\sigma_2}{\rho_2 + \rho_1} \right]^{-1/2}, \qquad (11.25)$$

which is often neglected in practical analyses.

The nature of the interface acceleration can be different. Some examples are

a) acceleration due to shock wave propagation which we call *global acceleration*, and
b) acceleration due to bubble collapse, which we call *local acceleration*.

The global acceleration causes changes of the average velocity differences $\frac{d}{d\tau}(\Delta V_{12})$, and the local acceleration causes changes in the local interface velocity, e.g. during bubble collapse $\left|\frac{d^2 R}{d\tau^2}\right|$.

For the discussion of the thermal fragmentation we will make use also of the velocity of the wave normal to the interface which we will call coolant micro jet velocity

$$V_{2,jets} = \left[2\pi a \lambda_{RT} \frac{\rho_2 - \rho_1}{\rho_2 + \rho_1}\right]^{1/2} = 2\pi \left[a \frac{3\sigma_2 (\rho_2 - \rho_1)}{(\rho_2 + \rho_1)^2}\right]^{1/4}. \tag{11.26}$$

11.4.1.2.1 Global coolant/vapor interface instability due to phase acceleration

If we consider the film surrounding the particle in the case of film boiling, only distortions having wavelengths shorter than the "bubble" diameter, $D_3 + 2\delta_{1F}$, are able to destroy the film. *Gelfand* et al. [41] (1977) analyzed experimentally and theoretically the distortion of bubbles with different densities behind the front of pressure waves. Idealizing the bubble as a cube, the authors applied to the surfaces normal to the pressure wave direction the *Rayleigh-Taylor*, and for the parallel surfaces the *Kelvin - Helmholtz* instability criteria to explain the distortion conditions. Extending this analysis to the gas-water interface in our case, we have a stable film if

$$D_3 + 2\delta_{1F} > \pi \left[24 \frac{\sigma_2}{\frac{d}{d\tau}(\Delta V_{12})(\rho_2 - \rho_1)}\right]^{1/2} \tag{11.27}$$

and

$$D_3 + 2\delta_{1F} > \frac{2\pi\sigma(\rho_2 + \rho_1)}{\Delta V_{12} \rho_2 \rho_1}. \tag{11.28}$$

Keeping in mind the simplified momentum difference Eq. (8.56),

$$\frac{d}{d\tau}(\Delta V_{21}) \approx -\left(\frac{1}{\rho_2} - \frac{1}{\rho_1}\right)\nabla p \approx \frac{\rho_2 - \rho_1}{\rho_2 \rho_1}\nabla p, \tag{11.29}$$

11.4 Thermo-mechanical fragmentation of liquid metal in water

the first *Taylor* instability criterion takes the form

$$D_3 + 2\delta_{1F} > \pi \left[24 \frac{\sigma_2 \rho_2 \rho_1}{(\rho_2 - \rho_1)^2 \nabla p} \right]^{1/2} \tag{11.30}$$

or

$$\nabla p > 24\pi^2 \frac{\rho_2 \rho_1}{(\rho_2 - \rho_1)^2} \frac{\sigma_2}{(D_3 + 2\delta_{1F})^2}. \tag{11.31}$$

If the pressure pulse propagation velocity through the mixture is a_m, the spatial component of the total pressure change with the time is

$$a_m \nabla p > 24\pi^2 a_m \frac{\rho_2 \rho_1}{(\rho_2 - \rho_1)^2} \frac{\sigma_2}{(D_3 + 2\delta_{1F})^2}. \tag{11.32}$$

The criterion given by this equation contains useful information:

1. For atmospheric pressure and $D_3 + 2\delta_{1F}$ of the order of 1 *mm*, small particles concentration for which the mixture sound speed is equal to liquid sound speed, $a_m = a_2$, we obtain

$$24\pi^2 a_m \frac{\rho_2 \rho_1}{(\rho_2 - \rho_1)^2} \frac{\sigma_2}{(D_3 + 2\delta_{1F})^2} \approx 1.5 \times 10^7 \, Pa/s, \tag{11.33}$$

   which explains the experimentally observed values given in Table 11.1.

2. The linear dependence of the spatial pressure change $a_m \nabla p$ on the surface tension given by Eq. (11.32) is also supported by the experiments of *Ando* and *Caldarola* [2].

3. The higher the surface tension, the higher the trigger pressure gradient. The linear dependence of the spatial pressure change $a_m \nabla p$ on the surface tension given by Eq. (11.32) is also supported by the experiments of *Ando* and *Caldarola* [2]. The increasing oxide content at increasing temperature for liquid copper droplets reduces the surface tension and therefore reduces the threshold pressure impulse required for thermal fragmentation.

4. The higher the system pressure, the higher $\dfrac{\rho_2 \rho_1}{(\rho_2 - \rho_1)^2}$ i.e. the steam to liquid acceleration is smaller for the same pressure gradient and therefore the higher the trigger pressure gradient.

5. The smaller the melt particle size, the higher the trigger pressure gradient (experimentally supported by *Fröhlich*, see fig. 5 in [39] (1991)).

6. The trigger pressure pulse increases with decreasing film thickness [experimentally confirmed by *Fröhlich* [39] (1991)].

7. Equation (11.29) explains why $|dp/d\tau|_{trigger}$ depends on the spatial form of the trigger pulse - the dependence on ∇p. Obviously the application of the condition (11.32) for practical computational analysis needs good spatial resolution in order to estimate accurately the local value of the pressure gradient.

11.4.1.2.2 Local coolant/vapor interface instability due to film collapse

As will be discussed in the chapter "Interface instabilities due to bubble collapse", the interface acceleration during film collapse in subcooled coolant leads to interface instabilities. The condition that the interface instability wavelength is smaller than the particle size leads to an interesting expression for the so called *threshold pressure*

$$p_{2,th} > p_1 + \frac{2\sigma_2}{R_3}\left(\pi^2 \frac{3}{2} \frac{\rho_2}{\rho_2 - \rho_1} - 1\right). \tag{11.34}$$

The threshold pressure condition contains an important message: The threshold pressure increases with decreasing particle size due to the term $\frac{2\sigma_2}{R_3}$, and with increasing system pressure due to the term $\frac{\rho_2}{\rho_2 - \rho_1}$.

The reestimation of the existing data and future experiments should be performed in order to establish a reliable database for the criterion (11.32).

11.4.2 Experimental observations

The internal triggers may originate due to different reasons classified as follows:

1. The reduction of the melt surface temperature below the minimum film boiling temperature leads to transition boiling. Fragmentation is caused due to

    a) instabilities induced by bubble formation and departure,

b) micro-shocks caused by liquid entering the volumes that are made free after the bubbles departure, and

c) local thermal stresses caused by the local differences in the temperature at the particle surface.

2. Spontaneous vapor condensation (cavitation) in subcooled water.

3. Geometry dependent pressure wave formation and interactions. In order to describe this process, adequate geometry description of the system is necessary associated with appropriate spatial resolution of the pressure waves. The computational models can be tested using well defined pressure wave propagation experiments not necessary associated with melt/water interactions. The mechanism of the thermo-mechanical fragmentation is then described with the information provided in chapter external triggers.

4. Particles reaching the bottom or an obstacle in a liquid state may serve as a trigger by the mechanism of the so called entrapment - see *Mitsumura* et al 1997 [72].

11.4.2.1 Interface solidification and availability of internal triggers

11.4.2.1.1 Solidification macro time scale of the system

We now consider the condition that the surface of the molten drop has to be in a liquid state in order to allow further fragmentation of the droplet. For such processes the external cooling due to film boiling and radiation causes solidification starting at the surface and penetrating into the drop. Therefore, if the drop possesses a specific entropy below s_3'', the crust on the surface is already formed and the probability that the droplet will be destroyed due to thermo-mechanical fragmentation is very low. Therefore the first order approximation for the condition allowing fragmentation is that the locally average field specific entropy is greater than the liquidus entropy of the molten drop,

$$s_3 > s_3''. \tag{11.35}$$

This approximate criterion is very convenient for application in computer codes. Note that there is a spatial temperature profile inside the drop which explains the term approximate. In addition, this criterion neglects the solidification delay.

Next we will estimate under which condition melt penetrating the water may create internal triggers due to entrapment using a simple lumped parameter approach. If particles of melt enter a water reservoir with water depth L_{pool_depth} one can estimate the time within the particle reaches the bottom of the reservoir

$$\Delta\tau_{life\_time} = \frac{L_{pool\_depth}}{\Delta w_{32}}.\tag{11.36}$$

Here Δw_{32} is the average velocity with which the droplet crosses the water. If there are some obstacles normal to the flow direction L_{pool_depth} should represent the distance between water level and these obstacles. Particles reaching the bottom or an obstacle in a liquid state may serve as a triggered for the mechanism of the so called entrapment - see *Mitsumura* et al. [72] 1997.

For nuclear reactor systems the melt possesses temperatures considerably higher than the critical temperature of water and therefore the film boiling and radiation are the heat transfer controlling mechanism at non-disturbed droplet interfaces. Thus the heat flux emitted from the surface is

$$\dot{q}_3''^{\sigma 2} = k_{SB}\left[T_3^{\sigma 4} - T'(p)^4\right] + h_{FB}\left[T_3^{\sigma} - T'(p)\right].\tag{11.37}$$

The characteristic time elapsed from the beginning of the melt-coolant mixing process to the formation of the crusts at the interface having the solidus-liquid temperature is computed from equating the heat flux released from the interface and the heat flux coming from the droplet bulk to the interface computed by the short time solution of the *Fourier* equation

$$\dot{q}_3''^{\sigma 2} = (T_3 - T_{3\sigma})\sqrt{\frac{3}{\pi}\frac{\lambda_3 \rho_3 c_{p3}}{\tau}}.\tag{11.38}$$

The result is

$$\Delta\tau_{crust} \approx \frac{3}{\pi}\lambda_3\rho_3 c_{p3}\left\{\frac{T_3 - T'''}{k_{SB}\left[T_3^{m4} - T'(p)^4\right] + h_{FB}\left[T_3^m - T'(p)\right]}\right\}.\tag{11.39}$$

Thus the approximate condition to expect entrapment triggers is

$$\Delta\tau_{crust} > \Delta\tau_{life\_time},\tag{11.40}$$

or

$$\frac{3}{\pi}\frac{\lambda_3 \rho_3 c_{p3}\Delta w_{32}}{L_{pool\_depth}}\left\{\frac{T_3 - T'''}{k_{SB}\left[T_3^{m4} - T'(p)^4\right] + h_{FB}\left[T_3^m - T'(p)\right]}\right\}^2 > 1.\tag{11.41}$$

11.4.2.1.2 Micro time scale of the system

As already mentioned, primary mechanical breakup of a non-stable particle in film boiling may lead to intimate melt/water contact and therefore introduce some thermal fragmentation which may serve as a local trigger.

Mechanical fragmentation of melt entering a water pool is possible if the time required for mechanical fragmentation is less than the time to reach surface solidification. As already discussed in Chapter 8, there are different regimes of mechanical fragmentation. The data collection of *Pilch* et al. [82] from 1981, see Fig.8.2, shows that the dimensionless fragmentation time

$$\Delta \tau_{br}^* = \frac{\Delta \tau_{br}}{\dfrac{D_3}{|\Delta V_{3c}|}\sqrt{\dfrac{\rho_3}{\rho_c}}} \qquad (11.42)$$

varies between 2 and 7. Note that for thin vapor layers in film boiling, which is usually the case, the continuum density is effectively equal to the coolant density. Thus the condition to have primary breakup before reaching the obstacle is

$$\Delta \tau_{crust} > \Delta \tau_{life\_time} > \Delta \tau_{br}. \qquad (11.43)$$

11.4.2.2 Transition boiling

The reduction of the melt surface temperature below the minimum film boiling temperature

$$T_3^\sigma < T_{FB,\min} \qquad (11.44)$$

leads to transition boiling. In addition, if the melt is in a liquid state, that is the melt solidification temperature is lower than the minimum film boiling temperature

$$T_3'' < T_{FB,\min}. \qquad (11.45)$$

violent explosion starts, *Board* et al. [14] (1972), *Frost* and *Ciccarelli* [40] (1988). One example is the experiment reported by *Board* et al. [14] (1972), in which tin with melting temperature 230°C was dropped in water at 60°C having minimum film boiling temperature about 250°C at atmospheric pressure. 2 to 5 mm droplets of tin with initial temperature of 800°C dropped in 20°C water are reported to reach these conditions. The first following thermal interaction happens within 7 *ms*. Generally if the tin temperature was above 400°C and the water temperature was below 60°C the droplets always undergo thermal interaction. An experimentally observed dependence of the projected area of the tin fragments (initial temperature 375 to 600°C) to the initial projected area of the molten drop is given by *Cho* and *Gunther,* Fig. 1 in [25]. It decreases starting at about 4-times of the initial area at about 5°C water temperature to 2-times at 70°C.

Instabilities caused by bubble formation, departure, and micro-shocks caused by liquid entering the volumes that are made free after the bubbles departure and local thermal stresses caused by the local differences in the temperature at the particle surface are the reason for the fragmentation. For larger subcooling of the coolant cyclic nature of the pressure spikes are reported by many researchers.

The minimum film boiling temperature is a function of the local pressure

$$T_{FB,\min} = T_{FB,\min}(p) < T_{2,cr} \quad \text{for} \quad p < p_{2,cr} \tag{11.46}$$

and is limited by the critical temperature,

$$T_{FB,\min} = T_{2,cr} \quad \text{for} \quad p \geq p_{2,cr}, \tag{11.47}$$

see *Schröder-Richter* [85] (1990). Theoretical dynamic stability condition is provided by *Matsumura* and *Nariai* [66] analysis.

11.4.2.3 Bubble collapse in subcooled liquid

Bubbles in subcooled *infinite* liquid having the same center of mass velocity as the liquid collapse *symmetrically*. This is not the case for *bubbles in asymmetric* flow field like

a) bubbles attached to surfaces or in the vicinity of surfaces,
b) bubbles attached to a droplet or in the vicinity of droplets,
c) bubbles having relative velocity with respect to the liquid which is always the case in multiphase flows.

Bubbles in asymmetric flow, e.g. in the vicinity of a wall, elongate in the direction normal to the wall and develop a jet towards the wall. The velocity, length and diameter are found to be

$$V_{2,jet} = 13\left(\frac{\Delta p}{\rho_2}\right)^{1/2}, \tag{11.48}$$

$$\ell_{2,jet} = 0.4929 R_{10}, \tag{11.49}$$

$$D_{2,jet} = 0.1186 R_{10}, \tag{11.50}$$

as reported by *Plesset* and *Chapman* in 1971 [83]. *Gibson* [42] (1968) found experimentally that the constant in Eq. (11.48) is 7.6. *Blake* and *Gibson* [43] (1987) indicate that this constant is a function of the ratio of the initial distance of the bubble center from wall to the initial radius. For ratio of 1.5 the constant is 11 and for 1, 8.6. *Voinov* and *Voinov* [95] (1979) found that the constant could be as high as 64 if the initial bubble had a slightly excentric shape. The pressure exerted by the jet is either the "water hammer pressure",

$$p_{2,jet,WH} = \rho_2 a_{2,sound} V_{2,jet} \frac{\rho_3 a_{3,sound}}{\rho_2 a_{2,sound} + \rho_3 a_{3,sound}} \quad (11.51)$$

or the stagnant pressure,

$$p_{2,jet} = \frac{1}{2} \rho_2 V_{2,jet}^2. \quad (11.52)$$

The duration of the water hammer pulse is of order of $D_{2,jet}/a_{2,sound}$ whilst that of the stagnant pressure is $\ell_{2,jet}/V_{2,jet}$ - see the discussion by *Buchman* 1973 [17]. It has been known for a long time that collapsing bubbles in the neighborhood of a solid surface can lead to pitting of the surface. That the damage is caused by liquid jets on the bubbles was first suggested by *Kornfeld* and *Suvorov* 1944 [58]. Experiments by *Benjamin* and *Ellis* 1966 [10] have confirmed that jets do indeed form on bubbles collapsing near a solid wall.

Note that a pressure excursion may bring initially saturated water to a subcooled state, which may result in self triggering of the system.

11.4.2.4 Entrapment

Melt reaching the bottom or an obstacle in a liquid state may serve as a trigger by the mechanism of the so called entrapment - see *Mitsumura* et al. [72] (1997). The entrapment is defined as inclusion of coolant into the melt. The intensive contact heat transfer leads to local explosive evaporation of the liquid. Depending on the amount of the water involved the entrapment may result in a very strong trigger.

Entrapment is the geometry typically considered as the initial state for volcano eruptions - see *Zimanowski* et al. [102]. Laboratory experiments with silicate melts and water injected into the melt are intensively studied. *Buetner* and *Zimanowski* reported in [20] a summary of the results of their own investigations. The authors reported that static water injection may lead to virtually stable layering. Imposing of pressure pulse - trigger - results in violent explosion. The energy gain depends on the pulse straight intensity. The authors reported experiments with triggers ranging between about 4×10^{12} and $10^{12} MPa/s$ with rise time between 1 and 7µs.

The maximum energy gain was obtained for about 2×10^{12} *MPa/s* with rise time about 4μs. Local thermal efficiencies of maximum 48.7% are reported. Because only one small part was able to react during the expansion process, the global efficiency resulted in 2%.

11.4.3 The mechanism of the thermal fragmentation

11.4.3.1 Film collapse dynamics

The arriving pressure wave has an effect on the vapor film depending on its *time signature*. If the time necessary for crossing the distance of the particle size is much smaller than the duration of the pressure pulse, we can assume that the film collapse is governed by the difference between the water bulk pressure and the vapor pressure. Different approaches are thinkable to compute the collapse time of a vapor film with thickness δ_{1F}. The contact discontinuity velocity for the case of negligible mass transfer is given by the mass and momentum jump conditions at the interface

$$V_{1\sigma} = V_1 + \frac{p_2 - p_1}{\rho_1 (V_2 - V_1)} \qquad (11.53)$$

valid for plane geometry. Even for plane geometry this approach gives constant velocity and no surface acceleration. It can, therefore, not explain the interfacial instabilities which are resulting from non-zero acceleration of the interface. That is why modeling the bubble collapse dynamics is required to compute the surface averaged interfacial acceleration. *Board* et al. [14] (1972) expressed the hypothesis that the cyclic nature of thermal fragmentation in subcooled water observed in the experiments is due to bubble growth and collapse. *Bankoff* et al. [6] succeeded in 1983 to reproduce very well the collapse time and the cyclic behavior of the process using the bubble collapse theory.

The following system of non-linear ordinary differential equations describes the bubble dynamics.

$$R\frac{d^2 R}{d\tau^2} + \frac{3}{2}\left(\frac{dR}{d\tau}\right)^2 = \frac{1}{\rho_2}\left(p_1 - p_2 - \frac{2\sigma}{R}\right), \quad R_3 \leq R \qquad (11.54)$$

$$\frac{dp_1}{d\tau} = \frac{3R^2}{\left(R^3 - R_3^3\right)} a_1^2 \left\{ \left[(\rho w)_{M21} - (\rho w)_{M12} \right] \left(\frac{1 + C_{n1} \frac{R_{M1} - R_{n1}}{R_1}}{+ \frac{c_{pM1}}{c_{p1}} \frac{T_2^{\sigma 1} - T_1}{T_1}} \right) - \rho_1 \frac{dR}{d\tau} \right\}, \tag{11.55}$$

$$\frac{dT_1}{d\tau} = \frac{1}{\rho_1 c_{p1}} \left\{ \frac{dp_1}{d\tau} + \frac{3R^2}{\left(R^3 - R_3^3\right)} \left[\dot{q}_1'' + (\rho w)_{M21} \left(h_{M1}^{\sigma 2} - h_{M1} \right) \right] \right\}, \tag{11.56}$$

$$\frac{dC_{n1}}{d\tau} = -\frac{C_{n1}}{\rho_1} \frac{3R^2}{\left(R^3 - R_3^3\right)} \left[(\rho w)_{M21} - (\rho w)_{M12} \right]. \tag{11.57}$$

The first equation is the momentum equation for incompressible fluid known as the *Rayleigh* equation. Note that for fast pressure change, which can be linearized by

$$p_2 = p_{20} + \frac{dp_2}{d\tau} \tau, \tag{11.58}$$

the *Rayleigh* equation is

$$R \frac{d^2 R}{d\tau^2} + \frac{3}{2} \left(\frac{dR}{d\tau} \right)^2 = \frac{1}{\rho_2} \left(p_1 - p_{20} - \frac{2\sigma}{R} - \frac{dp_2}{d\tau} \tau \right). \tag{11.59}$$

The second equation is the mass conservation equation of the gas film combined with the energy conservation equation of an ideal gas mixture. The third equation is the energy conservation equation of an ideal gas mixture. The last equation is the mass conservation equation of the non-condensable gas components. The dependent variable vector in this case is (R, p_1, T_1, C_{n1}).

For the case of no non-condensable gases in the film we have $C_{n1} = const = 0$ and the pressure and temperature equations simplify to

$$\frac{dp_1}{d\tau} = \frac{3R^2}{\left(R^3 - R_3^3\right)} a_1^2 \left\{ \left[(\rho w)_{M21} - (\rho w)_{M12} \right] \frac{T'(p_1)}{T_1} - \rho_1 \frac{dR}{d\tau} \right\}, \tag{11.60}$$

$$\frac{dT_1}{d\tau} = \frac{1}{\rho_1 c_{p1}}\left\{\frac{dp_1}{d\tau} + \frac{3R^2}{\left(R^3 - R_3^3\right)}\left[\dot{q}_1'' + (\rho w)_{M21} c_{pM1}\left(T'(p_1) - T_1\right)\right]\right\}. \quad (11.61)$$

Constitutive relations are required here in order to estimate the condensation and the evaporation mass flow rate from the liquid surface, and the total heat flux into the vapor. Thin thermal boundary layer solution of the *Fourier* equation is applicable to compute the heat flux from the interface into the water $\rho_2 c_{p2}\left(T_3^\sigma - T_2\right)\sqrt{\frac{3a_2}{\pi\tau}}$. The heat flux transferred from the hot melt surface through the vapor film is $\lambda_1 \frac{T_3 - T_3^\sigma}{R_1 - R_3}$. The net flux of evaporation is computed from the *Hertz* [47] equation modified by *Knudsen* [55] (1915) and *Langmuir* [61] (1913), *Langmuir* et al. [62] (1927).

$$(\rho w)_{M21} - (\rho w)_{M12} = c\left(\frac{p'(T_3^\sigma)}{\sqrt{2\pi R_{M1} T_3^\sigma}} - \frac{p_{M1}}{\sqrt{2\pi R_{M1} T_1}}\right) \quad (11.62)$$

In addition, for no sound physical reason, the assumption is made in the literature that the probability of escape for liquid molecule at the interface is equal to the probability of capture for a vapor molecule at the interface - the *c*-constant in the above equation. For pressures below 0.1 *bar* measurements for *c* reported by *Fedorovich* and *Rohsenow* [36] (1968) for liquid metals tightly clustered about unity. For larger pressure

$$c = 0.1 \text{ to } 1 \quad (11.63)$$

is a good approximation - see the discussion by *Mills* [71] and *Mills* and *Seban* [70]. For water the discussion and the measurements by [70, 73, 11] give

$$c = 0.35 \text{ to } 1. \quad (11.64)$$

Thus from the energy jump condition at the water surface the interface temperature is governed by

$$c\left(\frac{p'(T_3^\sigma)}{\sqrt{2\pi R_{M1} T_{di}}} - \frac{p_{M1}}{\sqrt{2\pi R_{M1} T'(p_1)}}\right)$$

$$= \frac{1}{h''(p_1) - h'(p_1)}\left[\lambda_1 \frac{T_3 - T_3^\sigma}{R_1 - R_3} - \rho_2 c_{p2}\left(T_3^\sigma - T_2\right)\sqrt{\frac{3a_2}{\pi\tau}}\right]. \quad (11.65)$$

It is interesting to note that for large film thickness, $R_1 - R_3$, the condensation predominates,

$$\lambda_1 \frac{T_3 - T_3^\sigma}{R_1 - R_3} < \rho_2 c_{p2} \left(T_3^\sigma - T\right) \sqrt{\frac{3a_2}{\pi \tau}} \ , \qquad (11.66)$$

and the pressure decreases. If the film thickness decreases sufficiently to have

$$\lambda_1 \frac{T_3 - T_3^\sigma}{R_1 - R_3} > \rho_2 c_{p2} \left(T_3^\sigma - T\right) \sqrt{\frac{3a_2}{\pi \tau}} \ , \qquad (11.67)$$

the pressure starts to increase. This physics combined with the momentum equation (11.54) (*Rayleigh*) describes the experimentally observed cyclic nature of the bubble collapse in subcooled liquid.

11.4.3.2 Interfacial instability due to bubble collapse

During the bubble collapse the vapor/water interface is subject to instability. As shown by the experiments performed by *Ciccarelli* [26], the pressure differences leading to surface oscillations do not lead to symmetric melt drop compression and successive hydraulic explosion as assumed by *Drumheller* [30], but to the following mechanism. The surface instabilities during the bubble collapse will cause melt-coolant contact at predominant places. The result is contact heat transfer, local explosion of highly superheated coolant pressing the melt at the initial contact points and causing them to erupt on the non-contact places. This mechanism leads to eruption of melt particles into the surrounding liquid causing a cloud of additional bubble production. The cloud is observed optically by several authors, e.g. [20] (1998). Successive condensation due to the resultant local pressure spike leads to collapse of the voided region for subcooled coolant. Then the process may repeat. This is known in the literature as a *cyclic nature of the thermal fragmentation*. The cyclic nature of the thermal fragmentation is observed in [53, 44, 75, 4, 99] (1974-1989). The droplet surface topology during the instability interactions is not directly visible from usual photographs, *Henry* and *Fauske* [46], but clearly demonstrated by the x-ray photographs by *Ciccarelli* [26]. The work by *Ciccarelli* provided the experimental evidence for the hypothesis which was asked for a long time. Analytical understanding of the process is provided by direct numerical simulation of the interaction and gives surprisingly similar results - see the work by *Koshizuka* et al. 1997 [59].

For saturated water only one cycle is usually experimentally observed. It is obvious that one cycle will cause fragmentation only of a surface droplet layer with limited thickness. Therefore this effect is a *surface entrainment effect*. The residual part of the hot droplets will remain not fragmented.

The most important part of this observation is that the whole process is driven by the energy transferred by the melt during the contact time at the contact melt/coolant spots.

At the beginning of the bubble collapse,

$$R\frac{d^2R}{d\tau^2} \gg \frac{3}{2}\left(\frac{dR}{d\tau}\right)^2, \qquad (11.68)$$

$$R \sim R_3, \qquad (11.69)$$

the interface acceleration is

$$\frac{d^2R}{d\tau^2} \approx \frac{1}{R_3\rho_2}\left(p_1 - p_2 - \frac{2\sigma}{R_3}\right) \qquad (11.70)$$

and therefore

$$\lambda_{RT} \sim \pi\left[\frac{12\sigma_2}{\frac{1}{R_3\rho_2}\left(p_1 - p_2 - \frac{2\sigma}{R_3}\right)(\rho_2 - \rho_1)}\right]^{1/2}, \qquad (11.71)$$

and

$$V_{2,jets} = 2\pi\left[3\sigma_2 \frac{1}{R_3\rho_2}\left(p_1 - p_2 - \frac{2\sigma}{R_3}\right)\frac{\rho_2 - \rho_1}{(\rho_2 + \rho_1)^2}\right]^{1/4}. \qquad (11.72)$$

11.4.3.3 Contact heat transfer

The purpose of this section is to estimate the local efficiency of possible contact heat transfer from fragmented melt to water after arrival of a trigger shock wave. After the pressure pulse collapses locally the vapor film, the wavy liquid interface touches the droplet. The heat transfer coefficients on those spots is much higher than the heat transfer in film boiling. *Inoue* and *Aritomi* [44] (1989) obtained the empirical expression for the local heat transfer coefficient as a function of the time elapsed from the contact initiation

$$h_{TB} \approx 10^5 \left[1-\left(2\tau/\Delta\tau_H\right)^2\right] \; W/(m^2 K), \qquad (11.73)$$

where $\Delta\tau_H \approx 30\times 10^{-6} s$. The average heat transfer coefficient within the contact time $\Delta\tau_H$ is $\overline{h_{TB}} \approx (2/3)\times 10^5 W/(m^2 K)$. It causes significant drop cooling at these spots and simultaneous superheating of the coolant liquid.

Analytical estimation of the local heat transfer as a function of time is possible in the following way.

11.4.3.3.1 Non-solidifying droplet

For contact spots much smaller than the droplet interface we can assume a plane geometry and use the solution of the *Fourier* equation

$$\frac{\partial T}{\partial \tau} = a\frac{\partial^2 T}{\partial x^2}, \qquad (11.74)$$

with initial conditions $T = T_3$ inside the particle and $T = T_2$ inside the coolant, requiring that T and $\lambda\frac{\partial T}{\partial x}$ be continuous at the interface. The spatial coordinate starts at the interface and points into the droplet. The short time solution of the temperature field inside the *non-solidifying* droplet is

$$T(x) = \frac{T_3\lambda_3\sqrt{a_2} + T_2\lambda_2\sqrt{a_3} + \lambda_2\sqrt{a_3}\left(T_3-T_2\right)erf\left(\frac{x}{2\sqrt{a_3\tau}}\right)}{\lambda_2\sqrt{a_3} + \lambda_3\sqrt{a_2}}. \qquad (11.75)$$

The instantaneous interface temperature is

$$T_{3\sigma} = \frac{T_3\sqrt{\rho_3 c_{p3}\lambda_3} + T_2\sqrt{\rho_2 c_{p2}\lambda_2}}{\sqrt{\rho_2 c_{p2}\lambda_2} + \sqrt{\rho_3 c_{p3}\lambda_3}}. \qquad (11.76)$$

This relation is sometimes written in the literature in the form

$$\frac{T_{3\sigma}-T_2}{T_3-T_{3\sigma}} = \sqrt{\frac{\rho_3 c_{p3}\lambda_3}{\rho_2 c_{p2}\lambda_2}}, \qquad (11.77)$$

see e.g. *Fauske* [35].

The instantaneous heat flux with corrector for spherical geometry is

$$\dot{q}''_{\sigma 2} = \rho_2 c_{p2} (T_{3\sigma} - T_2) \sqrt{\frac{3a_2}{\pi \tau}}$$

$$= \sqrt{\frac{3}{\pi}} \frac{1}{1/\sqrt{\rho_2 c_{p2} \lambda_2} + 1/\sqrt{\rho_3 c_{p3} \lambda_3}} \frac{T_3 - T_2}{\sqrt{\tau}}. \tag{11.78}$$

This is valid as far as the heat conduction within non-disintegrated water is possible. The water disintegrates due to homogeneous nucleation if the interface temperature is larger than the homogeneous nucleation temperature,

$$T_3^\sigma > T_{2,spin}. \tag{11.79}$$

That is why injecting water in liquid materials like steel or uranium dioxide always results in explosive interactions. The opposite, injection of molten materials in water not always results in explosive interaction.

For systems which have demonstrated propagating vapor explosions, the corresponding propagation velocity has been observed to be of the order of 100 m/s with a rise time for the shock wave of order of

$$\Delta \tau_{23} \approx 0.001s. \tag{11.80}$$

Obviously, the contact time cannot exceed this scale. It is interesting to know what the maximum of the energy is that can be transferred during this time. The heat flux averaged over this time is limited by

$$\dot{q}''_{32,average} = \sqrt{\frac{3}{\pi}} \frac{T_3 - T_2}{1/\sqrt{\rho_2 c_{p2} \lambda_2} + 1/\sqrt{\rho_3 c_{p3} \lambda_3}} \frac{1}{\Delta \tau_{12}} \int_0^{\Delta \tau_{23}} \frac{d\tau}{\sqrt{\tau}}$$

$$= 2\sqrt{\frac{3}{\pi}} \frac{T_3 - T_2}{1/\sqrt{\rho_2 c_{p2} \lambda_2} + 1/\sqrt{\rho_3 c_{p3} \lambda_3}} \frac{\sqrt{\Delta \tau_{23}}}{\Delta \tau_{23}}. \tag{11.81}$$

Over the period of time $\Delta \tau_{23}$ the ratio of the released energy

$$\pi D_3^2 2 \sqrt{\frac{3}{\pi}} \frac{T_3 - T_2}{1/\sqrt{\rho_2 c_{p2} \lambda_2} + 1/\sqrt{\rho_3 c_{p3} \lambda_3}} \sqrt{\Delta \tau_{23}}$$

to the available energy inside a single droplet

$$\frac{\pi}{6}D_3^3\rho_3\left\{c_{p3}\left[T_3-T'(p)\right]+h_3'-h_3'''\right\} \tag{11.82}$$

is

$$\eta_{efficency}=\chi\frac{\pi D_3^2 2\sqrt{\dfrac{3}{\pi}}\dfrac{T_3-T_2}{1/\sqrt{\rho_2 c_{p2}\lambda_2}+1/\sqrt{\rho_3 c_{p3}\lambda_3}}\sqrt{\Delta\tau_{23}}}{\dfrac{\pi}{6}D_3^3\rho_3\left\{c_{p3}\left[T_3-T'(p)\right]+h_3'-h_3'''\right\}}$$

$$=12\sqrt{\frac{3}{\pi}}\frac{T_3-T_2}{T_3-T'(p)+\dfrac{h_3'-h_3'''}{c_{p3}}}\frac{\chi\sqrt{\dfrac{\lambda_3}{\rho_3 c_{p3}}\Delta\tau_{23}}}{\left(1+\sqrt{\dfrac{\rho_3 c_{p3}\lambda_3}{\rho_2 c_{p2}\lambda_2}}\right)}\cdot\frac{1}{D_3}. \tag{11.83}$$

Here the χ is the ratio of the contact surface to the total surface of a single drop being between zero and one depending on the mode of contact disturbances. Comparing this expression with the one reported by *Henry* and *Fauske* in 1996 [46]

$$3\left[1+\frac{T'(p)-T_{2,spin}}{T_3-T'(p)}\right]\frac{\sqrt{\dfrac{\lambda_3}{\rho_3 c_{p3}}\Delta\tau_{12}}}{D_3}, \tag{11.84}$$

we see that the coolant properties are neglected by these authors relying on the assumption that the surface temperature is equal to the homogeneous nucleation temperature $T_{2,spin}$ and the latent heat of solidification is also neglected. These assumptions are not necessary and are omitted in our analysis.

11.4.3.3.2 Solidifying droplet

In the above derivation it is assumed that no freezing occurs at the surface, that is the surface temperature is always greater than the solidus temperature. Freezing at the surface is associated with release of the latent heat of melting. An approximate method to take this effect into account is to increase the effective temperature difference T_3-T_2 by $\dfrac{h_3'-h_3'''}{c_{p3}}$, as proposed by *Buchman* [17] in 1973. Thus the final expression for the efficiency is

$$\eta_{efficiency} = 12\sqrt{\frac{3}{\pi}} \frac{T_3 - T_2 + \frac{h_3' - h_3'''}{c_{p3}}}{T_3 - T'(p) + \frac{h_3' - h_3'''}{c_{p3}}} \frac{\chi \sqrt{\frac{\lambda_3}{\rho_3 c_{p3}} \Delta \tau_{23}}}{\left(1 + \sqrt{\frac{\rho_3 c_{p3} \lambda_3}{\rho_2 c_{p2} \lambda_2}}\right)} \frac{1}{D_3} \qquad (11.85)$$

$$\approx 12\sqrt{\frac{3}{\pi}} \frac{\sqrt{\frac{\lambda_3}{\rho_3 c_{p3}} \Delta \tau_{23}}}{1 + \sqrt{\frac{\rho_3 c_{p3} \lambda_3}{\rho_2 c_{p2} \lambda_2}}} \frac{\chi}{D_3}. \qquad (11.86)$$

For UO_2 $T_3 = 3000K$, $T_2 = 30°C$, $\chi = 1$, $\Delta \tau_{23} \approx 0.001s$ and atmospheric pressure we have

$$\eta_{efficency} = \frac{0.000133}{D_3}. \qquad (11.87)$$

For the same conditions for Al_2O_3 we have

$$\eta_{efficency} = \frac{0.000129}{D_3}. \qquad (11.88)$$

It interesting to note some valuable information contained in this expression:

a) The smaller the size of the particles before the thermal interaction the higher the efficiency, which is also confirmed by the experiments. The thermal efficiency of the heat transfer from particles with size smaller that 0.1 mm can be considered as 100%. The final size of the solidified fragments after steam explosion experiments is in order of $D_{3\infty} \approx (1 \text{ to } 100) \times 10^{-6} m$. It was measured by several experimentalists, e.g. *Kim* et. al. [53] (1989), (150 to 250) $\times 10^{-6} m$, see in *Corradini* [27] for Fe-Al_2O_3 systems, (1 to 700) $\times 10^{-6} m$, in KROTOS experiments, *Huhtiniemi* et al. [49]. Obviously the fine fragmented melt in steam explosion is able to release its thermal energy completely if the contact with the water is possible. Experimental data reported by *Buxton* et al. in 1979 [21] for Al_2O_3 and *Fe* thermite give thermal efficiency of the order of 0.1 to 1.4%. Recently *Corrardini* [28] (1996) summarized the Alumina-Water interaction data obtained from nine KROTOS explosion experiments [48] in confined geometry. Again all of the data give thermal efficiency between 0.1 and 3%. The reason for not having global thermal efficiency of 100% is the fact that only small part of the melt participates effectively in the interaction.

b) For particles larger than 0.1 *mm* the energy transferred during the liquid-liquid contact is limited by the amount stored inside the thermal boundary layer of the large premixed particles before the explosion. For 1 *mm* particle size and $\chi = 1$ the thermal efficiency for UO_2 is 13.3% and for Al_2O_3 is 12.9%. This demonstrates how important it is to know the particle size distribution during the interaction process. Non-explosive interactions produce debris in order of 1 to 3 *mm*, Corradini [27], 3.8 to 4.8 *mm*, FARO experiments with 80w%UO_2 and 20w% ZrO_2, *Magalon* et al [64]. This is in fact the range of particle size which should be used to estimate the efficiency of postulated non-explosive melt-water interactions for real reactor systems.

c) The water subcooling has virtually no effect on the local thermal efficiency.

d) The mode of contact, χ, greatly influences the local thermal efficiency. If there is locally no water, $\alpha_2 = 0$, no steam can be produced and therefore $\chi = 0$. Another interesting aspect resulting from this fact is that to transfer the thermal energy completely into evaporation of water in a control volume by some hypothetical mechanism, i.e.

$$\alpha_{2,st} \rho_2 \left(h_1'' - h'' \right) = \alpha_3 \rho_3 \left(h_3' - h''' \right), \tag{11.89}$$

we need a optimum liquid volume fraction of about

$$\alpha_{2,st} = \frac{1-\alpha_1}{1+\dfrac{\rho_2 \left(h_1'' - h'' \right)}{\rho_3 \left(h_3' - h''' \right)}}. \tag{11.90}$$

We call this liquid volume fraction "*stochiometric*" in analogy to the chemical combustion processes where the maximum of the released energy can not be increased by providing more fuel or more oxidizer than the stochiometric one. If the melt mixture is "*oversaturated*", that is

$$\alpha_2 > \alpha_{2,st}, \tag{11.91}$$

all the melt can virtually discharge its thermal energy. If the melt is "*undersaturated*", that is

$$\alpha_2 < \alpha_{2,st}, \tag{11.92}$$

only part of the melt,

$$\chi = \frac{\alpha_2}{\alpha_{2,st}} = \frac{\alpha_2}{1-\alpha_1}\left[1+\frac{\rho_2(h_1''-h'')}{\rho_3(h_3'-h''')}\right], \quad \text{for } \alpha_1 < 1 \qquad (11.93)$$

can virtually discharge its thermal energy. Both conditions can be written as

$$\chi = \min\left\{1, \frac{\alpha_2}{1-\alpha_1}\left[1+\frac{\rho_2(h_1''-h'')}{\rho_3(h_3'-h''')}\right]\right\} \quad \text{for } \alpha_1 < 1. \qquad (11.94)$$

For corium-water mixtures we obtain

$$\chi = \min\left\{1, \frac{\alpha_2}{1-\alpha_1}\left[1+\frac{1\,000 \times 2\,257\,200}{8\,105.9 \times 305\,350}\right]\right\} = \min\left(1, 1.912\frac{\alpha_2}{1-\alpha_1}\right). \qquad (11.95)$$

For a mixture consisting of melt and water only the *stochiometric* liquid volume fraction turns out to be 0.523. In other words, if the volume of the water in a control volume is less than 52% the complete release of the thermal energy is simply impossible. Increasing the vapor void fraction decreases this value to

$$0.523/(1-\alpha_1). \qquad (11.96)$$

Conclusions: There are three important thermal limitations on the fast melt-water heat transfer.

1) The contact time is limited;

2) The heat conduction in the thermal boundary layer is limited;

3) The local availability of water in the mixture is limited

Crust straight: If the contact temperature is lower than the solidus temperature the droplet starts to solidify at the surface. *Cronenberg* found in 1973 that the crystallization is much faster than heat transfer [29]. For plane geometry the growth of the crusts is governed by the solution obtained by *Carslaw* and *Jaeger* see Eq. (39) in [22], p 289,

$$\delta(\tau) = \psi 2\sqrt{a_3'''\tau}, \qquad (11.97)$$

where ψ satisfies the following equation

$$\frac{(T_3'''-T_2)\lambda_2\sqrt{a_3'''}\exp(-\psi^2)}{\lambda_3'''\sqrt{a_2}+\lambda_2\sqrt{a_3'''}erf(\psi)} - \frac{\lambda_3\sqrt{a_3'''}}{\lambda_3'''\sqrt{a_3}} \frac{(T_3-T_3''')\exp\left(-\psi^2\frac{a_3'''}{a_3}\right)}{erfc\left(\psi\frac{\sqrt{a_3'''}}{\sqrt{a_3}}\right)}$$

$$=\psi\sqrt{\pi}\frac{\Delta h_{3,melt}}{c_{p3}'''}.\qquad(11.98)$$

Thus any interactions leading to break up of the crust have to provide a local pressure larger than the ultimate pressive strength ($Al \approx 10^8 Pa$, $UO_2 \approx 2\times10^9 Pa$). If the break is caused by shearing, the shear stress should be larger than the ultimate shear strength.

It is important to note that for many systems of practical interest there is solidification at the cold spot. However, the explosion post-test debris analysis in KROTOS shows perfectly spherical fine debris which runs quickly on inclined surface of paper. This seems to contradict this prediction of local solidification. The surrounding of the micro-crust is liquid and the entrained droplets retain in fact the large part of the superheating energy so that after a while in a film boiling the complete melting is recovered and the further solidification mechanism is that of radiating micro-droplets in film boiling and having almost ideal spherical geometry.

11.4.3.4 Liquid coolant fragmentation

After the melt-coolant contact heat transfer the coolant boundary layer becomes metastable and expands dramatically due to homogeneous nucleation. This causes fragmentation of the coolant and highly turbulent oscillation of the liquid-liquid interface causing further fragmentation of the liquid metal droplet and of the coolant. *Henry* and *Fauske* [46] pointed out the importance of the fine *water* fragmentation at the water surface, the oscillation of the resulting fine spray, and the improvement of the heat transfer even under established vapor film conditions.

11.4.3.5 The maximum of the contact time

The purpose of this section is to estimate an upper limit of the possible contact time. We postulate that the absolute maximum is proportional to the time which is required to bring all of the liquid to homogeneous nucleation temperature by contact heat transfer.

Consider a water film surrounding the fragments, with a thickness proportional to the size of the fragments

$$\delta_{2F} \approx \delta_{2F}^* D_{3\infty},\qquad(11.99)$$

where $\delta_{2F}^{*} > 0$. The ratio of the volume of the sphere consisting of (a) one particle and (b) the surrounding film, to the volume of the particle itself is

$$(\alpha_3 + \alpha_{2F})/\alpha_3 = (D_3 + 2\delta_{2F})^3 / D_3^3 = (1 + 2\delta_{2F}^{*})^3. \tag{11.100}$$

Therefore the liquid volume fraction of the non-stable film is

$$\alpha_{2F} = \alpha_3 \left[(1 + 2\delta_{2F}^{*})^3 - 1 \right]. \tag{11.101}$$

In fact the film thickness is limited by the maximum packing density given by

$$\alpha_{2F} \leq 0.52 - \alpha_3 \tag{11.102}$$

or

$$\delta_{2F}^{*} \leq \delta_{2F,\max}^{*}, \tag{11.103}$$

where

$$\delta_{2F,\max}^{*} = \frac{1}{2}\left[\left(\frac{0.52}{\alpha_3}\right)^{1/3} - 1\right]. \tag{11.104}$$

An estimate of the time constant is obtained by the condition that the average temperature of the liquid equals the homogeneous nucleation temperature,

$$\frac{\pi}{6} D_3^3 \left[(1+\delta_{2F}^{*})^3 - 1\right] \rho_2 c_{p2} (T_{2,nc} - T_2) \frac{1}{\Delta \tau_{23}}$$

$$= \pi D_3^2 (1+\delta_{2F}^{*})^2 2\sqrt{\frac{3}{\pi}} \frac{T_3 - T_2}{1/\sqrt{\rho_2 c_{p2} \lambda_2} + 1/\sqrt{\rho_3 c_{p3} \lambda_3}} \frac{\sqrt{\Delta \tau_{23}}}{\Delta \tau_{23}}. \tag{11.105}$$

Solving with respect to the liquid *Fourier* number, we finally obtain

$$\frac{\Delta \tau_{23}}{D_3^2} \frac{\lambda_2}{\rho_2 c_{p2}} \approx \frac{\pi}{432} \left\{ \frac{\left[(1+\delta_{2F}^{*})^3 - 1\right]}{(1+\delta_{2F}^{*})^2} \left(1 + \sqrt{\frac{\rho_2 c_{p2} \lambda_2}{\rho_3 c_{p3} \lambda_3}}\right) \frac{T_{2,nc} - T_2}{T_3 - T_2} \right\}^2. \tag{11.106}$$

11.4 Thermo-mechanical fragmentation of liquid metal in water

Thus the maximum of the contact time is given by replacing of δ_{2F}^* with $\delta_{2F,\max}^*$. The real contact time will surely be smaller than the maximum one. To estimate it one needs additional mechanistic arguments.

11.4.3.6 Marangoni effect

Next we discuss the role of the *Marangoni* effect on the thermo - mechanical fragmentation. The surface tension for liquid metals increases with decreasing temperature, $d\sigma_3/dT_3 < 0$, in

$$\sigma_3 \approx \sigma_{30} + \frac{d\sigma_3}{dT_3}(T_{30} - T_3). \tag{11.107}$$

Examples of the temperature gradient of the surface tension for different substances $d\sigma/dT$ in $kg/(s^2 K)$ are given below, compare with *Brennen* [16], p. 132.

Water	2.02×10^{-4}	Oxygen	1.92×10^{-4}
Uranium Dioxide	1.11×10^{-4}	Methane	1.84×10^{-4}
Sodium	0.90×10^{-4}	Butane	1.06×10^{-4}
Mercury	3.85×10^{-4}	Carbon Dioxide	1.84×10^{-4}
Hydrogen	1.59×10^{-4}	Ammonia	1.85×10^{-4}
Helium-4	1.02×10^{-4}	Toluene	0.93×10^{-4}
Nitrogen	1.92×10^{-4}	Freon-12	1.18×10^{-4}

Therefore, the cooling of the spots causes significant tangential forces at the surface,

$$\tau_3^\sigma \approx \frac{\Delta \sigma_3}{\delta_C} = -\frac{d\sigma_3}{dT_3} \Delta T_3^\sigma / \delta_C. \tag{11.108}$$

δ_C ($\approx const\, D_3$, $const < 1$) is the distance between the neighboring hot and cold places, i.e. 1/2 of the surface wavelength along which the temperature change ΔT_3^σ takes place. These forces induce a thermo-capillary flow in the drop that is resisted by the viscosity of the drop, i.e.

$$\tau_3^\sigma = \eta_3 \frac{V_{Mar}}{\delta_3} = -\frac{d\sigma_3}{dT_3} \Delta T_3^\sigma / \delta_C, \tag{11.109}$$

δ_3 is some characteristic thickness of the wavy surface. *Ostrach* [80] (1982) solved with respect to V_{Mar} and obtained the characteristic *thermo-capillary velocity* caused by these forces

$$V_{Mar} \approx -\frac{d\sigma_3}{dT_3}\frac{\Delta T_3^\sigma}{\eta_3}(\delta_3/\delta_C). \qquad (11.110)$$

The surface wave needs the time

$$\tau_{br,TM} \approx \delta_C / V_{Mar} \qquad (11.111)$$

to collapse and to create a droplet (a kind of micro jet). The dimensional analysis provides an estimate for the thickness of the disturbed surface layer

$$\delta_3 \approx (v_3 \tau_{br,TM})^{1/2} \approx (v_3 \delta_C / V_{Mar})^{1/2}. \qquad (11.112)$$

After substituting δ_3 in Eq. (11.110) with the RHS of Eq. (11.112) and solving with respect to V_{Mar} one obtains

$$V_{Mar} \approx \left(-\frac{1}{\rho_3}\frac{d\sigma_3}{dT_3}\Delta T_3^\sigma\right)^{2/3}(v_3 \delta_C)^{-1/3}. \qquad (11.113)$$

Therefore the velocity induced by thermo-capillary forces can be significant. These forces transport material from the hot to the cold spots. This is an important example of the well known *Marangoni* effect, originally discovered by *Thomson* [92] (1855). *Marangoni* [65] (1871) found that the presence of a surface tension gradient along the interface causes convective currents or vortices close to the interface. The transported mass accelerates the growing of the surface waves induced by *Taylor* instability and causes fragmentation - see the discussion by *Henkel* [45] (1987) p 34. We call this mechanism *thermo-mechanical fragmentation* in order to distinguish it from the pure acceleration induced fragmentation called pure mechanical fragmentation.

11.4.3.7 Coolant interface classification

Injecting water in transparent liquid salt with temperature higher than the nucleation temperature of the water *Zimanovsky* et al. [100] observed classification of the interface of the salt and cracks characteristic for solid materials. The cracks are caused due to the enormous local pressure increase due to the water fragmentation by homogeneous nucleation. The explanation that the final fragments have smooth surface is that the energy release from the melt is not enough to solidify the melt fragments and they fly away from the interaction zone in liquid state.

11.4.3.8 Particle size after thermal fragmentation

The third condition for thermo-mechanical liquid metal fragmentation is that the initial droplet size, D_3, is considerably greater than the size of the final fragments, $D_{3\infty}$,

$$D_3 \gg D_{3\infty}. \tag{11.114}$$

As already mentioned before, the final size of the solidified fragments being of the order of $D_{3\infty} \approx (1 \text{ to } 100) \times 10^{-6} m$, is measured by several experimentalists, e.g. Kim et al. [53] (1989), $(150 \text{ to } 250) \times 10^{-6} m$, see in *Corradini* [28] for Fe-Al$_2$O$_3$ systems, $(1 \text{ to } 700) \times 10^{-6} m$, in KROTOS experiments, *Huhtiniemi* et al [49]. For comparison remember that non-explosive interactions produces debris in order of 1 to 3 mm, *Corradini* [28], 3.8 to 4.8 mm, FARO experiments with 80w%UO$_2$ and 20w% ZrO$_2$, *Magalon* et al [44]. *Inoue* and *Aritomi* [44] (1989) estimated $D_{3\infty} \approx 50 \text{ to } 60 \times 10^{-6} m$ assuming that the *Taylor* instability before solidification is responsible for the scale of the fragments which is of the order of magnitude observed experimentally. The fastest growing wavelength *Taylor* [91] (1950) is

$$D_{3\infty} \approx \delta_{3Taylor} = \pi \left[24 \frac{\sigma_3}{\frac{d}{d\tau}(\Delta V_{13})(\rho_3 - \rho_1)} \right]^{1/2}, \tag{11.115}$$

or

$$D_{3\infty} \approx \delta_{3Taylor} = \pi \left[24 \frac{\sigma_3}{\frac{d}{d\tau}(\Delta V_{23})(\rho_3 - \rho_2)} \right]^{1/2}, \tag{11.116}$$

where $\frac{d}{d\tau}(\Delta V_{13})$ and $\frac{d}{d\tau}(\Delta V_{23})$ are the accelerations of the unstable surfaces between gas and droplets and between liquid and droplet, respectively.

It appears that the *Taylor instability* starts the fragmentation and the *Marangoni effect* finishes it. Therefore we can assume that the final diameter of the newly created droplet, $D_{3\infty}$, should be between the *Taylor* wavelength and the dimension computed by the following consideration. The static pressure in the liquid droplets exceeds the environment pressure by approximately the pressure difference caused by the complete collapsed surface wave (the stagnant pressure of the wave) and therefore

$$\Delta p \approx \frac{1}{2}\rho_3 V_{Mar}^2. \qquad (11.117)$$

The static force balance on the newly formed droplet gives

$$D_{3\infty} \approx 4\sigma_3 / \Delta p \approx 4\sigma_3 / \left(\frac{1}{2}\rho_3 V_{Mar}^2\right). \qquad (11.118)$$

11.5 Particle production rate during the thermal fragmentation

Next we compute the particle production rate. If the liquid is saturated $T_2 \approx T'(p)$ no further fragmentation cycle can be expected in the microscopic time scale because the produced bubble did not collapse. If the liquid is sub-cooled, $T_2 < T'(p)$, the successive condensation causes the collapse of the so formed bubble and the residual hot fragment is subject to new fragmentation - new cycle of surface entrainment.

It is obvious that this mechanism will cause fragmentation only for a surface droplet layer with some thickness δ_3. Therefore, as already mentioned, this effect is a *surface entrainment effect*. The residual part of the hot droplets will remain not fragmented. This is observed by several authors, e.g. [53] (1989). The *cyclic nature of the fragmentation* is reported in [53, 44, 75, 4, 102] (1974-1989). *Inoue* and *Aritomi* estimated by comparisons with experiments that a fraction f_1 of the initial drop mass m_3 participates in the first fragmentation cycle, where $f_1 \approx 0.2$. Therefore, after the first cycle we have a number of $(D_3/D_{3\infty})^3 f_1$ fragments originating from one droplet and accelerated into the surrounding liquid. The second fragmentation is again a surface effect, as the first one, and takes only a part f_2 of the residual mass of the first fragmentation $(1-f_1)m_3$. *Inoue* and *Aritomi* [44] (1989) estimated this part as $f_2 \approx 0.8$ by comparison with experiments. Therefore, after the second fragmentation cycle we have $(D_3/D_{3\infty})^3 (1-f_1)f_2$ additional fragments per one particle. Obviously, after K cycles of fragmentation we have

$$n_{3\infty} = n_3^+ \left\{1 + (D_3/D_{3\infty})^3 \left[f_1 + (1-f_1)f_2 \sum_{k=2}^{K}(1-f_2)^{k-2}\right]\right\} \qquad (11.119)$$

particles per unit mixture volume, where K is the number of the finished fragmentation cycles. The production of particles per unit mixture volume rate for the time $\Delta \tau_{br}$ is therefore

$$\dot{n}_{3,sp} = (n_{3\infty} - n_3)/\Delta \tau_{br} = n_3 f_{3,frag}, \qquad (11.120)$$

where

$$f_{3,frag} = \left(f_3^+ \left\{ 1 + (D_3/D_{3\infty})^3 \left[f_1 + (1-f_1) f_2 \sum_{k=2}^{K} (1-f_2)^{k-2} \right] \right\} - 1 \right) \Big/ \Delta \tau_{br} \qquad (11.121)$$

is the fragmentation frequency of a single particle.

Most of the authors observe a delay time interval needed for the mechanical distortions of the film [31, 12, 13] (1974-1986). The time needed for the mechanical film distortion, $\Delta \tau_{mfd}$, is observed by *Kim* et al. [53] (1989) as

$$\Delta \tau_{mfd} < 200 \times 10^{-6} s, \qquad (11.122)$$

and the time needed for one fragmentation event is

$$\Delta \tau_{br} \approx 5 \times 10^{-6} s. \qquad (11.123)$$

The time needed for the following bubble collapse is obviously a function of the water subcooling

$$\Delta \tau_{collapse} \approx f\left[T_2 - T'(p)\right]. \qquad (11.124)$$

The time for one fragmentation cycle is the sum of the above times. It is estimated by *Inoue* and *Aritomi* in [44] (1989) to be

$$\Delta \tau_k = \Delta \tau_{collapse} + \Delta \tau_{fr} \approx 30 \times 10^{-6} s. \qquad (11.125)$$

Therefore the number of the finished fragmentation cycles is

$$K = \text{integer}\left(\Delta \tau - \Delta \tau_{mfd}\right)/\Delta \tau_k \quad \text{for} \quad T_2 < T'(p) \qquad (11.126)$$

$$K = 1 \quad \text{for} \quad T_2 \approx T'(p). \qquad (11.127)$$

Thus, the breakup time required to compute the production rate is

$$\Delta\tau_{br} = \Delta\tau_{mfd} + K\Delta\tau_{k}. \tag{11.128}$$

The thermo-mechanical fragmentation needs further attention. The experimental work provided up to now gives the *superposition of the effects of the mechanical and thermo-mechanical fragmentation but not quantitative information on the separate phenomena*. From the point of view of computational analysis, fine spatial discretization is necessary in order to model accurately the local pressure gradient and therefore to predict the trigger conditions.

11.6 *Tang*'s thermal fragmentation model

Tang [89] considered the thermal fragmentation as a process of surface entrainment of particles with entrainment mass flow rate

$$(\rho w)_{32,termal} = const\, \rho_3 V_{3,jets} F_1(\alpha_1) F_2(\tau_{br}), \tag{11.129}$$

governed by *Rayleigh-Taylor* instabilities at the water-vapor interface caused by the pressure difference at both sides of the interface, Δp. This pressure difference causes bubble collapse governed by the *Rayleigh* equation at the beginning of the bubble collapse where the velocity of the bubble surface is zero and the initial acceleration is

$$\frac{d^2 R}{d\tau^2} = \frac{p_1 - p_2}{R\rho_2}. \tag{11.130}$$

The interface is subject to instability with the growing rate

$$V_{2,jets} \approx const \left(\frac{d^2 R}{d\tau^2} \lambda_{RT} \right)^{1/2} \approx const \left(\frac{p_1 - p_2}{R\rho_2} \lambda_{RT} \right)^{1/2}. \tag{11.131}$$

The instability wavelength is assumed to be proportional to the particle size $\lambda_{RT} \approx R$ and therefore

$$V_{2,jets} \approx const \left(\frac{d^2 R}{d\tau^2} \lambda_{RT} \right)^{1/2} \approx const \left(\frac{p_1 - p_2}{\rho_2} \right)^{1/2}. \tag{11.132}$$

The reaction of the melt is again expressed in terms of surface instability with growing rate proportional to those of the water instability. Therefore melt jets erupt from the melt surface with

$$V_{3,jets} \approx const \; V_{2,jets}. \tag{11.133}$$

The entrainment mass source per unit time and unit mixture volume is consequently

$$\mu_{32,thermal} = (F/V)_3 (\rho w)_{32,thermal} = const \; \alpha_3 \rho_3 \left(\frac{\Delta p}{\rho_2 D_3^2}\right)^{1/2} F_1(\alpha_1) F_2(\tau_{br}). \tag{11.134}$$

The constant is proposed to take values smaller than 0.12. The function

$$F_2(\tau_{br}) = 1 - \Delta\tau/\tau_{br} \quad \text{for} \quad \Delta\tau < \tau_{br} \tag{11.135}$$

and

$$F_2(\tau_{br}) = 0 \quad \text{for} \quad \Delta\tau \geq \tau_{br} \tag{11.136}$$

limits the fragmentation process within the fragmentation time of $\tau_{br} \approx 0.002$. *Yerkess* [96] uses the *Kim* and *Corradini* model [52] and derived from the KROTOS experiments the following corrections

$$const = 0.002, \tag{11.137}$$

$$F_1(\alpha_1) = 0.5 - \arctan\left[\frac{100(\alpha_1 - 0.35)}{\pi}\right], \tag{11.138}$$

$$\tau_{br} = 0.0012 s. \tag{11.139}$$

For the prediction of the KROTOS experiments *Yerkess* assumed that the external trigger pressure pulse should be greater than some threshold value

$$p_2 > p_{2,th} \tag{11.140}$$

where

$$p_{2,th} \sim 2 \times 10^5 \; Pa. \tag{11.141}$$

The definition of the pressure difference is simple in the single drop case with stepwise pressure increase. This makes it difficult to apply this model in general purpose computer code where quite different pressure histories, resulting in different Δp, can be the reason for successive thermal fragmentation. *Brayer* et al. [15]

overcome this difficulty by using the time average pressure from many previous time steps as initial pressure.

Recently *Koshizuka* et al. [59] clarified the problem whether water jets can penetrate heavier melt droplets. Using direct numerical simulation the authors conclude that it depends on the density ratio. For heavier melt they found that penetration of water into the melt is impossible for velocities of 5 *m/s*. Note, that cavitation of bubbles in subcooled liquid may lead to considerably higher velocities.

Yuen et al. [99] noted that this approach does not explain how these microscopic jets can survive in the heating environment and penetrate the melt droplet surface. The *Marangoni* effect will counteract the water penetration into the melt at the contact point. The melt water contact will for sure transfer energy and cause nucleation explosion in the contact places. This will accelerate much more two-phase mist volume in all possible directions and aggravate the melt eruptions from contact surroundings as clearly demonstrated by the experiments of *Ciccarelli* [26]. Further, the volumetric displacement will be for sure associated with the number of contact points depending of the maximum growing wavelength during the *Rayleigh-Taylor* instabilities.

At least one improvement of this model is possible: There is no need to assume $\lambda_{RT} \approx R$. In this case setting again the melt jet velocity proportional to the coolant micro jet velocity

$$V_{3,jets} \approx const\ V_{2,jets}$$

$$= const\ 2\pi \left[3\sigma_2 \frac{1}{R_3 \rho_2} \left(p_1 - p_2 - \frac{2\sigma}{R_3} \right) \frac{\rho_2 - \rho_1}{(\rho_2 + \rho_1)^2} \right]^{1/4}, \qquad (11.142)$$

we obtain

$$\mu_{32,thermal} = (F/V)_3 (\rho w)_{32,thermal}$$

$$= const\ 12\pi 6^{1/4} \alpha_3 \rho_3 \left[\frac{\sigma_2}{\rho_2 D_3^5} \left(p_1 - p_2 - \frac{2\sigma}{R_3} \right) \frac{\rho_2 - \rho_1}{(\rho_2 + \rho_1)^2} \right]^{1/4}. \qquad (11.143)$$

Comparing this expression we see that $\mu_{32,thermal} \propto (p_1 - p_2)^{1/4}$ instead of $(p_1 - p_2)^{1/2}$ and $\mu_{32,thermal} \propto \frac{1}{D_3^{5/4}}$ instead of $\mu_{32,thermal} \propto \frac{1}{D_3}$ as obtained by *Tang* [89].

11.7 *Yuen*'s thermal fragmentation model

The thermal fragmentation model proposed by *Yuen* et al. [99] is applied by the authors to already existing mechanically pre-fragmented melt droplets. The further fragmentation is considered as a sheet stripping and wave crest stripping governed by the water-melt velocity difference and their properties. The latter is very important. The dependence on the pressure gradients is taken into account as a dependence on the resulting velocity differences. This is an implicit statement saying *what ever the reason for the origination of the relative velocity between water and melt corresponding to Weber number greater than 100 is, the thermal fragmentation takes place*. Note that in accordance with the *Ciccarelli* experiment this is correct [26] if the fragmentation time constant is shortened for the thermal fragmentation by factor of about two. This is also supported by the experiments performed in *Bürger* et al. [18, 19].

The new idea is that the fine fragments interact not with all of the water in the computational cell but with an amount proportional to the fine fragmentation rate. The proportionality factor was *empirically* chosen to be 7 based on volume. This amount is considered to immediately reach the fine particles/water thermal equilibrium, which may result in complete evaporation of the entrained water and superheating of the steam. Further interaction of the fine particles with the water is not considered. We call this phenomenon vapor shielding.

11.8 Oxidation

Liquid metals can oxidize in steam/water environment with different intensity. Three consequences of the oxidation are very important: a) Production of hydrogen; b) Heat release; c) Covering the fragment with oxide having given permeability for the reactant materials and melting temperature which may be higher than the metal melt temperature. As an example we consider the oxidation the aluminum next.

Aluminum: Molten aluminum interacts with water [63] as follows

$$2Al + 3H_2O \rightarrow Al_2O_3 + 3H_2.$$ (11.144)

The melting temperature of aluminum is 933.2 K [50] and the melting temperature of the aluminum oxide (alumina) is 2324.15 K [87]. Consequently, if the initial temperature of the molten aluminum is less than 2324.15 K oxide layer forms and crystallizes very quickly [93]. The speed of the chemical reaction is then controlled by the diffusion processes at both sides of the interaction front: diffusion of molten metal into the oxide and diffusion of steam through the steam - hydrogen boundary layer and again partially into the oxide. It is known that for this system the interaction front is close to the outer surface [69]. The participating thermal processes are the heat transfer by film boiling and radiation at the external surface and the heat transfer into the both materials oxide and metal. The film boiling at sphere is well understood as demonstrated in Chapter 21 and in [57] by comparison with about 2000 experimental data points in different regimes, as well the transient heat conduction in multi-layers sphere - see in [22]. Theoretical considerations containing the available models for the above discussed processes are presented in [94]. The most important result of this study was the ignition temperature of aluminum as a function of the particle size. Figure 16 in [94] shows in log-normal scale a linear dependence between about 1440 and 1720 K for particle sizes varying between 0.001 and 10 mm. These results are fairly close to the results obtained previously by *Epstein* [34]. These results are consistent with the experimental observation reported from the TREAT experiments (rapid heating) [32] in which temperatures of 1473 to 2273.15 K are necessary to burn the aluminum alloy under water. This result is also consistent with the experiments performed on the THERMIR facility at Winfrith which show that just pouring the 1123.15 K molten charge into subcooled water was not able to initiate a steam explosion [88].

11.9 Superposition of thermal fragmentation

11.9.1 Inert gases

Because the thermal fragmentation is associated with film condensation and with local melt/coolant contact any mechanism suppressing condensation in the film acts as a suppressive for steam explosion. There are two experimentally observed mechanisms:

a) entrapment of non-condensables if fragmented melt is falling through a atmosphere of non-condensable gases before entering the coolant as reported by *Akiyoshi* et al. [1] or artificially introduced bubbles of non-condensable gases *Zimanowski* et al. [100], and

b) melt oxidation leading to generation of inert gases inside the vapor film, *Nelson* et al. [78], *Corradini* [28].

Increasing the content of the non-condensable gases in the vapor film reduces the coolant/vapor interface oscillation, *Kim* and *Corradini* [51], and the reduces the maximum peak of the produced pulses *Akiyoshi* et al. - see fig. 8 in [51].

It is very interesting to note that recently the FARO experiments shows that heating UO_2 up to melting temperature makes the oxide under-stochiometric. During the melt-water interaction steam reduction is possible and H_2 production up to

0.150 *kg* per 175 *kg* melt is possible, releasing 300 to 500 *kJ* per mole of reacted UO_2 – see *Matzke* [67].

11.9.2 Coolant viscosity increase

Because the thermal fragmentation is associated with the stability of the coolant/vapor interface, increasing the viscosity of the coolant increases the stability of the coolant/vapor interface.

Bang and *Kim* [5] changed the water viscosity by adding *polyethylene oxide polymer*. 800*ppm* solutions have 2.7-times larger viscosity then pure water, see fig. 2 in [5]. The authors reported that for subcooled boiling the increase of the coolant viscosity decreases the minimum film boiling temperature and suppresses the instabilities caused by transition of the film boiling temperatures. With the used external trigger with amplitude 50 *kPa* 300*ppm* solution suppressed completely the explosion.

For water the viscosity, nominally at atmospheric pressure 0.001 *kg/(ms)*, can be increased to 0.04 and 0.24 *kg/(ms)* by adding *cellulose gum* into the water 0.1 and 0.4 *w/o*, respectively as reported by *Kim* et al. [53] (1989). The authors reported the following:

> "...The mechanism of suppressing steam explosion is the mechanism of suppressing the trigger. The increase of coolant viscosity >0.05 *kg/(ms)* at fixed trigger pressure (and impulse) results in complete explosion suppression. As the trigger pressure was increased the explosion could be triggered for intermediate viscosities, but suppression still remained for high viscosities >0.15 *kg/(ms)*. The explosion conversion ratio varies from 0 to 6 %."

This in fact confirms the findings reported previously by *Nelson* and *Guay* [76] in 1986. The latter authors found that adding *glycerol* in water requires 70 *w/o* in order to suppress explosions, which is rather large. The same effect can be reached by adding of 0.5 *w/o* cellulose gum.

11.9.3 Surfactants

Surfactants are substances changing the coolant/vapor interface properties. An examples of an-ionic surfactant is sodium- dodecyl- benzene- sulphonate $C_{12}H_{25}C_6H_4SO_3Na$ which is available as UFSAN-65. An example of a non-ionic surfactant is the ethoxilated- nonyle- phenol, $C_9H_{19}C_6H_4O(C_2H_4O)_9H$, available as Emulgator U-9. They possess long molecules being at the one end hydrophilic and at the other end hydrophobic. The hydrophilic end is always inside the bulk of the water and the hydrophilic tends to attach to the vapor side across the interface.

The major effects of the dense surfactant population (1 to 5 *ppm*) at the interface as reported in [8] are

a) decrease of surface tension (in small window of the concentration change),
b) local increase of water viscosity,
c) local increase of liquid density,
d) imposes a rigid surface on the bubble.

Becker and *Linland* [8] explained the stabilizing effect with the surface concentration *Marangoni* effect increasing the stability of the vapor/water interface. *Becker* and *Linland* drew the attention to these substances in [8] especially because of their property to suppress steam explosions reporting the following:

> "...As a matter of fact the surfactants have been used successfully in two metallurgical factories in Norway. In these factories copper granulate and ferrous alloys granulate are produced by pouring molten material into about 5 *m* deep pools of water, where the melts fragments hydrodynamically, sinks through the water, freezes and finally is collected as granulate at the bottom of the vessels. In the copper granulate factory 800 *kg* of copper is poured each time, while in the other factories the content of one batch is 8 000 *kg*. Before, strong steam explosions with severe damage to equipment and buildings occurred in both plants, but after the systematic use of surfactants was introduced no steam explosions have taken place in the copper factory, despite the fact that more than 10 000 pourings have been carried out until ...(April 1991). With regards to the ferrous alloy plant only mild interactions with water splashing out of the vessel have occurred on a few occasions. The number of pourings in this latter plants have now (April 1991) exceeded 20 000 since the surfactants were introduced."

Unfortunately small scale experiments performed later confirmed *statistically* the general trend by allowing occasionally steam explosions [8]. *Kowal* et al. [60] confirmed that a small concentration *on average* reduces the severity of steam explosions (*g*-quantities of 800°*C*-tin in water) by reducing the resulting pressure peak by 65% compared to pure water. There was no evidence in their study that surfactants can completely suppress steam explosions. *Chapman* et al. [23] performed experiments with *kg*-quantities of 1000°*C*-tin in water and again did not find the evidence that the surfactants mitigate steam explosions.

11.9.4 Melt viscosity

Since the pioneer work by *Ohnesorge* [79] in 1936 it is well known that increasing the liquid viscosity reduces the fragmentation. In several experiments particulate liquids with particle *Ohnesorge* number larger than 4 are not observed to fragment - see Fig. 8.8.

Nomenclature

Latin

a	normal interface acceleration pointing from the heavier to the lighter continuum, m/s^2
a_m	pressure pulse propagation velocity through the mixture, m/s
a_2	coolant sound velocity, m/s
C_{n1}	mass concentration of the non-condensable gases in the gas-vapor film, *dimensionless*
c_p	specific heat at constant pressure, $J/(kgK)$
D_3	hot spheres diameter, m
$D_{2,jet}$	jet diameter towards the wall in collapsing bubble, m
d	differential
f_3^+	part of the particles n_3 per unit mixture volume experiences melt/water fragmentation, *dimensionless*
g	gravitational acceleration, m/s^2
h	specific enthalpy, J/kg
h_{TB}	local heat transfer coefficient, $W/(m^2K)$
$\overline{h_{TB}}$	average heat transfer coefficient within the contact time $\Delta\tau_H$, $W/(m^2K)$
h_{FB}	Film boiling heat transfer coefficient, $W/(m^2K)$
k_{SB}	Stefan-Boltzman constant,
L_{pool_depth}	water reservoir water depth, m
$\ell_{2,jet}$	length jet towards the wall in collapsing bubble, m
n_3	hot spheres particle number, *dimensionless*
p	pressure, Pa
$p_{2,cr}$	critical pressure, Pa
$p_{2,th}$	threshold pressure, Pa
$p_{2,jet,WH}$	pressure exerted by the jet - "water hammer pressure", Pa
$p_{2,jet}$	$=\frac{1}{2}\rho_2 V_{2,jet}^2$ stagnant jet pressure, Pa
$\dot{q}_3^{"\sigma 2}$	heat flux emitted from the surface, W/m^2
$\dot{q}_{32,average}^{"}$	time-averaged heat flux, W/m^2
R	bubble radius, m
R_{10}	initial size of the collapsing bubble, m
R_{M1}	vapor gas constant, $J/(kgK)$
r	ratio of the film thickness for natural convection only, *dimensionless*

s_3	locally average field specific entropy, $J/(kgK)$
s''	liquidus entropy of the molten drop, $J/(kgK)$
$T'(p)$	saturation temperature as a function of pressure p, K
T_3^σ	droplet surface temperature, K
$T_{FB,\min}$	minimum film boiling temperature, K
$T_{2,cr}$	critical temperature, K
$T_{2,spin}$	homogeneous nucleation temperature, K
$V_{2,jets}$	coolant micro jet velocity, velocity jet towards the wall in collapsing bubble, m/s
V_{Mar}	characteristic thermo-capillary velocity, m/s
$\langle w_1 \rangle_1$	vapor velocity averaged over the film thickness computed for particle having large size compared to the film thickness, m/s

Greek

α	volume fraction, *dimensionless*
$\alpha_{2,st}$	optimum "*stochiometric*" liquid volume fraction, *dimensionless*
$\Delta \mathbf{V}_{23}$	relative velocity vector particle-water, m/s
ΔV_{32}	$= \sqrt{\Delta u_{32}^2 + \Delta v_{32}^2 + \Delta w_{32}^2}$, magnitude of the relative velocity vector particle-water, m/s
$\Delta u_{32}, \Delta v_{32}, \Delta w_{32}$	components of the relative velocity vector particle-water in the three coordinate directions, m/s
Δp_{pulse}	pulse pressure increase, Pa
$\Delta \tau_{pulse}$	pulse duration, s
$\Delta \tau_{life\_time}$	time necessary to reach the bottom of the reservoir, s
$\Delta \tau_{crust}$	time necessary to cool down the liquid particle up to formation of the surface crust, s
Δw_{32}	average velocity with which the droplet crosses the water, m/s
$\Delta \tau_{br}$	fragmentation time, s
$\Delta \tau_{br}^*$	$= \dfrac{\Delta \tau_{br}}{\dfrac{D_3}{|\Delta V_{3c}|}\sqrt{\dfrac{\rho_3}{\rho_c}}}$, fragmentation time, *dimensionless*
$\Delta \tau_H$	time elapsed from the contact initiation, s
δ_{1F}	film thickness at the equator, m
$\delta_{1F,nc}$	film thickness for natural convection only, m
$\delta_{1F,fc}$	film thickness for forced convection only, m

δ_{1F}^*	$=\dfrac{\delta_{1F}}{D_3}$	dimensionless film thickness
δ_C		distance between the neighboring hot and cold places, m
$\eta_{efficency}$		efficiency, *dimensionless*
λ_0		initial disturbance wavelength, m
λ		wavelength, m
λ_{RT}		wavelength of the fastest growing oscillation, m
λ		thermal conductivity, $W/(mK)$
$\Delta\tau_{RT}$		growing time constant of the fastest growing oscillation, s
μ_{21}		mass generated due to film boiling per unit mixture volume and unit time, $kg/(m^3s)$
ρ		density, kg/m^3
$(\rho w)_{M21}$		evaporation mass flow rate, $kg/(sm^2)$
$(\rho w)_{M12}$		condensation mass flow rate, $kg/(sm^2)$
$(\rho w)_{32,termal}$		entrainment mass flow rate due to thermo-mechanical fragmentation, $kg/(sm^2)$
σ		surface tension, N/m
τ		time, s
τ_3^σ		tangential shear at the surface, N/m^2
∇		gradient

Subscripts

1	gas
2	liquid
3	particles
1F	vapor (gas) film
m	mixture
M	vapor
n	inert gas

Superscripts

'	saturated steam
''	saturated liquid
'''	saturated solid phase

References

1. Akiyoshi R, Nishio S, Tanasawa I (1990) A study of the effect of non-condensable gas in the vapor film on vapor explosion, Int. J. Heat Mass Transfer, vol 33 no 4 pp 603-609
2. Ando M, Caldarola L (1982) Triggered fragmentation experiments at Karlsruhe, in Müller U and Günter C (eds), Post Accident Debris Cooling, Proc. of the Fifth Post Accident Heat Removal information Exchange meeting, NRC Karlsruhe, G. Braun Karlsruhe, pp 13-21
3. Ando K (1984) Experiment zur getriggerten Fragmentation an einem schmelzflüssigen Kupfertröpfen in Wasser, KfK 3667
4. Arakeri VH et al. (1978) Thermal interaction for molten tin dropped into water, Int. J. Heat Mass Transfer, vol 21 pp 325-333
5. Bang KH, Kim MK (1995) Boiling characteristics od delute polymer solutions and implications for the suppression of vapor explosions, Proceedings of the Seventh International Topical Meeting on Nuclear Reactor Thermal Hydraulics NURETH-7, New York, USA, NUREG/CP-0142, pp 1677-1687
6. Bankoff SG, Kovarik F, Yang JW (1983) A model for fragmentation of molten metal oxides in contact with water, Proc. Int. Mtg. on LWR Severe Accident Evaluation, Cambridge, MA pp TS-6.6-1 to 6.6-8
7. Bankoff SG, Yang JW (October 10-13, 1989) Studies Relevant to in-Vessel Steam Explosions, Proceedings Fourth International Topical Meeting on Nuclear Reactor Thermal-Hydraulics, Karlsruhe. Müller U, Rehme K and Rust K (eds), G. Braun, Karlsruhe, p 312
8. Becker KM, Linland KP (April 1991) The effect of surfactants on hydrodynamic fragmentation and steam explosions, KTK-NEL-50, Rev. Ed.
9. Belman R, Pennington RH (1954) Effect of surface tension and viscosity on Taylor instability, Quart. Appl. Maths, vol 12 pp 151-162
10. Benjamin TB, Ellis A T (1966) Phil. Trans. R. Soc. A, vol 260 pp 221-240
11. Berman LD (1961) Soprotivlenie na granize razdela fas pri plenochnoi kondensazii para nizkogo davleniya, Tr. Vses. N-i, i Konstrukt in-t Khim Mashinost, vol 36 p 66
12. Bjorkquist GM (1975) An experimental investigation of molten metal in water, TID-26820
13. Bjornard TA et al. (1974) The pressure behavior accompanying the fragmentation of tin in water, Transaction of American Nuclear Society, vol 29 p 247
14. Board S J, Farmer CL, Poole DH (October 1972) Fragmentation in thermal explosions, Berkeley Nuclear Laboratories, RD/B/N2423, CFR/SWP/P (72) p 81
15. Brayer C, Berthoud G (19th-21th May 1997) First vapor explosion calculations performed with MC3D thermal-hydraulic code, OECD/CSNI Specialist Meeting on Fuel Coolant Interactions, JAERI-Tokai Research Establishment, Japan
16. Brennen CE (1995) Cavitation and bubble dynamics. Oxford University Press, Oxford, Ney York
17. Buchman DJ (1973) Penetration of a solid layer by a liquid jet, J. Phys.D: Appl. Phys., vol 6 pp 1762-1771
18. Bürger M et al. (1991) Examination of thermal detonation codes and included fragmentation models by means of triggered propagation experiments in a tin/water mixture, Nucl. Eng. Des., vol 131 pp 61-70

19. Bürger M, Cho SH, von Berg E, Schatz A (January 5-8, 1993) Modeling of drop fragmentation inthermal detonation waves and experimental verification, Specialist's Meeting on Fuel-Coolant Interactions, Santa Barbara, California, USA
20. Buetner R, Zimanowski B (1998) Physics of thermohydraulic explosions, Physical Review E, vol 57 no 5
21. Buxton LD, Nelson LS, Benedick WB (1979) Steam explosion triggering in efficiency studies, Forth CSNI Specialists Meeting on Fuel-Coolant Interaction in Nuclear Reactor Safety, Bournemouth, UK, pp 387-408
22. Carslaw HS, Jaeger JC (1959) Conduction of heat in solids, Oxford Science Publications, Second edition, Oxford University Press
23. Chapman R, Pineau D, Corradini M (May 22-25, 1997) Mitigation of vapor explosions in one-dimensional large scale geometry with surfactant coolant additives, Proc. of the Int. Seminar on Vapor Explosions and Explosive Eruptions, Sendai, Japan, pp 47-58
24. Chen X, Yuen WW, Theofanous T (1995) On the constitutive description of micro-interactions concept in steam explosions, Proceedings of the Seventh International Topical Meeting on Nuclear Reactor Thermal Hydraulics NURETH-7, New York, USA, NUREG/CP-0142.
25. Cho DH, Gunther WH (June 1973) Fragmentation of molten materials dropped into water, Trans. of the Am. Nucl. Soc., vol 16 no 1 pp 185-186
26. Ciccarelli G (1992) Investigation of vapor explosions with single molten metal drops in water using x-ray, PhD Thesis, McGill University, Montreal, Quebec, Canada
27. Corradini M (1982) Analysis and Modelling of Large Scale Steam Explosion Experiments, Nucl. Sci. Eng., vol 82 pp 429-447
28. Corradini ML (March 10-14, 1996) Vapor explosion phenomena: scaling considerations, Proc. of the ASME-JSME 4th International Conference on Nuclear Engineering, New Orleans, Louisiana U.S.A., vol 1 Part A, ASME 1996, pp 309-316
29. Cronenberg AW (August 29, 1973) Solidification phenomena and fragmentation, in Sachs RG and Kyger JA, Reactor development program progress report, ANL-RDP-18, Liquid Metal Fast Breeder Reactors (UC-79), p 7.19
30. Drumheller DS (1979) The initiation of melt fragmentation in fuel-coolant interactions, Nucl. Sc. Engineering, vol 72 pp 347-356
31. Dullforce TA, Buchanan D, Peckover R S (1986) Self-triggering of small scale fuel-coolant interaction: I. Experiments, Journal of Physics D: Applied Physics, vol 9 pp 1295-1303
32. Ellison PG, Hyder ML, Monson PR, DeWald AB Jr, Long TA, and Epstein M (April-June 1993) Aluminium-uranium fuel-melt behavior during severe nuclear reactor accidents, Nuclear Safety, vol 34 no 2 pp 196-212
33. Epstein M, Fauske K (August 4-7, 1985) Steam film instability and the mixing of core-melt jets and water, ANS Proceedings, National Heat Transfer Conference, Denver, Colorado, pp 277-284
34. Epstein M (Aug. 1991) Underwater vapor phase burning of aluminium particles and on aluminium ignition during steam explosions, WSRC-RP-91-1001, Westinghouse Savannah River Co.
35. Fauske HK (1973) On the mechanism of uranium dioxide-sodium explosive interactions, Nuclear Science and Engineering, vol 51 pp95-101
36. Fedorovich ED, Rohsenow W M (1968) The effect of vapor subcooling on film condensation of metals, Int. J. of Heat Mass Transfer, vol 12 pp 1525-1529

37. Fletcher DF, Thyagaraja A (June 1989) A mathematical model of melt/water detonation, Appl. Math. Modelling, vol 13 pp 339-347
38. Fritz W, Ende W (1966) Über den Verdampfungsvorgang nach kinematographischen Aufnahmen an Dampfblasen. Phys. Z., vol 37 pp 391-401
39. Froehlich G (1991) Propagation of fuel-coolant interactions in multi-jet experiments with molten tin, Nuclear Engineering and Design, vol 131 pp 209-221
40. Frost D L, Ciccarelli G (July 24-27, 1988) Propagation of explosive boiling in molten tin - water mixtures, Proc. Nat. Heat Transfer Conference, HDT - 96, Vol.2, Ed. H. R. Jacobs, Houston, Texas, pp 539-574
41. Gelfand BE, Gubin SA et al. (1977) Influence of gas density on breakup of bubbles, Dokl. USSR Ac. Sci., vol 235 no 2 pp 292-294
42. Gibson DC (1968) Cavitation adjacent to plane boundaries. Proc. Australian Conf. On Hydraulic and Fluid Machinery, pp 210-214
43. Gibson DC and Blacke JR (1982) The growth and collapse of bubbles near deformable surface. App. Sci. Res., vol 38 pp 215-224
44. Inoue A, Aritomi M (October 10-13, 1989) An Analytical model on vapor explosion of a high temperature molten metal droplet with water induced by a pressure pulse, Proceedings Fourth International Topical Meeting on Nuclear Reactor Thermal-Hydraulics, Karlsruhe. Müller U, Rehme K and Rust K (eds), G. Braun, Karlsruhe, p 274
45. Henkel P (1987) Hüllrohrmaterialbewegung während eines Kühlmittelverluststörfalls in einem schnellen, natriumgekühlten Reaktor, Dissertation, Kernforschungszentrum Karlsruhe
46. Henry RE, Fauske HK (March 10-14, 1996) A diferent approach to fragmentation in steam explosions, Proc. of the ASME-JSME 4th International Conference on Nuclear Engineering, New Orleans, Louisiana U.S.A., vol 1 Part A, ASME 1996, pp 309-316
47. Hertz H (1882) Wied. Ann., vol 17 p 193
48. Hohmann H, Magalon D, Huhtiniemi I, Annunziato A, Yerkess A (October 1995) Recent results in FARO/KROTOS test series, Trans.23rd WRSIM, Bethesda MD
49. Huhtiniemi I, Magalon D, Hohmann H (19th-21th May 1997) Results of recent KROTOS FCI tests: alumna vs. corium melts, OECD/CSNI Specialist Meeting on Fuel Coolant Interactions, JAERI-Tokai Research Establishment, Japan
50. Kammer C (1995) Aluminium-Taschenbuch, Bd.1 Grundlagen und Wekstoffe, Herausgeber: Aluminium-Zentrale Düsseldorf, Aluminium-Verlag Düsseldorf
51. Kim BJ, Corradini ML (1986) Recent film boiling calculations: implication on fuel-coolant interactions, Int. J. Heat Mass Transfer, vol 29 pp 1159-1167
52. Kim B, Corradini M (1988) Modeling of small scale single droplet fuel/coolant interactions, Nucl. Sci. Eng., vol 98 pp 16-28
53. Kim H, Krueger J, Corradini ML (October 10-13, 1989) Single droplet vapor explosions: effect of coolant viscosity, Proceedings Fourth International Topical Meeting on Nuclear Reactor Thermal-Hydraulics, Karlsruhe. Müller U, Rehme K and Rust K (eds), G. Braun, Karlsruhe, p 261
54. Knowles JB (1985) A mathematical model of vapor film destabilization, Report AEEW-R-1933
55. Knudsen M (1915) Ann. Physik, vol 47 p 697
56. Kolev NI (1993) Fragmentation and coalescence dynamics in multi-phase flows, Experimental Thermal and Fluid Science, vol 6 pp 211 – 251

57. Kolev NI (19th-21st May 1997) Verification of the IVA4 Film boiling model with the data base of liu and theofanous, Proceedings of OECD/CSNI Specialists Meeting on Fuel-Coolant Interactions (FCI), JAERI-Tokai Research Establishment, Japan
58. Kornfeld M, Suvorov L (1944) J. Appl. Physics, vol 15 pp 495-506
59. Koshizuka S, Ikeda H, Oka Y (May 22-25, 1997) Effect on Spontaneous Nucleation on Melt Fragmentation in Vapor Explosions, Proc. of the Int. Seminar on Vapor Explosions and Explosive Eruptions, Sendai, Japan, pp 185-192
60. Kowal MG, Dowling MF, Abdel-Khalik S I (1993) An experimental investigation of the effects of surfactants on the severity of vapor explosions, Nuclear Science and Engineering, vol 115 pp 185-192
61. Langmuir I (1913) Physik. Z., vol 14 p 1273
62. Langmuir I, Jones HA, Mackay GMJ (1927) Physic. Rev., vol 30 p 201
63. Lide DR, Frederikse HPR (eds) (1997) CRC Handbook of chemistry and physics, 8th Edition, CRC Press, New York 1997
64. Magalon D, Huhtiniemi I, Hohmann H (19th-21th May 1997) Lessons learnt from FARO/TERMOS corium melt quenching experimnts, OECD/CSNI Specialist Meeting on Fuel Coolant Interactions, JAERI-Tokai Research Establishment, Japan
65. Marangoni CGM (1871) Über die Ausbreitung der Tropfen einer Flüssigkeit auf der Oberfläche einer anderen. Ann. Phys., vol 143 p 337
66. Matsumura K, Nariai H (March 10-14, 1996) The occurance condition of spontaneous vapor explosions, Int. Conf. on Nuclear Engineering ICONE-4, New Orleans, Louisiana, vol 1 - Part A, ASME, pp 325-332
67. Matzke H (28-29 September 1998) Status of FARO debris analysis by ITU, Karlsruhe, 9th FARO Expert Meeting, Ispra
68. Medhekar S, Amarasooriya WH, Theofanous TG (October 10-13, 1989) Integrated analysis of steam explosions, Proceedings Fourth International Topical Meeting on Nuclear Reactor Thermal-Hydraulics, Karlsruhe. Müller U, Rehme K and Rust K (eds), G. Braun, Karlsruhe, p 319-326
69. Merzhanov AG, Grigoriev YM, Gal'chenko YA (1977) Aluminium ignition, Combustion Flame, vol 29 p 1
70. Mills AF, Seban RA (1967) The condensation coefficient of water, J. of Heat Transfer, vol 10 pp 1815-1827
71. Mills AF (1967) The condensation of steam at low pressure, Techn. Report Series No. 6, Issue 39. Space Sciences Laboratory, University of California, Berkeley
72. Mitsumura K, Nariai H, Egashira Y, Ochimizu M (May 22-25, 1997) Experimental study on the base triggering spontaneous vapor explosions for molten tin-water system, Proc. of the Int. Seminar on Vapor Explosions and Explosive Eruptions, Sendai, Japan, pp 27-32
73. Nabavian K, Bromley LA (1963) Condensation coefficient of water; Chem. Eng. Sc., vol 18 pp 651-660
74. Naylor P (1985) Film boiling destabilization, Ph.D. Thesis, University of Exeter
75. Nelson LS, Duda PM (1981) Steam explosion experiments with single drops of iron-oxide melted with CO_2 laser, SAND81-1346, NUREG/CR-2295, Sandia National Laboratory
76. Nelson LS, Guay KP (1986) Suppression of steam explosion in tin and $FeAl_2O_3$ melts by increasing the viscosity of the coolant, High Temperature and High Pressures, vol 18 pp 107-111

77. Nelson LS, Buxton LD (1980) Steam explosion triggering phenomena: stainless steel and corium-e simulants studied with a floodable arc melting apparatus, SAN77 - 0998, NUREG/CR-01222, Sandia National Laboratories
78. Nelson LS, Hyndman DA, Duda PM (July 21-25, 1991) Steam explosions of single drops of core-melt simulants: Triggering, work output and Hydrogen generation, Proc. of the Int. Top. Meeting on Safety of Thermal Reactors, Portland, Oregon, pp 324-330
79. Ojnesorge W (1936) Die Bildung von Tröpfen an Düsen und die Auflüssiger Strahlen, Z. Angew. Math. Mech., vol 16 pp 335-359
80. Ostrach S (1982) Low gravity fluid flows, Ann. Rev. Fluid Mech., vol 14 pp 313-345
81. Peppler W, Till W, Kaiser A (Sept. 1991) Experiments on Thermal Interactions: Tests with Al_2O_3 Droplets and Water, Kernforschungszentrum Karlsruhe , KfK 4981
82. Pilch M, Erdman CA, Reynolds AB (Aug. 1981) Acceleration induced fragmentation of liquid drops, Charlottesville, VA: Department of Nucl. Eng., University of Virginia, NUREG/CR-2247
83. Plesset MS, Chapman RB (1971) Collapse of an initially spherical vapor cavity in the neighborhood of solid boundary, J. of Fluid Mechanics, vol 47 no 2 pp 283-290
84. Saito M, Sato K, Imahori S (July 24-July 27, 1988) Experimental study on penetration behaviours on water jet into Freon-11 and liquid nitrogen, ANS Proc. Nat. Heat Transfer Conference, HTC-vol 3. Jacobs HR (ed), Houston, Texas
85. Schröder-Richter D, Bartsch G (1990) The Leidenfrost phenomenon caused by a thermo-mechanical effect of transition boiling: A revisited problem of non-equilibrium thermodynamics, in Witte LC and Avedisian CT (eds) HTD-vol 136, Fundamentals of phase change: Boiling and condensation. Book No.H00589-1990, pp 13-20
86. Spencer BW et al. (August 4-7, 1985) Corium quench in deep pool mixing experiments, ANS Proceedings, National Heat Transfer Conference, Denver, Colorado, pp 267-276
87. Shpillrain EE, Yakimovich KA, Tsitsarkin AF (1973) Experimental study of the density of liquid alumina up to 2750 C, High Temperures - High Pressures, vol 5 pp 191-198
88. Taleyarkhan R (Febr. 1990) Steam-explosion safety consideration for the advanced neutron source reactor at the Oak Ridge National Laboratory, ORNL/TM-11324
89. Tang J (1993) A complete model for the vapor explosion process, PhD Thesis, University of Wisconsin, Madison WI
90. Taylor GI (1935) Proc. Roy. Soc. A, vol 151 p 429. See also (1950) The instability of liquid surface when accelerated in a direction perpendicular to their plane, Proc. Roy. Soc. A, vol 201 pp 192-196
91. Taylor G (1950) The instability of liquid surfaces when accelerated in a direction perpendicular to their planes. I., Proceedings of the Royal Society of London, Series A. Mathematical and Physical Sciences, vol 201 pp 192-196
92. Thomson J (July-December 1855) On certain curious motions observable at the surfaces of wine and other alcoholic liquors, Philosophical Magazine and Journal of Science, vol X - Fourth Series, pp 330-333
93. Turbill D, Fisher JC (1949) Rate of nucleation in condensed systems, J. Chem. Phys., vol 17 pp 71
94. Uludogan A, Corradini ML (Feb. 1995) Modeling of molten metal/water interactions, Nuclear Technology, vol 109 pp 171-186
95. Voinov OV and Voinov VV (1975) Numerical method of calculating non-stationary motions of ideal incompressible fluid with free surface. Sov. Phys. Dokl. Vol 20 pp 179-18091.

96. Yerkess (1997) TEXAS-IV Dynamic fragmentation model, private communication, 8th FARO Expert Meeting, Ispra, Italy
97. Young MF (Sept. 1987) IFCI: An integrated code for calculation of all phases of fuel coolant interaction, NUREG/CR-5084, SAND87-1048
98. Young MF (1990) Application of the IFCI Integrated fuel-coolant interaction code to a fits-type pouring mode experiment, SAND89-1692C
99. Yuen WW, Chen X, Theofanous TG (1994) On the fundamental micro-interactions that support the propagation of steam explosions, NED, vol 146 pp 133-146
100. Zimanowski B, Fröhlich G, Lorenz V (1995) Experiments on steam explosion by interaction of water with silicate melts, NED vol 155 pp 335-343
101. Zimmer HJ, Peppler W, Jacobs H (October 10-13, 1989) Thermal fragmentation of molten alumina in sodium, Proceedings Fourth International Topical Meeting on Nuclear Reactor Thermal-Hydraulics, Karlsruhe. Müller U, Rehme K and Rust K (eds) G. Braun, Karlsruhe, p 268
102. Zyskowski W (1975) Thermal interaction of molten copper with water, Int. J. of Heat and Mass Transfer, vol 18 pp 271-287

12. Nucleation in liquids

After reviewing the literature for description of the nucleation in superheated liquids the following conclusions and recommendations have been drawn. The maximum superheating in technical systems is a function of the depressurization velocity and of the produced turbulence. The maximum superheating can be predicted by the Algamir and Lienhard and by the Bartak correlations within an error band of 48.5%. Flashing in short pipes and nozzles leads to critical flows driven by the pressure difference equal to the entrance pressure minus the flashing inception pressure. For the prediction of the maximum achievable superheating which represents the Spinoidal line the Skripov correlation is recommended. The wetting angle is an important property of the polished surface characterizing its capability to activate nucleation sites. For the prediction of the activated nucleation sites the correlation obtained by Wang and Dhir is recommended. The establishing of a vapor film around a heated surface having temperature larger than the minimum film boiling temperature takes a finite time. Availability of small bubbles of non-condensing gases reduces the superheating re-quired to initiate evaporation. Evaporation at lower than the saturation temperature is possible.

12.1 Introduction

Liquids having temperature larger than the saturation temperature corresponding to the local pressure are called superheated liquids. Superheated liquids are unstable and start to disintegrate. This process is in generally called *flashing*. The process of rupturing a *continuous* liquid by decrease in pressure at roughly constant liquid temperature is often called *cavitation* - a word proposed by *Froude*. The process of rupturing a *continuous* liquid by increase the temperature at roughly constant pressure is often called *boiling*. The fluctuation of molecules having energy larger than that characteristic for a stable state causes the formation of clusters of molecules, which after reaching some critical size are called nuclei. The theory of the nucleation provides us with information about the generation of nuclei per unit time and unit volume of the liquid as a function of the local parameter.

12.2 Nucleation energy, equation of *Kelvin* and *Laplace*

Let us abstract from a superheated *non-stable liquid* a *spherical volume*, having an initial radius R_{10}, a volume $\frac{4}{3}\pi R_{10}^3$, pressure p, temperature T_2 and density $\rho_2 = \rho_2(p, T_2)$. After some time the selected sphere liquid volume increases due to the *evaporation* to the radius R_1 (respectively to the volume $\frac{4}{3}\pi R_1^3$) and reaches a pressure $p'(T_2)$. The pressure inside the bubble is assumed to be uniform because of the small bubble size. The density of the evaporated steam inside the sphere is $\rho'' = \rho''[p'(T_2)] = \rho''(T_2)$. The *initial* and the *end spheres* have the same *mass* per definition, therefore

$$\left(\frac{R_{10}}{R_1}\right)^3 = \frac{\rho''}{\rho_2}. \tag{12.1}$$

Consequently, the initial volume of the sphere is changed by

$$\frac{4}{3}\pi\left(R_1^3 - R_{10}^3\right) = \frac{4}{3}\pi R_1^3\left(1 - \frac{\rho''}{\rho_2}\right). \tag{12.2}$$

During this expansion a *mechanical work*

$$4\pi \int_{R_{10}}^{R_1}\left[p'(T_2) - p\right]r^2 dr \approx \frac{4}{3}\pi\left(R_1^3 - R_{10}^3\right)\left[p'(T_2) - p\right]$$

$$= \frac{4}{3}\pi R_1^3\left(1 - \frac{\rho''}{\rho_2}\right)\left[p'(T_2) - p\right] \tag{12.3}$$

is performed and transferred into total kinetic energy of the surrounding liquid [41]. In other words, this work is introduced into the liquid. For the creation of a sphere with a free surface additional work

$$\int_0^{R_1}\frac{2\sigma}{r}4\pi r^2 dr = 4\pi R_1^2 \sigma \tag{12.4}$$

is needed. The surface tension of water in N/m, σ, in contact with its vapor is given in *Lienhard* [31] with great accuracy by

12.2 Nucleation energy, equation of Kelvin and Laplace

$$\sigma = 0.2358\left[1-\frac{T'(p)}{T_c}\right]^{1.256}\left\{1-0.625\left[1-\frac{T'(p)}{T_c}\right]\right\}, \tag{12.5}$$

where the T_c is the thermodynamic critical temperature (for water $T_c = 647.2\ K$). For the region of 366 to 566 K the above equation can be approximated by

$$\sigma = 0.14783\left(1-T/T_c\right)^{1.053} \tag{12.6}$$

with an error of ±1%. The surface tension is a function of the surface temperature, which can strongly vary in transients. The surface tension is usually measured at macroscopic surfaces. Whether the so obtained information is valid for the microscopic metastable bubbles is not clear. Next we assume that this relationship holds also for microscopic surfaces.

Thus, the work necessary to create a single bubble with radius R_1 is

$$\Delta E_1 = 4\pi R_1^2\sigma - \left[p'(T_2)-p\right]\frac{4}{3}\pi R_1^3\left(1-\frac{\rho''}{\rho_2}\right) = 4\pi\sigma\left(R_1^2 - \frac{2}{3}\frac{R_1^3}{R_{1c}}\right),\ p'(T_2) > p \tag{12.7}$$

We see that this work depends on the bubble radius and has a maximum

$$\Delta E_{1c} = \frac{16\pi\sigma^3}{3\left[p'(T_2)-p\right]^2\left(1-\frac{\rho''}{\rho_2}\right)^2} = \frac{4}{3}\pi\sigma R_{1c}^2 \approx \frac{4}{3}\pi\sigma\left[\frac{T'(p)}{T_2-T'(p)}\frac{2\sigma}{\rho''(h''-h')}\right]^2 \tag{12.8}$$

for

$$R_{1c}^2 = \frac{3\Delta E_{1c}}{\sigma 4\pi} = \left\{\frac{2\sigma}{\left[p'(T_2)-p\right]\left(1-\rho''/\rho_2\right)}\right\}^2. \tag{12.9}$$

The corresponding bubble volume is then

$$V_{1c} = \frac{4}{3}\pi\left(\frac{3\Delta E_{1c}}{\sigma 4\pi}\right)^{3/2}. \tag{12.10}$$

The equation (12.9) is known as the *Laplace* and *Kelvin* equation. Gibbs [17] (1878) noted that the expression for the maximum " ... does not involve any geometrical magnitudes".

12.3 Nucleus capable to grow

The above consideration did not lead to any conclusion whether a bubble with size R_{1c} will further grow or collapse. It says only that at that size the mechanical energy needed to create a bubble at the initial state $R_1 = 0$ and at the final state $R_{1\infty} > R_{1c}$ possesses a maximum at R_{1c} and no more. There are many papers in which this size is taken to represent the bubble size, at which the bubble is further capable to grow, which as we will show below is not true.

Next we consider the conservation of the liquid mechanical energy in order to describe the bubble growth from the beginning through the critical radius. The mass conservation equation of the liquid can be approximated by

$$\frac{1}{r^2}\frac{\partial}{\partial r}\left(r^2 u_2\right) \approx 0, \qquad (12.11)$$

or integrating with the boundary condition

$$r = R_1, \quad u_2 = \frac{\partial R_1}{\partial \tau} \qquad (12.12)$$

$$u_2 = const/r^2 = \frac{\partial R_1}{\partial \tau}\left(R_1/r\right)^2. \qquad (12.13)$$

It gives the liquid velocity as a function of the radius if the liquid is assumed to be incompressible during the bubble expansion. The total kinetic energy of the liquid environment estimated using the above equation is therefore

$$\int_{R_1}^{\infty} \frac{1}{2}\rho_2 u_2^2 dVol_2 = \frac{1}{2}\rho_2\left(\frac{\partial R_1}{\partial \tau}\right)^2 R_1^4 4\pi \int_{R_1}^{\infty}\frac{dr}{r^2} = 2\pi\rho_2\left(\frac{\partial R_1}{\partial r}\right)^2 R_1^4 \int_{R_1}^{\infty} d\left(-\frac{1}{r}\right)$$

$$= 2\pi\rho_2\left(\frac{\partial R_1}{\partial \tau}\right)^2 R_1^3. \qquad (12.14)$$

During the bubble growth the work performed by the bubble expansion is transferred in total kinetic energy of the liquid environment, i.e.

$$2\pi\rho_2\left(\frac{\partial R_1}{\partial \tau}\right)^2 R_1^3 = -4\pi\sigma\left(R_1^2 - \frac{2}{3}\frac{R_1^3}{R_{1c}}\right) \equiv \Delta E_1$$

or

$$\left(\frac{\partial R_1}{\partial \tau}\right)^2 = \frac{2\sigma}{\rho_2}\left(\frac{2}{3}\frac{1}{R_{1c}} - \frac{1}{R_1}\right). \tag{12.15}$$

The constant 2/3 valid for bubble growth in a bulk liquid should be replaced by $\pi/7$ if a spherical bubble grows on a flat surface.

Equation (12.15) is a very important result. We see that real bubble growth is possible *if and only if*

$$R_1 > \frac{3}{2} R_{1c}. \tag{12.16}$$

For the case of $R_1 \gg R_{1c}$ Eq. (12.15) transforms into

$$\left(\frac{\partial R_1}{\partial \tau}\right)^2 = \frac{4}{3}\frac{\sigma}{\rho_2 R_{1c}}. \tag{12.17}$$

This mechanism of bubble growth is called *inertially controlled bubble growth*. For low pressure where the assumption $\rho''/\rho_2 = 0$ is reasonable the above equation reduces to one obtained for the first time by *Besand* [6] (1859) and in a more elegant way by *Rayleigh* [36] (1917). This mechanism controls the bubble growth within the first 10^{-8} s of the life of the stable bubble.

It follows from the above consideration that for the creation of a bubble with a critical unstable diameter $3/2\ R_{1c}$, a surplus of internal energy of the liquid is needed greater than E_{1c}, and for creation of bubble that is capable to grow, a surplus of internal energy of the liquid - greater than $(9/4)\ E_{1c}$.

12.4 Some useful forms of the *Clausius-Clapeyron* equation, measures of superheating

Usually the liquid superheating is expressed by the temperature difference $T_2 - T'(p)$. Sometimes the pressure difference $\Delta p = p'(T_2) - p$ corresponding to the superheating of the liquid with respect to the saturation temperature $T_2 - T'(p)$ is also used as a measure for the liquid superheating. Using the *Clausius - Clapeyron* equation

$$\frac{dT}{dp} = \frac{v''-v'}{s''-s'} = \frac{\rho'-\rho''}{s''-s'}\frac{1}{\rho'\rho''} = T'(p)\frac{\rho'-\rho''}{h''-h'}\frac{1}{\rho'\rho''} \tag{12.18}$$

integrated between the initial state $[p, T'(p)]$ and the final state $[p'(T_2), T_2]$ for

$$\frac{\rho' - \rho''}{h'' - h'} \frac{1}{\rho'\rho''} \approx const \qquad (12.19)$$

one obtains

$$p'(T_2) - p = (h'' - h') \frac{\rho'\rho''}{\rho' - \rho''} \ln\left[1 + \frac{T_2 - T'(p)}{T'(p)}\right]. \qquad (12.20)$$

With this result the critica bubbles size can be approximated by

$$R_{1c} \approx 2\sigma / \left\{ (h'' - h')\rho' \ln\left[1 + \frac{T_2 - T'(p)}{T'(p)}\right]\right\}. \qquad (12.22)$$

For $T_2 - T'(p) \ll T'(p)$ we have for the tensile pressure difference

$$p'(T_2) - p = (h'' - h') \frac{\rho'\rho''}{\rho' - \rho''} \frac{T_2 - T'(p)}{T'(p)}, \qquad (12.23)$$

and for the critical bubble size

$$R_{1c} \approx \frac{T'(p)}{T_2 - T'(p)} \frac{2\sigma}{\rho''(h'' - h')}. \qquad (12.24)$$

Thus one can use as a measure of liquid superheating either the *tension pressure difference* $p'(T_2) - p$ or the *liquid superheat* $T_2 - T'(p)$.

Equation (12.23) makes it possible to compute the liquid superheating corresponding to the radius R_{1c} from Eq. (12.9)

$$[T_2 - T'(p)] / T'(p) = \frac{2\sigma}{R_{1c}} \frac{(\rho' - \rho'')\rho_2}{(\rho_2 - \rho'')\rho'\rho''} \frac{1}{h'' - h'} \approx 2 \frac{\sigma}{R_{1c}} \frac{1}{\rho''(h'' - h')} \qquad (12.25)$$

or corresponding to $3 R_{1kr}/2$

$$[T_2 - T'(p)] / T'(p) = \frac{4}{3} \frac{\sigma}{R_{1c}} \frac{(\rho' - \rho'')\rho_2}{(\rho_2 - \rho'')\rho'\rho''} \frac{1}{h'' - h'} \approx \frac{4}{3} \frac{\sigma}{R_{1c}} \frac{1}{\rho''(h'' - h')}. \qquad (12.26)$$

For the atmospheric pressure Eq. (12.23) gives

$$T_2 - T'(p) = 6.4197 \times 10^{-5} / D_{1c}. \qquad (12.27)$$

12.4 Some useful forms of the Clausius-Clapeyron equation, measures of superheating 347

A nonlinear expression is obtained also for low pressure for which the density difference between liquid and vapor is very large. In this case Eq.(12.18) simplifies to $dT/dp = v''/(T\Delta h)$. This equation is known since 1828 in France as the *August equation*. Asuming the vapor behave as a perfect gas results in $dp/p = (\Delta h/R)dT/T^2$, where R is the vapor gas constant for the specific substance. Integrating between an initial state $[p, T'(p)]$ and a final state $[p'(T_2), T_2]$ and rearanging results in useful expresion for computing the relation between *tension pressure difference* $p'(T_2) - p$ and the *liquid superheat* $T_2 - T'(p)$ of metastable liquid,

$$\frac{p'(T_2)-p}{p} = \exp\left[\frac{\Delta h}{RT_2}\frac{T_2-T'(p)}{T'(p)}\right] - 1. \tag{12.28}$$

With this result the *Laplace* and *Kelvin* equation reads

$$R_{1c} \approx \frac{2\sigma}{\left\{\exp\left[\frac{\Delta h}{RT_2}\frac{T_2-T'(p)}{T'(p)}\right]-1\right\}p(1-\rho''/\rho_2)}. \tag{12.29}$$

One should carefully use the approximations of the *Laplace* and *Kelvin* equation. At low pressure the appoximations are good as shown in Fig.12.1. In this case at higher superheat Eq. (12.29) is much close to the acurate equation.

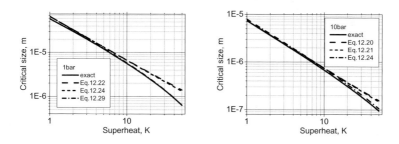

Fig. 12.1. Critical bubbles size at low pressure as a function of the liquid superheat

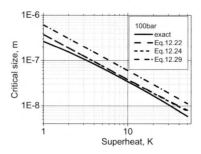

Fig. 12.2. Critical bubbles size at high pressure as a function of the liquid superheat

At high pressure the approximations may deviate substantially from the exact solution as shown in Fig. 12.2.

The dimensionless number constructed as follows

$$Gb_2 = \frac{\Delta E_{1c}}{kT_2} = \frac{\Delta S_{1c}}{k} = \frac{16\pi\sigma^3}{kT_2 3\left[p'(T_2)-p\right]^2 \left(1-\rho''(T_2)/\rho_2\right)^2} \qquad (12.30)$$

is called the *Gibbs* number. Here $k = 13.805 \times 10^{-24}$ J/K is the *Boltzmann* constant. The *Gibbs* number is frequently used as a dimensionless measure of superheating. The greater the superheating the smaller the *Gibbs* number. Gb_2 converging to infinity means no superheating. For $T_2 = T_c$, $\sigma = 0$ and $Gb_2 = 0$.

However, it should be noted that neither of the measures discussed above tells us at what ΔT liquid starts boiling or flashing.

12.5 Nucleation kinetics

12.5.1 Homogeneous nucleation

Now we discuss the dependence between liquid superheating and production rates of bubbles with critical size called nuclei. Consider a continuous liquid characterized by average state parameters p and T_2. An observer measures along long time $\Delta\tau$ the time $\Delta\tau_n$ in which a mass Δm_{2c} of the continuum departs from the normal, average, state. The entropy change necessary for this fluctuation is $\Delta m_{2c} \Delta s_{2c} = \Delta S_{2c}$. *Volmer* [45] (1939) p. 81 following *Boltzman* and *Einstein* assumed that

$$\Delta \tau_n / \Delta \tau \approx const\ e^{-\Delta S_{2c}/k} = const\ e^{-Gb_2} \qquad (12.31)$$

where k is the *Boltzmann* constant. It means, that the smaller the needed entropy change for the fluctuation the higher the probability of this fluctuation. This is the leading idea for describing such processes. It did not change with the years. Actually it is known from experimental observations that the greater the superheating the greater the probability of origination of bubbles with critical size. The probability of origination of a single bubble is defined as

$$\text{probability of origination of single bubble} = \frac{nucleation\ events}{molecular\ colisions}. \qquad (12.32)$$

Thus

$$\frac{nucleation\ events}{molecular\ colisions} = \exp(-Gb_2) \qquad (12.33)$$

Obviously for $T_2 = T_c$ and $\sigma_2 = 0$, the probability of origination of a single bubble $\exp(-Gb_2) = 1$. Nuclei with critical size can originate in the bulk liquid. This phenomenon is called *homogeneous* nucleation. Nuclei with critical size can originate also at the wall. This phenomenon is called *heterogeneous* nucleation. The theories modeling the nucleation in the bulk liquid are called homogeneous nucleation theories. The theories modeling the nucleation at surfaces are called heterogeneous nucleation theories. Next we give the main result of the homogeneous nucleation theory.

Kaishew and *Stranski* [25] computed in 1934 the number of the created nuclei per unit time in unit volume of the liquid as follows

$$\dot{n}_{1cin} = N_2 \left\{ \frac{6\sigma}{[2 + p/p'(T_2)]\pi m_2} \right\}^{1/2} \exp(-Gb_2)$$

$$= \exp\left\{ \ln\left[\rho_2 \left(\frac{N_A}{m_\mu} \right)^{3/2} \left\{ \frac{6\sigma}{[2 + p/p'(T_2)]\pi m_2} \right\}^{1/2} \right] - Gb_2 \right\}. \qquad (12.34)$$

Here $N_2 = \rho_2/m_2 = \rho_2 N_A/m_\mu$ ($\approx 3.3 \times 10^{28}$ for water) is the number of the molecules in one cubic meter, $m_2 = m_\mu/N_A$ is the mass of a single molecule, m_μ is the kg-mole mass (18 kg for water), $N_A = 6.02 \times 10^{26}$ (1/kg-mole) is the number of molecules in one kilogram-mole mass, the *Avogadro* number, and

$$\left\{\frac{6\sigma}{[2+p/p'(T_2)]\pi m_2}\right\}^{1/2} \qquad (12.35)$$

is the frequency with which each single liquid molecule interacts with its neighbors. The logarithmic expression is not very sensitive with respect to p and T_2 and ranges over

$$\ln\left[\rho_2\left(\frac{N_A}{m_\mu}\right)^{3/2}\left\{\frac{6\sigma}{[2+p/p'(T_2)]\pi m_2}\right\}^{1/2}\right] \approx 80 \text{ to } 83 \qquad (12.36)$$

for water.

12.5.2 Heterogeneous nucleation

12.5.2.1 Characteristics of the surfaces and of the liquid-surface contact

The technical surfaces possess roughness as a result of the manufacturing procedure. The structure of the roughness is an important characteristic of the nucleation processes at the surface. Another important characteristic is the molecular interaction between the surface and the liquid. It is usually characterized by the so called contact angle. The angle between the tangent to the interface and the wall θ is called contact angle. Hydrophobic surfaces, $\theta > 0$, causes heterogeneous nucleation at much reduced pressure difference (tensile strenght). In this section we give some examples of the characteristics of surfaces influencing the heterogeneous nucleation kinetics.

At real surfaces there are several cavities. For ordinary machined surfaces the cavities have sizes from 2 to 6 μm with density in the range of

$$n_w'' \approx (10 \text{ to } 250) \times 10^4 \, m^{-2} \qquad (12.37)$$

For mirror finished copper surfaces as described in [46] (1993) the physically existing cavities have been found to have surface densities depending on the equivalent cavity size in the range

$$n_w'' \approx 9 \times 10^{-9} / D_{cav}^2 \quad \text{for} \quad D_{cav} \geq 5.8 \times 10^{-6} \, m \qquad (12.38)$$

$$n_w'' \approx 10.3 \times 10^4 + 1.5 \times 10^{-25.2} / D_{cav}^{5.2} \quad \text{for} \quad 3.5 \times 10^{-6} \leq D_{cav} \leq 5.8 \times 10^{-6} \, m \qquad (12.39)$$

$$n_w'' \approx 2.2135 \times 10^7 + 3.981 \times 10^{-26.4} / D_{cav}^{5.4} \quad \text{for} \quad D_{cav} \leq 3.5 \times 10^{-6} m. \quad (12.40)$$

These are surface properties, which depend on the surface manufacturing only.

The following tables give some information on how the contact angle can change for different wall materials and different surface preparation procedures.

Table 12.1. Static contact angles θ for distilled water at polished surfaces.

Steel	$\pi/3.7$	*Siegel* and *Keshock* [39]
Steel, Nickel	$\pi/4.74$	*Bergles* and *Rohsenow* [53]
Nickel	$\pi/4.76$ to $\pi/3.83$	*Tolubinsky* and *Ostrovsky* [43]
Nickel	$\pi/4.74$ to $\pi/3.83$	*Siegel* and *Keshock* [39]
Chrome-Nickel Steel	$\pi/3.7$	*Arefeva* and *Aladev* [3]
Silver	$\pi/6$ to $\pi/4.5$	*Labuntsov* [29] for p =1 to 150 bar
Zinc	$\pi/3.4$	*Arefeva* and *Aladev* [3]
Bronze	$\pi/3.2$	*Arefeva* and *Aladev* [3]
Zr-4	$\pi/3.16$	*Basu, Warrier* and *Dhir* [48]
Note the contradictory data for copper in the literature		
Copper	$\pi/4$	*Arefeva* and *Aladev* [3]
Copper	$\pi/3$	*Gaertner* and *Westwater* [15]
Copper	$\pi/2$	*Wang* and *Dhir* [46]

Table 12.2. Static contact angle θ for distilled water at thermally or chemically treated polished surfaces

Copper heated to 525 K and exposed to air one hour:	$\pi/5.14$	*Wang* and *Dhir* [46]
Copper heated to 525 K and exposed to air two hour:	$\pi/10$	*Wang* and *Dhir* [46]
Chrome-Nickel Steel chemically treated:	$\pi/2.9$	*Arefeva* and *Aladev* [3]

In case of nucleation at walls the segment of a bubble attached at the cavity possesses

$$\text{surface area} = \varphi 4\pi R_{1c}^2 \quad (12.41)$$

which is less than the nucleation surface by a factor of φ. Therefore less energy is needed for the creation of bubbles at walls

$$\Delta E_{1c}^* = \varphi \Delta E_{1c}, \qquad (12.42)$$

than inside the liquid. Consequently, the superheating which can be achieved in technical systems is much smaller than in spontaneous bulk nucleation systems. φ is frequently called in the literature the work reduction factor.

Tolubinski [44] (1980) computed the work reduction factor, φ, for an idealized wall surface having *plane geometry* without cavities as follows

$$\varphi = \frac{1}{4}(1+\cos\theta)^2 (1-\cos\theta). \qquad (12.43)$$

Using the values from the Table 12.1 we have φ's in the range of 0.078 to 0.25.

The work reduction factor for surface with cavities calculated by *Kottowski* [27] (1973) is

$$\varphi = \frac{1}{4}\left[2 - 3\sin(\theta - \Phi) + \sin^3(\theta - \Phi)\right]. \qquad (12.44)$$

Here Φ is the cavity angle if the cavity is idealized as a cone.

12.5.2.2 Nucleation theories and experimental observations

Blander and *Katz* [7] (1975) computed for the heterogeneous nucleation the number of the nucleations per unit liquid volume and unit time in a pipe with wetted diameter D_{hy} as follows

$$\dot{n}_{1cin} = \frac{4}{D_{hy}} N_2^{2/3} \frac{1+\cos\theta}{2} \left\{\frac{6\sigma}{\left[2 + p/p'(T_2)\right]\pi\, m_2 \varphi}\right\}^{1/2} \exp\left(-\varphi\, Gb_2\right) =$$

$$\exp\left\{\ln\left[\frac{4}{D_{hy}} \rho_2^{2/3} \left(\frac{N_A}{m_\mu}\right)^{7/6} \frac{1+\cos\theta}{2} \left\{\frac{6\sigma}{\left[2 + p/p'(T_2)\right]\pi\, m_2 \varphi}\right\}^{1/2}\right] - \varphi\, Gb_2\right\}$$

$$(12.45)$$

Instead of N_2 in the *Kaishew* and *Stranski* equation $N^{2/3}$ is used because the nucleation occurs at the wall instead in the bulk.

The problem with the homogeneous and heterogeneous nucleation theories is that they did not describe appropriate nucleation rates in real liquid. *Rhohati* and *Reshotko* [38] and *Algamir* and *Lienhard* [2] (1981) data comparisons indicated substantial deviation from the above theories in a real technical systems.

Rhohati and *Reshotko* [38] (1975) tried to match the *Simoneau's* experimental data [40] (1975) with an expression

$$\dot{n}_{1cin} = (2.236 \text{ to } 4.472) \times 10^3 \left(\frac{2\sigma}{\pi m_2 \varphi}\right)^{1/2} \exp(-\varphi \, Gb_2) \qquad (12.46)$$

where

$$\varphi = 5 \times 10^{-6}. \qquad (12.47)$$

As it will be discussed below *Algamir* and *Lienhard* [2] (1981) found

$$\varphi = 0.055 \text{ to } 2 \times 10^{-7} \qquad (12.48)$$

for different depressurization speeds. *Algamir* and *Lienhard* [2] (1981) found that depending on the depressurization speed the nucleation process can be delayed. They use the work reduction factor φ to correlate experimental data with the following relationship

$$\varphi = 0.1058(T_2/T_c)^{28.46}\left\{1 + 14\left[\left(\gamma_v \frac{\delta p}{\delta \tau} + \mathbf{V}\gamma.\text{grad } p\right)/10^{11}\right]^{0.8}\right\} \qquad (12.49)$$

taking values as already mentioned of 0.055 to 2×10^{-7} for different depressurization rates. The correlation is valid for decompression speeds within

$$0.004 \leq (dp/d\tau)10^{-11} \leq 1.803 \qquad (12.50)$$

in channels with constant cross section in the temperature range

$$0.62 \leq T_2/T_c \leq 0.935. \qquad (12.51)$$

Here the decompression rate $dp/d\tau$ is measured in Pa/s.

For some practical applications it is useful to compute the time required to cover strongly superheated walls with nuclei. This is in fact the time which is required to establish film boiling at walls heated at temperatures above the film boiling temperature. Next we discuss briefly this problem. If the growth of the nucleus is neglected and the initial conditions are assumed to remain constant the nucleation number density originating during the time interval $\Delta\tau$ is

$$n_1 = \int_0^{\Delta\tau} \dot{n}_{1cin} d\tau = \dot{n}_{1cin}\Delta\tau. \qquad (12.52)$$

If the nuclei are spherical in shape and their centers form a triangular array at the wall the maximum number of nucleation sites per unit volume in a pipe is

$$n_{1,max} = \frac{4}{D_h} \frac{2}{\sqrt{3}} \frac{1}{D_{1c}^2} \qquad (12.53)$$

and the time required to reach it is

$$\Delta\tau_{max} = \frac{4}{D_h} \frac{2}{\sqrt{3}} / \left(D_{1c}^2 \dot{n}_{1cin} \right). \qquad (12.54)$$

If the cause for the superheating is depressurization with a $dp/d\tau$ starting from the saturation state the number of the produced particles is

$$n_1 = \int_0^{\Delta\tau} \dot{n}_{1cin} d\tau = n_1 = \frac{1}{dp/d\tau} \int_0^{\Delta p} \dot{n}_{1cin} dp \qquad (12.55)$$

If Δp corresponding to experimentally observed superheating is known and n_1 is set equal to $n_{1,max}$ the critical work reduction factor can be estimated. This procedure was used by *Algamir* and *Lienhard* [2] (1981). The authors found as already mentioned

$$\varphi = 0.055 \text{ to } 2 \times 10^{-7} \qquad (12.56)$$

for different depressurization speeds. Obviously only the functional form derived from the nucleation theories is useful to fit experimental data but not the theoretical formula itself. One of the reasons, why this is so may by the neglecting of nucleus growth of the already existing bubble generations during the nucleation.

Next we improve the computation by allowing each generation to grow. During a time interval $N \Delta\tau$, N-generation of nuclei, or $n_{1,n=1N}$ nuclei per unit flow volume are produced

$$n_{1,n} = \int_0^{\Delta\tau} \dot{n}_{1cin} d\tau, \quad n = 1, N. \qquad (12.57)$$

Each generation has its own bubble growth history. If we assume that the bubble growth is thermally controlled following the following relationship

$$\frac{dD_{1,n}}{d\tau} = \frac{B}{\sqrt{\tau}} \qquad (12.58)$$

where B is a function of local parameter the bubble size of the n-th generation after the time $N \Delta \tau$ is

$$D_{1,n} = D_{1c} + \sum_{k=1}^{N-n+1} \int_0^{\Delta \tau} \frac{dD_{1,n}}{d\tau} d\tau = D_{1c} + 2(N-n+1) B \Delta \tau^{1/2}. \qquad (12.59)$$

Thus the total number of nuclei and bubbles after N time steps of $\Delta \tau$ is

$$n_1 = \sum_{n=1}^{N} n_{1,n} = N \int_0^{\Delta \tau} \dot{n}_{1cin} d\tau. \qquad (12.60)$$

The cross section at the wall surface covered by attached bubbles per unit flow volume is

$$F_1/Vol = \frac{\pi}{4} \sum_{n=1}^{N} n_{1,n} D_{1,n}^2, \qquad (12.61)$$

which for deformable bubble can not be larger than $4/D_h$, where D_h is the hydraulic diameter of the pipe, i.e.

$$\frac{4}{D_h} \approx \frac{\pi}{4} \sum_{n=1}^{N} n_{1n} D_{1,n}^2. \qquad (12.62)$$

Analogous conditions can be derived for a hot sphere with diameter D_3 immersed in liquid

$$D_3^2 \approx \frac{1}{4} \sum_{n=1}^{N} D_{1,n}^2. \qquad (12.63)$$

By known nucleation and bubble growth laws and prescribed time step $\Delta \tau$, the above equation defines the number of bubble generation, N, and therefore the time needed to cover the surface with bubbles and nuclei, $N \Delta \tau$. At this time the maximum possible superheating is reached. At the end of this process the void fraction generated within $N \Delta \tau$ is

$$\alpha_1 = \frac{4}{3} \pi \sum_{n=1}^{N} n_{1,n} D_{1,n}^3 \qquad (12.64)$$

and volume-averaged bubble size is then

$$D_1^{*3} = \left(\frac{3}{4\pi}\alpha_1/n_1\right)^{1/3}. \tag{12.65}$$

After this moment the bubble production frequency is governed by combined thermal interaction and mechanical entrainment phenomena, which will be considered later.

12.6 Maximum superheat

Skripov et al. [41] (1980) p. 136, 143, found that rapid decompressions can bring the water into a state of maximum superheating defined by $Gb_2 = 6.8$ to 9.6, which means

$$p'(T_2^{spin}) - p = (1.57 \text{ to } 1.32) \frac{\sigma^{3/2}}{(kT_2)^{1/2}\left[1 - \rho''(T_2)/\rho'(T_2)\right]}. \tag{12.66}$$

Lienhard [52] found in 1976 the following explicit approximation of the homogeneous nucleation temperature which can be taken as a spinoidal temperature

$$T_2^{spin} = T'(p) + \left\{0.905 - \frac{T'(p)}{T_c} + 0.095\left[\frac{T'(p)}{T_c}\right]^8\right\}T_c. \tag{12.67}$$

Under rapid decompression we understand decompression with

$$dp/d\tau \geq 2\times 10^{11} \ Pa/s. \tag{12.68}$$

In addition we should mention that *Hutcherson* et al. [18] (1983) classified the boundary between "fast" and "slow" depressurization as about 400×10^6 *Pa/s*. The line in the *p-T* diagram, defined by the Eq. (12.66) or (12.67) is called the spinoidal line. The database defining the spinoidal line for water is given by *Skripov* [41] p. 144 Table 6.4. *Lienhard* [30] (1981) proposed for the constant the value 1.349 corresponding to $Gb_2 \approx 9.2$. For n-pentane, n-hexane, n-nepthane, ether and benzene *Eberhard* and *Schnydess* [12] reported in 1973 $Gb_2 \approx 11.5$. The authors observed experimentally negative pressure up to $\sim 0.4 p_c$.

In technical systems the superheating is a function of several parameters not considered in our previous discussion. Next we discuss some of them.

The question of the maximal achievable superheating in technical systems has attracted the attention of the engineers since 1947. Probably *Burnell* [10] was the first who investigated in 1947 experimentally the flow of boiling water through orifices. The inlet conditions are corresponding to saturated water at $p \approx (0.34$ to $12.7)\times 10^5$ *Pa*. *Burnell* found that a superheating of $\Delta p_{Fi}^* = p'(T_2) - p_{Fi}$ is necessary to observe flashing inception at the throat. The flashing inception at the throat reduces dramatically the mass flow rate compared to the prediction of the *Bernoulli* equation for subcooled water. *Burnell* found that the product of the initial saturated pressure multiplied by the critical bubble radius is constant

$$p'(T_2)R_{1c} \approx const \tag{12.69}$$

and therefore

$$\Delta p^*_{Fi} = p'(T_2) - p_{Fi} = 2\sigma_2(T_2)/R_{1c} \approx 2\sigma_2(T_2)p'(T_2)/const$$

$$\approx f(\sigma_2)p'(T_2) \tag{12.70}$$

and

$$f(\sigma_2) = 0.264 \frac{\sigma_2[T'(p)]}{\sigma'_2[T'(12.066 \times 10^5 \, Pa)]} \tag{12.71}$$

where the surface tension of the saturation condition was computed as follows

$$\sigma'_2(T_2) = 0.07548[1 - 1.85(T_2 - 273.15)] \text{ in } N/m. \tag{12.72}$$

Thus the critical bubble radius in this region is

$$R_{1c} = 2\sigma_2(T_2)/\Delta p^*_{Fi}. \tag{12.73}$$

Algamir and *Lienhard* correlated several experimental data for the reached water superheating with a standard deviation of ± 10% with the following value of the dimensionless critical work $Gb_2 = 28.2 \pm 5.8$ which after using Eqs. (12.30) and (12.49) results in the following relationship for flashing inception,

$$\Delta p^*_{Fi} = \left\{ \frac{16\pi\sigma^3}{3kT_c[1 - \rho''(T_2)/\rho'(T_2)]^2} \frac{T_2}{T_c} \frac{Gb_2}{\varphi} \right\}^{1/2}$$

$$= \Delta p^o_{Fi} \sqrt{1 + 14[(\gamma_v \, \partial p/\partial \tau + \mathbf{V}\gamma.\mathrm{grad} \, p)/10^{11}]^{0.8}} \tag{12.74}$$

where

$$\Delta p_{Fi}^o = 0.253 \frac{\left[\sigma(T_2)\right]^{3/2} (T_2/T_c)^{13.73}}{(kT_c)^{1/2} \left[1 - \rho''(T_2)/\rho'(T_2)\right]} \qquad (12.75)$$

is the pressure difference corresponding to the maximum possible superheating in an isobar system and T_2 is the initial temperature before the pressurization. For

$$T_2 = 273.15 + 100 \ K, \ \Delta p_{Fi}^* > 0.2 \times 10^5 \ Pa, \qquad (12.76)$$

and for

$$T_2 \approx 273.15 + 300 \ K, \ \Delta p_{Fi}^* < 9.5 \times 10^5 \ Pa \qquad (12.77)$$

For $T_2 > 273.15 + 300K$ Δp_{Fi}^o decreases to zero due to vanishing surface tension as the critical point is approached. In fact Eq. (12.75) is the modified *Skripov* equation. Instead of the constant the expression $0.253(T_2/T_c)^{13.73}$ is used.

Bartak [5] (1990) correlated the data of several authors including his own data with the following correlation

$$\Delta p_{Fi}^* = \left\{ \frac{16\pi\sigma^3}{3kT_2 \left[1 - \rho''(T_2)/\rho'(T_2)\right]^2 \frac{T_2}{T_c} \frac{Gb_2}{\varphi}} \right\}^{1/2} \qquad (12.78)$$

where

$$\frac{Gb_2}{\varphi} = f_1 f_2, \qquad (12.79)$$

$$\log f_1 = 11 - 0.0274(T_2 - 273.15), \qquad (12.80)$$

$$f_2 = 36 / \left[(\gamma_v \, \partial p/\partial \tau + \mathbf{V}\gamma.\text{grad } p)/10^6 \right]^{0.37}, \qquad (12.81)$$

with a mean relative error of 16%. It is valid for water in the temperature range from 373.15 to 583.15 K and depressurization rates of 4×10^8 to 2×10^{11} Pa/s. Bartak reported that the *Algamir* and *Lienhard* correlation 65 approximates his own data with 21% mean relative error. In fact *Baratak's* correlation predict more then 30% higher values than the *Algamir* and *Lienhard* correlation. Consequently, it can be concluded that the achievable prediction accuracy is within the error band of about 48.5%.

If the maximum possible superheating under given conditions is reached, the maximum nucleation density is reached too. Note that the reduction factor, Eq.

(12.49) obtained by *Algamir* and *Lienhard* can be used to compute the heterogeneous nucleation densities using the *Blander* and *Katz* equation (12.45) only together with the assumption of no bubble growth during the nucleation process.

Jones [21] found in 1979 that the turbulization of the flow could reduce the maximum superheating

$$\Delta p_{Fi} = \Delta p_{Fi}^* - 27\tfrac{1}{2}\rho_2 V_2'^2, \qquad (12.82)$$

where the turbulent pulsation velocity in a straight pipe is 7.2 % from the local flow velocity

$$\sqrt{V'^2} = 0.072 V_2 \text{ or } V_2'^2 = 5.184 \times 10^{-3} V_2^2 \qquad (12.83)$$

For a converging nozzle the local fluctuation velocity can be expressed as a function of the entrance velocity V_{2o} using the continuity equation

$$V_2'^2 = 5.184 \times 10^{-3} V_{2o}^2 (F_o/F)^2. \qquad (12.84)$$

Comparing with experimental data *Abuaf* et al. [1] (1982) found that the turbulent pulsation velocity varies in accordance with

$$V_2'^2 = 5.184 \times 10^{-3} V_{2o}^2 (F_o/F)^n, \qquad (12.85)$$

where

$$n = 1.75 \text{ for } F/F_o \geq 1/6, \qquad (12.86)$$

and for stronger contraction

$$n = 1.4 \text{ for } F/F_o < 1/6. \qquad (12.87)$$

Jones [22] (1982) shows that for $G > 15\,000$ kg/(m²s) no superheating is observed. To what extent the superheating reduction explained by *Jones* [21] with turbulization is separable from the convective transport of nuclei by high-liquid velocities is not clear.

12.7 Critical mass flow rate in short pipes, orifices and nozzles

The most important outcome of the heterogeneous nucleation theory for the engineering design practice is the possibility to compute in a simple manner the critical mass flow rate in short pipes, orifices and nozzles for saturated and subcooled water. Probably *Burnell* was the first approximating critical flow in nozzles with saturated and subcooled inlet condition modifying the *Beronulli* equation as follows

$$G^* = \sqrt{2\rho_2(p_o - p_{Fi})} \qquad (12.88)$$

where the critical pressure is set to the flashing inception pressure

$$p_{Fi} = p'(T_2) - \Delta p_{Fi}. \qquad (12.89)$$

Jones [22] (1982) described successfully critical mass flow rate in nozzles modifying Eq. (12.79) to

$$G^* = (0.93 \pm 0.04)\sqrt{2\rho_2(p_o - p_{Fi})} \qquad (12.90)$$

The error of this approach was reported to be ± 5% for inlet conditions of $p = (28$ to $170) \times 10^5 \, Pa$ and $T_2 = 203$ to $288°C$ and use of the *Algamir* and *Lienhard* correlation. *Fincke* [14] (1984) found for the discharge coefficient instead of 0.93 the value 0.96 and reported that this coefficient does not depend on the *Reynolds* number.

12.8 Nucleation in the presence of non-condensable gases

In the liquids usually used in technology there are *dissolved inert gases* and *microbubbles*. It is known that at $p = 10^5 Pa$, $T_{2o} = 298.15 \, K$ the amount of dissolved gases and microbubbles in the coolant is $\alpha_{1o} \cong 0.005$, for boiling water reactors, and $\alpha_{1o} \cong 0.001$, for pressurized water reactor *Malnes* and *Solberg* [32] (1973). *Brennen* [9] (1995) p. 20 reported that it takes weeks of deaeration to reduce the concentration of air in the water tunnel below $3ppm$ (saturation at atmospheric pressure is about $15ppm$).

This amount is dissolved in a form of

$$n_{1o} = \alpha_{1o} / (\pi D_{1o}^3 / 6) \qquad (12.91)$$

bubbles per unit volume, so that before starting the nucleation n_{1o} nucleation sites already exist. Here the initial bubble diameter is in any case less than the bubble diameter computed after equating the buoyancy force and the surface force

$$D_{1o} < \sqrt{6}\lambda_{RT}. \qquad (12.92)$$

Note that the so called free stream nuclei number density is subject to distribution depending on the nucleation size, e.g. $n_{10} \sim R_{10}^{-4}$, $R_{10} > 5 \times 10^{-6}$ as reported by Brennen [12] (1995).

Let us suppose that at p_o, T_{2o} we have no water steam in the single bubble [6] (1980) and therefore the mass of the single bubble is

$$m_{\sum nlo} = p_{\sum nl} \frac{\frac{4}{3}\pi R_{1o}^3}{R_{\sum nl} T_{2o}} = (p + \frac{2\sigma}{R_{1o}}) \frac{\frac{4}{3}\pi R_{1o}^3}{R_{\sum nl} T_{2o}} = (p + \frac{2\sigma}{R_{1o}}) \frac{\alpha_{1o}/n_{1o}}{R_{\sum nl} T_{2o}}. \qquad (12.93)$$

Further let us assume that the steam evaporating from the surface into the bubble is saturated. The initial internal bubble pressure $\frac{3}{4}\frac{m_{\sum nlo} R_{\sum nl} T'}{\pi R_1^3}$ increases with $p'(T_2)$ and equals the external pressure $p + \frac{2\sigma}{R_1}$

$$p'(T_2) - p = \frac{2\sigma}{R_1} - \frac{3}{4}\frac{m_{\sum nlo} R_{\sum nl} T'}{\pi R_1^3}, \qquad T_2 > T'(p). \qquad (12.94)$$

With increasing liquid superheating the bubble radius increases to its critical size

$$R_{1c} = 3\sqrt{\frac{m_{\sum nlo} R_{\sum nl} T'(p)}{8\pi\sigma}}, \qquad (12.95)$$

Blake [8], Neppiras and Noltingk [47] (1951), corresponding to the superheating

$$T_2 - T'(p) = \frac{2^{5/2}}{3\sqrt{\pi}}\sigma^{3/2}\sqrt{T'(p)}\frac{(\rho'-\rho'')\rho_2}{(\rho_2-\rho'')\rho'\rho''}\frac{1}{h''-h'}\frac{1}{\sqrt{m_{\sum nlo} R_{\sum nl}}}. \qquad (12.96)$$

This is the maximum possible superheating. Thereafter the bubble loses its stability and starts to increase in size. We see that the presence of dissolved gases in the liquid decreases the superheating necessary for starting the intensive evaporation.

A considerable amount of dissolved gases can lead to flashing initiation at temperatures *lower than the saturation temperature*.

For the first time steps, where the nucleation occurs, the instantaneous source terms are

$$\mu_{21} = \rho'' V_{1c} \dot{n}_{1cin}, \qquad (12.97)$$

$$\dot{q}_2^{m1\sigma} = -\mu_{21}(h'' - h_{M2}). \qquad (12.98)$$

12.9 Activated nucleation site density – state of the art

Not all cavities at the wall serve as a nucleation sites. It is experimentally observed that the increase of the wall superheating activates an increasing number of nucleation sites. Many authors tried to represent this observation by setting

$$n''_{1w} \approx const / D_{1c}^n \qquad (12.99)$$

where the exponent n varies from 2 to 6. If nucleation sites forms a triangular array and are touching each other their maximum number per unit surface corresponding to the local superheating is

$$n''_{1w,\max} = 2/(\sqrt{3} D_{1c}^2). \qquad (12.100)$$

Note that at atmospheric pressure this expression reduces to

$$n''_{1w,\max} = 0.20 \times 10^9 \Delta T^2. \qquad (12.101)$$

Labuntsov [29] assumed in 1963 that the active nucleation sites density is a part of this value

$$n''_{1w,\max} = const\, 2/(\sqrt{3} D_{1c}^2) \qquad (12.102)$$

Later in 1977 this equation was used by *Avdeev* et al. [4] for modeling of flashing flows

$$n''_{1w,\max} = 20 \times 10^{-5} / D_{1c}^2 \qquad (12.103)$$

which for atmospheric pressure gives an activated site density four order of magnitude lower than the thinkable maximum. The constant 20×10^{-5} was estimated by comparison of the critical mass flow rates with those experimentally measured. For atmospheric pressure Eq. (12.92) results in

$$n''_{1w,\max} = 4.85 \times 10^4 \Delta T^2 . \tag{12.104}$$

Johov [20] (1969) plotted data for water and non-water liquids for the nucleation site density as a function of the critical diameter. The data have been in the region $400 \le n''_{1w} \le 4 \times 10^8$. The data correlated with

$$n''_{1w} = 4 \times 10^{-12} / D^3_{1c} , \tag{12.105}$$

$\left[= 15.12 \Delta T^3 \right]$ for $p = 0.1$ *MPa*, with considerable data spreading within two orders of magnitude.

Mikic and *Rohsenow* [33] (1969) proposed the expression

$$n''_{1w} = c \left(D^*_1 / D_{1c} \right)^{3.5} , \tag{12.106}$$

where D^*_1 is the diameter for which n''_{1w} would be 1 per unit area, and c is a dimensional constant ($1/m^2$) depending on the cavity size distribution. Here n''_{1w} is interpreted as the number of the active bubble sites, per unit area, having sizes greater than D^*_1. For a triangular array

$$D^*_1 = \left(2 / \sqrt{3} \right)^{1/2} . \tag{12.107}$$

Hutcherson et al. [18] (1983) uses instead of the critical nucleation size the bubble departure size and obtained good agreement with experimental data for the initial depressurization of a vessel blow down assuming

$$n''_{1w} D^2_{1d} \approx 0.01 \times 2 / \sqrt{3} \tag{12.108}$$

using the *Fritz* equation for the bubble departure diameter D_{1d}.

$$D_{1d} \equiv D_{1d,Fritz} = 1.2\theta \left\{ \frac{\sigma}{g \left[\rho_2 - \rho''(p) \right]} \right\}^{1/2} . \tag{12.109}$$

For the case of nucleate boiling at the wall, *Kocamustafaogullari* and *Ishii* [26] (1983) proposed the following correlation to predict the active nucleation sites density

$$n''_{1w}D_{1d}^2 = \left[\left(D_{1d}/D_{1c}\right)^{4.4} 2.157 \times 10^{-7} \left(1+0.0049\bar{\rho}\right)^{4.13} / \bar{\rho}^{3.2}\right] \quad (12.110)$$

where

$$\bar{\rho} = (\rho' - \rho'')/\rho'', \quad (12.111)$$

$$D_{1d} = 0.0012\bar{\rho}^{0.9} D_{1d,Fritz} \quad (12.112)$$

and the size of the activated sites, D_{1c}, is computed using Eq. (12.9) with a temperature difference $S[T_w - T'(p)]$. Here S is the so called *Chen* superposition factor for taking into account the effective superheating of the wall boundary layer. $S = 1$ for free convection pool boiling and ≤ 1 for forced convection boiling. The authors reported good agreement with data for pressures $p = 0.1$ to $19.8\ MPa$ and for both subcooled and pool boiling.

Jones [23, 24] (1992) found

$$n''_{1w}D_{1d}^2 = 10^{-7} \left(D_{1d}/D_{1c}\right)^4 \quad (12.113)$$

for critical flashing flow in a nozzle using the equilibrium between drag and surface forces in the viscous boundary layer to compute departure diameter. *Riznic* and *Ishii* [37] (1989) demonstrated the applicability of this correlation for flashing adiabatic flow using D_{1c} computed with the equation by *Kocamustafaogullari* and *Ishii* for the temperature difference $T_2 - T'(p)$.

Cornwell and *Brown* [11] (1978) reported experimental data for pool boiling of saturated water at atmospheric pressure on a copper surface correlated by

$$n''_{1w} \approx \Delta T^{4.5} \approx 1.36 \times 10^{-19} / D_{1c}^{4.5}. \quad (12.114)$$

Plotting the available data for nucleation site density as a function of superheating for boiling water at atmospheric pressure, Fig. 12.1, gives an average trend that can be represented by

$$n''_{1w} \approx \Delta T^{4.18} \approx 2.988 \times 10^{-18} / D_{1c}^{4.18}. \quad (12.115)$$

The minimum of the nucleation site density can be represented by

$$n''_{1w} \approx \left(\Delta T / 3.729\right)^{2.593} \text{ for } \Delta T \leq 21.1, \quad (12.116)$$

and

12.9 Activated nucleation site density – state of the art

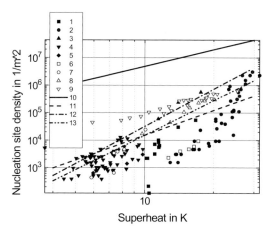

Fig. 12.3. Active nucleation site density as a function of superheat. Saturated water at 0.1 *MPa*. Data: 1) Gaertner [16] 1965, 4/0 polished copper, 2) Gaertner and Westwater [15] 1960, 4/0 polished copper, 20 p.c. nickel salt-water solution, 3) Sultan and Judd [42] 1978, diamond grid 600 polished copper, 4) Yamagata et al. [47] 1955, fine polished brass, 5) Jakob and Linke [19] 1933, polished steel, 6) Cornwell and Brown [11] 1978, 4/0 polished copper, 7) Kurihara and Myers [28] 1960, 4/0 polished copper, 8) Rallis and Jawurek [35] 1964, nickel wire, 9) Faggani et al. [13] 1981, polished 316 steel horizontal cylinder. Prediction with correlations proposed by 10) Avdeev et al. [4], 11) Johov [20], 12) Carnwell and Brown [11] and 13) Kocamustafaogullari and Ishii [26]

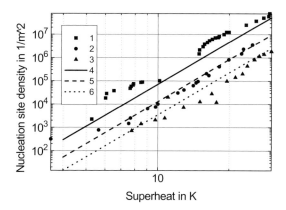

Fig. 12.4. Active nucleation site density as a function of superheating. Saturated water at 0.1 *MPa*. Wang and Dhir [46] data for three different static contact angles 1) 90°, 2) 35° and 3) 18°. Prediction of the same data with their correlation 4), 5), and 6), respectively. Larger static contact angle results of larger active nucleation site density by the same superheating

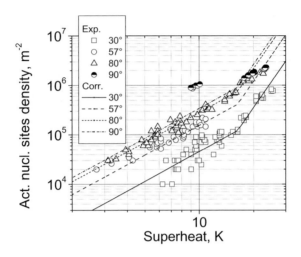

Fig. 12.5. Active nucleation site density as a function of superheating. Saturated water at 0.1 MPa. Basu, Warrier and Dhir [48] data for different static contact angles: 30°,57°,80°, 90°. $124 < G < 886 kg/(m^2 s)$, $6.6 < \Delta T_{sub} < 52.5 K$, $25 < \dot{q}''_{w2} < 960 kW/m^2$. Prediction of the same data with their correlation respectively

$$n''_{1w} \approx 100(\Delta T/16.135)^{16.53} \text{ for } 21.19 \leq \Delta T \leq 40. \qquad (12.117)$$

Comparing the predictions for n''_{1w} by the different above mentioned methods with each other shows discrepancies of several orders of magnitude in different regions which leads to the conclusion that the knowledge in this field is unsatisfactory for practical application.

An important step in this field was made by *Wang* and *Dhir* [46]. *Wang* and *Dhir* [46] (1993) succeeded to quantify the effect of the static contact angle θ, and correlated their data for active nucleation sites density as a function of the wall superheating as follows

$$n''_{1w} = 5 \times 10^{-27} (1 - \cos\theta)/D_{1c}^6 = 714.3 \Delta T^6, \qquad (12.118)$$

where the constant has a dimension m^4 - see Fig. 12.4. The uncertainty in measuring the static contact angle was $\pm 1/60$ *rad*; the uncertainty of the measured nucleation site density was estimated to ± 20 %. The data are collected for atmospheric pool boiling of water with static contact angle of π 90/180, π 35/180 and π 18/180 *rad*. The authors discovered that only cavities having a nearly spherical form and mouth angle less than the static contact angle serve as nucleation sites. The mouth angle is the angle between the boiling surface and the internal surface of the cavity forming the mouth. The region of changing the heat flux covers the entire nucleate boiling region.

Note that Eq. (12.118) should be used together with the experimentally observed relationship between the averaged nearest-neighbor distance and the nucleation site density,

$$\ell \approx 0.84 / n''_{1w}.$$
(12.119)

In 2002 *Basu, Warrier* and *Dhir* [48] proposed a replace for the Eq. (12.118)

$$n''_{1w} = 3400(1 - \cos\theta)\Delta T^2 = 2.183 \times 10^{-5} / D_{1c}^2$$
(12.120)

for $\Delta T_{inb} < \Delta T < 16.298K$, and

$$n''_{1w} = 0.34(1 - \cos\theta)\Delta T^{5.3} = 2.048 \times 10^{-23}(1 - \cos\theta) / D_{1c}^{5.3}$$
(12.121)

for $16.298K \leq \Delta T$. Here ΔT_{inb} is the superheat required for initiation of the nucleate boiling. The correlation reproduces data given Fig. 12.5 for $\theta = \pi/2$ and $\theta = \pi/6$ within $\pm 40\%$. The correlation is valid in the range $124 < G < 886 kg/(m^2s)$, inlet subcooling $6.6 < \Delta T_{sub} < 52.5K$, $25 < \dot{q}''_{w2} < 960 kW/m^2$, and $\pi/6 < \theta < \pi/4$.

Benjamin and *Balakrishnan* [49] reported in 1997 the only correlation taking into account the thermal properties of the wall and its roughness as follows

$$n''_{1w} = 218.8 \frac{\Pr_2^{1.63}}{\gamma \delta_w^{*0.4}} \Delta T^3,.$$
(12.122)

The liquid *Pradtl* number, the dimesinless rougness and the dimensionless factor defining the transient interface temperature are defined as follows

$$\Pr_2 = c_{p2}\eta_2 / \lambda_2,$$
(12.123)

$$\delta_w^* = 14.5 - 4.5 \frac{\delta_w p}{\sigma_2} + \left(\frac{\delta_w p}{\sigma_2}\right)^{0.4},$$
(12.124)

$$\gamma = \left[\lambda_w \rho_w c_{pw} / (\lambda_2 \rho_2 c_{p2})\right]^{1/2}.$$
(12.125)

δ_w is the arithmetic averaged deviation of the roughness surface countor from the averaged line. The correlation is valid within $1.7 < \Pr_2 < 5$, $4.7 < \gamma < 93$, $0.02 < \delta_w < 1.17 mm$, $5 < \Delta T < 25 K$, $13 \times 10^{-3} < \sigma < 59 \times 10^{-3} N/m$, $2.2 < \delta_w^* < 14$. Table 12.3 provides usefull information about the rougness of differently polished materials.

Table 12.3. Rougness of differently polished materials

Material	Finish	Roughness in μm	Refference
Cooper	3/0 emery paper	0.14	*Grifith* and *Wallis* [55]
Cooper	4/0 emery paper	0.07	*Kurihara* and *Myers* [28]
Cooper	4/0 emery paper	0.07	*Gaertner* and *Westwater* [15]
Cooper	4/0 emery paper	0.07	*Gaertner* [16]
Cooper	Mirror finish	<0.02	*Wang* and *Dhir* [46]
Nickel	4/0 emery paper	0.045	*Zuber* [54]
Stainless steel	1/0 emery paper	0.2	*Benjamin* et al [49]
Aluminium	2/0 emery paper	1.17	*Benjamin* et al [49]
Aluminium	3/0 emery paper	0.89	*Benjamin* et al [49]
Aluminium	4/0 emery paper	0.52	*Benjamin* et al [49]

There is no attempt to describe theoretically the activated nucleation sites per unit surface as a function of the local superheating.

12.10. Conclusions and recommendations

1. The maximum superheating in technical systems is a function of the depressurization velocity and of the produced turbulence.
2. The maximum superheating can be predicted by the *Algamir* and *Lienhard* and by the *Bartak* correlations within an error band of 48.5%.
3. Flashing in short pipes and nozzles leads to critical flows driven by the pressure difference equal to the entrance pressure minus the flashing inception pressure.
4. For the prediction of the maximum achievable superheating which represents the spinoidal line the *Skripov* correlation is recommended.
5. The wetting angle is an important property of the polished surface characterizing its capability to activate nucleation sites.
6. For the prediction of the activated nucleation sites the correlation obtained by *Wang* and *Dhir* is recommended.
7. The establishing of a vapor film around a heated surface having temperature larger than the minimum film boiling temperature takes a finite time.

8. Availability of small bubbles of non-condensing gases reduces the superheating required to initiate evaporation. Evaporation at lower than the saturation temperature is possible.

Nomenclature

B	thermally controlled bubble growth parameter
D_{cav}	equivalent cavity size, m
D_h	hydraulic diameter, m
$D_{1,n}$	bubble diameter belonging to n-th bubble generation, m
D_{1c}	critical bubble diameter, m
D_{1d}	bubble departure diameter, m
d	differential, *dimensionless*
F_1/Vol	cross section at the wall surface covered by attached bubbles per unit flow volume, $1/m$
G^*	critical mass flow rate, $kg/(m^2s)$
Gb_2	$= \dfrac{\Delta E_{1kr}}{kT_2}$, Gibbs number, *dimensionless*
h	specific enthalpy, J/kg
k	$= 13.805 \times 10^{-24}$, Boltzmann constant, J/K
m_2	$= m_\mu/N_A$, mass of single molecule, kg
m_μ	kg-mole mass (18 kg for water), $kg/mole$
N_2	$= \rho_2/m_2 = \rho_2 N_A/m_\mu$ ($\approx 3.3 \times 10^{28}$ for water), number of the molecules per unit volume, $1/m^3$
N_A	$= 6.02 \times 10^{26}$, Avogadro number, number of molecules in one kilogram-mole mass, $1/kg\text{-}mole$
$\dot{n}_{1cin}$	number of the created nuclei per unit time in unit volume of the liquid, $1/(m^3s)$
n_w''	cavity per unit surface, $1/m^2$
n_{1w}''	bubble generating (active) cavity per unit surface, $1/m^2$
n_1	bubble number density, $1/m^3$
$n_{1,n}$	bubble number density belonging to n-th bubble generation, $1/m^3$
$n_{1,\max}$	maximum number of nucleation sites per unit volume in a pipe, $1/m^3$
p	pressure, Pa
$\dot{q}_2^{m1\sigma}$	thermal energy consumed from the liquid per unit time and unit flow volume, W/m^3
R	gas constant, $J/(kgK)$
R_1	bubble radius, m

R_{10}	initial bubble radius, m
R_{1c}	critical bubble radius, m
$R_{1\infty}$	final radius of the bubble, m
r	radius, m
s	specific entropy, $J/(kgK)$
T	temperature, K
u	radial velocity, m/s
V	velocity vector, m/s
V_{1c}	volume of the bubble with critical size
v	specific volume, m^3/kg
Vol	volume, m^3

Greek

α_1	bubble volumetric fraction, m^3/m^3
γ_v	volume occupied by the flow divided by the volume of the system, *dimensionless*
γ	cross section flowed by the flow divided by the total cross section of the system, *dimensionless*
Δ	finite difference
ΔE_1	work necessary to create a single bubble with radius R_1, J
ΔE_{1c}	work necessary to create a single bubble with radius R_{1c}, J
ΔE_{1c}^*	$= \varphi \Delta E_{1c}$, energy for the creation of a single bubble at a wall, J
Δp_{Fi}^*	$= p'(T_2) - p_{Fi}$, flashing inception pressure difference, Pa
$\Delta \tau$	time interval, s
θ	static contact angle, rad
λ_{RT}	Rayleigh-Tailor wavelength, m
μ_{21}	evaporating mass per unit time and unit flow volume, $kg/(m^3s)$
ρ	density, kg/m^3
σ	surface tension, N/m
τ	time, s
Φ	cavity angle if the cavity is idealized as a cone, rad
φ	work reduction factor, *dimensionless*

Superscripts

$'$	saturated liquid
$''$	saturated vapor
spin	spinoidal line

Subscripts

1	vapor, gas
2	liquid
c	critical state
s	at constant entropy
Fi	flashing inception
$\sum_{nlo}$	sum of all non-condensing gases
M	evaporating chemical component

References

1. Abuaf N, Jones OC Jr, Wu BJC (May 1983) Critical flashing flows in nozzles with subcooled inlet conditions, J. of Heat Transfer, vol 105 pp 379-383
2. Algamir M, Lienhard JH (1981) Correlation of pressure undershoot during hot-water depressurisation. J. of Heat Transfer, vol 103 no 1 p 60
3. Arefeva EI, Aladev IT (July 1958) O wlijanii smatchivaemosti na teploobmen pri kipenii, Injenerno - Fizitcheskij Jurnal, vol 1 no 7 pp 11 - 17, in Russian
4. Avdeev AA, Maidanik VN, Selesnev LI, Shanin VK (1977) Calculation of the critical flow rate with saturated and subcooled water flashing through a cylindrical duct, Teploenergetika, vol 24 no 4 pp 36-38
5. Bartak J (1990) A study of the rapid depressurization of hot water and the dynamics of vapor bubble generation in superheated water, Int. J. Multiphase Flow, vol 16 no 5 pp 789-798
6. Besand WH (1859) Hydrostatics and hydrodynamics, Deighton Bell, Cambridge, p 170
7. Blander M, Katz JL (1975) Bubble nucleation in liquids. AIChE, vol 21 pp 833-838
8. Blake FG (1949) The onset of cavitation in liquids. I. Acoustic Res. Lab., Harvard Univ., Tech. Memo. no 12
9. Brennen CE (1995) Cavitation and bubble dynamics. Oxford University Press, Oxford, Ney York
10. Burnell JG (Dec.12, 1947) Flow of boiling water through nozzles, orifices and pipes, Engineering, pp 572-576
11. Cornwell K, Brown RD (1978) Boiling surface topology, Proc. 6th Int. Heat Transfer Conf. Heat Transfer, Toronto, vol 1 pp 157-161
12. Eberhard JG and Schnyders MC (1973) Application of the mechanical stability condition to the prediction of the limit of superheat for normal alkanes, ether and water. J. Phys. Chem. Vol 77 no 23 pp 2730-2736
13. Faggiani S, Galbiati P, Grassi W (1981) Active site density, bubble frequency and departure on chemically etched surfaces, La Termotechnica, vol 29 no 10 pp 511-519
14. Fincke JR (August 5 - 8, 1984) Critical flashing flow of subcooled fluids in nozzles with contour discontinuities, Basic Aspects of Two Phase Flow and Heat Transfer, 22nd Nat. Heat Transfer Conference and Exhibition, Niagara Falls, New York, HTD – vol 34 p 85-93
15. Gaertner RF, Westwater JW (1960) Population of active sites in nucleate boiling heat transfer, Chem. Eng. Progr. Symp. Ser., vol 30 no 30 pp 39-48

16. Gaertner RF (Feb. 1965) Photographic study of nucleate pool boiling on a horizontal surface, transaction of the ASME, Journal of Heat Transfer, vol 87 pp17 - 29
17. Gibbs JW (1878) Thermodynamische Studien (Leipzig 1892), Amer. J. Sci. and Arts, vol 16 pp 454-455
18. Hutcherson MN, Henry RE, Wollersheim DE (Nov. 1983) Two-phase vessel blowdown of an initially saturated liquid - Part 2: Analytical, Trans. ASME, J. Heat Transfer, vol 105 pp 694-699
19. Jakob M, Linke W (1933) Der Wärmeübergang von einer waagerechten Platte an siedendes Wasser, Forsch. Ing. Wes., vol 4 pp 75-81
20. Johov KA (1969) Nucleations number during steam production, Aerodynamics and Heat Transfer in the Working Elements of the Power Facilities, Leningrad, in Russian, Proc. CKTI, vol 91 pp 131- 135
21. Jones OC Jr (1979) Flashing inception in flowing liquids, non-equilibrium two-phase flow, in Chan JC and Bankoff SG (eds), ASME, New York, pp 29-34
22. Jones OC Jr (1982) Towards a unified approach for thermal non-equilibrium in gas-liquid systems, Nucl. Eng. and Design, vol 69 pp 57-73
23. Jones OC (1992) Non-equilibrium phase change 1. Flashing inception, critical flow, and void development in ducts, Boiling Heat Transfer, Lahey RT Jr (ed) Elsevier Science Publishers BV, pp 189-234
24. Jones OC (1992) Nonequilibrium phase change 2. Relaxation models, general applications, and post heat transfer, in Lahey RT Jr (ed) Boiling heat transfer, Elsevier Science Publishers BV, pp 447-482
25. Kaishew R, Stranski IN (1934) Z. Phys. Chem., vol 26 p 317
26. Kocamustafaogullari G, Ishii M (1983) Interfacial area and nucleation site density in boiling systems, Int. J. Heat Mass Transfer, vol 26 pp 1377-1389
27. Kottowski HM (1973) Nucleation and superheating effects on activation energy of nucleation, in Progress in Heat and Mass Transfer, Dwyer OE (ed) Pergamon, New York, vol 7 pp 299-324
28. Kurihara HM, Myers JE (March 1960) The effect of superheat and surface roughness on boiling coefficients, AIChE Journal, vol 6 no 1 pp 83 - 91
29. Labuntsov DA (1963) Approximate theory of heat transfer by developed nucleate boiling (Russ.), Izvestiya AN SSSR, Energetika i transport, no 1
30. Lienhard JH (1981) Homogeneous nucleation and the spinoidal line, J. of Heat Transfer, vol 103 p 72
31. Lienhard JH, A Heat Transfer Textbook, Prentice-Hall, Inc., Engelwood Cliffts, New Jersey 07632
32. Malnes D, Solberg K (May 1973) A fundamental solution to the critical two-phase flow problems, applicable to loss of coolant accident analysis. SD-119, Kjeller Inst., Norway
33. Mikic BB, Rohsenow WM (May 1969) A new correlation of pool-boiling data including the effect of heating surface characteristics, Transactions of the ASME, J. Heat Transfer, vol 91 pp 245-250
34. Neppiras EA and Noltingk BE (1951) Cavitation by ultrasonic: theoretical conditions for onset of cavitation. Proc. Phys. Soc., London, vol 64B pp 1032-1038
35. Rallis CJ, Jawurek HH (1964) Latent heat transport in saturated nucleate boiling, Int. J. Heat Transfer, vol 7 pp 1051-1068
36. Rayleigh L (1917) On the pressure developed in a liquid during the collapse of spherical cavity, Phil. Mag., vol 34 p 94

37. Riznic J, Ishii M (1989) Bubble number density and vapor generation in flashing flow, Int. J. Heat Mass Transfer, vol 32 pp 1821-1833
38. Rohatgi US, Reshotko E (1975) Non-equilibrium one dimensional flow in variable area channels. In Lahey RT and Wallis GB (eds) Non-equilibrium two phase flows, ASME, New York
39. Siegel R, Keshock EG (July 1964) Effects of reduced gravity on nucleate boiling bubble dynamics in saturated water, AIChE Journal, vol 10 no 4 pp 509-517
40. Simoneau RJ (1975) Pressure distribution in converging diverging nozzle during two phase choked flow of subcooled nitrogen. Lahey RT and Wallis GB (eds) Non-equilibrium two phase flows, ASME, New York
41. Skripov VP et al. (1980) Thermophysical properties of liquids in meta - stable state, Moskva - Atomisdat, in Russian
42. Sultan M, Judd RL (Feb. 1978) Spatial distribution of active sites and bubble flux density, Transactions of the ASME, Journal of Heat Transfer, vol 100 pp 56-62
43. Tolubinsky VI, Ostrovsky JN (1966) On the mechanism of boiling heat transfer (vapor bubbles growth rate in the process of boiling in liquids, solutions, and binary mixtures), Int. J. Heat Mass Transfer, vol 9 pp 1463-1470
44. Tolubinski IS (1980) Boiling heat transfer, Kiev, Naukova Dumka, in Russian
45. Volmer M (1939) Kinetik der Phasenbildung, Dresden und Leipzig, Verlag von Theodor Steinkopf
46. Wang CH, Dhir VK (Aug. 1993) Effect of surface wettability on active nucleation site density during pool boiling of water on a vertical surface, ASME Journal of Heat Transfer, vol 115 pp 659-669
47. Yamagata K, Hirano F, Nishikawa K, Matsuoka H (1955) Nucleate boiling of water on the horizontal heating surface, Mem. Fac. Engng; Kyushu Univ., vol 15 p 97
48. Basu N, Warrier GR and Dhir VK (2002) Onset of nucleate boiling and active nucleation site density during subcooled flow boiling, J. Heat Transfer, vol 124 pp 717–728
49. Benjamin RJ and Balakrishnan AR (1997) Nucleation site density in pool boiling of saturated pure liquids: effect of surface microroughness and surface and liquid physical properties, Exp. Thermal Fluid Sci., vol 15 pp 32–42
50. Benjamin RJ and Balakrishnan AR (1997) Nucleation site density in pool boiling of binary mixtures: effect of surface microroughness and surface and liquid physical properties, Can. J. Chem. Eng., vo 75 pp 1080–1089
51. Hibiki T and Ishii M (2003) Active nucleation site density in boiling systems, International Journal of Heat and Mass Transfer, vol 46 pp 2587–2601
52. Lienhard JH (1976) Correlation for the limiting liquid superheat, Chem. Eng. Sci., vol 31 pp 847-849
53. Bergles AE and Rohsenow WM (1964) The determination of forced convection surface-boiling heat transfer, ASME J. Heat Transfer, vol 1 pp 365-372
54. Zuber N (1963) Nucleate boiling: The region of isolated bubbles and the similarity with natural convection, Int. J. Heat Mass Transfer, vol 6 pp 53-79
55. Griffith P and Wallis GB (1960) The role of the surface conditions in nucleate boiling, Chem. Eng. Prog. Symp., vol 56 pp 49-63

13. Bubble growth in superheated liquid

This review presents the achievements of the theory of bubble growth in superheated liquids. The thermally controlled bubble growth solutions are summarized and the derivation of the Mikic equation is discussed in some more details. Then the link between the solutions for the bubble growth and the mass source terms for the averaged conservation equations for two-phase flow are presented. The way to derive non-averaged mass source terms and time-averaged mass source terms is given. The effect of the steam superheating is discussed. Brief discussion of diffusion controlled bubble growth initially containing non-condensable gases is also given.

13.1 Introduction

The *spontaneous flashing of superheated liquid* is associated with the *bubble growth* in the liquid. The mechanism of the *bubble growth* is described already by *Jakob* [10] in 1932 in the following way:

"*...it can be imagined that during the small explosion which starts the growth of a bubble, the interface temperature, because of the consumed heat for vaporization, drops immediately from the superheat temperature to the saturation temperature, for example from 110°C to 100°C. As a consequence of the heat transfer from the liquid to the vapor bubble the liquid envelope is being cooled progressively from the inside toward the outer boundary; a temperature boundary layer is created with a constantly decreasing temperature drop. This thermal boundary layer increases in thickness until the thermal wave, which advances from the vapor bubble interface into the liquid, has reached the outer limit of the hydrodynamics boundary layer. The decrease in thickness of the hydrodynamics boundary layer because of the evaporation at the interface is, initially, a small fraction of the total thickness...*".

The rigorous mathematical description of the processes in a system of spherical coordinates with the origin equivalent with the center of the mass of the nucleation site, needs the solution of a system of mass, momentum, and energy conservation equations for the corresponding initial and boundary conditions, see by *Nigmatulin* [18] (1978), *Beylich* [2] (1991) among others. An important result of such a solution is the *bubble radius as a function of time* and of the local thermodynamic parameters describing the non-equilibrium of the mixture. A general analytical solution of these problems is not known to the author. For practical applications there are several approximate solutions obtained under different simplifying assumptions. The most successful compromise between strict formulation and sound simplifying assumption was made by *Mikic* et al. [17] (1970). In what follows we present the arguments used by *Mikic*.

13.2 The thermally controlled bubble growth

Consider infinite liquid with temperature T_2 which is superheated with respect the saturation temperature by the system pressure p, i.e.

$$T_2 > T'(p). \qquad (13.1)$$

The liquid side bubble interface temperature is $T_2^{1\sigma}$.

$$T_2^{1\sigma} < T_2, \qquad (13.2)$$

Due to the intensive steam production from the bubble surface and the fact that vapor sound velocity is larger then the bubble interface velocity, it can be assumed that the pressure inside the bubble is uniform and that the *vapor temperature in the bubble is equal to the surface temperature*

$$T_1 = T_2^{1\sigma}. \qquad (13.3)$$

The pressure inside the bubble can be assumed to be the saturation pressure at the vapor temperature and therefore

$$p'(T_1) = p'\left(T_2^{1\sigma}\right). \qquad (13.4)$$

The change of the bubble mass with the time is equal to the mass evaporating from the surface per unit time, i.e.

$$\frac{d}{d\tau}\left(\frac{4}{3}\pi R_1^3 \rho_1\right) = 4\pi R_1^2 (\rho w)_{21}, \qquad (13.5)$$

see *Bosnjakovic* [3] (1930). This is the vapor mass conservation equation. Here $(\rho w)_{21}$ is the mass flow rate from the surface into the bubble. This equation is used in the literature in different forms e.g.

$$\frac{1}{3}\frac{d}{d\tau}\left[\left(R_1 \rho_1^{1/3}\right)^3\right] = R_1^2 (\rho w)_{21} \qquad (13.6)$$

or

$$\frac{d}{d\tau}\left[\left(R_1 \rho_1^{1/3}\right)\right] = (\rho w)_{21} / \rho_1^{2/3} \tag{13.7}$$

or in integrated form

$$R_1 = \left(\rho_{10} / \rho_1\right)^{1/3} R_{10} \left\{1 + \frac{1}{R_{10}\rho_{10}} \int_0^{\Delta\tau} \left[(\rho w)_{21} / \rho_1^{2/3}\right] d\tau\right\}, \tag{13.8}$$

which is appropriate to take into account the density change with the time, or

$$\frac{dR_1}{d\tau} = (\rho w)_{21} / \rho_1 - \frac{R_1}{3\rho_1} \frac{d\rho_1}{d\tau}. \tag{13.9}$$

The enthalpy of the liquid experiencing evaporation before crossing the interface is $h'(T_2^{1\sigma})$ and after crossing the interface $h''(T_2^{1\sigma})$. Thus the energy jump condition at the interface, assuming no heat transfer between the bubble and the interface, $\dot{q}_1''^{2\sigma} = 0$, is

$$(\rho w)_{21}\left[h''(T_2^{1\sigma}) - h'(T_2^{1\sigma})\right] = -\dot{q}_2''^{1\sigma}. \tag{13.10}$$

Solving with respect to the evaporation mass flow rate and substituting into the mass conservation equation (13.9) we obtain

$$\frac{dR_1}{d\tau} = -\dot{q}_2''^{1\sigma} / \left\{\rho_1\left[h''(T_2^{1\sigma}) - h'(T_2^{1\sigma})\right]\right\} - \frac{R_1}{3\rho_1}\frac{d\rho_1}{d\tau} \tag{13.11}$$

The heat transferred from the superheated bulk liquid to the surface is controlled by the *Fourier* equation, in spherical coordinates (see *J. Fourier*, [7] (1822))

$$\frac{\partial T}{\partial \tau} = -a_2 \frac{1}{r^2}\frac{\partial}{\partial r}\left(r^2 \frac{\partial T}{\partial r}\right) \tag{13.12}$$

solved for the initial and boundary conditions

$$\tau = 0, \ r = R_1, \ T = T_2^{1\sigma}, \tag{13.13}$$

$$\tau = 0, \ r > R_1, \ T = T_2, \tag{13.14}$$

$$r = R_1,\ T = T_2^{1\sigma},\tag{13.15}$$

$$r = \infty,\ T = T_2\ .\tag{13.16}$$

The text book solution for *thin thermal boundary layer*, Glasgow and Jager [4] (1959), is used to compute the temperature gradient at the bubble surface and to compute the resulting heat flux as a function of time

$$\dot{q}_{i2}'' = -\lambda_2 \left.\frac{\partial T}{\partial r}\right|_{r=R_1} = -\rho_2 c_{p2}\left(T_2 - T_2^{1\sigma}\right)\sqrt{\frac{3a_2}{\pi\tau}}$$

$$= -\rho_2 c_{p2}\left[T_2 - T'(p)\right]\sqrt{\frac{3a_2}{\pi\tau}}\left[1 - \frac{T_2^{1\sigma} - T'(p)}{T_2 - T'(p)}\right].\tag{13.17}$$

Usually the so obtained heat flux is substituted in Eq. (13.11) and the resulting equation is rewritten in the form

$$\frac{dR_1}{d\tau} = \frac{1}{2}\frac{B}{\sqrt{\tau}}\left[1 - \frac{T_2^{1\sigma} - T'(p)}{T_2 - T'(p)}\right] - \frac{R_1}{3\rho_1}\frac{d\rho_1}{d\tau}\tag{13.18}$$

where

$$B = Ja\sqrt{\frac{12 a_2}{\pi}}\ .\tag{13.19}$$

The group

$$Ja = \frac{\rho_2 c_{p2}\left[T_2 - T'(p)\right]}{\rho_{10}\left[h''(T_2^{1\sigma}) - h'(T_2^{1\sigma})\right]}\tag{13.20}$$

is named *Jakob* number in honor of the German scientist considering for the first time this problem in [10] (1932). *Fritz* and *Ende* [8] obtained in 1936 Eq. (13.18) for $\rho_1 \approx const$ and plane geometry. For processes with dramatic pressure change the vapor compressibility is important and $d\rho_1/d\tau$ cannot be neglected. Equation (13.18) controls the bubble growth if the surface temperature $T_2^{1\sigma}$ is known. If the surface temperature is assumed to be saturation temperature at the system pressure

$$T_2^{1\sigma} = T'(p),\tag{13.21}$$

the obtained solution for $\rho_1 \approx const$ is the so called thermally controlled bubble growth solution obtained by *Plesset* and *Zwick* [21] in 1954

$$\frac{dR_1}{d\tau} = \frac{1}{2}\frac{B}{\sqrt{\tau}}.\qquad(13.22)$$

There are several authors contributing to this topic. Their results are summarized in Appendix 13.1.

13.3 The *Mikic* solution

Note that in accordance with the assumption that the pressure inside the bubble is equal to the saturation pressure as a function of the interface temperature, there is no pressure difference between bubble and environment, which is not true. Improved solution can be obtained if one allows a surface temperature different from the saturation temperature at the system pressure. In this case one needs one more equation to close the description. *Mikic* uses the following way: The mass conservation equation of the liquid can be approximated by

$$\frac{1}{r^2}\frac{\partial}{\partial r}(r^2 u_2) \approx 0,\qquad(13.23)$$

or integrating with the boundary condition

$$r = R_1,\; u_2 = \frac{\partial R_1}{\partial \tau},\qquad(13.24)$$

$$u_2 = const/r^2 = \frac{\partial R_1}{\partial \tau}(R_1/r)^2.\qquad(13.25)$$

It gives the liquid velocity as a function of the radius if the liquid were incompressible during the bubble expansion. During the bubble growth the work performed by the bubble expansion

$$4\pi \int_0^{R_1}(p_1 - p)r^2 dr \approx \frac{4}{3}\pi R_1^3 (1 - \frac{\rho''}{\rho_2})(p_1 - p),\qquad(13.26)$$

is transferred in total kinetic energy of the liquid environment

$$\int_{R_1}^{\infty} \frac{1}{2}\rho_2 u_2^2 dVol_2 = \frac{1}{2}\rho_2 \left(\frac{dR_1}{d\tau}\right)^2 R_1^4 4\pi \int_{R_1}^{\infty} \frac{dr}{r^2} = 2\pi\rho_2 \left(\frac{dR_1}{d\tau}\right)^2 R_1^4 \int_{R_1}^{\infty} d(-\frac{1}{r})$$

$$= 2\pi\rho_2 \left(\frac{dR_1}{d\tau}\right)^2 R_1^3 \,,$$

i.e.

$$\left(\frac{dR_1}{d\tau}\right)^2 = \frac{2}{3}\left(1 - \frac{\rho''}{\rho_2}\right)(p_1 - p)/\rho_2 = \frac{2}{3}\left(1 - \frac{\rho''}{\rho_2}\right)\left(p'(T_2^{1\sigma}) - p\right)/\rho_2 = \frac{2}{3}\frac{2\sigma}{R_{1c}}/\rho_2\,,$$
(13.27)

which is the second needed equation. The constant 2/3 valid for bubble growth in a bulk liquid should be replaced by π/7 if a spherical bubble growing on a flat surface is considered. In fact this is the mechanical energy conservation equation. If one assumes that the interface temperature is equal to the liquid temperature

$$T_2^{1\sigma} = T_2\,,$$ (13.28)

and therefore the bubble pressure is the saturation pressure at the liquid temperature the above equation describes alone the bubble growth. This is the limiting case giving the fastest bubble growth with the liquid superheating. This solution is named inertially controlled solution and was obtained for the first time by *Besand* [1] (1859) and in the above described more elegant way by *Rayleigh* [23] (1917) for $\rho''/\rho_2 = 0$. The solution is valid within the first 10^{-8} s of the life of the bubble.

In reality the bubble in a superheated liquid behaves between the both above described limiting mechanisms, so the bubble growth is governed by the equations (13.22) and (13.27) simultaneously. To transfer the pressure difference $p'(T_2^{1\sigma}) - p$ in the last equation into a temperature difference $T_2^{1\sigma} - T'(p)$ one can use the linearized equation of *Clausius* and *Clapeyron*

$$\frac{p'(T_2^{1\sigma}) - p}{T_2^{1\sigma} - T'(p)} = \left(\frac{dp}{dT}\right)_{sat},$$ (13.29)

where

$$\left(\frac{dp}{dT}\right)_{sat} = \frac{1}{T'(p)}\frac{h''(p) - h'(p)}{v''(p) - v'(p)},$$ (13.30)

or

$$\left[p'(T_2^{1\sigma}) - p\right]/\rho_2 = \left[T_2^{1\sigma} - T'(p)\right]\frac{1}{\rho_2}\left(\frac{dp}{dT}\right)_{sat}. \qquad (13.31)$$

The result is

$$\left(\frac{dR_1}{d\tau}\right)^2 = \frac{2}{3}\left(1 - \frac{\rho''}{\rho_2}\right)\left[T_2^{1\sigma} - T'(p)\right]\frac{1}{\rho_2}\left(\frac{dp}{dT}\right)_{sat}$$

$$= \frac{2}{3}\frac{1}{\rho_2}\left(\frac{dp}{dT}\right)_{sat}\left(1 - \frac{\rho''}{\rho_2}\right)\left[T_2 - T'(p)\right]\frac{T_2^{1\sigma} - T'(p)}{T_2 - T'(p)} \qquad (13.32)$$

or

$$\left(\frac{dR_1}{d\tau}\right)^2 = A^2 \frac{T_2^{1\sigma} - T'(p)}{T_2 - T'(p)}, \qquad (13.33)$$

where

$$A^2 = \frac{2}{3}\frac{1}{\rho_2}\left(\frac{dp}{dT}\right)_{sat}\left(1 - \frac{\rho''}{\rho_2}\right)\left[T_2 - T'(p)\right]. \qquad (13.34)$$

Eliminating the temperature difference from the both equations, Eqs. (13.22) and (13.33), *Mikic* et al. obtained the following final form of the equation describing best the available experimental data

$$\frac{1}{A^2}\left(\frac{dR_1}{d\tau}\right)^2 + \frac{2\sqrt{\tau}}{B}\frac{dR_1}{d\tau} - 1 + \frac{2\sqrt{\tau}}{B}\frac{R_1}{3\rho_1}\frac{d\rho_1}{d\tau} = 0, \qquad (13.35)$$

or after solving for $dR_1/d\tau$ and substituting $(dR_1/d\tau)/A = dR_1^+/d\tau^+$

$$\frac{dR_1^+}{d\tau^+} = \left(\tau^+ + 1 - \frac{2}{3}R_1^+ \frac{\tau_1^{+1/2}}{\rho_1^+}\frac{d\rho_1^+}{d\tau^+}\right)^{1/2} - \left(\tau^+\right)^{1/2}, \qquad (13.36)$$

where

$$R^+ = AR_1/B^2, \qquad (13.37)$$

$$\tau^+ = A^2\tau/B^2. \qquad (13.38)$$

Substituting the time derivative of the bubble radius from Eq. (13.36) into Eq. (13.33) we compute the surface temperature

$$T_2^{1\sigma} = T'(p) + [T_2 - T'(p)] \left[\left(\tau^+ + 1 - \frac{2}{3} R_1^+ \frac{\tau_1^{+1/2}}{\rho_1^+} \frac{d\rho_1^+}{d\tau^+} \right)^{1/2} - (\tau^+)^{1/2} \right]^2 \quad (13.39)$$

which is necessary to compute the enthalpies of the transferred mass before and after crossing the interface.

Integrating the equation (13.35) for $R_1^+ = R_{1o}$ at $\tau^+ = 0$, and $\rho_1 = const$, Mikic et al. obtained the general bubble growth relation

$$R_1^+ - R_{1o}^+ = \frac{2}{3} \left[(\tau^+ + 1)^{3/2} - (\tau^+)^{3/2} - 1 \right], \quad (13.40)$$

which for $\tau^+ \ll 1$ simplifies to the *Rayleigh* solution, and for $\tau^+ \gg 1$ to the thermally controlled bubble growth solution.

The validity of the Eq. (13.40) is confirmed experimentally by *Lien* and *Griffith* [15] (1969) for superheated water over the pressure range 0.0012 to 0.038 *MPa*, superheating range 8 to 15 *K*, and $58 \le Ja \le 2690$. The comparison with experimental data made by *Tolubinskii* [25] (1980) shows that in a earlier stadium of the bubble growth $\tau^+ < 10^{-3} (\tau \le 10^{-8} s)$ the bubble growth is inertially dominated. For $\tau^+ > 10^{-3}$ the *Mikic* et al. solution is the best one. For the time $\tau^+ > 10$ all of the heat diffusion controlled solutions tend to the asymptotic solution, the so called thermally controlled bubble growth.

Olek et al. [19] (1990) obtained a general counterpart solution to that of *Mikic* et al. [17] (1970) using the hyperbolic heat conduction equation, instead of the classical *Fourier* (parabolic) equation. The new solution is important for bubble growth in fluids like Helium II.

13.4 How to compute the mass source terms for the averaged conservation equations?

13.4.1 Non-averaged mass source terms

In what follows we show how to link the results obtained in the previous chapter with the macroscopic flow description from Volume 1 of this monograph.

The evaporated mass per unit mixture volume and unit time is equal to the product of the number of the bubbles per unit mixture volume and the mass change of a single bubble. Therefore

13.4 How to compute the mass source terms for the averaged conservation equations?

$$\mu_{21} = \rho''(T_2^{1\sigma})n_1 dV_1/d\tau = \rho''(T_2^{1\sigma})n_1 4\pi R_1^2 dR_1/d\tau, \qquad (13.41)$$

or replacing

$$R_1 = \left(\frac{3}{4\pi}\frac{\alpha_1}{n_1}\right)^{1/3}, \qquad (13.42)$$

$$\mu_{21} = 3^{1/3}(4\pi)^{2/3}\rho''(T_2^{1\sigma})n_1^{2/3}\alpha_1^{1/3}R_1 dR_1/d\tau. \qquad (13.43)$$

Replacing $dR_1/d\tau$ with one of the expressions from Appendix 13.1 we have for the non-averaged mass source term for the thermally controlled bubble growth

$$\mu_{21} = C\rho''(T_2^{1\sigma})n_1^{2/3}\alpha_1^{1/3}B^2 = C\rho''(T_2^{1\sigma})n_1^{2/3}\alpha_1^{1/3}\left(\frac{B}{Ja}\right)^2\left[\frac{\rho_2 c_{p2}}{\rho''(h''-h')}\right]^2[T_2 - T'(p)]^2, \qquad (13.44)$$

with constant C ranging from 2.2 to 14.89. It is very interesting to note that the evaporation source term depends on the void fraction, on the bubble number density, on the liquid temperature, and on the system pressure. All of these values change during a single time interval.

The heat taken away from the liquid and used for the evaporation per unit mixture volume and unit time is

$$\dot{q}_2^{m1\sigma} = -\mu_{21}\left[h''(T_2^{1\sigma}) - h'(T_2^{1\sigma})\right], \qquad (13.45)$$

or for the thermally controlled bubble growth

$$\dot{q}_2^{m1\sigma} = -const\, n_1^{2/3}\alpha_1^{1/3}\frac{\rho_2 c_{p2}\lambda_2}{\rho''(h''-h')}[T_2 - T'(p)]^2. \qquad (13.46)$$

We see that it is nearly proportional to the square root of the difference between the liquid temperature and the saturation temperature. This is the non-averaged energy source term. The use of the non-averaged source terms is justified for time steps considerably smaller then the characteristic time interval needed to reach the equilibrium. Otherwise the averaged source terms have to be used. Further we discuss how to compute the averaged source terms.

13.4.2 The averaged mass source terms

For a constant number of bubbles during the time step, the *volume difference* between the end and the beginning of the time step, *multiplied* by the *steam density* and the *bubble number per unit mixture volume*, gives the *integral mass evaporating* during the considered time step per unit mixture volume. Dividing this mass by the time step we obtain the averaged mass source term

$$\mu_{21} = \rho'' n_{1o} V_1 / \Delta\tau, \qquad \alpha_1 = 0, \tag{13.47}$$

for the first integration step when the bubble growth just starts and

$$\mu_{21} = n_{1o} \frac{\rho_1 V_1 - \rho_{1o} V_{1o}}{\Delta\tau} = \frac{\rho_{1o} n_{1o} V_{1o}}{\Delta\tau} \left(\frac{\rho_1 V_1}{\rho_{1o} V_{1o}} - 1 \right) = \frac{\rho_{1o} \alpha_{1o}}{\Delta\tau} \left[\frac{\rho_1}{\rho_{1o}} \left(\frac{R_1}{R_{1o}} \right)^3 - 1 \right], \quad \alpha_1 > 0 \tag{13.48}$$

for the next time steps. Using Eq. (13.40) we obtain for $\alpha_{1a} = 0$

$$\mu_{21} = \frac{\rho'' n_{1o}}{\Delta\tau} \left(\frac{4}{3} \pi R_1^3 \right) = \frac{\rho'' n_{1o}}{\Delta\tau} \frac{4}{3} \pi \left\{ \frac{B^2}{A} \frac{2}{3} \left[\left(\tau^+ + 1 \right)^{3/2} - \left(\tau^+ \right)^{3/2} - 1 \right] \right\}^3 \tag{13.49}$$

and for $\alpha_{1a} > 0$

$$\mu_{21} = \frac{\rho'' \alpha_{1o}}{\Delta\tau} \left[\left\{ 1 + \frac{2}{3} \left[\left(\tau^+ + 1 \right)^{3/2} - \left(\tau^+ \right)^{3/2} - 1 \right] / R_{1o}^+ \right\}^3 - 1 \right] \tag{13.50}$$

where

$$\tau^+ = \frac{\Delta\tau}{B^2 / A^2}. \tag{13.51}$$

For the thermally controlled bubble growth the third power of the radii ratio reduces to

$$\left(\frac{R_1}{R_{1o}} \right)^3 = \left[1 + \left(\Delta\tau / \tau^* \right)^{1/2} \right]^3. \tag{13.52}$$

The time interval measured from the beginning of the bubble growth to the beginning of the time step can easily be calculated

$$\tau^* = \left(R_{10}/B\right)^2 = \left[\left(\frac{3\alpha_{10}}{4\pi n_{10}}\right)^{1/3}\Big/B\right]^2. \tag{13.53}$$

Having in mind, that for the spontaneous evaporation $\mu_{12} = 0$ and $\dot{q}_1^{"2\sigma} = 0$, we obtain the corresponding energy source term from the energy jump condition on the bubble surface, i.e. Eq. (13.45).

13.5. Superheated steam

The above derived formalism is strictly valid if the bubble surface temperature is equal to the gas temperature T_1,

$$T_2^{1\sigma} = T_1. \tag{13.54}$$

In the following we denote the evaporation rate for this case with μ'_{21}. In general it is possible to have superheated or subcooled steam with respect to the surface temperature due to the history of the gas - liquid mixing process. In such cases there is a small amount of heat transferred from the bubble to the surface if $T_1 > T_2^{1\sigma}$ or from the surface to the bubble, if $T_1 < T_2^{1\sigma}$, due to a mechanism similar to natural convection inside the cavity. In this case the above theory should be revised. An approximate correction of the theory can be introduced as follows:

$$\dot{q}_2^{m1\sigma} = -\mu'_{21}\left[h"(T_2^{1\sigma}) - h'(T_2^{1\sigma})\right], \tag{13.55}$$

$$\mu_{12} = 0, \tag{13.56}$$

$$\dot{q}_1^{m2\sigma} = a_{21}h_{NC}\left(T_2^{1\sigma} - T_1\right) \tag{13.57}$$

$$\mu_{21} = \mu'_{21} - \dot{q}_1^{m2\sigma}/\left[h"(T_2^{1\sigma}) - h'(T_2^{1\sigma})\right], \tag{13.58}$$

where the heat transfer coefficient due to natural e.g. convection in a cavity is

$$h_{NC} = const\frac{\lambda_1}{D_1}(Gr_1 \Pr_1)^m, \tag{13.59}$$

$$Gr_1 = g\left|\rho_1 - \rho_1"(T_2^{1\sigma})\right|\rho_1 D_1^3/\eta_1^2, \tag{13.60}$$

$$\Pr_1 = \eta_1 c_{p1}/\lambda_1, \tag{13.61}$$

$$const \approx 0.59 \div 0.9, \tag{13.62}$$

$$m \approx 1/4. \tag{13.63}$$

The second term in Eq. (13.58) is usually neglected as very small compared to the first one. This term becomes important only for mass transfer processes on the bubble surface in saturated or nearly saturated liquid.

13.6 Diffusion controlled evaporation into mixture of gases inside the bubble

Consider a bubble consisting of a mixture of vapor and non-condensable gases. The partial pressure of the vapor is p_{M1}. If the interface pressure $p'(T_2)$ is larger than the partial vapor pressure there is an evaporation mass flux from the interface into the bubble. The emitted steam molecules from the surface enter the boundary layer of the bubble by diffusion. The non-averaged instantaneous mass source term is

$$\mu_{21} = a_{21}(\rho w)_{21} = -a_{12}\beta \frac{C_{M1}\rho_1}{p_{M1}}[p_{M1} - p'(T_2)]$$

$$= -a_{21}\alpha_c \frac{M_{M1}}{M_1} \ln \frac{p - p'(T_2)}{\sum p_{n1}}/(c_{p1}Le^{2/3}) \quad \text{see in [26] (1984)} \tag{13.64}$$

where

$$\beta = \frac{p_{M1}}{C_{M1}\rho_1[p'(T_2) - p_{M1}]}\alpha_c \frac{M_{M1}}{M_1} \ln \frac{p - p'(T_2)}{\sum p_{n1}}/(c_{p1}Le_1^{2/3}), \tag{13.65}$$

$$p_{M1} < p'(T_2) < p, \tag{13.66}$$

$$Le_1 = \lambda_1/(\rho_1 c_{p1} D_{M \to \sum n}) \tag{13.67}$$

$$M_1 = \frac{\sum p_{n1}}{p}\sum M_{n1} + \frac{p - \sum p_{n1}}{p} M_{M1} \tag{13.68}$$

($\sum M_{n1} = 28.96$ kg for air, $M_{M1} = 18.96$ kg for water steam) \hfill (13.69)

p' is the partial pressure of the steam in the boundary layer, where the steam is supposed to be saturated at a temperature nearly equal to the liquid temperature

$$p' = p'(T_2).\tag{13.70}$$

During the evaporation the liquid is cooled by

$$\dot{q}_2^{m1\sigma} = -\mu_{21}(h'' - h_{M2}).\tag{13.71}$$

Because this process is relatively slow in most of the practical cases averaging in the time steps of the source terms is not necessary.

13.7 Conclusions

1. After the origination of bubbles with stable bubble size the evaporation of the superheated liquids is controlled by the bubble growth. The bubble growth is over a very short time controlled by the inertia and is well described by the *Besand* and *Rayleigh* solution. For the time after the inertia-controlled period the thermally controlled bubble growth controls the evaporation. From nine known models for this region the model by *Mikic* is recommended. The *Mikic* model interpolates properly between inertia controlled and thermally controlled regions.

2. Thermally controlled bubble growth at heated walls is described by modifying the solutions for thermally controlled bubble growth in the bulk.

3. Evaporation processes are the driving forces for pressure excursions. The pressure change has an important feedback on the bubble growth by changing the driving temperature difference controlling the energy transfer between the bulk and the interface. For pressure change with semi-constant pressure change velocity over a given time step the bubble growth is best approximated by the solution of *Jones* and *Zuber*.

Nomenclature

Latin

a	liquid thermal diffusivity, m^2/s
a_{21}	interfacial area density between the liquid 2 and the vapor 1, $1/m$
C	mass concentration, kg/kg

13. Bubble growth in superheated liquid

c_p	specific heat at constant pressure, $J/(kgK)$
$D_{M \to \sum n}$	diffusion constant for the species M into the mixture $\sum n$, $1/m$
D_{cav}	equivalent cavity size, m
D_h	hydraulic diameter, m
$D_{1,n}$	bubble diameter belonging to n-th bubble generation, m
D_{1c}	critical bubble diameter, m
D_{1d}	bubble departure diameter, m
d	differential, *dimensionless*
F_1/Vol	cross section at the wall surface covered by attached bubbles per unit flow volume, $1/m$
g	gravitational acceleration, m/s^2
Ja	$= \dfrac{\rho_2 c_{p2}[T_2 - T'(p)]}{\rho_{10}\left[h''(T_2^{1\sigma}) - h'(T_2^{1\sigma})\right]}$, Jakob number, *dimensionless*
G^*	critical mass flow rate, $kg/(m^2 s)$
Gb_2	$= \dfrac{\Delta E_{1kr}}{kT_2}$, Gibbs number, *dimensionless*
Gr_1	$= g\left|\rho_1 - \rho_1''(T_2^{1\sigma})\right|\rho_1 D_1^3 / \eta_1^2$, bubble *Grashoff* number, *dimensionless*
h	specific enthalpy, J/kg
h_{NC}	heat transfer coefficient by natural circulation, $W/(m^2 K)$
k	$= 13.805 \times 10^{-24}$, Boltzmann constant, J/K
Le_1	$= \lambda_1 /(\rho_1 c_{p1} D_{M \to \sum n})$, gas *Lewis* number, *dimensionless*
m_2	$= m_\mu / N_A$, mass of single molecule, kg
m_μ	kg-mol mass (18 kg for water), $kg/mole$
N_2	$= \rho_2 / m_2 = \rho_2 N_A / m_\mu$ ($\approx 3.3 \times 10^{28}$ for water), number of the molecules per unit volume, $1/m^3$
N_A	$= 6.02 \times 10^{26}$, Avogadro number, number of molecules in one kilogram-mole mass, $1/kg$-mole
$\dot{n}_{1cin}$	number of the created nuclei per unit time in unit volume of the liquid, $1/(m^3 s)$
n_w''	cavity per unit surface, $1/m^2$
n_{1w}''	bubble generating (active) cavity per unit surface, $1/m^2$
n_1	bubble number density, $1/m^3$
$n_{1,n}$	bubble number density belonging to n-th bubble generation, $1/m^3$
$n_{1,max}$	maximum number of nucleation sites per unit volume in a pipe, $1/m^3$
Pr_1	$= \eta_1 c_{p1} / \lambda_1$, gas *Prandtl* number, *dimensionless*
p	pressure, Pa

$\dot{q}_2^{m1\sigma}$	thermal energy flow density coming from the vapor interface and introduced into the bulk liquid, W/m^3
$\dot{q}_1^{n2\sigma}$	thermal energy flux coming from the liquid interface and introduced into the bulk vapor, W/m^2
$\dot{q}_2^{n1\sigma}$	thermal energy flux coming from the vapor interface and introduced into the bulk liquid, W/m^2
R	gas constant, $J/(kgK)$
R_1	bubble radius, m
R_{10}	initial bubble radius, m
R_{1c}	critical bubble radius, m
$R_{1\infty}$	final radius of the bubble, m
r	radius, m
s	specific entropy, $J/(kgK)$
T	temperature, K
u	radial velocity, m/s
$\mathbf{V}$	velocity vector, m/s
V_{1c}	volume of the bubble with critical size
v	specific volume, m^3/kg
Vol	volume, m^3

Greek

α_1	bubble volumetric fraction, m^3/m^3
β	diffusion coefficient
γ_v	volume occupied by the flow divided by the volume of the system, *dimensionless*
γ	cross section flowed by the flow divided by the total cross section of the system, *dimensionless*
Δ	finite difference
ΔE_1	work necessary to create a single bubble with radius R_1, J
ΔE_{1c}	work necessary to create a single bubble with radius R_{1c}, J
ΔE_{1c}^*	$= \varphi \Delta E_{1c}$, energy for the creation of a single bubble at a wall, J
Δp_{Fi}^*	$= p'(T_2) - p_{Fi}$, flashing inception pressure difference, Pa
$\Delta \tau$	time interval, s
η	dynamic viscosity of liquid, $kg/(ms)$
θ	static contact angle, *rad*
λ_{RT}	*Rayleigh-Tailor* wavelength, m
λ	thermal conductivity, $W/(mK)$

μ_{21}	evaporating mass per unit time and unit flow volume, $kg/(m^3s)$
μ_{12}	condensing mass per unit time and unit flow volume, $kg/(m^3s)$
ρ	density, kg/m^3
$(\rho w)_{21}$	evaporation mass flow rate, $kg/(m^2s)$
σ	surface tension, N/m
M_{M1}	kg-mole mass of the condensing gases, kg/mol
$\sum M_{n1}$	kg-mole mass of the non condensing gases, kg/mol
M_1	kg-mole mass of the gas mixture, kg/mol
τ	time, s
Φ	cavity angle if the cavity is idealized as a cone, rad
φ	work reduction factor, *dimensionless*

Superscripts

$'$	saturated liquid
$''$	saturated vapor
$spin$	spinoidal line
1σ	interface with vapor
2σ	interface with liquid

Subscripts

1	vapor, gas
2	liquid
c	critical state
s	at constant entropy
sat	saturated
Fi	flashing inception
$\sum_{n1o}$	sum of all non-condensing gases
M	evaporating or condensing chemical component
n	inert (non condensing) chemical component

References

1. Besand WH (1859) Hydrostatics and hydrodynamics, Deighton Bell, Cambridge, p 170
2. Beylich AE (January 1991) Dynamics and thermodynamics of spherical vapor bubbles, International Chemical Engineering, vol 31 no 1 pp 1-28
3. Bosnjakovic F (1930) Techn. Mech. Thermodyn, vol 1 p 358
4. Carlsow HS, Jaeger JC (1959) Condution of heat in solids, Oxford
5. Cole R, Schulman HL (1966) Bubble growth rates of high Jacob number, Int. J. Heat Mass Transfer, vol 9 pp 1377-1390

6. Foster HK, Zuber N (1954) J. Appl. Physics, vol 25 pp 474-478
7. Fourier J (1822) Theory Analytique de la Chaleur
8. Fritz W, Ende W (1936) Ueber den Verdampfungsvorgang nach kinematographischen Aufnahmen an Dampfblasen. Phys. Z. , vol 37 pp 391-401
9. Hutcherson MN, Henry RE, Wollersheim DE (Nov. 1983) Two - phase vessel blow down of an initially saturated liquid - Part 2: Analytical, Trans. ASME, vol. 105 pp 694-699
10. Jakob M (1932) Z. d. Ver. Dtsch. Ing., vol 76 p 1161
11. Jones OC Jr, Zuber N (Aug. 1978) Bubble growth in variable pressure fields, Journal of Heat Transfer, vol 100 pp 453-458
12. Kroshilin AE, Kroshilin VE, Nigmatulin BI (May - June 1986) Growth of a vapor bubble in a superheated liquid volume for various laws governing the pressure variation in the liquid, Teplofizika Vysokih Temperatur, vol 24 no 3 pp 533 - 538
13. Kutateladze SS (1982) Analiz podobija v teplofisike (Similarity analysis in tharmal physics), Nauka, (in Russian)
14. Labuntsov DA, Kol'chugin VA, Golovin VS et al (1964) Teplofiz. Vys. Temp., vol 3 pp 446 - 453
15. Lien Y, Griffith P (1969) Bubble growth in reduced pressure, Ph. D. Thesis, Massachusetts Institute of Technology, Cambridge
16. Mikic BB, Rohsenow WM (1969) Bubble growth rates in non uniform temperature fields, Progr. Heat Mass Transfer, vol 2 pp 283-293
17. Mikic BB, Rohsenhow WM, Griffith P (1970) On bubble growth rates, Int. J. Heat Mass Transfer, vol 13 pp 657-666
18. Nigmatulin RI (1978) Basics of the mechanics of the heterogeneous fluids, Moskva, Nauka, In Russian.
19. Olek S, Zvirin Z, Elias E (1990) Bubble growth prediction by the hyperbolic and parabolic heat conduction equations, Waerme- and Stoffuebertragung, vol 25 pp 17-26
20. Piening J (1971) Der Wärmeübergang an eine an der Heizwand wachsende Dampfblase beim Sieden. Dissertation, Berlin TU 1971
21. Plesset MS, Zwick S (April 1954) The growth of bubbles in superheated liquids, J. of Applied Physics, vol 25 no 4 pp 493-500
22. Prisnjakov VF (January 1970) Bubble growth in liquids, Journal of Engineering Physics, vol 18 no 1 pp 584-588
23. Rayleigh L (1917) On the pressure developed in a liquid during the collapse of spherical cavity, Phil. Mag., vol 34 p 94
24. Scriven LE (1959) On the dynamics of phase growth, Chem. Engng. Sci., vol 10 no 1 pp 113
25. Tolubinskii IS (1980) Boiling heat transfer, Kiev, Naukova Dumka, in Russian
26. VDI-Waermeatlas, 4. Auflage 1984, VDI-Verlag
27. Wang Z (December 1989) Transient level swell and liquid carryover phenomena of a vapor-liquid pool, PhD Dissertation, Northwestern University, Evenston, Illinois, USA

Appendix 13.1 Radius of a single bubble in a superheated liquid as a function of time

Thermally controlled bubble growth: $\dfrac{dR_1}{d\tau} = \dfrac{1}{2}\dfrac{B}{\sqrt{\tau}}$, $R_1 = B\sqrt{\tau}$, $R_1 \dfrac{dR_1}{d\tau} = B^2/2$

where $Ja = \dfrac{\rho_2 c_{p2}(T_2 - T')}{\rho''(h'' - h')}$, $a_2 = \dfrac{\lambda_2}{\rho_2 c_{p2}}$ and B is given below.

Bosnjakovic [3] (1930), *Fritz* and *Ende* [8] (1936) for plane geometry:

$$B = Ja\sqrt{\dfrac{4a_2}{\pi}}.$$

Plesset and *Zwick* [21] (1954) for $Ja < 100$:

$$B = Ja\sqrt{\dfrac{12a_2}{\pi}}.$$

Foster and *Zuber* [6] (1954), *Dergarabedijan* see in [25] (1967) for $Ja < 100$:

$$B = Ja\sqrt{\pi a_2}.$$

Labunzov [14] (1964) approximate the analytical solution obtained by *Scriven* [24] (1959) for bubble growth in uniformly heated liquid with an error of 2%

$$B = Ja\sqrt{\dfrac{12a_2}{\pi}}\left[1 + \dfrac{1}{2}\left(\dfrac{\pi}{6Ja}\right)^{2/3} + \dfrac{\pi}{6Ja}\right]^{1/2}.$$

Olek et al. [19] (1990):

$$B = \sqrt{\dfrac{a_2}{\pi}}Ja\left[1 + \left(1 + \dfrac{2\pi}{Ja}\right)^{1/2}\right].$$

Cole and *Schulman* [5] (1966):

$$B = Ja\sqrt{\dfrac{\pi a_2}{4}}.$$

Cooper - see in [27] (1969) for H_2O:

$$B = 1.57\, Ja\sqrt{\frac{a_2}{Pr_2}}.$$

Mikic and *Rohsenow* [16] (1963)

$$B = 0.83\, Ja\sqrt{\pi a_2}.$$

Prisnjakov [22] (1970) for $Ja < 500$:

$$B = Ja\sqrt{\frac{16 a_2}{9\pi}}.$$

Bubble growth at heated wall

Labuntsov et al. [14] (1964), validated for water boiling at silver surface at $p = 1$ to 100 bar.

$$B = \sqrt{2\beta\, a_2 Ja},$$

where

$$\beta = \frac{2(\cos\frac{\theta}{2})\ln(\delta_v/\delta_m)}{(1+\cos\theta)(2-\cos\theta)},$$

$\theta \approx \pi 35/180$, δ_m is the averaged distance between the liquid molecules, $\approx 10^{-9}$ to $10^{-10}\,m$, and δ_v is the thickness of viscous boundary layer of the boiling surface $\approx 10^{-4}\,m$. $\beta \approx 5.6$ to 6.7, $\beta = 6$ recommended.

Piening [20] (1971) for $8 \times 10^{-2} < Ja < 500$:

$$B = \sqrt{a_2 Ja}.$$

Kutateladze [13] (1982) p.120 for bubble growth at heated wall:

$$B \approx const \sqrt{a_2 Ja}.$$

Hutcherson et al. [9] (1983):

$$B = \sqrt{12 a_2 Ja / \pi} \ .$$

Wang [27] (1989) for $p \neq const$:

$$R_1 = (\rho_{10}/\rho_1)^{1/3} R_{10} \left[1 + \frac{1}{R_{10}\rho_{10}} \int_0^{\Delta\tau} \left[(\rho w)_{21} / \rho_1^{2/3} \right] d\tau \right],$$

where

$$(\rho w)_{21,wall} = \beta (\rho w)_{21,\text{inf init liquid see}},$$

$$\beta = \frac{2(1+\cos\theta)+\sin^2\theta}{2+\cos\theta(2+\sin^2\theta)}, \quad \theta = 1.05 \text{ to } 1.31 \ rad \ .$$

Cole and *Shulman* [5]:

$$B = \frac{5}{2} Ja^{3/4} \sqrt{a_2} \ .$$

Jones and *Zuber* [11] (1978), *Kroshilin* et al. [12] (1986) for $p \neq const$ in uniformly heated liquid

---

The heat flux at the bubble surface $\dot{q}_2^{"1\sigma}$ is computed using the *Carlsow* and *Jager* [4] (1959) solution of the *Fourier* equation for heat slab with variable temperature difference at the one boundary

$$T_2 = const, \quad \Delta T(\tau) = T_2 - T_2^{1\sigma}(\tau), \quad \Delta T(0) = T_2 - T'(p_0),$$

$$\dot{q}_2^{"1\sigma} = -k_s \frac{\lambda_2}{\sqrt{\pi a_2}} \left[\frac{\Delta T(0)}{\sqrt{\tau}} + \int_0^\tau \frac{\Delta T(\eta)}{\sqrt{\tau-\eta}} d\eta \right]$$

k_s is the sphericity correction = $\pi/2$ for *Foster - Zuber*, $\sqrt{3}$ for *Plesset - Zwick*. For linear pressure decrease during the time step $\Delta\tau$,

$$\Delta T(\eta) = b\eta,$$

where

Appendix 13.1 Radius of a single bubble in a superheated liquid as a function of time

$$b = -\left(\frac{dT}{dp}\right)_{sat} \frac{dp}{d\tau} = -\left(\frac{dT}{dp}\right)_{sat} \frac{p - p_a}{\Delta \tau},$$

the heat flux is

$$\dot{q}_2^{"\sigma} = -\frac{k_s}{\sqrt{\pi}} \sqrt{\lambda_2 \rho_2 c_{p2}} \left[\frac{\Delta T(0)}{\sqrt{\tau}} + \frac{4}{3} b \tau^{3/2}\right].$$

Neglecting the expression containing the density change and substituting into the mass transport equations with

$$(\rho w)_{21} = -\dot{q}_2^{"\sigma} / (h'' - h').$$

we obtain

$$\frac{R_1}{R_{10}} = 1 + \frac{k_s}{\sqrt{\pi}} \frac{1}{\rho_1 (h'' - h') R_{10}} \sqrt{\lambda_2 \rho_2 c_{p2}} \int_0^{\Delta \tau} \left[\frac{\Delta T(0)}{\sqrt{\tau}} + \frac{4}{3} b \tau^{3/2}\right] d\tau.$$

After the integration the result is

$$\frac{R_1}{R_{10}} = 1 + \frac{2 k_s}{\sqrt{\pi}} \frac{1}{\rho_1 (h'' - h') R_{10}} \sqrt{\lambda_2 \rho_2 c_{p2}} \left(\Delta \tau^{1/2} + \frac{4}{15} b \Delta \tau^{5/2}\right).$$

The equation rewritten in dimensionless form is

$$R_1 = R_{10} \left[1 + \frac{2 k_s}{\sqrt{\pi}} \left(Ja_T Fo^{1/2} + \frac{4}{15} \frac{R_{10}^2}{a_2} Ja_p Fo^{5/2}\right)\right],$$

where

$$Ja_T = \frac{\rho_2 c_{p2}}{\rho_{10} (h'' - h')} [T_2 - T'(p_0)],$$

$$Ja_p = b \frac{\rho_2 c_{p2}}{\rho_{10} (h'' - h')} \frac{R_{10}^2}{a_2},$$

$$Fo = a_2 \tau / R_{10}^2.$$

The result obtained by *Jones* and *Zuber* is slightly different

$$R_1 = (\rho_{10}/\rho_1)^{1/3} R_{10} \left\{1 + \frac{2k_s}{\sqrt{\pi}}\left[Ja_T Fo^{1/2} + \frac{2}{3}Ja_p Fo^{3/2}\right]\right\}.$$

The time-averaged evaporation mass per unit time and unit mixture volume is therefore

$$\mu_{21} = \frac{\rho_{1o}\alpha_{1o}}{\Delta\tau}\left\{\left\{1 + \frac{2k_s}{\sqrt{\pi}}\left[Ja_T Fo^{1/2} + \frac{2}{3}Ja_p Fo^{3/2}\right]\right\}^3 - 1\right\} \quad \text{for} \quad \alpha_1 > 0$$

Inertia controlled bubble growth, $\Delta\tau \leq 10^{-8}$ s, *Besand* [1] (1859), *Rayleigh* [23] (1917).

$$\frac{dR_1}{d\tau} = \left\{\frac{2}{3}[p'(T_2) - p]/\rho_2\right\}^{1/2} = \left\{\frac{2}{3}\frac{1}{\rho_2}\left(\frac{dp}{dT}\right)_{sat}[T_2 - T'(p)]\right\}^{1/2}$$

$$R_1 = \left\{\frac{2}{3}\frac{1}{\rho_2}\left(\frac{dp}{dT}\right)_{sat}[T_2 - T'(p)]\right\}^{1/2}\tau$$

Inertia and thermal controlled bubble growth, *Mikic* et al. [17] (1970):

$$R_1^+ - R_{1o}^+ = \frac{2}{3}\left[(\tau^+ + 1)^{3/2} - \tau^{+3/2} - 1\right]$$

where

$$R_1^+ = R_1(B^2/A), \quad \tau^+ = \frac{\tau}{B^2/A^2}$$

$$Ja = \frac{\rho_2 c_{p2}[T_2 - T'(p)]}{\rho_1\left[h''(T_2^{1\sigma}) - h'(T_2^{1\sigma})\right]}, \quad B = \left(\frac{12}{\pi}a_2\right)^{1/2} Ja$$

$$A^2 = c\frac{1}{\rho_2}\left(\frac{dp}{dT}\right)_{sat}[T_2 - T'(p)],$$

$$\left(\frac{dp}{dT}\right)_{sat} = \frac{1}{T'(p)}\frac{h''(p) - h'(p)}{v''(p) - v'(p)}.$$

$c = 2/3$ for bubble growth in infinite continuum,

$c = \pi/7$ for bubble growth on a wall surface.

Instantaneous mass transfer per unit time and unit mixture volume:

$$\mu_{21} = (36\pi)^{1/3} n_1^{1/3} \alpha_1^{2/3} \rho'' A \left[(\tau^+ + 1)^{1/2} - (\tau^+)^{1/2} \right].$$

14. Condensation of a pure steam bubble in a subcooled liquid

This is a review Chapter about the different methods for description of bubble collapse due to vapor condensation in subcooled liquid. First we consider the case of stagnant bubbles. The solutions for moving bubbles are then presented. The available experimental data are then discussed. Then the link between the solutions for the bubble growth and the mass source terms for the averaged conservation equations for two-phase flow is presented. The way to derive non-averaged mass source terms and time-averaged mass source terms is given. The very interesting case of disappearance of the bubbles from the size distribution spectrum due to condensation is discussed in some details. The influence of the liquid turbulence on the bubble condensation in pipes is discussed too. Then a brief description of the vapor condensation from gas mixtures inside bubbles surrounded by subcooled liquid is given for the two limiting cases a) thermally controlled collapse and b) diffusion controlled collapse.

14.1 Introduction

Injecting steam in subcooled water causes condensation. If a pressure wave crosses a water-steam system in thermodynamic equilibrium condensation also happens. One of the basic condensation mechanisms in multi-phase systems is the condensation of a single bubble in subcooled water. Several authors study condensation of a saturated bubble in a subcooled liquid. As for the evaporation, the theoretical results of the approximate integration of the process governing a system of PDEs describe the radius as a function of time. Unlike the evaporation, the condensation is associated with a given *initial bubble size*. There are two classes of solutions: a) for a stagnant bubble and b) for a bubble in motion with respect to the surrounding liquid. The motion increases the heat transfer between the bubble interface and the bulk liquid and therefore increases the condensation.

14.2 Stagnant bubble

Table 14.1 contains some of the results known from the literature, obtained with the assumption that there is no relative motion of the bubble with respect to the surrounding liquid.

Table 14.1. Size of a bubble condensing in a subcooled liquid as a function of time by $\Delta V = 0$

$$\tau_H = \frac{4}{\pi} Ja^2 Fo, \quad Ja = \frac{\rho_2}{\rho''} \frac{c_{p2}\left[T'(p) - T_2\right]}{h'' - h'}, \quad Fo = \frac{a_2 \tau}{R_{lo}^2}$$

Forschuetz and *Chao* [8] (1965) for $\tau \leq \dfrac{\pi R_{lo}^2}{4 Ja^2 a_2}$

$$\frac{R_1}{R_{lo}} = 1 - \sqrt{\tau_H}.$$

Plesset and *Zwick* [20] (1954):

$$\tau_H = \frac{2}{3}\left(\frac{R_{lo}}{R_1}\right) + \frac{1}{3}\left(\frac{R_1}{R_{lo}}\right)^2 - 1, \quad \tau_{99\%} = 65 \frac{\pi R_{lo}^2}{4 Ja^2 a_2}.$$

Prisnjakov [21] (1970) for $\tau \leq \dfrac{1}{4\varepsilon^2} \dfrac{\pi R_{lo}^2}{4 Ja^2 a_2}$:

$$\frac{R_1}{R_{lo}} = 1 - 2\varepsilon\sqrt{\tau_H}, \quad \varepsilon = 1\Big/\left[1 - \frac{\rho''}{\rho_2} + 2\frac{h_2}{h'' - h'}\right].$$

Zuber [26] (1961) for $\tau \leq 4\tau_m$:

$$\frac{R_1}{R_{1m}} = \left(\frac{\tau}{\tau_m}\right)^{1/2}\left[2 - \left(\frac{\tau}{\tau_m}\right)^{1/2}\right].$$

τ_m time when the bubble reaches its maximum radius R_{1m}. For bubble growing and collapsing on a heated wall in subcooled water

$$\tau_m = \frac{1}{\pi a_2}\left(\lambda_2 \frac{T_w - T'}{\dot{q}_w''}\right)^2, \quad R_m = \frac{b}{\pi} Ja\sqrt{\pi a_2 \tau_m}, \quad b \approx 1 \div \sqrt{3},$$

$$Ja = \frac{(T_w - T')c_{p2}\rho_2}{(h'' - h')\rho''}.$$

Buher and *Nordman* [4] (1978) show that the best fitting of the experimental data is obtained with *Zuber's* equation from Table 14.1.

14.3 Moving bubble

The influence of the relative velocity on the temperature profile around the bubble and therefore on the condensation process can be taken into account by using the results of several authors describing the heat transfer around a solid sphere as summarized in Table 14.2.

Table 14.2. Averaged heat transfer coefficient on the surface of solid sphere moving in a liquid

$$Pe_2 = \frac{D_1 \Delta V_{12}}{a_2}, \quad a_2 = \frac{\lambda_2}{\rho_2 c_{p2}}, \quad Pr_2 = \frac{\eta_2}{\rho_2 a_2}, \quad Re_1 = \frac{D_1 \rho_2 \Delta V_{12}}{\eta_2}, \quad Nu_1 = \frac{D_1 h_k}{\lambda_2},$$

$$Fo = \frac{a_2 \tau}{R_{1o}^2}, \quad Re_{1o} = \frac{D_{1o} \rho_2 \Delta V_{12}}{\eta_2}, \quad Fo \le 1/(Nu_1 Ja)$$

Potential flow $\left(\dfrac{\partial R_1}{\partial \tau} \ll \Delta V \right)$, $Re_1 \ll 1$:

Soo [22] (1969) for $Pe_2 \ll 1$:

$$Nu_1 = 2 + \frac{9}{16} Pe_2 + \frac{9}{64} Pe_2^2 + ...$$

Nigmatulin [17] (1978) for $Pe_2 \gg 1$:

$$Nu_1 = \frac{0.65}{\left(1 + \dfrac{\eta_1}{\eta_2}\right)^{1/2}} Pe_2^{1/2}, \quad \eta_1/\eta_2 \le 1.$$

Hunt [10] (1970) for a single-component system:

$$Nu_1 = \frac{2}{\pi^{1/2}} (cPe_2)^{1/2},$$

where $c = 1$.

For a two-component system *Isenberg* and *Sideman* [11] (1970) take into account viscous effects in the *Hunt* equation using

$$c = 0.25 \, Pr_2^{-1/3}.$$

Kendouch [13] (1976) for bubbles in a swarm:

$$Nu_1 = \frac{2}{\pi^{1/2}} Pe_2^{1/2} (1-\alpha_1)^{-1/2}.$$

Wilson [24] (1965) for $Re_1 < 1$:

$$Nu_1 = 2 + 0.37 Re_1^{3/5} Pr_2^{1/3}.$$

Nigmatulin [17] (1978) for $Re_1 < 1$ or $Pe_2 < 10^3$ (e.g. for $D_{1H2O} < 0.1$ mm):

$$Nu_1 = 2 + \frac{0.65 Pe_2^{1.7}}{1 + Pe_2^{1.3}}, \; \eta_1/\eta_2 \cong 0.$$

Brauer et al. [3] (1976) for bubbles without internal circulation:

$$Nu_1 = 2 + \frac{0.65 Pe_2^{1.7}}{\left[1+(0.84 Pe_2^{1.6})^3\right]^{1/3} (1+Pe_2^{1.2})}.$$

Soo [22] (1969) for $1 < Re_1 < 7 \times 10^4$, $0.6 < Pr_2 < 400$:

$$Nu_1 = 2 + (0.55 \text{ to } 0.7) Re_1^{1/2} Pr_2^{1/3}.$$

There are a limited number of correlations, approximating directly experimental data for a *condensing moving bubble in a subcooled liquid*. Some of them are summarized in Table 14.3. Next we first show the analytical way to derive such an expression which can be successfully modified to correlate the data. Then, to illustrate the idea how the approximate form of the correlations in Table 14.3 is obtained we apply the theory using the heat transfer mechanism described by the correlation by *Hunt* and *Isenberg*.

The energy conservation equation with neglected convection and diffusion from the surrounding computational cells gives

$$\rho'' \frac{dR_1}{d\tau} = -h_c \frac{T'-T_2}{h''-h_2}, \tag{14.1}$$

or

$$\frac{dR_1}{d\tau} = -\frac{h_c}{\rho''}\frac{T'-T_2}{h''-h_2} = -\frac{Nu_1 \lambda_2 (T'-T_2)}{R_1 2\rho''(h''-h_2)} = -c/R_1 = -\frac{a_2 Nu_1 Ja}{2R_1}, \quad (14.2)$$

or

$$\frac{d}{dFo}\left(\frac{R_1}{R_{1o}}\right)^2 = -Nu_1 Ja. \quad (14.3)$$

Integrating the above equation for $Nu_1 = const$ we obtain

$$\frac{R_1}{R_{1o}} = \left(1 - \frac{2c\Delta\tau}{R_{1o}^2}\right)^{1/2} = \sqrt{1 - \Delta\tau/\Delta\tau_o} = (1 - Nu_1 JaFo)^{1/2} = (1 - Fo/Fo_o)^{1/2}. \quad (14.4)$$

The last form of the above notation contains the important information for the dimensionless time, Fo_o, within which the bubble collapses completely if the mixture still does not reached the saturation. The bubble collapses entirely

$$R_1 = 0, \quad (14.5)$$

within

$$\Delta\tau_o = R_{1o}^2/(2c), \quad [Fo_o = 1/(Nu_1 Ja)]. \quad (14.6)$$

After this time interval Eq. (14.4) is no longer valid. For the third power of the radii ratio we have

$$\left(\frac{R_1}{R_{1o}}\right)^3 = (1 - \Delta\tau/\Delta\tau_o)^{3/2} = (1 - Fo/Fo_o)^{3/2} \quad (14.7)$$

If we drop the assumption $Nu_1 = const$ and use the *Hunt* and *Isenberg*'s equation from Table 14.2 we obtain

$$\left(\frac{R_1}{R_{1o}}\right)^3 = (1 - Fo/Fo_o)^2, \quad (14.8)$$

where

$$Fo_o = 1/\left(\frac{3}{2}\frac{1}{2\sqrt{\pi}} Re_{1o}^{1/2} Pr_2^{1/3} Ja\right), \quad (14.9)$$

see the first equation in Table 14.3. It differs from the *Moalem* and *Sideman* correlation only in the constant.

As seen from Table 14.3 all the correletions are based on the above form modifying the exponents except the correlation by *Dushkin* and *Kolomenzev*.

Table 14.3. Bubble size of saturated steam condensing in a subcooled liquid $\Delta V \neq 0$. Approximation of experimental data. Fo_o is the dimensionless lifetime of condensing bubble

Theory
$Nu_1 \neq f(r_1)$, and Eq. (14.3)
$\dfrac{R_1}{R_{1O}} = (1 - Nu_1 JaFo)^{1/2} = (1 - Fo/Fo_o)^{1/2}, Fo = 1/(Nu_1 Ja)$
Nu_1 = *Hunt* and *Isenberg* Table 14.2, and Eq. (14.3) + $\Delta V \neq f(R_1)$
$\dfrac{R_1}{R_{1O}} = (1 - \dfrac{3}{2}\dfrac{1}{2\sqrt{\pi}} Re_{1o}^{1/2} Pr_2^{1/3} JaFo)^{2/3}$
$= (1 - 0.423 Re_{1o}^{1/2} Pr_2^{1/3} JaFo)^{2/3} = (1 - Fo/Fo_o)^{2/3}$,
$Fo_o = 1/(0.423 Re_{1o}^{1/2} Pr_2^{1/3} Ja)$.

Experiment
Maiynger and *Nordmann* [15] (1979) for $4 < Ja < 300$, $p = 0.2$ to 5 bar:
$\dfrac{R_1}{R_{1O}} = \left[1 - \dfrac{5}{3}\dfrac{10^3}{2}\left(\dfrac{\rho_1}{\rho_2}\right)^{0.79} Re_{1o}^{1/3} Pr_2^{2/3} Ja^{1.27} Fo\right]^{3/5}$
$= \left[1 - 0.833 \times 10^3 \left(\dfrac{\rho_1}{\rho_2}\right)^{0.79} Re_{1o}^{1/3} Pr_2^{2/3} Ja^{1.27} Fo\right]^{3/5} = (1 - Fo/Fo_o)^{3/5}$,
$Fo_o = 1/\left[0.833 \times 10^3 \left(\dfrac{\rho_1}{\rho_2}\right)^{0.79} Re_{1o}^{1/3} Pr_2^{2/3} Ja^{1.27}\right]$.

Condensation of bubbles in a pool behind nozzles

Akiyama [1] (1973) for $\Delta V_{12} = 1.18 \left[\dfrac{\sigma g(\rho_2 - \rho'')}{\rho_2^2} \right]^{1/4}$

$$\frac{R_1}{R_{1o}} = \left(1 - 0.259 Re_{1o}^{0.6} Pr_2^{1/3} JaFo\right)^{1/1.4} = \left(1 - Fo/Fo_o\right)^{1/1.4},$$

$$Fo_o = 1/\left(0.259 Re_{1o}^{0.6} Pr_2^{1/3} Ja\right).$$

Brucker and *Sparrow* [5] (1977) for $\Delta T = 15$ to $100K$, $p = (10.3$ to $62.1) \times 10^5 Pa$, $Pe_2 = 2000$ to 3000, $Nu_1 \pm 50\%$:

$$\frac{R_1}{R_{1o}} = \left(1 - 0.259 Re_{1o}^{0.6} JaFo\right)^{1/1.4} = \left(1 - Fo/Fo_o\right)^{1/1.4},$$

$$Fo_o = 1/\left(0.259 Re_{1o}^{0.6} Ja\right).$$

Moalem and *Sideman* [16] (1973):

$$\frac{R_1}{R_{1O}} = \left(1 - \frac{2}{3}\frac{1}{\sqrt{\pi}} Re_{1o}^{1/2} Pr_2^{1/2} JaFo\right)^{2/3}$$

$$= \left(1 - 0.846 Re_{1o}^{1/2} Pr_2^{1/2} JaFo\right)^{2/3} = \left(1 - Fo/Fo_o\right)^{2/3},$$

$$Fo_o = 1/\left(0.846 Re_{1o}^{1/2} Pr_2^{1/2} Ja\right).$$

Dushkin and *Kolomenzev* [7] (1989) for $0.8 \times 10^6 \leq p \leq 2 \times 10^6$ Pa, $15 \leq T' - T_2 \leq 25$ K, $2 \times 10^{-3} \leq D_{1o} \leq 4 \times 10^{-3}$ m, or $0.1 < Ja < 20$, $0 < Fo < 0.2$, $10^3 < Pe_2 < 10^5$, $0.1 \leq \dfrac{R_1}{R_{1o}} \leq 1$, $\alpha_1 = 0.2$ to 0.7, $Re_1 = 10^4$ to 10^6, $566 \leq \alpha_1 \rho_1 w_1 \leq 1\ 698$ kg/(m²s), $11\ 317 \leq \alpha_2 \rho_2 w_2 \leq 56\ 588$ kg/(m²s) approximate experimental data of several authors for a vertical upwards flow, and their own data for downward flow of Freon-22 in a tube with $D_h = 0.015$ with

$$\frac{R_1}{R_{1o}} = \exp\left[-\frac{4}{3}\left(Pe_2^{0.43} + 1\right) JaFo\right].$$

Chen and *Myinger* [6] (1989), experiment $2 < Pr_2 < 15$, $1 < Ja < 120$.

$$\frac{R_1}{R_{1o}} = \left(1 - 0.56 Re_{1o}^{0.7} Pr_2^{1/2} JaFo\right)^{0.9} = \left(1 - Fo/Fo_o\right)^{0.9},$$

$$Fo_o = 1/\left(0.56 Re_{1o}^{0.7} Pr_2^{1/2} Ja\right).$$

At *Jacob* numbers above 70 inertia controlled phenomena start to dominate.

Bucher and *Nordman* [4] (1978) compared experimental data with the predictions of the equations from Table 14.3 and found that the *Akiyama* equation shows the best results up to 1978.

14.4 Non-averaged source terms

The non-averaged source terms required for microscopic flow description are given below. First the evaporation mass per unit volume and unit time is equal to zero,

$$\mu_{21} = 0. \tag{14.10}$$

There is no heat transfer between the bubble interface and the vapor

$$\dot{q}_1^{m2\sigma} = 0. \tag{14.11}$$

Only the heat transfer between the interface and the bulk liquid

$$\dot{q}_2^{m1\sigma} = a_{12} h_c \left[T'(p) - T_2\right], \tag{14.12}$$

controls the condensed mass per unit mixture volume and unit time

$$\mu_{12} = \dot{q}_2^{m1\sigma} / \left[h_{M1} - h'(p)\right]. \tag{14.13}$$

Here the instantaneous heat transfer coefficient between the interface and the bulk liquid is

$$h_c = \frac{\lambda_2}{D_1} Nu_1, \tag{14.14}$$

where the *Nusselt* number is a function of the local parameter of the flow.

In what follows we show how to link the empirically obtained results for the change of the radius of the condensing bubble with the macroscopic flow description. The condensed mass per unit mixture volume and unit time is equal to the product of the number of the bubbles per unit mixture volume and the mass change of a single bubble.

$$\mu_{12} = -\rho''(p) n_1 dV_1 / d\tau = -\rho''(p) n_1 4\pi R_1^2 dR_1 / d\tau \qquad (14.15)$$

or replacing

$$R_1 = \left(\frac{3}{4\pi} \frac{\alpha_1}{n_1} \right)^{1/3}, \qquad (14.16)$$

$$\mu_{12} = -3^{2/3} (4\pi)^{1/3} \rho''(p) n_1^{1/3} \alpha_1^{2/3} \frac{dR_1}{d\tau} = 4.84 \rho''(p) n_1^{1/3} \alpha_1^{2/3} \frac{dR_1}{d\tau}, \qquad (14.17)$$

where for $dR_1/d\tau$ one can use one of the expressions from Table 14.2 or 14.3. The heat absorbed by the liquid during the condensation per unit mixture volume and unit time is

$$\dot{q}_2'''^{1\sigma} = \mu_{12} \left[h''(p) - h'(p) \right]. \qquad (14.18)$$

14.5 Averaged source terms

In order to calculate the averaged mass source term we proceed as follows.

The integral expression for the averaged condensation within the time interval $\Delta\tau$, where $\Delta\tau < \Delta\tau_o$, is

$$\mu_{12} = \rho'' n_1 \frac{V_{1o} - V_1}{\Delta\tau} = \frac{\rho'' n_1 V_{1o}}{\Delta\tau} \left(1 - \frac{V_1}{V_{1o}} \right) = \frac{\rho'' \alpha_1}{\Delta\tau} \left[1 - \left(\frac{R_1}{R_{1o}} \right)^3 \right]. \qquad (14.19)$$

If the time interval is selected larger than the lifetime of the bubble, $\Delta\tau \geq \Delta\tau_o$, we have

$$\mu_{12} = \rho'' \alpha_1 / \Delta\tau. \qquad (14.20)$$

The last relationship is practically the condition for the entire condensation of the available steam within the time step $\Delta\tau$. Having in mind that during the condensation

$$\mu_{21} = 0 \tag{14.21}$$

$$\dot{q}_1'''^{2\sigma} = 0, \tag{14.22}$$

we obtain from the energy jump condition on the bubble surface the averaged energy source term

$$\dot{q}_2'''^{1\sigma} = \mu_{12}(h_{M1} - h'). \tag{14.23}$$

The above relationship is strictly valid if the bubble surface temperature

$$T_2^{1\sigma} = T'(p_{M1}) \tag{14.24}$$

is equal to the gas temperature T_1. We note the condensation rate in this case with μ_{12}'. In general there is a small amount of heat transfer from the bubble to the surface if $T_1 > T'(p_{M1})$ or from the surface to the bubble if $T_1 < T'(p_{M1})$ due to a mechanism similar to natural convection inside the cavity. In this case we have

$$\dot{q}_2'''^{1\sigma} = \mu_{12}'[h_{M1} - h'(p)]. \tag{14.25}$$

$$\mu_{21} = 0, \tag{14.26}$$

$$\dot{q}_1'''^{2\sigma} = a_{21} h_{NC}[T'(p_{M1}) - T_1]. \tag{14.27}$$

$$\mu_{12} = \mu_{12}' + \dot{q}_1'''^{2\sigma} / [h_{M1} - h'(p)], \tag{14.28}$$

where h_{NC} is the heat transfer coefficient due to natural convection in a cavity. The second term in Eq. (14.28) is usually neglected as very small compared to the first one. This term becomes important only for mass transfer processes on the bubble surface in saturated or nearly saturated liquid.

14.6 Change of the bubble number density due to condensation

Now let as answer the question what happens to the particle number density during the condensation of bubbles? Suppose the bubbles size distribution obeys the *Nukiama-Tanasava* law [18] (1938)

$$P(D_1) = 4\left(\frac{D_1}{D_1'}\right)^2 e^{-2(D_1/D_1')}. \qquad (14.29)$$

Here $P(D_1)$ is the probability that a bubble has its size between D_1 and $D_1 + \delta D_1$. D_1' is the most probable particle size, i.e. the size where the probability function has its maximum value. The particle sizes may take values between zero and a maximum value

$$0 < D_1 < D_{1max}. \qquad (14.30)$$

Thus, if we know D_1' and D_{1max} the particle distribution is uniquely characterized. *Mac Vean*, see in *Wallis* [23] (1969), found that a great deal of data could be correlated by assuming that

$$D_1' = D_1/2 \qquad (14.31)$$

where D_d is the volume averaged particle size. The relationship between D_d and the maximum particle size, D_{1max}, is reported as

$$D_{1max} \approx (2.04 \text{ to } 3.13)\, D_1, \qquad (14.32)$$

see *Pilch* et al. [19] (1981), *Kataoka* et al. [12] (1983), among others. Suppose that the distribution remains unchanged during the time step $\Delta \tau$. For that time the volume-averaged diameter changes as

$$\Delta D_1 = D_{10} - D_1 = D_{10}(1 - R_1/R_{10}) = D_{10}\left[1 - \left(1 - \frac{\mu_{12}\Delta\tau}{\rho_1 \alpha_1}\right)^{1/3}\right] \qquad (14.33)$$

and all particles having sizes

$$0 \le D_1 \le 2\Delta D_1 \qquad (14.34)$$

namely

$$n_1 \int_0^{2\Delta D_1} P(D_1)dD_1 = 2n_1 \left\{ 1 - e^{-4\Delta D_1/D_1'} \left[1 + 4\frac{\Delta D_1}{D_1'}\left(1 + 2\frac{\Delta D_1}{D_1'}\right) \right] \right\} \qquad (14.35)$$

disappear. The averaged particle sink per unit time and unit mixture volume is consequently

$$\dot{n}_1 \approx -(n_1/\Delta\tau)2\left\{ 1 - e^{-4\Delta D_1/D_1'}\left[1 + 4\frac{\Delta D_1}{D_1'}\left(1 + 2\frac{\Delta D_1}{D_1'}\right)\right]\right\}, \qquad (14.36)$$

for $\Delta D_1 \leq D_1'$. The latter condition means that approximately within the interval $2\,\Delta D_1$ the averaged bubble size is ΔD_1.

14.7 Pure steam bubble drifting in turbulent continuous liquid

In a bubble flow or in a churn-turbulent flow with considerable turbulence, the bubbles are moving practically with the *same* velocity as the liquid. The mechanism governing the condensation is quite different compared to the mechanism described in the previous Sections. There are several authors assuming that the heat released during the condensation is transported *by exchange of turbulent eddies* between the boundary layer and the bulk liquid. The characteristic time of one cycle, estimated by dimensional analysis of the turbulent characteristics of the continuous velocity field, is of the order of

$$\Delta\tau = const\,(l_{e2}^2/\varepsilon_2)^{1/2} \qquad (14.37)$$

for high frequency pulsations in accordance with the *statistical theory of turbulence*. Replacing the characteristic size of the turbulent eddies in the liquid with

$$l_{e2} = const\,(v_2^3/\varepsilon_2)^{1/4} \qquad (14.38)$$

we obtain for the time constant

$$\Delta\tau = const(v_2/\varepsilon_2)^{1/2}. \qquad (14.39)$$

After replacing the dissipated specific kinetic energy of the turbulent pulsations with

$$\alpha_2 p_2 \varepsilon_2 = \frac{1}{2} \frac{\rho^2}{\rho_2} \left[u^3 \Phi_{2ou}^2 \left(\frac{\lambda_{Ru}}{D_{hu}} + \frac{\zeta_u}{\Delta r} \right) + v^3 \Phi_{2ov}^2 \left(\frac{\lambda_{Rv}}{D_{hv}} + \frac{\zeta_\theta}{r^\kappa \Delta \theta} \right) + w^3 \Phi_{2ow}^2 \left(\frac{\lambda_{Rw}}{D_{hw}} + \frac{\zeta_w}{\Delta z} \right) \right]$$
(14.40)

we obtain

$$\Delta \tau = const \frac{\rho_2}{\rho} \left(2\alpha_2 v_2 \bigg/ \left[\begin{array}{c} u^3 \Phi_{2ou}^2 \left(\frac{\lambda_{Ru}}{D_{hu}} + \frac{\zeta_u}{\Delta r} \right) + v^3 \Phi_{2ov}^2 \left(\frac{\lambda_{Rv}}{D_{hv}} + \frac{\zeta_\theta}{r^\kappa \Delta \theta} \right) \\ + w^3 \Phi_{2ow}^2 \left(\frac{\lambda_{Rw}}{D_{hw}} + \frac{\zeta_w}{\Delta z} \right) \end{array} \right]^{1/2} \right)$$
(14.41)

The heat flux on the bubble surface can be determined to the accuracy of a constant as

$$\dot{q}_2^{\prime\prime\sigma} = const(T' - T_2) \sqrt{\frac{\lambda_2 \rho_2 c_{p2}}{\Delta \tau}}, \quad (const = \frac{2}{\sqrt{\pi}} \text{ plane}),$$
(14.42)

where $\Delta \tau$ is the time interval in which the high frequency eddy is in contact with the bubble surface. During this time, the heat is transported from the surface to the eddy by molecular diffusion. Thereafter the eddy is transported into the bulk flow again, and its place on the surface is occupied by another one. In this way the heat released during the condensation is transported from the bubble to the turbulent bulk liquid. We substitute $\Delta \tau$ from Eq. (14.41) into Eq. (14.42) and obtain

$$\dot{q}_2^{\prime\prime\sigma} = const(T' - T_2) \lambda_2 Pr_2^{1/2} \sqrt{\frac{\rho}{\eta_2 v_2^{1/2} \alpha_2^{1/2}}}$$

$$\times \left[\begin{array}{c} u^3 \Phi_{2ou}^2 \left(\frac{\lambda_{Ru}}{D_{hu}} + \frac{\zeta_u}{\Delta r} \right) + v^3 \Phi_{2ov}^2 \left(\frac{\lambda_{Rv}}{D_{hv}} + \frac{\zeta_\theta}{r^\kappa \Delta \theta} \right) \\ + w^3 \Phi_{2ow}^2 \left(\frac{\lambda_{Rw}}{D_{hw}} + \frac{\zeta_w}{\Delta z} \right) \end{array} \right]^{1/4}.$$
(14.43)

The constant can be determinate by comparison with a result obtained by *Avdeev* in [2] (1986) for one-dimensional flow without local resistance ($\xi = 0$). *Avdeev* used the known relationship for the *friction coefficient of turbulent flow*

$$\lambda_{Rw} = 0.184 Re_2^{-0.2} \qquad (14.44)$$

where

$$Re_2 = \frac{D_{hw} w_2}{v_2}, \qquad (14.45)$$

compared the so obtained equation with experimental data, and estimated the constant in

$$\dot{q}_2''^{1\sigma} = const\ 0.184^{1/4} (T'-T_2) \frac{\lambda_2}{D_{hw}} Pr_2^{1/2} Re_2^{0.7} (\Phi_{2ow}^2 / \alpha_2)^{1/4} \qquad (14.46)$$

as

$$const\ 0.184^{1/4} = 0.228 \qquad (14.47)$$

or

$$const = 0.348. \qquad (14.48)$$

The final relationship recommended by *Avdeev* for one-dimensional flow

$$Nu = 0.228 Pr_2^{1/2} Re_2^{0.7} (\Phi_{2ow}^2 / \alpha_2)^{1/4} \qquad (14.49)$$

describes his own data within ± 30% error band for $D_1/D_{hw} > 80/Re_2^{0.7}$. Note that in the *Avdeev* equation ρ/ρ_2 is set to one. Thus, the so estimated constant can be successfully applied to three-dimensional flows. For comparison let us write the relationships obtained by *Hancox* and *Nikol* (see in [9] (1981))

$$Nu = 0.4 Pr_2 Re_2^{2/3} \qquad (14.50)$$

and *Labunsov* (see in [14])

$$Nu = \frac{\lambda_{Rw}/8}{1-12\sqrt{\frac{\lambda_{Rw}}{8}}} Pr_2 Re_2 \frac{D_1}{D_h} \cong \frac{0.023}{1-1.82 Re_2^{-0.1}} \frac{D_1}{D_h} Pr_2 Re_2^{0.8}. \qquad (14.51)$$

We see that in the three equations obtained independently from each other the dependence on Re_2 is $Re_2^{0.7\ to\ 0.8}$, and the dependence on Pr_2 is $Pr_2^{0.5\ to\ 1}$. In case of $Re_2 \approx 0$ the energy dissipated behind the bubbles should be taken into account.

14.8 Condensation from a gas mixture in bubbles surrounded by subcooled liquid

In the *presence of a non-condensing gas*, the temperature of the bubble surface T'_k is not equal to the saturation temperature T' at the system pressure p. During the condensation the partial pressure of the inert components increases with the decreasing bubble size due to the simultaneously decreasing partial steam pressure until T'_k becomes equal to the liquid temperature and the condensation ceases. Note that in this case the *bubble remains with a stable equilibrium volume*. Because in accordance with *Dalton*'s law the non-condensing mixture occupies the whole bubble volume at the beginning of the condensation as well at its end the ratio of the end volume to the initial volume is inversely proportional to the densities of the non-condensing mixture in both states, respectively

$$\left(\frac{R_{1\infty}}{R_1}\right)^3 = \frac{\sum \rho_{nlo}}{\sum \rho_{nl\infty}} = \frac{1}{\sum \rho_{nl\infty}} \rho_{lo}(1-C_{Mlo}) = \frac{R^*_{n1}T_{lo}}{p-p'(T_2)} \rho_{lo}(1-C_{Mlo}) \approx \frac{\sum \rho_{nlo}}{p-p'(T_2)} \tag{14.52}$$

This expression is necessary for the estimation of the *maximum duration of the bubble collapse*.

There are different approaches to estimate the condensing mass per unit mixture volume and unit time.

14.8.1 Thermally controlled collapse

One of them assumes that the non-condensing components are *uniformly* distributed through the whole bubble volume and the condensation is controlled, as in one-component bubble condensation, by the heat transport from the bubble surface to the bulk liquid. The only difference is that the surface temperature, assumed in the single-component case to be equal to the saturation temperature at the steam partial pressure, is not equal to the saturation temperature at the system pressure. Thus, the equation describing the change of the bubble radius for a one-component bubble can simply be corrected as follows

$$\frac{dR_1}{d\tau} = -\frac{h_c}{\rho''} \frac{T'_c - T_2}{h'' - h_2} = -\frac{Nu_1 \lambda_2 (T'_c - T_2)}{R_1 2 \rho''(h'' - h_2)} = -c_T \frac{Nu_1 \lambda_2 (T' - T_2)}{R_1 2 \rho''(h'' - h_2)}$$

$$= -\frac{a_2 Nu_1 Ja}{2R_1} = -c_T c / R_1, \tag{14.53}$$

where

$$c_T = \frac{T'_c - T_2}{T' - T_2} \tag{14.54}$$

is the dimensionless surface temperature. Thus, the derivation of the averaged source terms is similar to the derivation shown in Section 14.4 for single-component bubble condensation.

14.8.2 Diffusion controlled collapse

Another approach considers the *concentration profile* in the neighborhood of the bubble surface and models the condensation as a diffusion process. The non-averaged mass condensing per unit mixture volume and unit time is

$$\mu_{12} = a_{12}(\rho w)_{12} = -a_{12}\beta \frac{C_{M1}\rho_1}{p_{M1}}[p_{M1} - p'(T_2)]$$

$$= -a_{21}\alpha_c \frac{M_{M1}}{M_1} \ln \frac{[p - p'(T_2)]}{\sum p_{n1}} / (c_{p1}Le^{2/3}), \text{ see in [25] (1984)} \tag{14.55}$$

where

$$\beta = \frac{p_{M1}}{C_{M1}\rho_1[p'(T_2) - p_{M1}]} \alpha_c \frac{M_{M1}}{M_1} \ln \frac{p - p'(T_2)}{\sum p_{n1}} / (c_{p1}Le^{2/3}), \tag{14.56}$$

$$p'(T_2) < p_{M1}, \tag{14.57}$$

$$Le = \lambda_1 / (\rho_1 c_{p1} D_{M \to \sum n}), \tag{14.58}$$

$$M_1 = \frac{\sum p_{n1}}{p} \sum M_{n1} + \frac{p - \sum p_{n1}}{p} \sum M_{M1}, \tag{14.59}$$

p' is the partial steam pressure in the boundary layer, where the steam is supposed to be saturated, having a temperature nearly equal to the liquid temperature

$$p' = p'(T_2). \tag{14.60}$$

The released heat during the condensation is transported into the liquid

$$\dot{q}_2^{\prime\prime\prime 1\sigma} = \mu_{12}\left[h_{M1} - h'(p)\right].\tag{14.61}$$

Nomenclature

a	thermal diffusivity, m^2/s
a_2	$= \dfrac{\lambda_2}{\rho_2 c_{p2}}$, liquid thermal diffusivity, m^2/s
a_{21}	interfacial area density between the liquid 2 and the vapor 1, $1/m$
C	mass concentration, kg/kg
c_p	specific heat at constant pressure, $J/(kgK)$
$D_{M \to \sum n}$	diffusion constant for the species M into the mixture $\sum n$, $1/m$
D_{cav}	equivalent cavity size, m
D_h	hydraulic diameter, m
$D_{1,n}$	bubble diameter belonging to n-th bubble generation, m
D_{1c}	critical bubble diameter, m
D_{1d}	bubble departure diameter, m
D_1	bubble diameter, m
$D_{1\max}$	maximum bubble diameter, m
D_1'	most probable bubble size, m
d	differential, *dimensionless*
F_1/Vol	cross section at the wall surface covered by attached bubbles per unit flow volume, $1/m$
Fo	$= \dfrac{a_2 \tau}{R_{1o}^2}$, *Fourier* number – condensation time scale, *dimensionless*
Fo_o	time within which the bubble collapses completely, *dimensionless*
g	gravitational acceleration, m/s^2
Ja	$= \dfrac{\rho_2 c_{p2}\left[T_2 - T'(p)\right]}{\rho_{10}\left[h''(T_2^{1\sigma}) - h'(T_2^{1\sigma})\right]}$, evaporation *Jacob* number, *dimensionless*
Ja	$= \dfrac{\rho_2}{\rho''} \dfrac{c_{p2}\left[T'(p) - T_2\right]}{h'' - h'}$, condensation *Jacob* number, *dimensionless*
G^*	critical mass flow rate, $kg/(m^2s)$
Gb_2	$= \dfrac{\Delta E_{1kr}}{kT_2}$, *Gibbs* number, *dimensionless*
Gr_1	$= g\left|\rho_1 - \rho_1''(T_2^{1\sigma})\right|\rho_1 D_1^3/\eta_1^2$, bubble *Grashoff* number, *dimensionless*

14. Condensation of a pure steam bubble in a subcooled liquid

h	specific enthalpy, J/kg
h_{NC}	heat transfer coefficient by natural circulation, $W/(m^2K)$
h_c	heat transfer coefficient by forced convection, $W/(m^2K)$
k	$= 13.805 \times 10^{-24}$, Boltzmann constant, J/K
Le_1	$= \lambda_1 /(\rho_1 c_{p1} D_{M \to \sum n})$, gas Lewis number, dimensionless
l_{e2}	$= const\, (v_2^3 / \varepsilon_2)^{1/4}$, characteristic size of the turbulent eddies in the liquid, m
m_2	$= m_\mu / N_A$, mass of single molecule, kg
m_μ	kg-mol mass (18 kg for water), $kg/mole$
Nu_1	$= \dfrac{D_1 h_k}{\lambda_2}$, bubble Nusselt number, dimensionless
N_2	$= \rho_2 / m_2 = \rho_2 N_A / m_\mu$ ($\approx 3.3 \times 10^{28}$ for water), number of the molecules per unit volume, $1/m^3$
N_A	$= 6.02 \times 10^{26}$, Avogadro number, number of molecules in one kilogram mole mass, $1/kg\text{-}mole$
$\dot{n}_{1cin}$	number of the created nuclei per unit time in unit volume of the liquid, $1/(m^3s)$
n''_w	cavity per unit surface, $1/m^2$
n''_{1w}	bubble generating (active) cavity per unit surface, $1/m^2$
n_1	bubble number density, $1/m^3$
$\dot{n}_1$	bubble number density change per unit time, $1/(m^3s)$
$n_{1,n}$	bubble number density belonging to n-th bubble generation, $1/m^3$
$n_{1,max}$	maximum number of nucleation sites per unit volume in a pipe, $1/m^3$
$P(D_1)$	probability that a bubble has its size between D_1 and $D_1 + \delta D_1$, dimensionless
Pr_1	$= \eta_1 c_{p1} / \lambda_1$, gas Prandtl number, dimensionless
Pr_2	$= \dfrac{\eta_2}{\rho_2 a_2}$, liquid Prandtl number, dimensionless
Pe_2	$= \dfrac{D_1 \Delta V_{12}}{a_2}$, bubble Peclet number, dimensionless
p	pressure, Pa
Re_1	$= \dfrac{D_1 \rho_2 \Delta V_{12}}{\eta_2}$, bubble Reynolds number, dimensionless
Re_{1o}	$= \dfrac{D_{1o} \rho_2 \Delta V_{12}}{\eta_2}$, bubble Reynolds number based on the initial bubbles size, dimensionless

$\dot{q}_2^{m1\sigma}$ thermal energy flow density coming from the vapor interface and introduced into the bulk liquid, W/m^3

$\dot{q}_1^{m2\sigma}$ thermal energy flow density coming from the liquid interface and introduced into the bulk vapor, W/m^3

$\dot{q}_1^{\prime\prime 2\sigma}$ thermal energy flux coming from the liquid interface and introduced into the bulk vapor, W/m^2

$\dot{q}_2^{\prime\prime 1\sigma}$ thermal energy flux coming from the vapor interface and introduced into the bulk liquid, W/m^2

R gas constant, $J/(kgK)$

R_1 bubble radius, m

R_{10} initial bubble radius, m

R_{1c} critical bubble radius, m

$R_{1\infty}$ final radius of the bubble, m

r radius, m

s specific entropy, $J/(kgK)$

T temperature, K

u radial velocity, m/s

$\mathbf{V}$ velocity vector, m/s

V_{1c} volume of the bubble with critical size, m^3

V_1 volume of the bubble, m^3

V_{10} initial volume of the bubble, m^3

v specific volume, m^3/kg

Vol volume, m^3

Greek

α_1 bubble volumetric fraction, m^3/m^3

α_2 liquid volumetric fraction, m^3/m^3

β diffusion coefficient

γ_v volume occupied by the flow divided by the volume of the system, *dimensionless*

γ cross section flowed by the flow divided by the total cross section of the system, *dimensionless*

Δ finite difference

ΔE_1 work necessary to create a single bubble with radius R_1, J

ΔE_{1c} work necessary to create a single bubble with radius R_{1c}, J

ΔE_{1c}^* $= \varphi \Delta E_{1c}$, energy for the creation of a single bubble at a wall, J

Δp_{Fi}^* $= p'(T_2) - p_{Fi}$, flashing inception pressure difference, Pa

$\Delta\tau$	time interval, s
$\Delta\tau_o$	time within which the bubble collapses completely, s
ε	dissipation rate for kinetic energy from turbulent pulsations, irreversibly dissipated power by the viscous forces due to turbulent pulsations, W/kg
ξ	local friction coefficient
η	dynamic viscosity of liquid, $kg/(ms)$
θ	static contact angle, rad
λ_{RT}	*Rayleigh-Taylor* wavelength, m
λ	thermal conductivity, $W/(mK)$
λ_R	*Darsy* friction coefficient
μ_{21}	evaporating mass per unit time and unit flow volume, $kg/(m^3s)$
μ_{12}	condensing mass per unit time and unit flow volume, $kg/(m^3s)$
ρ	density, kg/m^3
$(\rho w)_{21}$	evaporation mass flow rate, $kg/(m^2s)$
σ	surface tension, N/m
M_{M1}	kg-mole mass of the condensing gases, $kg/mole$
$\sum M_{n1}$	kg-mole mass of the non-condensing gases, $kg/mole$
M_1	kg-mole mass of the gas mixture, $kg/mole$
τ	time, s
τ_H	$=\dfrac{4}{\pi}Ja^2 Fo$, condensation time scale , *dimensionless*
Φ	cavity angle if the cavity is idealized as a cone, rad
Φ_{2o}^2	two-phase friction multiplier, *dimensionless*
φ	work reduction factor, *dimensionless*

Superscripts

'	saturated liquid
"	saturated vapor
spin	spinoidal line
1σ	interface with vapor
2σ	interface with liquid

Subscripts

1	vapor, gas
2	liquid
c	critical state

s	at constant entropy
sat	saturated
Fi	flashing inception
$\sum_{nlo}$	sum of all non-condensing gases
M	evaporating or condensing chemical component
n	inert (non condensing) chemical component
u,v,w	in the positive direction of the velocity components u,v and w
R	friction

References

1. Akiyama A (1973) Bubble collapse in subcooled boiling, Bul. JSME, vol 16 no 93 pp 570-575
2. Avdeev AA (1986) Growth and condensation velocity of steam bubbles in turbulent flow, Teploenergetika, in Russian, vol 1 pp 53 - 55
3. Brauer et al H (1976) Chem. Ing. Tech., vol 48 pp 737 - 741
4. Bucher B, Nordman D (1978) Investigations of subcooled boiling problems in two-phase transport and reactor safety, Veziroglu RN, Kakac S (eds) Hemisphere, Washington, D.C., vol 1 pp 31-49
5. Brucker GG, Sparrow EM (1977) Direct contact condensation of steam bubbles in water of high pressure, Int. J. Heat Mass Transfer, vol 20 pp 371-381
6. Chen YM, Maynger F (August 6 - August 9, 1989) Measurement of heat transfer at the phase interface of condensing bubbles, ANS Proceedings Nat. Heat Transfer Conferecence, Philadelphia, Pensilvania, vol 4 pp 147-152
7. Dushkin AL, Kolomenzev AI (1989) Steam bubble condensation in subcooled liquid, High Temperature Physics, , in Russian, vol 27 no 1 pp 116-121
8. Forschuetz L, Chao BT (May 1965) On the mechanics of vapor bubble collapse, Transactions of the ASME, Journal of Heat Transfer, pp 209-220
9. Hughes ED, Paulsen MP, Agee LJ (Sept. 1981) A drift-flux model of two-phase flow for RETRAN. Nuclear Technology, vol 54 pp 410-420
10. Hunt DL (1970) The effect of delayed bubble growth on the depressurization of vessels Containing high temperature water. UKAEA Report AHSB(S) R 189
11. Isenberg J, Sideman S (June 1970) Direct contact heat transfer with change of phase: Bubble condensation in immiscible liquids, Int. J. Heat Mass Transfer, vol 13 pp 997-1011
12. Kataoka I, Ishii M, Mishima K (June 1983) Transactions of the ASME, vol 105 p 230-238
13. Kendouch AA (1976) Theoretical and experimental investigations into the problem of transient two-phase flow and its application to reactor safety, Ph.D.Thesis, Department of Thermodynamics and Fluid Mechanics, University of Strathclyde, U.C.
14. Labunzov DA (1974) State of the art of the nuclide boiling mechanism of liquids, Heat Transfer and Physical Hydrodynamics, Moskva, Nauka, in Russian, pp 98 - 115
15. Mayinger F, Nordmann D (1979) Temperature, pressure and heat transfer near condensing bubbles, Proc.of the Heat and Mass Transfer, Dubrovnik, Hemisphere Publ.Corp., New York, vol 1

16. Moalem D, Sideman S (1973) The effect of motion on bubble collaps, Int. J. Heat Mass Transfer, vol 16 pp 2321-2329
17. Nigmatulin RI (1978) Basics of the mechanics of the heterogeneous fluids, Moskva, Nauka, in Russian
18. Nukiama S, Tanasawa, Y (1938) Trans. Soc. Mech. Engrs. (Japan), vol 4 no 14 p 86
19. Pilch M, Erdman CA, Reynolds AB (August 1981) Acceleration induced fragmentation of liquid drops, Charlottesville, VA: Department of Nucl. Eng., University of Virginia, NUREG/CR-2247
20. Plesset MS, Zwick SA (April 1954) The growth of bubbles in superheated liquids, J. of Applied Physics, vol 25 no 4 pp 493-500
21. Prisnjakov VF (January 1970) Bubble growth in liquids, Journal of Engineering Physics, vol 18 no 1 pp 584-588
22. Soo SL (1969) Fluid dynamics of multiphase systems, Massachusetts, Woltham
23. Wallis GB (1969) One-dimensional two-phase flow, McGraw-Hil, New York
24. Wilson JF (1965) Primary separation of steam from water by natural separation. US/EURATOM Report ACNP-65002
25. VDI-Waermeatlas, 4. Auflage 1984, VDI-Verlag
26. Zuber N (1961) The dynamics of vapor bubbles in non uniform temperature fields, Int. J. Heat Mass Transfer, vol 2 pp 83-98

15. Bubble departure diameter

The tangential shear due to the volume- and time-averaged pulsation velocity caused by the thermally controlled bubble growth and successive cyclic departure is found to be responsible for the natural switch from the regime of isolated bubble growth and departure into the regime of mutual interaction bubble growth and departure as the temperature difference between the bubble interface and the surrounding liquid increases. The proposed theoretical model, based only on the first principles, agrees well with the experimental data for boiling water to which it was compared. The new model successfully predicts the isolated and the mutual interaction bubble departure size, and the quantitative description of the natural transition between the two regimes is the essentially new feature of this work. This Chapter is a slightly abbreviated version of the work published primarily in [21].

15.1 How accurately can we predict bubble departure diameter for boiling?

After 60 years of world wide research on boiling there are still "classical" problems of practical importance that have been not satisfactorily resolved. Such a problem is the analytical description of the bubble departure diameter during boiling or flashing. We demonstrate this problem as follows. Consider saturated water pool boiling at atmospheric pressure on a heated horizontal surface. The experimental data from the literature [5, 6, 36, 32, 33, 30] are depicted on Figure 15.1 as a function of the wall superheating. Note that data for bubble sizes measured away from the surface and sometimes reported as bubble departure diameters are excluded. Now we try to predict the data behavior. For the prediction of the bubble departure diameter we use some of the frequently cited theories from [4, 23, 17, 2, 27, 16]. The results presented in Figs. 15.1 and 15.2 are disappointing. None of these theories predicts even an appropriate data trend. In particular, none of the existing theories gives an explanation of why the bubble departure diameter starts to decrease after a certain superheating is exceeded. The interested reader will find reviews on modeling by *Hsu* and *Graham* [8] up to 1975, by *van Stralen* and *Cole* [34] up to 1979, and by *Klausner* et al; *Zeng* et al. [16, 42, 43] for more recent work. These works and the references given there present the state of the art in this field and will not be repeated here.

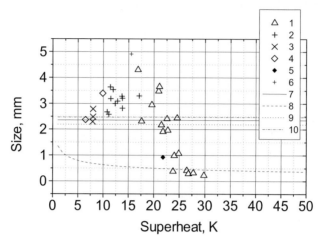

Fig. 15.1. Bubble departure diameter as a function of superheating. Saturated water pool boiling at 0.1 MPa pressure. Data: 1 Gaertner and Westwater [5], 2 Gaertner [6], 3 Tolubinsky and Ostrovsky [36], 4 Siegel and Keshok [32], 5 van Stralen et al. [33], 6 Roll and Mayers [30]. Theories: 7 Fritz [4], 8 van Krevelen and Hoftijzer [23], 9 Kocamostafaogullari and Ishii [17], 10 Cole and Rohsenow [2]

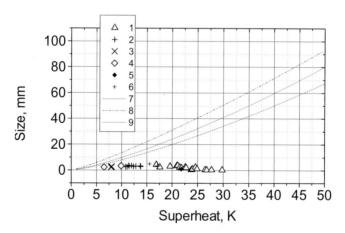

Fig. 15.2. Bubble departure diameter as a function of superheating. Saturated water pool boiling at 0.1 MPa pressure. Data: 1 Gaertner and Westwater [5], 2 Gaertner [6], 3 Tolubinski and Ostrovsky [36], 4 Siegel and Keshok [32], 5 van Stralen et al. [33], 6 Roll and Mayers [30]. Theories: 7 Moalem et al. [27], 8 Klausner et al. $n = 1/2$ [16], 9 Klausner et al. $n=1/3$ [16]

We present in this Section a new model for the prediction of the bubble departure diameter as a function of the local flow parameter, which agrees well with data for

15.2 Model development

Consider bubble growth at a heated surface as shown in Fig. 15.3. The surface vector and the upwards-directed vertical form an angle φ. There is a flow from left to right with an average velocity V_2. The derived expressions rely on the following assumptions:

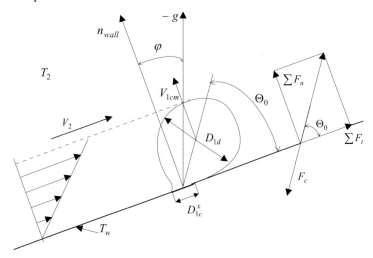

Fig. 15.3. Bubble detachment in heated wall in flow boiling

1) The bubble is capable to grow after reaching the size given by Eq. (12.16)

$$D_{1c}^* = \frac{3}{2} D_{1c} \tag{15.1}$$

where D_{1c} is the critical bubble size.

2) The bubble growth is thermally controlled. The expression governing the bubble growth is of the type given in Appendix 13.1

$$D_1 = 2B\tau^n \tag{15.2}$$

or

$$dD_1/d\tau = 2nB\tau^{n-1}. \tag{15.3}$$

It is generally agreed that the exponent n is of order of 1/2 for saturated liquids. For the purpose of data comparison we use in this work *Labuntsov*'s approximation of the *Scriven* solution [25], $n = 1/2$ and

$$B = c\, Ja\, a_2^{1/2}, \tag{15.4}$$

where

$$c = (12/\pi)^{1/2} \left[1 + \frac{1}{2}\left(\frac{\pi}{6Ja}\right)^{2/3} + \frac{\pi}{6Ja} \right]^{1/2}. \tag{15.5}$$

3) The base bubble diameter at the wall is equal to D_{1c}^* for high pressure. For low pressure, the base diameter initially grows and thereafter collapses, so that at the moment of detachment it again approximates D_{1c}^*.

4) At the moment of detachment the bubble is deformed such that its diameter measured parallel to the surface is larger than the volume-averaged bubble departure diameter D_{1d}.

5) The presence of the wall increases the drag coefficient of the bubble in the tangential drag force compared to the case of a bubble in an infinite medium.

6) The bubble axis inclination, θ_0, coincides with the main force inclination with respect to the wall.

7) At the moment of departure, the elongation of the bubble base to a cylindrical form with the axis coincident with the bubble axis is the reason for the tangential surface tension force component.

Some useful consequences of assumption 2 which will be used later are given below. At the moment of detachment, $\Delta\tau_{1d}$, the bubble radius is

$$R_{1d} = B\,\Delta\tau_{1d}^n \tag{15.6}$$

the bubble center of mass (c.m.) velocity normal to the wall is

$$V_{1cm} = dR_{1d}/d\tau = nB\Delta\tau_{1d}^{n-1} \tag{15.7}$$

and the c.m. temporal acceleration normal to the wall is

$$dV_{1cm}/d\tau = n(n-1)B\Delta\tau_{1d}^{n-2} \tag{15.8}$$

Consequently the total acceleration is

$$\left(\frac{dV_1}{d\tau} + V_1\frac{dV_1}{dy}\right)_{cm} = dV_{1cm}/d\tau + V_{1cm}^2/R_{1d} = n(n-1)B\Delta\tau_{1d}^{n-2} + n^2 B\Delta\tau_{1d}^{n-2}$$

$$= nB\Delta\tau_{1d}^{n-2}(2n-1). \tag{15.9}$$

For $n = 1/2$, which, as already mentioned, is the case for bubble growth in saturated liquid, the total bubble acceleration is zero, and therefore the bubble does not exert any dynamic forces like inertia and virtual mass force. This is a very surprising result. It simply means that at the moment of departure only the static mechanical equilibrium governs the bubble departure.

Before writing the momentum equation we compute the force components in normal and tangential directions. The buoyancy force components are

$$F_{b,n} = \frac{\pi}{6}D_{1d}^3(\rho_2 - \rho_1)g\sin\varphi \tag{15.10}$$

and

$$F_{b,t} = \frac{\pi}{6}D_{1d}^3(\rho_2 - \rho_1)g\cos\varphi. \tag{15.11}$$

For computation of the normal drag force resisting the bubble growth, we modify the Eq. (2.20) derived by *Ishii* and *Zuber* [10] for bubbles in infinite liquid

$$F_{d,n} = 3\pi\eta_2 c_{form}D_{1d}V_{1cm}\left[1 + 0.1(c_{form}D_{1d}V_{1cm}\rho_2/\eta_2)\right]^{3/4}. \tag{15.12}$$

The product

$$D_{1d}V_{1cm} = 2nB^2\Delta\tau_{1d}^{2n-1}, \tag{15.13}$$

which for $n = 1/2$

$$D_{1d}V_{1cm} = B^2 \tag{15.14}$$

is obviously a function of superheating only. Consequently

$$F_{d,n} = 3\pi\eta_2 c_{form} B^2 [1 + 0.1(c_{form} B^2 \rho_2 / \eta_2)]^{3/4}. \qquad (15.15)$$

The form coefficient, c_{form}, takes into account the ratio of the bubble size parallel to the wall to the volume-averaged bubble size as well the departure of the c.m. bubble velocity normal to the wall due to complicated bubble growth and deformation at the moment of detachment. It was found through the data comparison for pool boiling only that

$$c_{form} \approx 2. \qquad (15.16)$$

The drag force parallel to the wall is

$$F_{d,t} = 0.3\pi\rho_2 (c_{wall} D_{1d} V_{21d})^2. \qquad (15.17)$$

Here V_{21d} is the boundary layer velocity at distance R_{1d} from the wall. We compute this velocity using the *Reinhard* universal velocity profile in exactly the same manner as described in [16]. The deviation from the idealized profile in the real flow, caused by the bubbles themselves, the bubble deformation, and the wall influence are pooled into the wall correction coefficient, which was found to be

$$c_{wall} = 2 \qquad (15.18)$$

through comparison with flow boiling data for negligible mutual interaction. The lift force caused by the rotation of the non-uniform velocity field at the wall is

$$F_{l,n} = \pi \frac{1}{2} \rho_2 (D_{1d} V_{21d})^2 c_{lift}, \qquad (15.19)$$

where c_{lift} is computed in accordance with [16].

Finally we consider a force never before considered in the literature in the momentum equation describing bubble departure - the drag force caused by the growth of the neighboring bubbles

$$F'_{d,t} = 0.3\pi\rho_2 (D_{1d} \overline{\overline{V'_2}})^2. \qquad (15.20)$$

For an incompressible liquid denoted with subscript 2 the mass conservation equation for the liquid velocity gives $u_2 = (dR_1/d\tau)(R_1/r)^2$ in the region between R_1 and $R_{2,inf}$, where $R_{2,inf}$ is half of the average center-to-center spacing. *Zuber* [44] (1963) obtained the volume average liquid velocity as follows

$$\overline{V}_2' = \frac{1}{R_{2,\text{inf}} - R_1} \int_{R_1}^{R_{2,\text{inf}}} R_1^2 \frac{dR_1}{d\tau} \frac{dr}{r^2} = \frac{R_1}{R_{2,\text{inf}}} \frac{dR_1}{d\tau} = 2R_1 \frac{dR_1}{d\tau} / D_{2,\text{inf}} = B^2 / D_{2,\text{inf}}$$
(15.21)

where

$$B^2 = 2R_1 \frac{dR_1}{d\tau}.$$
(15.22)

for $n = 1/2$. For the idealized case of a triangular array at a plane surface, the influence circle has a diameter $D_{2,\text{inf}} = \left(\frac{2}{\sqrt{3}} \frac{1}{n_{1w}''}\right)^{1/2} \approx 1.074 / n_{1w}^{1/2}$, but the experimentally observed distribution of the active nucleation sites is a *Poisson* distribution [7, 35, 38]. It gives an average distance between two neighboring bubbles sites of

$$\left(D_{2,\text{inf}}^2\right)^{1/2} = \left(\frac{1}{\pi n_{1w}''}\right)^{1/2} \approx 0.56 / n_{1w}^{1/2}$$
(15.23)

More recent measurements by *Wang* and *Dhir* [38] (1993) give $\left(D_{2,\text{inf}}^2\right)^{1/2} \approx 0.84 / n_{1w}^{1/2}$. For the purpose of the data comparisons we use in this work Eq. (15.23). The *time - averaged fluctuation velocity* (called sometimes micro-convection velocity) during $\Delta\tau_w + \Delta\tau_d$ is therefore

$$\overline{\overline{V}}_2' = (B^2 / D_{2,\text{inf}}) \frac{\Delta\tau_d}{\Delta\tau_w + \Delta\tau_d} = B^2 \frac{\Delta\tau_d}{\Delta\tau_w + \Delta\tau_d} (\pi n_{1w}'')^{1/2},$$
(15.24)

which is generally applicable to different situations of bubble production at walls such as boiling, flashing, and gas injection through perforated plates. Note that nucleation at inner and outer walls of a cylinder with diameter D_h is characterized by

$$D_{2,\text{inf}} = \left(\frac{1}{\pi n_{1w}''}\right)^{1/2} (1 \pm D_{1d} / D_h),$$
(15.23a)

where the minus sign is valid for the inner surface. Consequently

$$\overline{\overline{V}}_2' = (B^2 / D_{2,\text{inf}}) \frac{\Delta\tau_d}{\Delta\tau_w + \Delta\tau_d} = B^2 \frac{\Delta\tau_d}{\Delta\tau_w + \Delta\tau_d} (\pi n_{1w}'')^{1/2} / (1 \pm D_{1d} / D_h)$$

$$= D_{1d}^2 f_{1w} (\pi n_{1w}'')^{1/2} / (1 \pm D_{1d} / D_h) \ . \tag{15.24a}$$

Consequently, with increased wall superheating, bubbles in a pipe interact more strongly than bubbles at a plane, and bubbles at an outer cylinder wall interact less intensively than bubbles at a plane. The higher the value of D_{1d}/D_h the stronger the influence of the geometry. For large values of superheating, $\Delta \tau_{1d} / (\Delta \tau_{1w} + \Delta \tau_{1d})$ approximates 1. For small values, this force is not important. Thus the approximation of the exact equation

$$\overline{\overline{V'}} = B^2 (\pi n_{1w}'')^{1/2} \tag{15.25}$$

can be used to compute the bubble departure diameter. Now we compute the bubble inclination angle using assumption 6

$$\theta_0 = \arctan\left[\left(F_{b,n} - F_{d,n} + F_{1,n} \right) / \left(F_{b,t} + F_{d,t} + F_{d,t}' \right) \right] \tag{15.26}$$

and write the static force balance along the axis of the bubble divided by π

$$\frac{1}{6} D_{1d}^3 (\rho_2 - \rho_1) g \sin(\varphi + \theta_0) + \frac{1}{2} \rho_2 (D_{1d} V_{21d})^2 c_{lift} \sin \theta_0$$

$$+ 0.3 \rho_2 \left(c_{wall} D_{1d} V_{21d} \right)^2 \cos \theta_0 + 0.3 \rho_2 \left(D_{1d} \overline{\overline{V_2'}} \right)^2 \cos \theta_0$$

$$= D_{1c}^* \sigma + 3 \eta_2 c_{form} B^2 \left[1 + 0.1 \left(c_{form} B^2 \rho_2 / \eta_2 \right) \right]^{3/4} \sin \theta_0$$

or simply

$$\left(D_{1d} / D_{1d,nc} \right)^3 + \left(D_{1d} / D_{1d,fc} \right)^2 = 1 \tag{15.27}$$

where

$$D_{1d,nc} = \left\{ 6A / \left[(\rho_2 - \rho_1) g \sin(\varphi + \theta_0) \right] \right\}^{1/3}, \tag{15.28}$$

$$D_{1d,fc} = \left\{ A / \left[\rho_2 V_{21d}^2 \left(\frac{1}{2} c_{lift} \sin \theta_0 + 0.3 c_{wall}^2 \cos \theta_0 \right) + 0.3 \rho_2 \overline{\overline{V_2'}}^2 \cos \theta_0 \right] \right\}^{1/2}, \tag{15.29}$$

and

$$A = D_{1c}^* \sigma + 3\eta_2 c_{form} B^2 \left[1 + 0.1(c_{form} B^2 \rho_2 / \eta_2)\right]^{3/4} \sin\theta_0 . \quad (15.30)$$

Note that the inclination angle for forced convection is a real one. For pool boiling it is the cone angle within which the fluctuation of the spatial inclination occurs. The angular position within this angle changes chaotically. That is why one experimentally sees fluctuation but never inclination of the vapor column. It explains the well known oscillating of the bubble columns. Note that the bubbles at a surface never start and grow simultaneously - such an idealized picture should give zero interaction force, which is never the real case.

Obviously we have two transcendental equations with two unknowns D_{1d} and θ_0 which are solved by iteration. It is important to note that after each iteration the new inclination is computed as an average value between the previous and the new computed values. This ensures a stable algorithm that converges to the desired accuracy after a maximum of seven iterations. For low superheating and low flow velocities, D_{1d} approximates $D_{1d,nc}$, and for large superheating and large flow velocities, D_{1d} approximates $D_{1d,fc}$. It is important to note that the surface tension and the latent heat of vaporization are functions of the wall temperature for boiling and of the liquid temperature for flashing.

Note that the contribution of the macro-convection in boiling heat transfer is generally accepted. Consequently we have to take it into account also in the momentum balance.

15.3 Comparison with experimental data

We have confined our attention to water data only because this model is operating in computer code IVA in all of its variants, which has water as one of the three flow components [18, 19]. For further information on this code see the references given in [18, 19]. The prediction of bubble departure diameter is a part of the analytical modeling of transient sources and sinks of bubbles and droplets in addition to the information presented in [20], which is used in IVA3.

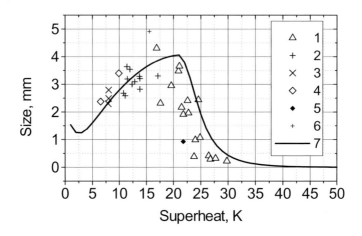

Fig. 15.4. Bubble departure diameter as a function of superheating. Saturated water pool boiling at 0.1 *MPa* pressure. Data 1: Gaertner and Westwater [5], 2 Gaertner [6], 3 Tolubinski and Ostrovsky [36], 4 Siegel and Keshok [32], 5 van Stralen et al. [33], 6 Roll and Mayers [30], 7 IVA5 model with bubble interactions

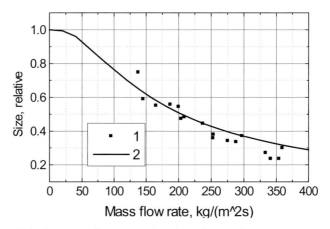

Fig. 15.5. Bubble departure diameter as function of mass flow rate. Saturated water flow boiling at 0.1 *MPa* pressure. D_{hy} = 0.019 *m*, $T_w - T_2$ = 15 *K*. 1 Data of Koumoutsos et al. [22], 2 IVA5 model without bubble interaction

Figure 15.4 presents a comparison of the prediction of the new theory with the already discussed data for saturated water pool boiling at a horizontal surface. Without considering the mutual bubble interaction the trend of the curve for superheatings lower then 15 K continuous to an asymptotic value of about 5 mm as shown in Fig. 4 in [21]. The line on Fig. 15.4 represents the new model considering the mutual bubble interaction. The particle number density for the *Gaertner* and *Westwater* [5] data, presented in Fig. 12.1 and approximated by Eqs. (12.107) and (12.108) is used for this comparison. In the original publication, *Gaetner* and *Westwater* [5] called the bubble departure diameter "lift off diameter", and in a later work by *Gaetner* [7] it is called "bubble column stem" diameter. However, the proportionality constant between both diameters is around one. Note that the *Gaetner* and *Westwater* data for particle number density and bubble departure diameter as a function of the wall superheating are confirmed also by the later measurements by *Iida* and *Kobayasi* [9], who measured particle number density and the thickness of the thermal boundary layer for pool boiling. Figure 12.1 presents the nucleation site density as a function of superheating also by other authors [6, 25, 40, 12, 3, 34, 29]. This is done to classify the *Gaertner* and *Westwater* data as the data giving the lowest possible nucleation site density for saturated pool boiling of water. The consideration of the mutual bubble interaction has a striking effect that can not be overlooked. This interaction is responsible for the decrease of the bubble diameter when the nucleation site density exceeds some value, that is, when the wall superheating exceeds some value. This regime causes a change of the character of the *Nukiama* diagram containing data for saturated water pool boiling at atmospheric pressure on polished surfaces [6, 34, 40, 35, 39, 5, 1, 11, 3, 37, 28] as shown in Fig. 15.7 and is in fact quite different from boiling with negligible interaction, which in the boiling literature is called the regime of isolated bubbles. Our theory predicts a smooth transition between the two regimes, which is the expected behavior in nature.

In Fig. 15.5 we compare the prediction of our theory with *flow boiling* data given in [22]. On the abscissa is entered the bubble departure diameter divided by the bubble departure diameter for pool boiling. The trend of the data is reasonably reproduced. Figure 8 in [21] shows the same data compared also with predictions using theories of other authors [26, 41, 14, 15] and with prediction of our theory neglecting the normal force components. We find from this comparison that neglecting the normal force component is allowed for mass flow rates larger than 150 $kg/m^2 s$. The theory of *Jones* [14, 15] gives unbounded values for small mass flow rate, and the *Yang* and *Weisman* modification [41] of *Levy*'s theory [26] underpredicts the data as shown in Fig. 8 in [21].

Figure 15.6 gives a comparison with data at elevated pressures reported in Refs. [36, 31]. In this case also, the trend of the data is properly reproduced by our theory.

From the above discussion the following conclusions are in order:

1. The mutual interaction of the growing and departing bubbles causes significant shear at the surface and forces the bubbles to detach at an earlier stage of their growth.

2. The proposed theory successfully describes quantitatively the natural transition between the isolated bubble boiling and boiling with mutual bubble interaction with changing temperature difference.

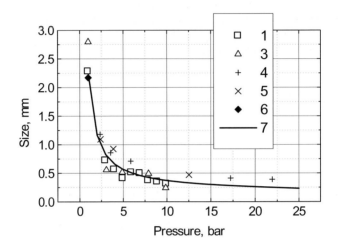

Fig. 15.6. Bubble departure diameter as a function of pressure. Saturated water pool boiling, superheat = 7.2 *K*. Data of Tolubinsky and Ostrovsky: 1 Permalloy, 2 brass, 3 copper. Data of Semeria: 4 wire D = 0.8*mm*, 5 plate. 6 Model IVA5 with bubble interaction

15.4 Significance

The proposed method can be used in various fields of engineering practice, a few of them being

1. Flashing in channel flow due to depressurization and critical flow - Chapter 17.
2. Boiling in pools and channels - Chapter 16.
3. Modeling of multi-phase flow in system computer codes.
4. Reexamination of all approaches for analytical description of boiling heat transfer using the bubble departure diameter as an input variable.
5. Reexamination of all approaches for analytical description of bubble diameter produced by gas injection into a liquid through perforated plates - Section 10.7.

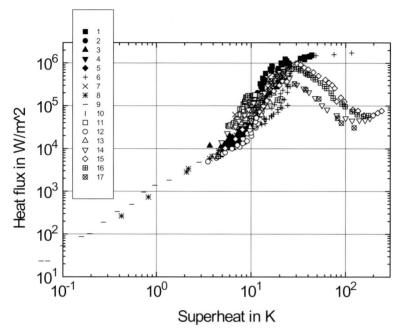

Fig. 15.7. Heat flux as a function of superheat. Saturated water at 0.1 MPa. 1) Gaertner [6] 1965, 4/0 polished copper, 2) Kurihara and Myers [24] 1960, 4/0 polished copper, 3) Yamagata et al. [40] 1955, fine polished brass, 4) Sultan and Judd [35] 1978, diamond grid 600 polished copper, 5) Wiebe [39] 1970, diamond 600 polished copper, 6) Gaertner and Westwater [5] 1960, 4/0 polished copper, 20 p.c. nickel salt-water solution, 7) Borishanskii et al. [1] 1961, steel, 5 to 8 p.c. error, 8) Fritz [4] 1931, polished steel, 9) Jakob and Linke [13] 1933, polished steel, 10) Cornwell and Brown [4] 1978, 4/0 polished cooper, 11) Vachon et al. [37] 1968, emery grid 600 polished 304 stainless steel, 12) Nishikawa et al. [28] 1984, emery grid No.0/10 polished copper, 13) Rallis and Jawurek [29] 1964, nickel wire, 14) Wang and Dhir [38] nuclide and transition boiling. Contact angle 90 *deg*, 15) Wang and Dhir [38] nuclide and transition boiling. Contact angle 18 *deg*, 16) Wang and Dhir [38] nuclide and transition boiling. Contact angle 35 *deg*, 17) Wang and Dhir [38] nuclide and transition boiling. Contact angle 35 *deg*. Liaw data.

15.5 Summary and conclusions

The *tangential shear* due to the volume- and time- averaged *pulsation velocity* caused by the thermally controlled bubble growth and successive cyclic departure is found to be responsible for the natural switching from the regime of *isolated* bubble growth and departure into the regime of *mutually interacting* bubble growth and departure as the temperature difference between the bubble interface and the surrounding liquid increases. The proposed theoretical model, based only on the first principles, agrees well with the experimental data for boiling water to which it was compared. The new model, which successfully predicts the isolated and the mutually interacting bubble departure sizes, and the quantitative description of the natural transition between the two regimes are the essentially new features of the work presented in this Chapter.

Nomenclature

Latin

a_2	liquid thermal diffusivity, m^2/s
B^2	$= 2R_1 dR_1/d\tau$, m^2/s
D_1	bubble diameter, m
D_{1c}	critical diameter, m
D_{1d}	bubble departure diameter, m
$D_{2,inf}$	$= 2R_{2,inf}$, m
$D_{1d,fc}$	bubble departure diameter for strongly predominant forced convection, $V_{21d} \gg \overline{\overline{V_2'}}$, m
$D_{1d,nc}$	bubble departure diameter for natural circulation, $V_{21d} \ll \overline{\overline{V_2'}}$, m
D	total differential, $dimensionless$
G	mass flow rate, $kg/(m^2 s)$
g	gravitational acceleration, m/s^2
h'	saturated liquid specific enthalpy, $J/(kgK)$
h''	saturated steam specific enthalpy, $J/(kgK)$
Ja	$= \dfrac{[T_2 - T'(p)]c_{p2}}{h''(p) - h'(p)} \dfrac{\rho'}{\rho''}$, Jacob number, $dimensionless$
n_{1w}'''	active nucleation site density, $1/m^2$
p	pressure, Pa
p_c	critical pressure, Pa
R_1	bubble radius, m
$R_{2,inf}$	half of the average center-to-center spacing, m
$T'(p)$	saturation temperature at system pressure p, K
T_2	liquid temperature, K
T_w	wall temperature, K
V_{1cm}	center-of-mass bubble velocity at the moment of detachment, m/s
V_{21d}	tangential velocity in the boundary layer of thickness D_{1d}, m/s
u_2	liquid velocity in the region between R_1 and $R_{2,inf}$, m/s
$\overline{V_2'}$	volume-averaged fluctuation velocity, m/s
$\overline{\overline{V_2'}}$	time- and volume-averaged fluctuation velocity over $\Delta\tau_w + \Delta\tau_d$, m/s

Greek

∂	partial differential, *dimensionless*
$\Delta \rho_{21}$	$= \rho_2 - \rho_1$, kg/m^3
$\Delta \tau_d$	time need from the origination of bubble with critical size to the bubble departure from the wall, s
$\Delta \tau_w$	delay time, s
η_2	dynamic viscosity of liquid, $kg/(ms)$
φ	angle between the flow direction and the upwards-directed vertical, - $\mathbf{n}_g$, *rad*
ρ_1	gas density, kg/m^3
ρ_2	liquid density, kg/m^3
ρ''	saturated steam density, kg/m^3
ρ'	saturated liquid density, kg/m^3
$\bar{\rho}$	$= (\rho' - \rho'')/\rho$, *dimensionless*
σ	surface tension, N/m
τ_{2w}	$= \dfrac{\xi}{8} G^2 / \rho$ shear stress, N/m^2
τ	time, s
θ_0	angle between the bubble axis and the wall, *rad*

References

1. Borishanskii V, Bobrovich G, Minchenko F (1961) Heat transfer from a tube to water and to ethanol in nucleate pool boiling, Symposium of Heat Transfer and Hydraulics in Two-Phase Media, Kutateladze SS (ed) Gosenergoizdat, Moscow
2. Cole R, Rohsenow WM (1969) Correlation of bubble departure diameters for boiling of saturated liquids, Chem. Eng. Prog. Symp. Ser. , vol 65 no 92 pp 211-213
3. Cornwell K, Brown RD (1978) Boiling surface topology, Proc. 6th Int. Heat Transfer Conf., Toronto, vol 1 pp 157-161
4. Fritz W (1935) Berechnung des maximalen Volumens von Dampfblasen, Phys. Z., vol 36 no 11 pp 379-384.
5. Gaertner RF, Westwater JW (1960) Population of active sites in nucleate boiling heat transfer, Chem. Eng. Progr. Symp. Ser, vol 30 pp 39-48
6. Gaertner RF (Feb. 1965) Photographic study of nucleate pool boiling on a horizontal surface, Transaction of the ASME, Journal of Heat Transfer, pp 17- 29
7. Gaertner RF (1963) Distribution of active sites in the nucleate boiling of liquids, Chem. Eng. Prog. Symp. Series, no 41 vol 59 pp 52-61

8. Hsu YY, Graham RW (1976) Transport processes in boiling and two - phase systems, Hemisphere Publishing Corporation, Washington - London, Mc Graw - Hill Book Company, New York
9. Iida Y, Kobayasi K (1970) An experimental investigation on the mechanism of pool boiling phenomena by a probe method, 4th Int. Heat Transfer Conf., Paris – Versailles, vol 5 no 3 B13 pp 1-11
10. Ishii M, Zuber N (1978) Relative motion, interfacial drag coefficients in dispersed two - phase flow of bubbles, drops and particles, Paper 56a, AIChE 71st Ann. Meeting, Miami
11. Jakob M, Fritz W (1931) Forsch. Ing.-Wes., vol 2 p 435
12. Jakob M (1932) Kondensation und Verdampfung, Zeitschrift des Vereins deutscher Ingenieure, vol 76 no 48 pp 1161-1170
13. Jakob M, Linke W (1933) Der Waermeuebergang von einer waagerechten Platte an siedendes Wasser, Forsch. Ing. Wes., vol 4 pp 75-81
14. Jones OC (1992) Nonequilibrium phase change --1. Flashing inception, critical flow, and void development in ducts, in Lahey RT Jr (ed) Boiling Heat Transfer, Elsevier Science Publishers B.V., pp 189 - 234
15. Jones OC (1992) Nonequilibrium Phase Change --2. Relaxation models, general applications, and post heat transfer, in Lahey RT Jr (ed) Boiling Heat Transfer, Elsevier Science Publishers B.V., pp 447-482
16. Klausner JF, Mei R, Bernard DM, Zeng LZ (1993) Vapor bubble departure in forced convection boiling, Int. J. of Heat and Mass Transfer, vol 36 no 3 pp 651-662
17. Kocamustafaogullari G, Ishii M (1983) Interfacial area and nucleation site density in boiling systems, Int. J. Heat Mass Transfer, vol 26 no 9 pp 1377-1389
18. Kolev NI (October 5-8 1993) IVA3 NW: Computer code for modeling of transient three phase flow in complicated 3D geometry connected with industrial networks, Proc. of the Sixt International Topical Meeting on Nuclear Reactor Thermal Hydraulics, Grenoble, France
19. Kolev NI (1993) The code IVA3 for modeling of transient three-phase flows in complicated 3D geometry, Kerntechnik, vol 58 no 3 pp 147-156
20. Kolev NI (1993) Fragmentation and coalescence dynamics in multi-phase flows, Experimental Thermal and Fluid Science, Elsevier, vol 6 pp 211-251
21. Kolev NI (1994) The influence of mutual bubble interaction on the bubble departure diameter, Experimental Thermal and Fluid Science, Elsevier, vol 8 pp 167-174
22. Koumoutsos N, Moissis R, Spyridonos A (May 1968) A study of bubble departure in forced-convection boiling, Journal of Heat Transfer, Transactions of the ASME pp 223-230
23. van Krevelen DW, Hoftijzer PJ (1950) Studies of gas- bubble formulation, calculation of interfacial area in bubble contactor, Chem. Eng. Progr. Symp. Ser., vol 46 no 1 pp. 29-35
24. Kurihara HM, Myers J E (March 1960) The effect of superheat and surface roughness on boiling coefficients, A. I. Ch. E. Journal, vol 6 no 1 pp 83-91
25. Labuntsov DA (1974) State of the art of the nucleate boiling mechanism of liquids, Heat Transfer and Physical Hydrodynamics, Moskva, Nauka, in Russian, pp. 98- 115
26. Levy S (1967) Forced convection subcooled boiling prediction of vapor volume fraction, Int. J. Heat Mass Transfer, vol 10 pp 951-965
27. Moalem D, Yijl W, van Stralen SJD (1977) Nucleate boiling at a liquid-liquid interface, letters heat and mass transfer, vol 4 pp 319-329

28. Nishikawa K, Fujita Y, Uchida S, Ohta H (1984) Effect of surface configuration on nucleate boiling heat transfer, Int. J. Heat and Mass Transfer, vol 27 no 9 pp 1559-1571
29. Rallis C J, Jawurek HH (1964) Latent heat transport in saturated nucleate boiling, Int. J. Heat Transfer, vol 7 pp 1051-1068
30. Roll JB, Mayers JC (July 1964) The effect of surface tension on factors in boiling heat transfer, A.I.Ch.E.Journal, pp 330-344
31. Semeria RF (1962) Quelques resultats sur le mechanisme de l'ebullition, 7, J. de l'Hydraulique de la Soc. Hydrotechnique de France
32. Siegel R, Keshock EG (July 1964) Effects of reduced gravity on nucleate boiling bubble dynamics in saturated water, A.I.Ch.E. Journal, vol 10 no 4 pp 509-517
33. van Stralen SJD, Sluyter WM, Sohal MS (1975) Bubble growth rates in nucleate boiling of water at subatmospheric pressures, Int. J. Heat and Mass transfer, vol 18 pp 655-669
34. van Stralen S, Cole R (1979) Boiling Phenomena, Hemisphere, USA
35. Sultan M, Judd RL (Feb. 1978) Spatial distribution of active sites and bubble flux density, Journal of Heat Transfer, Transactions of the ASME, vol 100 pp 56-62
36. Tolubinsky VI, Ostrovsky JN (1966) On the mechanism of boiling heat transfer (vapor bubbles growth rise in the process of boiling in liquids, solutions, and binary mixtures), Int. J. Heat Mass Transfer, vol 9 pp 1463-1470
37. Vachon RI, Tanger GE, Davis DL, Nix GH (May 1968) Pool boiling on polished chemically etched stainless-steel surafces, Transactions of ASME, Journal of Heat Transfer, pp 231-238
38. Wang CH, Dhir VK (Aug. 1993) Effect of surface wettability on active nucleation site density during pool boiling of water on a vertical surface, ASME Journal of Heat Transfer, vol 115 pp 659-669
39. Wiebe JR (1970) Temperature profiles in subcooled nucleate boiling, M. Eng. thes., Mechanical Engineering Department, McMaster University, Canada
40. Yamagata K, Hirano F, Nishikawa K, Matsuoka H (1955) Nucleate boiling of water on the horizontal heating surface, Mem. Fac. Engng, Kyushu Univ, vol 15 p 97
41. Yang JY, Weisman J (1991) A phenomenological model of subcooled flow boiling in the detached bubble region, Int. J. Multiphase Flow, vol 17 no 1 pp 77-94
42. Zeng LZ, Klausner JF, Mei R (1993) A unified model for the prediction of bubble detachment diameters in boiling systems - 1. Pool boiling, Int. J. of Heat and Mass Transfer, vol 36 no 9 pp 2261 - 2270
43. Zeng L Z, Klausner JF, Bernard DM, Mei R (1993) A unified model for the prediction of bubble detachment diameters in boiling systems - 2. Flow boiling, Int. J. of Heat and Mass Transfer, vol 36 no 9 pp 2271 - 2279
44. Zuber N (1963) Nucleate boiling. The region of isolated bubbles and the similarity with natural convection, Int. J. Heat and Mass Transfer, vol 6 pp 53-78

16. How accurately can we predict nucleate boiling?

In this Chapter we present a new theory of nucleate pool boiling based on turbulence in the boundary layer induced by bubble growth and departure which incorporates the effect of the static contact angle and explains the large spread of the available experimental data. In addition an exiting feature of the new theory is demonstrated - the capability to predict the critical heat flux with the same correlation which was never reported before in the literature as far as the author knows. This is a slightly modified version of the work previously published in [26].

16.1 Introduction

Nucleate boiling is probably the most investigated phenomenon in the thermal sciences over 60 years. Data are collected, e.g. [16, 18, 50, 30, 11,13, 3, 37, 44, 49, 41, 7, 35], see also Appendix 16.2, and many theoretical models exist, e.g. [46, 39, 10, 45, 51, 34, 6, 54, 55, 24, 57, 47, 36, 4, 9, 52, 5, 19, 27, 43]. It seems that the question of how accurately we can predict the heat flux as a function of the wall superheating for simple pool boiling is banal and boring. However, let us consider this problem in more detail.

Figure 16.1 depicts the heat flux as a function of the wall superheating for water at atmospheric pressure at plane surfaces taken from [16, 18, 50, 30, 11,13, 3, 37, 44, 49, 41, 7, 35] for mirror polished surfaces (data for specifically prepared rough surfaces are excluded). By taking a close look at these data we see that while the reported error for temperature measurements is about 5 to 6 % and for heat transfer measurements about 1 to 14 % in each particular experiment the spreading of different authors data is over two orders of magnitude which cannot be explained with measurements error. This spreading corresponds to the data for the nucleation site density (reported measurement error about 1 to 20 %) as a function of wall superheat presented in Fig. 16.2 for the same data set as far as available [2, 4, 5, 6, 8, 11, 12]. The state of the art of the modeling of this phenomenon is well represented by [46, 39, 10, 45, 51, 34, 6, 54, 55, 24, 57, 47, 36, 4, 9, 52, 5, 19, 27, 43]. Examining the main basic physical ideas on which these models rely (see Appendix 16.1) we surprisingly find that none of them predicts the main reason for the observed data spreading.

440 16. How accurately can we predict nucleate boiling?

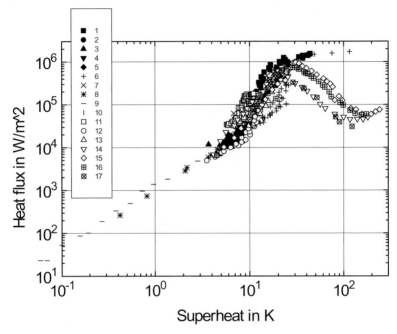

Fig. 16.1. Heat flux as a function of superheat. Saturated water at 0.1 *MPa*. 1) Gaertner [13] 1965, 4/0 polished copper, 2) Kurihara and Myers [30] 1960, 4/0 polished copper, 3) Yamagata et al. [50] 1955, fine polished brass, 4) Sultan and Judd [41] 1978, diamond grid 600 polished copper, 5) Wiebe [49] 1970, diamond 600 polished copper, 6) Gaertner and Westwater [11] 1960, 4/0 polished copper, 20 p.c. nickel salt-water solution, 7) Borishanskii et al. [3] 1961, steel, 5 to 8 p.c. error, 8) Jakob and Fritz [16] 1931, polished steel, 9) Jakob and Linke [18] 1932, polished steel, 10) Cornwell and Brown [7] 1978, 4/0 polished cooper, 11) Vachon et al. [45] 1968, emery grid 600 polished 304 stainless steel, 12) Nishikawa et al. [35] 1984, emery grid No.0/10 polished copper, 13) Rallis and Jawurek [37] 1964, nickel wire, 14) Wang and Dhir [48] nuclead and transition boiling. Contact angle 90 *deg*, 15) Wang and Dhir [48] nuclead and transition boiling. Contact angle 18 *deg*, 16) Wang and Dhir [48] nuclead and transition boiling. Contact angle 35 *deg*, 17) Wang and Dhir [48] nuclead and transition boiling. Contact angle 35 *deg*. Liaw data.

There are many qualitative hypotheses in the literature for the reason of the observed data spreading but none of them are used to quantitatively improve our predictive capability. *Wang* and *Dhir* [48] published interesting experimental work showing clearly the effect of the static contact angle on the active nucleation site density as a function of the wall superheat. We transform the new data in the coordinates of Fig. 16.2 and plot them in Fig. 16.3. Comparing Figs. 16.2 and 16.3 it is obvious that variation of the static contact angle could account for a substantial part of the observed data spread in Fig. 16.2 and consequently in Fig. 16.1. This conclusion is supported by the following consideration. Nucleation site density is reported to be measured with about 1 to 20 % error. Static contact angle (if measured at all) is measured with $\pm 3\ rad$ error. Distilled water being in contact with different polished solid surfaces not especially chemically treated possesses different static contact angles as indicated in Tables 12.1 and 12.2. Since in Fig. 16.1 we have data sets for different static contact angles varying between very small values, as is the case with the *Gaertner* and *Westwater*'s data, and $\pi/2$, as it seems to be the case for copper-water data, the data sets must differ from each other.

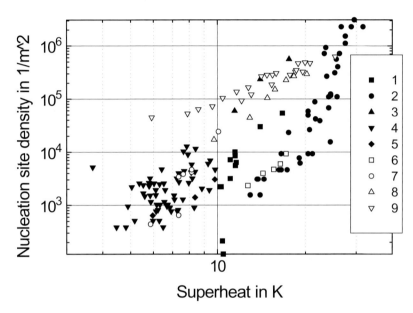

Fig. 16.2. Active nucleation site density as a function of superheat. Saturated water at 0.1 MPa. 1) Gaertner [13] 1965, 4/0 polished copper, 2) Gaertner and Westwater [11] 1960, 4/0 polished copper, 20 p.c. nickel salt-water solution, 3) Sultan and Judd [41] 1978, diamond grid 600 polished copper, 4) Yamagata et al. [50] 1955, fine polished brass, 5) Jakob and Linke [18] 1932, polished steel, 6) Cornwell and Brown [7] 1978, 4/0 polished cooper, 7) Kurihara and Myers [30] 1960, 4/0 polished copper, 8) Rallis and Jawurek [37] 1964, nickel wire, 9) Faggani et al. [8] 1981, polished 316 steel horizontal cylinder.

Wang and *Dhir* correlated their data for active nucleation sites density as a function of the wall superheat as follows

$$n''_{1w} = 5 \times 10^{-27}(1-\cos\varphi)/D_{1c}^6 \qquad (16.1)$$

where the constant has a dimension m^4, the critical diameter is a function of the local superheating

$$D_{1c} = \frac{4\sigma}{[p'(T_2)-p]\left(1-\dfrac{\rho''}{\rho_2}\right)} = \frac{4\sigma}{(h''-h')\rho'\ln\left[1+\dfrac{T_w-T'(p)}{T'(p)}\right]}$$

$$\approx \frac{T'(p)}{T_w-T'(p)}\frac{4\sigma}{\rho''(h''-h')} \qquad (16.2)$$

and $\sigma = \sigma(T_w)$.

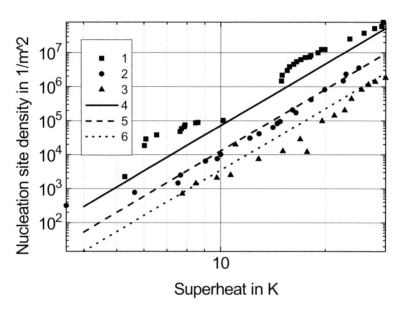

Fig. 16.3. Active nucleation site density as a function of superheat. Saturated water at 0.1 MPa. Wang and Dhir [48] data for three different static contact angles 1) 90, 2) 35 and 3) 18 deg. Prediction of the same data with their correlation 4), 5), and 6), respectively. Larger static contact angle results of larger active nucleation site density by the same superheating.

Note that all saturation properties in this paper are computed as a function of the wall temperature. The results of the prediction of n''_{1w} for the three different static contact angles ($\varphi = 90, 35$ and 18 deg) as function of the wall superheating and the experimental data which they correlate are presented in Fig. 16.3. For the prediction we use the measured wall superheating $T_w - T'(p)$. For comparison see the previous correlations not taking into account the influence of the static contact angle published in [7, 32, 34, 2, 15, 27, 38, 21, 22, 23] and summarized in Appendix 16.2 - some of them are depicted together with data in Fig 16.4. The quantitative dependence on the static contact angle reported by *Wang and Dhir* was not previously known. It strongly supports the qualitative arguments published in 1958 by *Arefeva* and *Aladev* [1] p.12 which are briefly summarized below:

" ... Boiling of liquid is associated with liquid fragmentation by liquid detachment from the wall. Consequently steam must appear for lower wall superheats at places where the work for liquid detachment is lower. Quantitative measure for the influence of the wall on heat transfer is the rate of wetting i.e. the static contact angle. Therefore, (a) the nucleation site density and consequently the heat transfer coefficient increases with increasing static contact angle, and (b) increasing static contact angle decreases the critical heat flux... "

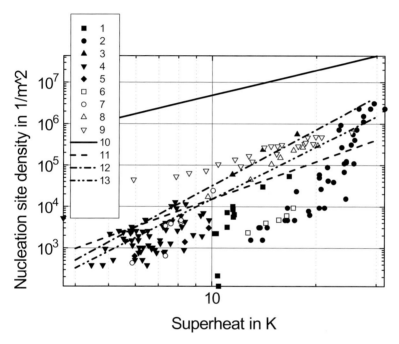

Fig. 16.4. Active nucleation site density as a function of superheat. Saturated water at 0.1 MPa. 1) to 9) Data from Fig. 16.2. Prediction with correlations proposed by 10) Avdeev et al. [2], 11) Johov [21], 12) Cornel and Brown [7] and 13) Kocamustafaogullari and Ishii [27].

Another important result of the measurements by *Wang* and *Dhir* is the relationship between the averaged nearest - neighbor distance and the nucleation site density

$$\left(D_{2,\inf}^2\right)^{1/2} \approx 0.84 / n_{1w}^{\prime\prime\,1/2} \tag{16.3}$$

which is obtained with better statistics than the earlier relation reported by *Gaertner* [12] and *Sultan* and *Judd* [41] $\approx 1/\left(\pi n''_{1w}\right)^{1/2}$. In what follows we shall demonstrate the importance of the experimental findings of *Wang* and *Dhir* for the prediction of heat transfer in nucleate boiling.

In this Chapter, we present a new theory of nucleate pool boiling based on bubble growth and departure induced turbulence in the boundary layer which incorporates the effect of the static contact angle and explains the large spreading of the available experimental data. In addition an exciting feature of the new theory is demonstrated - the capability to predict the critical heat flux with the same correlation. This was not previously reported in the literature as far as the author knows.

16.2 New phenomenological model for nucleate pool boiling

16.2.1 Basic assumptions

1. Wall superheating larger than a given value is necessary to initiate pool boiling. For smaller wall superheating the heat is transferred by natural convection, *Jakob* [17]

$$\dot{q}''_{w2,nc} = 2.417 \, \lambda_2 \left(\Pr_2 g\beta_2 / v_2^2 \right)^{1/4} \left(T_w - T_2 \right)^{21/16} \tag{16.4}$$

Here indices "2" indicates bulk liquid values.

2. The heat flux from the wall into the boiling mixture during nucleate pool boiling is caused by boundary layer turbulence due to bubble growth and departure.

3. The turbulent length scale is of the order of the *Rayleigh - Taylor* instability wavelength

$$\ell'_2 = \pi \, \lambda_{RT}, \tag{16.5}$$

where $\lambda_{RT} = \left\{ \sigma / \left[g(\rho' - \rho'') \right] \right\}^{1/2}$ is the capillary *Laplace* constant. The imagination that boiling at a heated wall is in fact counter-current gas liquid flow led *Zuber* [52] and *Zuber* et al. [53, 56] to the conclusion that the instability of such flow is controlled by the *Rayleigh-Taylor* instability wavelength. This imagination led to a successful theory for critical heat flux. *Rohsenow* [39] and *Tolubinsky* [43] used λ_{RT} as a length scale to correlate data also for sub-critical nucleate boiling heat transfer. Thus, we proceed with the imagination that not only the continuous steam stream counter current to the liquid but also the train of bubbles counter current to the liquid produce macroscopic instability of order of the *Rayleigh-Taylor* instability wavelength. As will be demonstrated later by comparison with data this assumption seems quite appropriate. We found that neither the average distance between two departing bubbles, nor the bubble departure diameter is appropriate as a macroscopic length scale of turbulence for this process.

4. The nucleation site distribution obeys the *Poisson* distribution law and the experimentally observed averaged site to site space $D_{2,inf}$ is given by Eq. (16.3).

5. The bubble growth at the wall is thermally controlled. The expression governing the bubble growth is

$$D_1 = 2B\tau^n \tag{16.6}$$

As shown in Appendix 13.1 it is generally agreed that the exponent is of the order of 1/2 for saturated liquids. For the purpose of data comparison we use in this work the *Labuntsov* approximation [31, 32] of the *Scriven* solution,

$$B = cJa'a'^{1/2} \tag{16.7}$$

where

$$c = (12/\pi)^{1/2}\left[1 + \frac{1}{2}\left(\frac{\pi}{6Ja'}\right)^{2/3} + \left(\frac{\pi}{6Ja'}\right)^{1/2}\right] \tag{16.8}$$

and $n = 1/2$.

6. The order of magnitude of the minimum waiting time will be computed in accordance with the proposal by *Han* and *Griffith* [14]

$$\Delta\tau_{1w} \approx \Delta\tau_{1w,min} = \delta_{2,min}^2/(\pi a') \tag{16.9}$$

where

$$\delta_{2,min} = 12(T_w - T_2)T'(p)\sigma/\{[T_w - T'(p)]^2 \rho''(h'' - h')\} \tag{16.10}$$

is the minimum of the thermal boundary layer thickness.

7. The bubble departure diameter D_{1d} and consequently the bubble departure time

$$\Delta\tau_{1d} = \frac{1}{4} D_{1d}^2 / B^2, \qquad (16.11)$$

are controlled by the force balance in accordance with the model

$$\left(D_{1d}/D_{1d,nc}\right)^3 + \left(D_{1d}/D_{1d,fc}\right)^2 = 1 \qquad (16.12)$$

presented recently in [25]. The reader will find in [25] the detailed derivation of Eq. (16.12) and the method for computation of the bubble departure diameter for natural convection only, $D_{1d,nc}$, and for forced convection only, $D_{1d,fc}$, which will not be repeated here. Further the model was compared in [25] with the available data for saturated water for pool and flow boiling at low and elevated pressures and good agreement was found. The most exiting feature of this model is the taking into account of the influence of the mutual bubble interaction on the bubble departure diameter, something which was experimentally observed but never theoretically explained before.

The frequency of bubble departure from the wall is therefore

$$f_{1w} = 1/\left(\Delta\tau_{1d} + \Delta\tau_{1w}\right), \qquad (16.13)$$

It should not be confused with the average frequency of the turbulent pulsation in the wall boundary layer.

16.2.2 Proposed model

The proposed model is strictly valid for pool boiling of saturated liquid. Nevertheless the main ideas used here can be easily extended for nucleate flow boiling of saturated liquid and pool and flow boiling of subcooled liquid as it will be discussed in Section 16.4.

Before jumping apart from the wall, the turbulent eddies stay at the wall during the time $1/f_{1w}^t$ and receives heat from the wall by heat conduction. Therefore, the average heat flux at the wall follows the analytical solution of the *Fourier* equation averaged over the period $\Delta\tau = 1/f_{1w}^t$

$$\dot{q}''_{w2,nb}(\Delta\tau) = \frac{1}{\Delta\tau} \int_0^{\Delta\tau} \dot{q}''_w(\tau) d\tau = 2\left(\frac{\lambda'\rho'c_p'}{\pi\Delta\tau}\right)^{1/2} (T_w - T_2)$$

$$= \left(2/\pi^{1/2}\right)\left(f_{1w}^t a'\right)^{1/2} \rho'c_p'(T_w - T_2) \qquad (16.14)$$

The idea to use the time-averaged heat flux at the wall stems from *Mikic* and *Rohsenow* [34]. In contrast to these authors we use here the turbulence renewal period rather than the bubble departure time. The time- and space-averaged pulsation velocity is

$$\overline{\overline{V_2'}} = (B^2/D_{2,\inf}) \frac{\Delta\tau_d}{\Delta\tau_w + \Delta\tau_d} = \frac{1}{0.84} B^2 n_{1w}^{\prime\prime 1/2} \frac{\Delta\tau_{1d}}{\Delta\tau_{1w} + \Delta\tau_{1d}}. \quad (16.15)$$

The above time- and space-averaged pulsation velocity is in fact the time-averaged micro-convection velocity, first computed by *Forster* and *Zuber* [9] or see also *Zuber* [54], p. 12. The most unexpected effect introduced by the additional time averaging is the capability of the finally obtained equation to predict the critical flux as it will be demonstrated later. The fluctuation frequency is therefore a function of the bubble departure frequency

$$f_{1w}^t = \overline{\overline{V_2'}}/\ell_2' = \frac{1}{0.84} B^2 n_{1w}^{\prime\prime 1/2} \frac{\Delta\tau_{1d}}{\Delta\tau_{1w} + \Delta\tau_{1d}} / \ell_2' \quad (16.16)$$

Substituting Eq. (16.16) in (16.14) and using Eq.(16.7) we obtain

$$\dot{q}_{w2,nb}'' = \frac{c_1 c}{\left[\ell_2'(1+\Delta\tau_{1w}/\Delta\tau_{1d})\right]^{1/2}} \frac{\lambda'\rho'c_p'}{\rho''\Delta h} n_{1w}^{\prime\prime 1/4} (T_w - T_2)^2, \quad (16.17)$$

where $c_1 = 2/(\pi 0.84)^{1/2}$ is of order of unity. For the data comparison the value of

$$c_1 = 1.4626 \quad (16.18)$$

was used. Because $\Delta\tau_{1w}$ is a very rough estimate for the averaged waiting time and therefore the ratio $\Delta\tau_{1w}/\Delta\tau_{1d}$ is only approximately analytically known we introduce the empirical constant $c_2\Delta\tau_{1w}/\Delta\tau_{1d}$

$$\dot{q}_{w2,nb}'' = \frac{c_1 c}{\left[\ell_2'(1+c_2\Delta\tau_{1w}/\Delta\tau_{1d})\right]^{1/2}} \frac{\lambda'\rho'c_p'}{\rho''\Delta h} n_{1w}^{\prime\prime 1/4} (T_w - T_2)^2. \quad (16.19)$$

The effect of c_2 will be demonstrated by data comparison. As will be shown

$$c_2 = 0.3 \quad (16.20)$$

is appropriate. Finally, replacing Eq. (16.1) in the form

$$n''_{1w} = 5\times 10^{-27}(1-\cos\varphi)/D_{1c}^6 = 5\times 10^{-27}(1-\cos\varphi)\left[\frac{\rho''\Delta h}{4\sigma T'(p)}(T_w-T_2)\right]^6$$

(16.21)

we obtain

$$\dot{q}''_{w2,nb} = \frac{c_1 c\left[5\times 10^{-27}(1-\cos\varphi)\right]^{1/4}}{\left[\ell'_2(1+c_2\Delta\tau_{1w}/\Delta\tau_{1d})\right]^{1/2}}\left[\frac{\rho''\Delta h}{4\sigma T'(p)}\right]^{3/2}\frac{\lambda'\rho'c'_p}{\rho'\Delta h}(T_w-T_2)^{7/2}.$$

(16.22)

The mass of steam generated per unit time and unit flow volume is therefore

$$\mu_{21,gen} = \frac{4}{D_h}\dot{q}''_{w2,b}/(h''-h_2).$$

(16.23)

The corresponding number of bubbles generated with bubble departure diameter per unit flow volume is

$$\dot{n}_{1w} = \mu_{21}/\left(\rho_1\frac{\pi}{6}D_{1d}^3\right).$$

(16.24)

From that moment the bubbles start their own history inside the bulk flow and undergo condensation or further evaporation as well as collisions, coalescence or fragmentation.

16.3 Data comparison

Comparison of the prediction of Eq. (16.22) with all data available to the author for atmospheric pool boiling of water at a horizontal surface is shown on Figs. 16.5, 16.6. We see from Figs. 16.5 and 16.6 that the data in the region of pool boiling are well reproduced by Eq. (16.22) including the important effect of the static contact angle. In order to demonstrate the effect of the additional time averaging we set

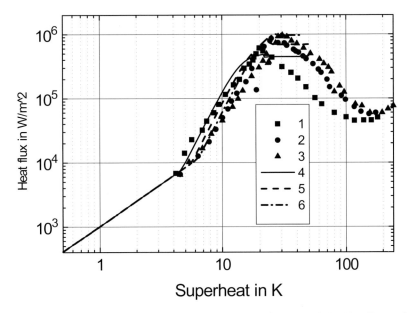

Fig. 16.5. Heat flux as a function of superheat. Saturated water at 0.1 *MPa*. Comparison of the prediction of the new theory with the experimental data by Wang and Dhir for three different static contact angles: 1) Exp.- 90 *deg*; 2) Exp.-35 *deg*; and 3) Exp.-18 *deg*, 4) Kolev-90 *deg*; 5) Kolev-35 *deg*; and 6) Kolev-18 *deg*. The larger the static contact angle the smaller the critical heat flux.

$$\Delta \tau_{1d} / (\Delta \tau_{1d} + \Delta \tau_{1w}) \approx 1 \tag{16.25}$$

which is equivalent to $c_2 = 0$,

$$\dot{q}''_{w2,nb} = c_1 c \left[5 \times 10^{-27} (1 - \cos \varphi) \right]^{1/4} \ell_2^{t-1/2} \left[\frac{\rho'' \Delta h}{4 \sigma T'(p)} \right]^{3/2} \frac{\lambda' \rho' c'_p}{\rho' \Delta h} (T_w - T_2)^{7/2} . \tag{16.26}$$

and compare only with the original data set reported by *Wang* and *Dhir*. The result is shown in Fig. 16.7.

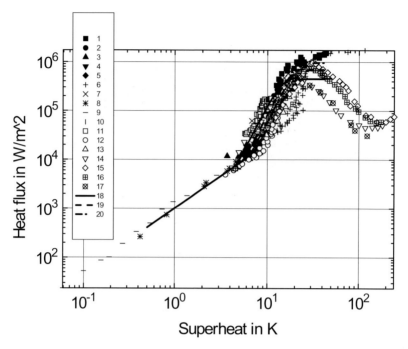

Fig. 16.6. Experimental data for heat flux as a function of superheat for saturated water at 0.1 *MPa* - 1) to 17) as in Fig. 16.1. Prediction of the new theory for three different static contact angles: 18) Kolev-90 *deg*; 19) Kolev-35 *deg*; and 20) Kolev-18 *deg*.

From the comparison of Fig. 16.7 with Fig. 16.5 we see also that the ratio $\Delta \tau_{1w} / \Delta \tau_{1d}$ introduced by the time average is getting important in the region of high superheatings where the predicted heat flux no longer increases with increasing superheating. Surprisingly, the obtained equation seems to predict critical heat flux without additional correlation. The mechanism which works in this case is the dramatic increase of the nucleation site density for high temperature differences which causes a dramatic decrease of the bubble departure diameter and therefore a dramatic decrease of the bubble departure time. As a result, we reach the critical heat flux, a critical heat flux that does not increase any more with increasing temperature difference. The surprising effect here is that we predict accurately the nucleate boiling heat transfer and the critical heat flux with a *single theory*. The theory predicts also the observed decreasing critical heat flux with increasing contact angle that is with decreasing wetability. This effect is observed also by *Ramilison*,

$$\dot{q}''_{CHF} / \dot{q}''_{CHF,Zuber} = 0.0336(\pi - \theta)^3 k^{0.125} \tag{16.27}$$

where k is the RMS surface roughness.

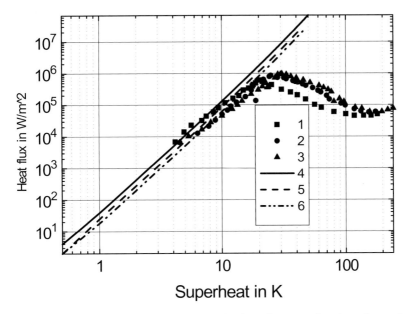

Fig. 16.7. Wang and Dhir experimental data for heat flux as a function of superheat for saturated water at 0.1 *MPa* and for three different static contact angles: 1) 90 *deg*; 2) 35 *deg*; and 3) 18 *deg*. 4), 5), 6) Prediction of the new theory for the corresponding static contact angles without taking into account the time averaging that is with $c_2 = 0$.

Note the difference to the so called "hydrodynamic theory of critical heat flux" proposed first by *Kutateladse* [28, 29] and later by *Zuber* [52], *Zuber* and *Tribus* [53] and *Zuber* et al. [56]. A brief summary of this theory is given in Section 20.4.1. The hydrodynamic theory of critical heat flux is intended to predict only the critical heat flux up to which the nucleate boiling regime is possible but it was not aimed to predict heat transfer in nucleate boiling itself. Note the two completely different approaches to the boiling crises by comparing this theory with the theory presented in Section 20.4.1. We showed here that the micro-scale of the nucleation processes, bubble growth and departure, and the turbulence generated by this process naturally lead to self saturation of the boundary layer with bubbles and dramatically changes of the phenomenon. In contrast, *Kutateldse* and *Zuber* analyzed the macro-scale of the instability above the layer where the microphysics happens. That is the reason why their analysis gives an absolute upper limit which is about 5 times larger then the observed critical heat flux. This is reflected in the necessity of introducing an empirical constant of about 1/5 – see Section 20.4.1.

16.4 Systematic inspection of all the used hypotheses

1) The used hypotheses for the development of the bubble departure theory presented in Chapter 15 are checked against data for saturated water pool boiling for

   a) wall superheating up to 40 K,
   b) mass flow rates up to 360 $kg/(m^2s)$ and
   c) pressure up to 24 bar,

and good agreement was obtained. The trends of the dependence of the bubble departure diameter on superheating, mass flow rate and pressure are also well predicted. Thus, I could expect that the theory will hold also outside this region.

2) There are only two constants in Eq. (16.17):

   a) The theoretical value of $c_1 = 1.23$ is corrected with 19 % to 1.4626 after comparison with experimental data, which is not unusual in the heat transfer theory at all.

   b) The c_2 constant is empirically introduced to account for uncertainties in the prediction of the ratio $\Delta\tau_{1w}/\Delta\tau_{1d}$. It was demonstrated that for superheatings giving heat flux up ≈ 70 % from the critical heat flux this correction is unimportant. It becomes important if one uses the new theory to improve heat flux predictions for superheatings giving heat fluxes ≈ 70 % from the critical heat flux.

3) The new heat transfer theory is strictly valid for pool boiling of saturated liquid. The bubble departure theory and therefore the theory for prediction of the fluctuation velocity in the boundary layer are correct also for flow boiling at saturated liquid. The ideas used in this Chapter to reach Eq. (16.17) can be also used to extend its validity to flow boiling. The main questions which should be addressed in this case is whether the characteristic length scale of turbulence in the wall boundary layer will very much change for flow boiling with respect to $\pi\lambda_{RT}$. If no, the same theory should be used. If yes, additional investigation is *necessary* in which the flow induced fluctuations in the wall boundary layer should be superimposed to the bubble departure generated fluctuations. This remains to be done.

4) The extension of the theory for subcooled nucleate boiling requires modification of the bubble departure model published in [25]. The road is: replace the bubble growth model for saturated water with a bubble growth model for subcooled water. This remains to be done.

16.5 Significance

1. The economical design of heat exchangers working with pool boiling depends on the accuracy of the prediction models. I believe that the proposed method will find application in this field.

2. The appropriate quantitative separation of the effects of single sub-mechanisms into a single correlation for the prediction of pool boiling has never succeeded before in spite of the fact that it was frequently tried. The prediction of the heat transfer in pool boiling together with the prediction of the critical heat flux for water boiling at atmospheric pressure seems to be indicative that Eq. (16.22) quantitatively predicts the appropriate separate effects.

16.6 Conclusions

1. The turbulence induced by the bubble growth and departure is the main heat transfer mechanism during pool boiling.

2. The static contact angle is a very important parameter explaining the large spread of the experimental data for pool boiling.

3. The new theory consists of sound verified sub theories and models

- the bubble departure diameter, confirmed by data comparison Chapter 15,
- the active nucleation site density provided by *Wang* and *Dhir* [48],
- the mean site to site distance provided by *Wang* and *Dhir* [48],

and the hypothesis of bubble growth and departure introduced turbulence in the boundary layer as the main heat transfer mechanism. These are new features that do not follow from the preceding works [46, 39, 10, 45, 51, 34, 6, 54, 55, 24, 57, 47, 36, 4, 9, 52, 5, 19, 27, 43].

4. Equation (16.22) predicts the existing experimental data including the effect of the static contact angle with sufficient accuracy. It is recommended for general use. Surprisingly, Eq. (16.22) also seems to predict the critical heat flux, which was demonstrated for water at atmospheric pressure.

5. Increasing static contact angle leads to decreasing critical heat flux.

6. From the simplified Eq. (16.26) one should easily estimate that if the measurement error for wall superheat is of order of 5 to 6 % the error of the comparison with the experimental data with known static contact angle will be about 21 %. Heat transfer data with unknown static contact angle are not appropriate for validation of any nucleate boiling heat transfer theory.

From the above conclusions we can derive some practical recommendations for the application of the new theory.

1. Equating the predictions of Eq. (16.4) and (16.22) computes the wall temperature $T_{w,NBI}$ necessary to initiate nucleate boiling

$$T_{w,NBI} = T_2 + \left\{ \frac{2.417 \lambda_2 \left(Pr_2 g \beta_2 / v_2^2 \right)^{1/4}}{c_1 c \left[5 \times 10^{-27} (1 - \cos \varphi) \right]^{1/4} \ell_2^{t-1/2} \left[\frac{\rho'' \Delta h}{4 \sigma T'(p)} \right]^{3/2} \frac{\lambda' \rho' c_p'}{\rho'' \Delta h}} \right\}^{8/35}$$

(16.22)

2. If the actual wall temperature is less that $T_{w,NBI}$ we have natural circulation heat transfer. In this case Eq. (16.4) describes the heat transfer number.

3. If the wall temperature is larger than $T_{w,NBI}$ we have two possibilities to compute the heat flux:

a) Use existing correlation for prediction of the critical heat flux, predict the heat flux by the simplified theory, Eq. (16.26), if the predicted heat flux is below the critical one the prediction of Eq. (16.26) is the desired one. Note that in this case the error is increasing in the region between 70 and 100 % of the critical heat flux.

b) Use the complete theory for prediction of the heat flux, that is Eqs. (16.19) and (16.21) together with the complete algorithm described in Chapter 15 for prediction of the bubble departure diameter.

4) In accordance with the measurements by *Arefeva* and *Aladev* [1] the influence of the contact angle on the heat flux is observed in the region between $\varphi = 0$ and $\pi/2$. Beyond this contact angle nucleate boiling heat transfer does not depend on the contact angle. That is why we recommend to use the present theory in the region $\varphi = 0$ and $\pi/2$. For larger contact angles we recommend to use $\varphi = \pi/2$ until data for the nucleation site density as a function of superheating are available for static contact angles larger $\pi/2$.

Nomenclature

a_2 liquid thermal diffusivity, m^2/s
a' saturation liquid thermal diffusivity, m^2/s

Nomenclature

B^2	$= 2R_1 dR_1/d\tau$, m^2/s
c_{p2}	specific heat of liquid, $J/(kgK)$
D_1	bubble diameter, m
D_{1c}	critical size, m
D_{1d}	bubble departure diameter, m
$D_{1d,fc}$	bubble departure diameter for strongly predominant forced convection, m
$D_{1d,nc}$	bubble departure diameter for natural circulation, m
$D_{2,inf}$	average center to center spacing, m
f_{1w}	bubble departure frequency, $1/s$
f_{1w}^t	boundary layer turbulence fluctuation frequency, $1/s$
G	mass flow rate, $kg/(m^2 s)$
g	gravitational acceleration, m/s^2
h', h''	specific saturation enthalpies for water and steam, $J/(kgK)$
Δh	$= h'' - h'$, J/kg
Ja	$= \rho_2 c_{p2}[T_2 - T'(p)]/(\rho_1 \Delta h)$, Jacob number, dimensionless
Ja'	$= \rho' c_p'[T_2 - T'(p)]/(\rho'' \Delta h)$, Jacob number, dimensionless
n_{1w}''	Active nucleation site density, $1/m^2$
p	pressure, Pa
p'	saturated pressure, Pa
Pr_2	$= \eta_2/(\rho_2 a_2)$, Prandtl number, dimensionless
$\dot{q}_{w2,nc}''$	heat flux from the wall into the liquid during natural convection without boiling, W/m^2
$\dot{q}_{w2,nb}''$	heat flux from the wall into the liquid during pool boiling, W/m^2
R_1	bubble radius, m
$R_{2,inf}$	$= D_{2,inf}/2$, m
$T'(p)$	saturation temperature at system pressure p, K
T_2	liquid temperature, K
T_w	wall temperature, K
$\overline{V_2'}$	the volume-averaged fluctuation velocity, m/s
$\overline{\overline{V_2'}}$	the volume-averaged fluctuation velocity time averaged over $\Delta\tau_w + \Delta\tau_d$, m/s

Greek

β_2	$= -\dfrac{1}{\rho_2}\left(\dfrac{\partial \rho_2}{\partial T_2}\right)_p$, thermal expansion coefficient, $1/K$
$\delta_{2,min}$	minimum of the thermal boundary layer thickness, m
$\Delta\rho_{21}$	$= \rho_2 - \rho_1$, kg/m^3

$\Delta \tau_d$ time needed from the origination of a bubble with critical size to the bubble departure from the wall, s

$\Delta \tau_w$ delay time, s

η_2 dynamic viscosity of liquid, $kg/(ms)$

λ_2 thermal conductivity of liquid, $W/(mK)$

$\lambda_{RT} = \left(\dfrac{\sigma}{g\Delta\rho_{21}} \right)^{1/2}$, *Rayleigh - Taylor* instability wavelength, m

v_2 cinematic viscosity of liquid, m^2/s

ρ_1 gas density, kg/m^3

ρ_2 liquid density, kg/m^3

ρ'' saturated steam density, kg/m^3

ρ' saturated liquid density, kg/m^3

σ surface tension, N/m

τ time, s

θ_0 angle between the bubble axis and the wall, *rad*

φ static contact angle between "liquid drop" and the wall, *rad*

References

1. Arefeva EI and Aladev IT (July 1958) O wlijanii smatchivaemosti na teploobmen pri kipenii, Injenerno - Fizitcheskij Jurnal, in Russian, vol 1 no 7 pp 11-17
2. Avdeev AA, Maidanik VN, Selesnev LI and Shanin VK (1977) Calculation of the critical flow rate with saturated and subcooled water flushing through a cylindrical duct, Teploenergetika, vol 24 no 4 pp 36-38
3. Borishanskii, V, Bobrovich G, Minchenko F (1961) Heat transfer from a tube to water and to ethanol in nucleate pool boiling, Symposium of Heat Transfer and Hydraulics in Two-Phase Media, Kuteladze SS (ed) Gosenergoizdat, Moscow, pp 75-93
4. Brauer H (1971) Stoffaustausch, Verlag Sauerländer
5. Chen JC (July 1966) Correlation for boiling heat transfer to saturated fluids in convective flow, Ind. & Eng. Chem. Progress Design and Development, vol 5 no 3 pp 322- 329
6. Cole R, Rohsenow WM (1969) Correlation of bubble departure diameters for boiling of saturated liquids, Chem. Eng. Prog. Symp. Ser., no 92 vol 65 pp 211-213
7. Cornwell K, Brown RD (1978) Boiling surface topology, Proc. 6th Int. Heat Transfer Conf. Heat Transfer 1978 - Toronto, vol 1 pp 157-161
8. Faggiani S, Galbiati P and Grassi W (1981) Active site density, bubble frequency and departure on chemically etched surfaces, La Termotechnica, vol 29 no 10 pp 511-519
9. Forster HK, Zuber N (1955) Dynamics of vapor bubbles and boiling heat transfer, AIChE J., vol 1 no 4 pp 531-535
10. Fritz W (1935) Berechnung des maximalen Volumens von Dampfblasen, Phys. Z., vol 36 no 11 pp 379-384

11. Gaertner RF, Westwater JW (1960) Population of active sites in nucleate boiling heat transfer, Chem. Eng. Progr. Symp. Ser., no 30 vol 30 pp 39-48
12. Gaertner RF (1963) Distribution of active sites in the nucleate boiling of liquids, Chem. Eng. Prog. Symp. Series, no 41 vol 59 pp 52-61
13. Gaertner RF (Feb. 1965) Photographic study of nucleate pool boiling on a horizontal surface, Transaction of the ASME, Journal of Heat Transfer, vol 87 pp 17 - 29
14. Han CY, Griffith P (1965) The mechanism of heat transfer in nucleate pool boiling, Part I, Bubble initiation, growth and departure, Int. J. Heat Mass Transfer, vol 8 pp 887-904
15. Hutcherson MN, Henry RE and Wollersheim DE (Nov. 1983) Two-phase vessel blowdown of an initially saturated liquid - Part 2: Analytical, Trans. ASME, J. Heat Transfer, vol 105 pp 694-699
16. Jakob M and Fritz W (1931) Versuche ueber den Verdampfungsvorgang, Forsch. Ing. - Wes., vol 2 p 435
17. Jakob M (1932) Kondensation und Verdampfung, Zeitschrift des Vereins Deutscher Ingenieure, vol 76 no 48 pp 1161-1170
18. Jakob M and Linke W (1933) Der Wärmeübergang von einer waagerechten Platte an siedendes Wasser, Forsch. Ing. Wes., vol 4 pp 75-81
19. Jakob M (1949) Heat transfer, Wiley, New York, vol 1 ch 29
20. Jens WH and Lottes PA (1951) Analysis of heat transfer, burnout, pressure data and density data for high pressure water. USAEC Rep. ANL-4627
21. Johov KA (1969) Nucleations number during steam production, Aerodynamics and Heat Transfer in the Working Elements of the Power Facilities, Proc. CKTI, Leningrad, in Russian, vol 91 pp 131- 135
22. Jones OC (1992) Nonequilibrium phase change --1. Flashing inception, critical flow, and void development in ducts, in boiling heat transfer, Lahey RT Jr (ed) Elsevier Science Publishers B.V., pp 189 - 234
23. Jones OC (1992) Nonequilibrium phase change --2. Relaxation models, general applications, and post heat transfer, Lahey RT Jr (ed) Boiling Heat Transfer, Elsevier Science Publishers B.V., pp 447-482
24. Judd RL, Hwang KS (Nov. 1976) A comprehensive model for nucleate pool boiling heat transfer including microlayer evaporation, Transaction of the ASME, Journal of Heat Transfer, vol 98 pp 623-629
25. Kolev NI (1994) The influence of mutual bubble interaction on the bubble departure diameter, Experimental Thermal and Fluid Science, vol 8 pp 167-174
26. Kolev NI (1995) How accurate can we predict nucleate boiling, Experimental Thermal and Fluid Science, Experimental Thermal and Fluid Science, vol 10 pp 370-378
27. Kocamustafaogullari G and Ishii M (1983) Interfacial area and nucleation site density in boiling systems, Int. J. Heat Mass Transfer, vol 26 pp 1377-1389
28. Kutateladse SS (1954) A hydrodynamic theory of changes in the boiling process under free convection conditions, Izv. Akad. Nauk SSSR, Otd. Tech. Nauk, vol 4 pp 529 - 536, 1951; AEC-tr-1991
29. Kutateladse SS (1962) Basics on heat transfer theory, in Russian, Moscow, Mashgis, p 456
30. Kurihara HM, Myers JE (March 1960) The effect of superheat and surface roughness on boiling coefficients, AIChE Journal, vol 6 no 1 pp 83-91
31. Labuntsov DA (1974) State of the art of the nucleate boiling mechanism of liquids, Heat Transfer and Physical Hydrodynamics, Moskva, Nauka, in Russian, pp 98-115

32. Labuntsov DA (1963) Approximate theory of heat transfer by developed nucleate boiling (Russ.), Izvestiya AN SSSR, Energetika i transport no 1
33. Miheev MA and Miheeva IM (1973) Basics of heat transfer, Moskva, Energija, in Russian, p 320
34. Mikic BB, Rohsenow WM (May 1969) A new correlation of pool-boiling data including the effect of heating surface characteristics, Transactions of the ASME, J. Heat Transfer, vol 91 pp 245-250
35. Nishikawa K, Fujita Y, Uchida S, Ohta H (1984) Effect of surface configuration on nucleate boiling heat transfer, Int. J. Heat and Mass Transfer, vol 27 no 9 pp 1559-1571
36. Pohlhausen K (1921) Zur nährungsweisen Integration der Differentialgleichung der laminaren Grenzschicht, Z. angew. Math. Mech., vol 1 pp 252-268
37. Rallis CJ, Jawurek HH (1964) Latent heat transport in saturated nucleate boiling, Int. J. Heat Transfer, vol 7 pp 1051-1068
38. Riznic J and Ishii M (1989) Bubble number density and vapor generation in flushing flow, Int. J. Heat Mass Transfer, vol 32 pp 1821-1833
39. Rohsenow WM (1952) A method of correlating heat transfer data for surface boiling of liquids, Trans. ASME, vol 74 pp 969-975
40. Siegel R and Keshock EG (July 1964) Effects of reduced gravity on nucleate boiling bubble dynamics in saturated water, AIChE Journal, vol 10 no 4 pp 509-517
41. Sultan M, Judd RL (Feb. 1978) Spatial distribution of active sites and bubble flux density, Transactions of the ASME, Journal of Heat Transfer, vol 100, pp 56-62
42. Thom IRS et. al. (1966) Boiling in subcooled water during up heated tubes or annuli. Proc. Instr. Mech. Engs., vol 180 3C pp 1965-1966
43. Tolubinsky VI, Ostrovsky JN (1966) On the mechanism of boiling heat transfer (vapor bubbles growth rate in the process of boiling in liquids, solutions, and binary mixtures), Int. J. Heat Mass Transfer, vol 9 pp 1463-1470
44. Vachon RI, Tanger GE, Davis DL, Nix GH (May 1968) Pool boiling on polished chemically etched stainless-steel surfaces, Transactions of ASME, Journal of Heat Transfer, vol 90 pp 231-238
45. Vachon RI, Nix GH, Tanger GH (May 1968) Evalution of the constants for the Rohsenow pool - boiling correlation, Transactions of the ASME Journal of Heat Transfer, vol 90 pp 239-247
46. van Stralen S, Cole R (1979) Boiling Phenomena, Hemisphere, USA
47. von Karman T (1921) Über laminare und turbulente Reibung, Z. angew. Math. Mech., vol 1 pp 233-252
48. Wang CH, Dhir VK (Aug. 1993) Effect of surface wettability on active nucleation site density during pool boiling of water on a vertical surface, ASME Journal of Heat Transfer, vol 115 pp 659-669
49. Wiebe JR (1970) Temperature profiles in subcooled nucleate boiling, M. Eng. thesis, Mechanical Engineering Department, McMaster University, Canada
50. Yamagata K, Hirano F, Nishikawa K, Matsuoka H (1955) Nucleate boiling of water on the horizontal heating Surface, Mem. Fac. Engng; Kyushu Univ, vol 15 p 97
51. Yang JK, Weisman J (1991) A phenomenological model of subcooled flow boiling in the detached bubble region, Int. J. Multiphase Flow, vol 17 no 1 pp 77 94
52. Zuber N (April 1958) On the stability of boiling heat transfer, Transactions of the ASME, vol 80 pp 711-720
53. Zuber N and M Tribus (1958) Further remarks on the stability of boiling heat transfer, report 58-5, Department of Engineering, University of California, Los Angeles

54. Zuber N (1959) Hydrodynamic aspect of boiling heat transfer, U.S. Atomic Energy Commission Rept., AECU - 4439, Tech. Inf. Serv. Oak Ridge, Tenn
55. Zuber N (Jan. 1960) Hydrodynamic aspect of nucleate pool boiling, Part I - The region of isolated bubbles, Research Laboratory Ramo - Wooldridge, RW-RL-164, 27
56. Zuber N, Tribus M and Westwater JW (1961) The hydrodynamic crisis in pool boiling of saturated and subcooled liquids, International Developments in Heat Transfer, Proc. Int. Heat Transfer Conf., Boulder, Colorado, Part 2, no 27 pp 230-236
57. Zuber N (1963) Nucleate boiling, The region of isolated bubbles and the similarity with natural convection, Int. J. Heat Mass Transfer, vol 6 pp 53-78
58. Borishanskii V, Kozyrev A and Svetlova L (1964) Heat transfer in the boiling water in a wide range of saturation pressure, High Temperature, vol 2 no 1 pp 119-121

Appendix 16.1 State of the art of nucleate pool boiling modeling

Detailed information on this topic can be found in [46] (1979). We confine our attention to some basic ideas on which some of the frequently used correlations in the literature rely.

Rohsenow [39] (1952) computed the heat flux from the wall during nucleate boiling of saturated liquid as $\dot{q}''_{w2} = G_{1d}[h'' - h'(T_w)]$, where G_{1d} is the mass flow rate of departing bubbles from the surface. $G_{1d} = \dot{q}''_{w2}/[h'' - h'(T_w)]$ is used in the *Reynolds* number defined as $Re_{1d} = G_{1d}D_{1d}/\eta_2 = D_{1d}\dot{q}''_{w2}/\{[h'' - h'(T_w)]\eta_2\}$ for postulating the heat transfer dependence $St_{1d} = Nu_{1d}/(Pr_2 Re_{1d}) \approx f(Re_{1d}, Pr_2)$, where the *Nusselt* number is defined as $Nu_{1d} = h_w D_{1d}/\lambda_2$. *Rohsenow* uses the *Fritz* equation [10] (1935) $D_{1d} = 1.2\,\varphi\lambda_{RT}$ for the estimation of the bubble departure size. *Rohsenow* finally obtained

$$1/St_{1d} = c_{p2}[T_w - T'(p)]/\Delta h \approx const\left[\dot{q}''_w \lambda_{RT}/(\eta_2 \Delta h)\right]^{1/3} Pr_2^m$$

where the constant and the exponent are fitted against experimental data for water: $m = 1$, $const = 0.006$ to 0.013, for other liquids: $m = 1.7$, $const = 0.0025$ to 0.015. Additional values for $const$ and m are reported by *Vachon* et al. [44, 45] (1968) for different fluid - surface combinations and different surface preparations.

The definitions by *Rohsenow* are used later by *Kurihara* and *Myers* [30] (1960) in the form $h_{w2}D_{1d}/\lambda_2 = const(G_{1d}D_{1d}/\eta_2)^m Pr_2^n$, where for saturated pool boiling the authors assumed that all produced steam is removed from the wall boundary layer, i.e. $G_{1d} = (\pi/6)D_{1d}\rho_1 f_{1w} n''_{1w}$. The authors assumed further $D_{1d}f_{1w} \approx const$. The comparison with experimental data for a constant pressure gives $m \approx 1/3$, which cancels the effect of the bubble departure diameter. Finally the authors obtained

$$h_{w2} = 820 \lambda_2 \left(\rho_1 / \eta_2 \right)^{1/3} n_{1w}''^{1/3} \text{Pr}_2^{-0.89}$$

Note that $D_{1d}f_{1w}$ is not a constant as assumed by *Kurihara* and *Myers* but a function of the pressure.

Mikic and *Rohsenow* [34] (1969) computed the temperature as a function of time and distance y from the wall

$$T = T_2 + (T_w - T_2) \text{erfc}\left\{ y / \left[2(a_2 \tau)^{1/2} \right] \right\}$$

and the heat flux at the wall in the liquid

$$\dot{q}_w''(\tau) = -\lambda_2 \left. \frac{dT}{dy} \right|_{y=0} = \lambda_2 (T_w - T_2) / \left[(\pi a_2 \tau)^{1/2} \right]$$

Thereafter they performed time averaging of the heat flux during the time interval $\Delta \tau$

$$\dot{q}_w''(\Delta \tau) = \frac{1}{\Delta \tau} \int_0^{\Delta \tau} \dot{q}_w''(\tau) d\tau = 2 \left(\frac{\lambda_2 \rho_2 c_{p2}}{\pi \Delta \tau} \right)^{1/2} (T_w - T_2)$$

The averaged heat flux over the total surface is computed assuming that the heat conduction from the wall into the liquid happens only inside a circle around the nucleation site with the diameter as the bubble departure diameter

$$\dot{q}_w'' = \frac{\pi}{4} D_{1d}^2 n_{1w}'' \dot{q}_w''(\Delta \tau)$$

In addition, the authors computed the bubble departure diameter in accordance with the correlation by *Cole* and *Rohsenow* [6] (1969) $D_{1d} = c_2 \lambda_{RT} Ja^{*5/4}$, where $Ja^* = \rho_2 c_{p2} T'(p) / (\rho_1 \Delta h)$, and $c_2 = 1.5 \times 10^{-4}$ for water, and 4.65×10^{-4} for other liquids. The time averaging is performed during the bubble departure period only $\Delta \tau = \Delta \tau_{1d} = 1 / f_{1w}$. Here the relationship obtained by *Zuber* [54, 55] (1959) for free bubble rise in a pool, $f_{1d} D_{1d} \approx 0.6 \left[\sigma g (\rho_2 - \rho_1) / \rho_2^2 \right]^{1/4}$, was used, where f_{1d} is the bubble removal frequency from the wall boundary layer, rather than the bubble departure frequency. The bubble number density was postulated to be $n_{1w}'' = c(D_1^* / D_{1c})^{3.5}$ where the critical bubble size is

$$D_{1c} = 4\sigma T'(p) / \{\rho_1 \Delta h [T_w - T'(p)]\}.$$

The constants $cD_1^{*3.5}$ are estimated by fitting experimental data.

Judd and *Hwang* [24] (1976) extended this model by postulating that the area of influence of a single bubble is $C \approx 1.7$ times the projected bubble area at departure and the heat transfer outside the area of influence is due to natural convection. The final expression by *Judd* and *Hwang* is

$$\dot{q}''_{w2} = C \frac{\pi}{4} D_{1d}^2 n''_{1w} \dot{q}''_{w2}(\Delta \tau) + (1 - C \frac{\pi}{4} D_{1w}^2 n''_{1w}) \dot{q}''_{w2,nc},$$

or

$$\dot{q}''_{w2} - \dot{q}''_{w2,nc} = C \frac{\pi}{4} D_{1d}^2 n''_{1w} \left[\dot{q}''_{w2}(\Delta \tau) - \dot{q}''_{w2,nc} \right],$$

where the natural convection heat flux is

$$\dot{q}''_{w2,nc} = 0.18 \lambda_2 \left[Pr_2 g \beta_2 / v_2^2 \right]^{1/3} (T_w - T_2)^{4/3},$$

and $\beta_2 = -\frac{1}{\rho_2} \left(\frac{\partial \rho_2}{\partial T_2} \right)$ is the volumetric thermal expansion coefficient. The data of of several investigators can be approximated using 0.272 instead 0.18 as proposed by *Zuber* [57] (1962). Note that *Jacob* used earlier the equation

$$\dot{q}''_{w2,nc} \delta_{2,th} / \left[\lambda_2 (T_w - T_2) \right] = const \left[(T_w - T_2) \delta_{2,th}^3 g \beta_2 / v_2^2 Pr_2 \right]^{1/4},$$

where the thermal boundary layer thickness was assumed to be

$$\delta_{2,th} \approx const / \Delta T^4,$$

and

$$\dot{q}''_{w2,nc} = 2.417 \lambda_2 \left[Pr_2 g \beta_2 / v_2^2 \right]^{1/4} (T_w - T_2)^{21/16}.$$

Bubble growth and departure causes liquid fluctuation parallel to the wall, the so called micro-convection. This leads some authors to the idea to modify the well-known *Karman - Pohlhausen* equation [47, 36] (1921)

$$Nu = h_{w2}\ell/\lambda_2 = 0.664\,\mathrm{Re}^{1/2}\,\mathrm{Pr}_2^{1/3}$$

for computing the averaged heat transfer at the wall for viscous flow. Note that the heat flux in the above equation is averaged along the distance length ℓ, and that ℓ is used as a scale in the *Nusselt* and *Reynolds* numbers. The velocity used in the Reynolds number is the bulk velocity of the flow parallel to the horizontal plate. The corresponding equation for turbulent flow is

$$Nu = h_{w2}\ell/\lambda_2 = 0.037\,\mathrm{Re}^{4/5}\,\mathrm{Pr}_2^{1/3}$$

see [4] (1971), p.182, Eq. 3.38.

Forster and *Zuber* [9] (1955) applied the *Pohlhausen* equation using as averaging length scale the bubble departure diameter D_{1d}

$$h_{w2,mic}\,D_{1d}/\lambda_2 = const\,\mathrm{Re}^m\,\mathrm{Pr}_2^n$$

The authors averaged the bubble growth velocity along an averaged distance between two neighboring bubbles, $D_{2,inf}$, for thermally controlled bubble growth and obtained

$$\bar{V}' = 2R_1\frac{dR_1}{d\tau} = B^2/D_{2,inf}.$$

where $B = \sqrt{\pi a_2}\,Ja_2$. This velocity is called micro convection velocity. The *Reynolds* number was computed to

$$Re = \bar{V}'D_{1d}/v_2 = \left(B^2/v_2\right)\left(D_{1d}/D_{2,inf}\right)$$

and substituted into the *von Karman - Pohlhausen* equation with

$$Re = B^2/v_2$$

setting $D_{1d}/D_{2,inf} \approx 1$. The ratio $D_{1d}/D_{2,inf} < 1$ was taken into account by the constant

const = 0.003

for the polar fluids *n*-pentane, benzene, ethanol and for water. The exponents are found to be $m = 0.62$, $n = 1/3$. For the bubble departure diameter the authors used the thermally controlled bubble growth mechanism

$$D_{d1} \approx 2B\Delta\tau_{1d}^{1/2},$$

where the order of magnitude of the departure time

$$\Delta\tau_{1d} \approx \{\sigma/[p'(T_{2i})-p]\}\{\rho_2[p'(T_{2i})-p]\}^{1/2}$$

is estimated from the mechanical energy equation setting the bubble growth work equal to the kinetic energy of the bubble environment

$$(dR_1/d\tau)^2 \approx (R_{1d}/\Delta\tau_{1d})^2 = (2/3)(1-\rho''/\rho_2)\Delta\rho/\rho_2.$$

The authors themselves considered the use of this time constant as "not the only one that may be significant". The verification of this equation is made with data for superheat and heat flux at the point of burn out for different pressure. The exponents were questioned later by *Katz*, see the *Katz* comments to *Zuber*'s work in [52] (1958), and new exponents are proposed in order to obtain agreement with experimental data for ethanol/chromium with *const* = 300, benzene with *const* = 1, and *n*-pentane with *const* = 0.3, where $m = 1.05$ and $n = -7$ was used for all three cases. Nevertheless, in a later work *Chen* [5] (1966) uses the *Forster* and *Zuber* equation with $m = 1/2$, $n = 1/3$ to successfully describe by the part of heat transfer due to nucleate boiling in forced convection.

In a later work *Zuber* [55] (1960) comes back to the *von Karman - Pohlhausen* equation using again D_{1d} as a length scale but replacing the average distance between the neighboring bubbles by assuming a quadratic arrangement $1/D_{2,\inf} = n_{1w}^{"1/2}$. Here, $n_{1w}^{"}$ is the nucleation site density in sites per square meter. Thus *Zuber* finally writes $h_{w,mic}D_{1d}/\lambda_2 = const\, Re^{1/2}Pr_2^{1/3}$ where $Re = B^2D_{1d}/(v_2D_{2,\inf}) = B^2D_{1d}n^{"1/2}_{1w}/v_2$. This time, the $B^2 = \dfrac{4}{\pi}a_2 Ja_2^2$ function is computed in accordance with *Bosnjakovic* [19] (1930). The intention of *Zuber* in [54] was not to derive a correlation but to show the dependence $h_{w,mic} \approx n_{1w}^{"1/4}$ which was experimentally observed. Nevertheless, *Zuber* computed the heat flux at the wall during nucleate boiling as

$$\dot{q}_w'' = \dot{q}_{w,mic}''\left(1-\frac{\pi}{4}D_{1d}^2 n_{1w}^{1/2}\right) + \dot{n}_{1w}'' f_{1w}\Delta h \frac{\pi}{6}D_{1d}^3\rho_1$$

In more recent work, *Kocamustafaogullari* and *Ishii* [27] (1983) use the idea of micro-convection in the following way

$$h_{w,mic} D_{2,\inf} / \lambda_2 = 13.18 Re^m Pr_2^n \left(D_{1d} / D_{2,\inf} \right)^p ,$$

where $m = 1/4$, $n = -0.14$, $p = -1/4$. The distance between the neighboring bubbles is used as a characteristic which is consistent with the length along which the fluctuation velocity was averaged. It gives a simple expression for the Reynolds number $Re = \overline{V_2'} D_{2,\inf} / v_2 = B^2 / v_2$. The authors used again the Bosnjakovic model for thermally controlled bubble growth at a wall $B^2 = \dfrac{4}{\pi} a_2 Ja_2^2$ and quadratic nucleation arrangement $1/D_{2,\inf} = n_{1w}^{"\ 1/2}$. Substituting into the starting equation and rearranging the authors came to

$$h_{w,mic} = 14 Ja^{1/2} Pr_2^{-0.39} \dfrac{\lambda_2}{D_{1d}^{1/4}} n_{1w}^{"\ 3/8}$$

The constant and the powers are determined by comparison with experimental data where the bubble departure diameter was computed using a modified form of the *Fritz* equation.

Tolubinsky and *Ostrogradsky* [43] (1966) generalized great number of data for water and other liquids by the correlation

$$h_w \lambda_{RT} / \lambda_2 = 75 \left[\dfrac{\dot{q}_{w1}'' / (\Delta h \rho_1)}{D_{1d} f_{1w}} \right]^{0.7} Pr_2^{-0.2}$$

λ_{RT} where the bubble departure velocity $D_{1d} f_{1w} = 0.36 \times 10^{-3} (p_c / p)^{1.4}$. The data are fitted within an error band of $\pm 25\%$.

Some conclusions seem to be in order after knowing the above discussed approaches to the modeling of nucleate pool boiling:

a) Obviously, heat conduction and micro-convection simultaneously exist in pool boiling.

b) The appropriate scale in the *von Karman - Pohlhausen* equation for micro-convection is the average distance between two neighboring bubbles. Nevertheless, the micro-convection is much more similar to turbulent heat transfer than to forced convection.

c) The bubble removal frequency from the wall boundary layer is not necessarily equal to the bubble departure frequency.

d) Besides the spatial averaging of the pulsation velocity, a time averaging during the bubble generating period $1/f_{1w}$ is necessary.

e) None of the theories takes into account the influence of the static contact angle on the heat transfer.

f) All models put the unknown phenomena in empirical constants so that none of the theories is able to predict separate phenomena like bubble departure diameter, nucleation site density etc. appropriately.

Appendix 16.2 Some empirical correlations for nucleate boiling

In the engineering practice typical correlations used are those by *Borishanskii* et al, *Jens* and *Lottes*, *Thom*, and *Micheev* and *Micheeva*. In these correlations the heat flux at the wall is proportional to the wall superheating with respect to the saturation temperature to a power n

$$\dot{q}''_{NB} = h_{NB}(p)\left[T_w - T'(p)\right]^n$$

which varies between 2 and 4. The coefficient h_{NB} is found to be mainly a function of the system pressure.

$h_{NB}(p)$	n	References
$h_{NB\_Thom} = 1942 \exp(p/4.35 \times 10^6)$	2	*Thom* [42] (1966)
$h_{NB\_Micheev} = \dfrac{0.07845 p^{0.53}}{(1 - 4.5 \times 10^{-8} p)^3}$	3	*Micheev* and *Micheeva* [33] (1973) $1 \times 10^5 \leq p \leq 200 \times 10^5 \, Pa$
$h_{NB\_Jens-Lottes} = 2.567 \exp\left(p/1.55 \times 10^6\right)$	4	*Jens* and *Lottes* [20] (1951) $p \leq 140 \times 10^5 \, Pa$ $\dot{q}''_{NB} < 1.1 \times 10^6 \, W/m^2$ $G < 10^4 \, kg/(m^2 s)$
$h_{NB\_Borishanskii} = \left(6.28 p^{0.14} + 5.86 p^2\right)^{10/3}$	10/3	*Borishanskii* et al [58] (1964), $\mp 5\%$, steel/water, $9.81 \times 10^4 \leq p \leq 196.2 \times 10^5 \, Pa$

$h_{NB}(p)$	n	References
$h_{NB\_Borishanskii} = \left\{ 190 \dfrac{p_c^{1/3}}{T_c^{5/6} M^{1/6}} \left(\dfrac{p}{p_c}\right)^{0.1} \right\}^3 \Big/ \left[1 + 4.65 \left(\dfrac{p}{p_c}\right)^{1.16} \right]$	3	Borishanskii et al [58] (1964), for different fluids, M - molar weight, c - critical state
$h_{NB\_Arefeva-Aladev} = (2.559\theta - 0.727)^4$	4	Arefeva and Aladev [1] (1958), $0.7 < \theta < 1.08$, $1 bar$
$h_{NB\_Arefeva-Aladev} = 3287.26$	3/2	Arefeva and Aladev [1] (1958), $1.6 < \theta < 2.41$, $1 bar$

17. Heterogeneous nucleation and flashing in adiabatic pipes

A new model of heterogeneous nucleation at walls during flashing and boiling is presented. The model is based on a new method for computation of the bubble departure diameter in flows accounting for mutual bubble interactions and the limitation of the bubble production due to the limited amount of energy supplied into the boundary layer by turbulence induced by bubble growth and departure. The final model together with fragmentation and coalescence models is incorporated into the system code IVA and verified by comparison with data. This Chapter is a short version of the primary published work in [8].

17.1 Introduction

Modern theories of multi-phase flows account for the dynamic generation of particles of one phase inside the other. Bubbles generation at walls during boiling or flashing is a typical example discussed in Chapter 16. The number of the bubbles generated per unit time and unit surface, and the bubble departure diameter, are the most important parameters in this process, which influence the later process of bubble growth and/or collapse into the bulk liquid. There have been some publications on this field in the last decade whose main result is the so called figure-of-merit model. This model produces agreement with data by introducing fitting parameters in the general model without verifying the sub-models. Examples are Refs. [10, 3], where the model used for bubble departure diameter is not appropriate for strong superheating [7]. In this Chapter we extend the validity of the results obtained in Chapters 15 and 16 to adiabatic flashing of superheated liquids in pipes. The final model together with fragmentation and coalescence models is incorporated into the system code IVA4 and verified by comparison with data for flashing flows in converging-diverging nozzles and pipe blow-down of water. IVA4 is computer code describing multi-phase non-equilibrium flows by means of three velocity fields in industrial networks and/or three-dimensional facilities. The system of partial differential equations, 21 PDE's for 3D and 15 PDE's for 1D network, solved by this code contains Eq. (1.62) from Volume 1 which is the local volume-averaged mass conservation equation subsequently time averaged

$$\frac{\partial}{\partial \tau}(\alpha_\ell \rho_\ell \gamma_v) + \nabla \cdot (\alpha_\ell \rho_\ell \gamma \mathbf{V}_\ell) = \mu_\ell \qquad (17.1)$$

and Eq. (1.102) from Volume 1 which is the local volume- and time-averaged conservation equation for the number of the particles per unit flow volume without diffusion terms

$$\frac{\partial}{\partial \tau}(n_\ell \gamma_v) + \nabla \cdot (n_\ell \gamma \mathbf{V}_\ell) = \dot{n}_{\ell,kin} + \dot{n}_{\ell,frag} - \dot{n}_{\ell,coal}. \qquad (17.2)$$

Knowing the local volume- and time-averaged particle number density n_l and the volumetric fraction α_ℓ, the characteristic local volume- and time-averaged particle size D_ℓ is computed from the equation

$$n_\ell \frac{\pi}{6} D_\ell^3 = a_\ell \qquad (17.3)$$

that is strictly valid for spherical bubble. Next we confine our attention only to origination of the steam velocity field, designated by 1, in water, designated by 2.

17.2 Bubbles generated due to nucleation at the wall

Heterogeneous nucleation theory is sometimes recommended to describe nucleation at walls [11]. This theory gives the probability of nucleation for given local conditions [5 and 12], but does not predict these conditions. As far the author knows there is no other attempt to describe the nucleation at walls theoretically as those presented in Chapters 15 and 16. The turbulence in the boundary layer is an important energy exchange mechanism also during flashing of superheated liquids in adiabatic pipes. This follows from the fact that no more energy for bubble generation at the wall can be consumed than is supplied by the turbulence transfer from the bulk into the boundary layer. That is why we propose to use Eqs. (16.22) to (16.24) also for flashing in adiabatic pipes with driving temperature difference $T_2 - T'$ instead $T_w - T'$:

$$\dot{q}''_{w2,b} = ...\text{Eq. (16.22)}, \qquad (17.4)$$

$$\mu_{21,gen} = \frac{4}{D_h} \dot{q}''_{w2,b} / (h'' - h_2), \qquad (17.5)$$

and

$$\dot{n}_{1w} = \mu_{21} / \left(\rho_1 \frac{\pi}{6} D_{1d}^3 \right). \qquad (17.6)$$

From the moment of bubble detachment from the wall the bubble starts its own history inside the bulk flow and undergoes condensation or further evaporation as well as collisions, coalescence or fragmentation. In the next section we consider an example of such history for superheated liquids.

17.3 Bubble growth in the bulk

Consider a constant number of bubbles per unit mixture volume, n_1, during the time step $\Delta\tau$. As discussed in Chapter 13, the mass difference between the end and the beginning of the time step in a single bubble multiplied by the bubble number per unit mixture volume, gives the mass evaporated in the bulk per unit mixture volume and unit time.

$$\mu_{21,bulk} = n_1 \frac{\pi}{6}\left(\rho_1 D_1^3 - \rho_{10} D_{1o}^3\right)/\Delta\tau = \frac{\rho_{1o}\alpha_{1o}}{\Delta\tau}\left[\frac{\rho_1}{\rho_{1o}}\left(\frac{R_1}{R_{1o}}\right)^3 - 1\right] \quad \alpha_{1o} > 0 \tag{17.7}$$

As already mentioned from the sixteen analytical solutions of the bubble growth problem obtained for different degrees of simplicity known to the author - see Appendix 13.1, the bubble growth model proposed by *Mikic* et al [9] is used as the best one:

$$\mu_{21,bulk} = \frac{\rho''\alpha_{1o}}{\Delta\tau}\left[\left\{1+\frac{2}{3}\left[\left(\tau^+ +1\right)^{3/2} - \left(\tau^+\right)^{3/2} - 1\right]/R_{1o}^+\right\}^3 - 1\right] \tag{17.8}$$

where

$$\tau^+ = \Delta\tau/\left(B^2/A^2\right), \tag{17.9}$$

$$A^2 = \frac{2}{3}\frac{1}{\rho_2}(dp/dT)_{sat}\left(1-\rho''/\rho_2\right)\left[T_2 - T'(p)\right], \tag{17.10}$$

$$R^+ = AR_1/B^2. \tag{17.11}$$

Thus the steam generation is a superposition of the nucleation at the wall and of the bubble growth inside the bulk flow

$$\mu_{21} = \mu_{21,bulk} + \mu_{21,gen}. \tag{17.12}$$

17.4 Bubble fragmentation and coalescence

A review of the existing fragmentation and coalescence models for gases and liquids was given in Chapters 7 to 10. For the computation presented below we use the following gas fragmentation model. The volume and time average number of generated particles per unit time and unit mixture volume due to dynamic fragmentation is

$$\dot{n}_{1,frag} = (n_{1\infty} - n_1)/\Delta\tau_{fr} \quad for \ n_{1\infty} > n_1 \ and \ \Delta\tau \leq \Delta\tau_{fr}, \tag{17.13}$$

$$\dot{n}_{1,frag} = (n_{1\infty} - n_1)/\Delta\tau \quad for \ n_{1\infty} > n_1 \ and \ \Delta\tau > \Delta\tau_{fr}, \tag{17.14}$$

$$\dot{n}_{1,frag} = 0 \quad for \ n_{1\infty} \leq n_1, \tag{17.15}$$

where $\Delta\tau$ is the time step, n_1 is local bubble density concentration, and

$$n_{1\infty} = 6\alpha_1/(\pi D_{1\infty}^3) \tag{17.16}$$

is the stable bubble number density concentration after fragmentation. Equation (17.14) gives in fact steady state fragmentation for large time steps. The stable bubble diameter is computed in accordance with the *Achmad's* equations (8.173) and (8.172),

$$D_{1\infty} = \lambda_{RT} \left[9\alpha_1/(1-\alpha_1) \right]^{1/3} \tag{17.17}$$

for $\alpha_1 > 0.1$, and

$$D_{1\infty} = 0.9\lambda_{RT} / \left\{ 1 + 1.34 \left[(1-\alpha_1)V_2 \right]^{1/3} \right\} \tag{17.18}$$

for $\alpha_1 < 0.1$, see in [2].

The fragmentation time is set equal to the natural fluctuation period of bubble flow, Eq. (8.177)

$$\Delta\tau_{fr} = \left(\frac{6}{\pi\sqrt{2}} \right)^{1/3} \alpha_1^{1/3} / \Delta V_{12}^*, \tag{17.19}$$

where the relative velocity is a superposition of the difference between the mean velocity and the turbulent fluctuation velocity,

$$\Delta V_{12,}^* = |V_1 - V_2| + V_2^t \qquad (17.20)$$

$$V_2^t \approx 0.3 |V_2| \quad \text{for} \quad \alpha_1 < 0.3 \qquad (17.21)$$

and

$$V_2^t \approx 0.7 |V_2| \quad \text{for} \quad \alpha_1 \geq 0.3 \qquad (17.22)$$

The volume and time average bubble disappearance per unit time and unit mixture volume due to collision and coalescence is computed in accordance with Eq. (7.22),

$$\dot{n}_{coal} = n_1 \left[1 - 1/\exp(f_{coal}/\Delta \tau) \right]/\Delta \tau, \qquad (17.23)$$

where $\Delta \tau$ is the time averaging period. Here the instantaneous coalescence frequency of a single bubble is

$$f_{coal} = P_{coal} f_{col} \qquad (17.24)$$

where the collision frequency is computed in accordance with Eq. (7.29),

$$f_{col} = 4.5 \alpha_1^{1/2} \Delta V_{12}^* / D_1 \qquad (17.25)$$

and the coalescence probability P_{coal} is set to 1 for superheated liquid in order to approximate the effect of non-uniform particle distribution in the cross section. In case of no liquid superheating we use Eq. (7.45),

$$P_{coal} \approx 0.032 \left(\Delta \tau_{col} / \Delta \tau_{coal} \right)^{1/3} \quad \text{for} \quad \Delta \tau_{col} / \Delta \tau_{coal} \geq 1, \qquad (17.26)$$

where

$$\Delta \tau_{col} / \Delta \tau_{coal} = 1.56 \left(\frac{D_1 \sigma_2}{3\rho_1 + 2\rho_2} \right)^{1/2} / V_2^t. \qquad (17.27)$$

17.5 Film flashing bubble generation in adiabatic pipe flow

Boiling and flashing are governed by the same physical phenomena. Having accepted that there is departure from nucleate boiling in heated channels we have also to accept that the departure from nucleate bubble generation is also possible in adiabatic pipe flow. The steam volumetric fraction in the boundary layer with a thickness D_{1d} is

$$\alpha_{1d} = (\pi/6) n''_{1w} D_{1d}^2 \qquad (17.28)$$

If $\alpha_{1d} > 0.52$ the bubbles touch each other and a film is formed. We call this bubble generation mechanism *film* flashing because it is controlled by bubble entrainment from the vapor layer. If there is no flow velocity imposed, we have

$$D_{1d} \approx 2\pi \lambda_{RT}, \qquad (17.29)$$

$$f_{1w} \approx V_{1Ku}/D_{1d}, \qquad (17.30)$$

where

$$V_{1Ku} = 1.41\left[g\sigma(\rho_2 - \rho_1)/\rho_2^2\right]^{1/4}, \qquad (17.31)$$

is the free rising bubble velocity (*Kutateladze* velocity). For forced convection we assume that the stability criterion

$$We_{1d} \approx 12 \qquad (17.32)$$

controls the bubble entrainment size, and therefore

$$D_{1d} \approx 12\sigma/(\rho_2 \Delta V_{21}^2), \qquad (17.33)$$

$$f_{1w} \approx \Delta V_{21}/D_{1d}. \qquad (17.34)$$

For both cases we have

$$\dot{n}_{1w} = \frac{4}{D_h} n_{1w} f_{1w}, \qquad (17.35)$$

$$\mu_{21,gen} \approx \rho_1 \dot{n}_{1w} (\pi D_{1d}^3/6), \qquad (17.36)$$

and

$$\dot{q}'''_{\sigma 2,gen} = -\mu_{21,gen}(h'' - h_2). \qquad (17.37)$$

17.6 Verification of the model

As already mentioned, the model developed in the preceding Sections was incorporated into the code IVA4. Its predicting capabilities will now be verified taking the flashing flow in converging - diverging nozzles and the blow-down in pipes as examples. It is very important in such calculations to ensure that the code properly predicts the single-phase steady state and transient flows.

The results of three BNL experiments for water flow in a converging-diverging nozzle [1] are presented in Fig. 17.1, together with the results calculated with our model. We see good agreement for pressure distribution as well as for the predicted mass flow rate as function of inlet/outlet pressure difference.

The *Kellner* and *Gissler* experiment [6] used for the next comparison produced pressure wave propagation in single pipes with five bends and one dead end. The imposed pressure function at the entrance of the pipe is shown in Fig. 17.2a. The comparison of the predicted and the measured responses presented on Figs. 17.2b through 17.2d shows good agreement.

Now we proceed to the BNL converging-diverging nozzle experiment [1] with an entrance condition which causes flashing flow behind the nozzle throat. The results of the computation are presented in Figs. 17.3 through 17.6. We see that our model

a) predicts the data for pressure distribution data well,
b) slightly over predicts the data for the volumetric void fraction distribution, resulting in
c) some under prediction of the mass flow as a function of the measured pressure difference.

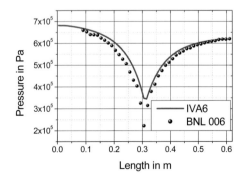

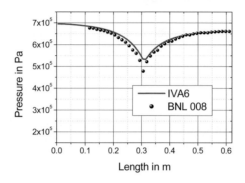

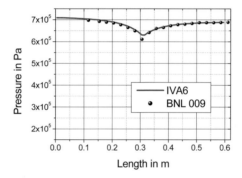

Fig. 17.1. Comparison of IVA4-NW predictions with BNL experimental data for pressure distribution. Single phase flow. Runs 6, 8, 9. $p_{in} = 682, 695, 709$ kPa, $p_{out} = 627, 665, 692$ kPa, $T_{in} = 27°C$, $G_{exp} = 7010, 4710, 3130$ kg/$(m^2 s)$, $G_{com} = 6319, 4362, 2991$ kg/$(m^2 s)$.

Next we simulate the pipe blow-down experiment documented in [4,8]. A horizontal pipe is initially filled with water slightly subcooled under the initial pressure. One end of the pipe is quickly opened. The pressure drops below the saturation pressure at the water temperature and the heterogeneous nucleation starts. After producing enough bubble sizes the pressure increases up to the saturation pressure corresponding to the initial temperature for a long time due to intensive flashing inside the pipe. Then it starts to decrease to the external pressure. The results are given in Figs. 17.7 through 17.11. We see that the nucleation theory applied here results in an appropriate pressure prediction (Fig. 17.7). The somewhat faster emptying of the pipe in the computation in the later stage of the process (Fig. 17.8) is caused by the several interactions in the computer code like drag forces, flow pattern recognition, etc., which will not be discussed here. We see also that the computed water temperature (Fig. 17.10) decreases faster than in reality. This is evidence that more energy is released from the liquid for evaporation in the model than in reality. This is consistent with the predicted void fraction at a given position (Fig. 17.9). The large void fraction increase in the first 0.1 s is not predicted by the model. This experimental observation is in contrast with the behavior observed behind the throat of the NL - nozzle.

The IVA5 force processor computes pipe forces due to hydrodynamics only as discussed in Chapter 8 of Volume 1. The result is presented in Fig. 17.11a. It corresponds to the pressure history. The oscillating character of the measurements, (Fig. 17.11b), is due to the fluid structure interactions. These interactions are not simulated by our computation.

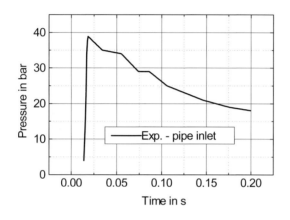

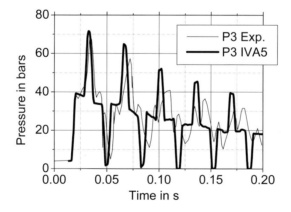

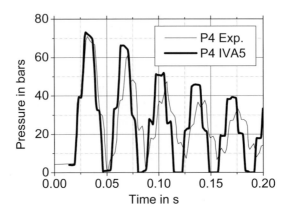

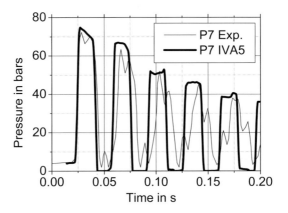

Fig. 17.2. a) Pressure at the pipe entrance as a function of time as measured by Kellner and Gissler [6]. Pipe length 13.12 m, diameter 0.1 m, 4 bends at 4.8, 10.2, 12.6, 13.64 m from the entrance. Initial conditions: water at atmospheric temperature and pressure. b) at 4.08 m from the pipe entrance. Comparison of IVA5 prediction with the experimental data by Kellner and Gissler [6]. c) at 8.18 m from the pipe entrance. d) at 12.5 m from the pipe entrance.

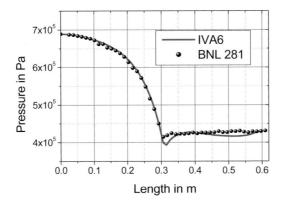

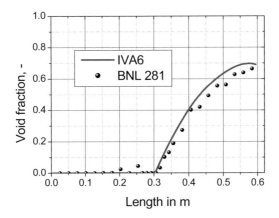

Fig. 17.3. Comparison of IVA4 prediction with BNL experimental data for pressure distributions and area averaged void fractions for Runs 281 - 283. $p_{in} = 688$ kPa, $T_{in} = 148.8°C$, $G_{in} = 5730$ $kg/(m^2s)$, $p_{out} = 431$ kPa, $p_c = 452$ kPa, $T_c = 148.8°C$, $G_{com} = 5533$ $kg/(m^2s)$.

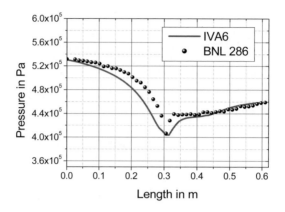

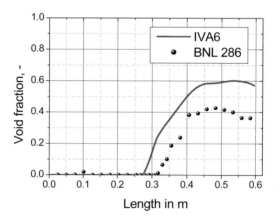

Fig. 17.4. Comparison of IVA4 predictions with BNL experimental data for pressure distributions and area average void fractions for Runs 286 - 288. p_{in} = 530 kPa, T_{in} = 149.2°C, G_{in} = 3580 kg/(m²s), p_{out} = 459 kPa, p_c = 456 kPa, T_c = 149.2°C, G_{com} = 2706 kg/(m²s).

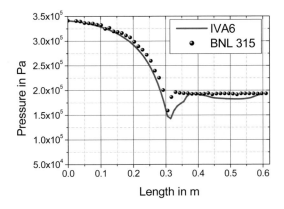

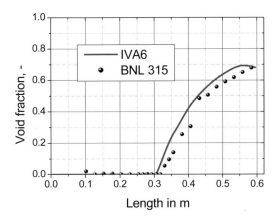

Fig. 17.5. Comparison of IVA4 predictions with BNL experimental data for pressure distributions and area average void fractions for Runs 313 - 315. $p_{in} = 341\ kPa$, $T_{in} = 121°C$, $G_{in} = 4410\ kg/(m^2 s)$, $p_{out} = 193\ kPa$, $p_c = 200\ kPa$, $T_c = 119°C$, $G_{com} = 4202\ kg/(m^2 s)$.

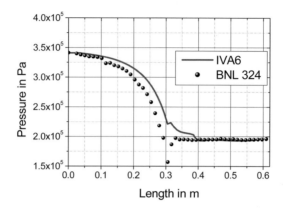

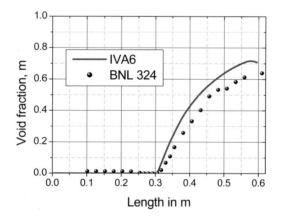

Fig. 17.6. Comparison of IVA4 predictions with BNL experimental data for pressure distributions and area average void fractions for Runs 324 - 325. p_{in} = 341 kPa, T_{in} = 121°C, G_{in} = 4410 kg/(m²s), p_{out} = 196 kPa, p_c = 200 kPa, T_c = 121.8°C, G_{com} =4004 kg/(m²s).

17.6 Verification of the model 481

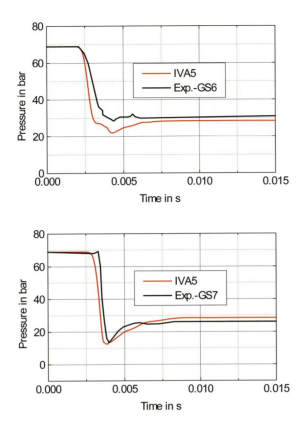

Fig. 17.7. Pressure at 0.072 and 0.914 *m* from the dead end of pipe as a function of time. Comparison IVA5 prediction with experimental data by Edwards and O'Brien. Pipe diameter 0.0732 *m*, length 4.096 *m*.

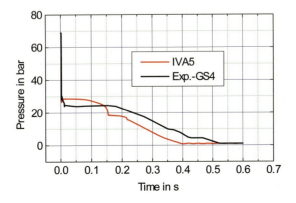

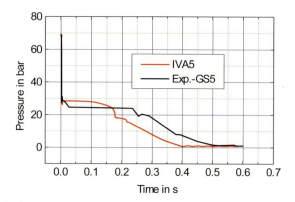

Fig. 17.8. Pressure at 2.024 and 2.469 *m* from the dead end of pipe as a function of time. Comparison IVA5 prediction with experimental data by Edwards and O'Brien. Pipe diameter 0.0732 *m*, length 4.096 *m*.

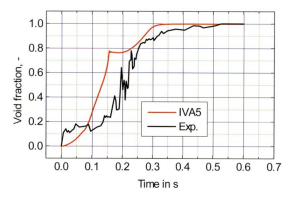

Fig. 17.9. Steam volume fraction at 1.469 *m* from the dead end of the pipe as a function of time. Comparison IVA5 prediction with experimental data by Edwards and O'Brien. Pipe diameter 0.0732 *m*, length 4.096 *m*.

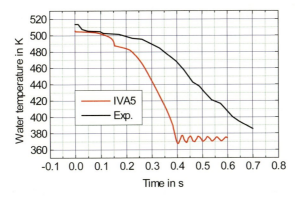

Fig. 17.10. Water temperature at 1.469 *m* from the dead end of the pipe as a function of time. Comparison IVA4 prediction with experimental data by Edwards and O'Brien. Pipe diameter 0.0732 *m*, length 4.096 *m*.

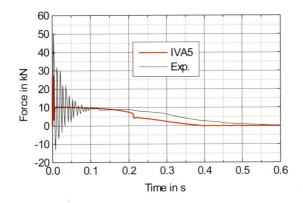

Fig. 17.11. Force as a function of time: a) IVA4 prediction without fluid-structure interactions; b) Edwards - O'Brien measurements reflecting fluid structure interactions.

17.9 Significance and conclusions

The proposed nucleation and flashing model consists of sub-models, which where verified separately with the available data sets for each particular sub-phenomenon like

- the bubble departure diameter, taking into account the mutual bubble interactions at high superheatings, confirmed by data comparison in Chapter 15, [19],
- the active nucleation site density, provided by *Wang* and *Dhir* [34] measurements,
- the mean site-to-site distance provided by *Wang* and *Dhir* [34] measurements,
- heat transfer between wall-fluid interface and bulk liquid caused mainly by bubble-growth and departure

introduced turbulence in the boundary layer verified in Chapter 16, [21] etc. As far the author knows this is the first boiling and flashing model which gives reasonable agreement with data for complex flashing flows in addition to the good agreement in separated tests. The model has the advantage to predict properly the separation between steam generation due to bubble production and bubble growth. I believe that it will be successfully applied in multi-phase flow modeling.

Nomenclature

a_2	liquid thermal diffusivity, m^2/s
B^2	$= 2R_1 dR_1/d\tau$, m^2/s
D_h	hydraulic diameter, m
D_l	particle size of velocity field l, m
D_1	bubble diameter, m
$D_{1\infty}$	stable bubble diameter under the local flow conditions, m
D_{1c}	critical size, m
D_{1d}	bubble departure diameter, m
$D_{1d,fc}$	bubble departure diameter for strongly predominant forced convection, m
$D_{1d,nc}$	bubble departure diameter for natural convection, m
$D_{2,inf}$	average center-to-center spacing, m
f_{1w}	bubble departure frequency, $1/s$
f_{1w}^t	boundary layer turbulence fluctuation frequency, $1/s$
f_{coal}	coalescence frequency, $1/s$
f_{col}	collision frequency of single bubble, $1/s$
P_{coal}	coalescence probability, dimensionless
G	gravitational acceleration, m/s^2
h', h''	specific saturation enthalpies for water and steam, J/kg
Δh	$= h'' - h'$, J/kg
Ja_2	$= \rho_2 c_{p2}[T_2 - T'(p)]/(\rho_1 \Delta h)$, Jacob number, *dimensionless*
n_l	particle number density of velocity field l, time averaged, $1/m^3$
$n_{l\infty}$	stable particle number density concentration after fragmentation, $1/m^3$
n_{1w}''	active nucleation site density, $1/m^2$

$\dot{n}_{l,kin}$ particles of velocity field *l* originating per unit time and unit mixture volume due to change of the state of aggregate, local volume and time averaged, $1/(s\ m^3)$

$\dot{n}_{1w}$ bubbles originating at the wall per unit time and unit mixture volume, local volume and time averaged, $1/(s\ m^3)$

$\dot{n}_{1,frag}$ particles of velocity field *l* originating per unit time and unit mixture volume due to fragmentation, local volume and time averaged, $1/(s\ m^3)$

$\dot{n}_{l,coal}$ particles of velocity field *l* disappearing per unit time and unit mixture volume due to coalescence, local volume and time averaged, $1/(s\ m^3)$

p pressure, *Pa*

p' saturated pressure, *Pa*

Pr_2 $= \eta_2/(\rho_2 a_2)$, Prandtl number, *dimensionless*

$\dot{q}'''_{2\sigma,gen}$ the energy consumed for the bubble generation until departure per unit mixture volume and unit time, W/m^3

$\dot{q}''_{2\sigma,gen}$ heat flux from the wall into the liquid during film flashing bubble generation in adiabatic pipe flow boiling, W/m^2

$\dot{q}''_{w2,b}$ heat flux from the wall into the liquid during pool boiling, W/m^2

R_1 bubble radius, *m*

R_{1o} bubble radius at the beginning of the time step $\Delta\tau$, *m*

$R_{2,inf}$ $= D_{2,inf}/2$, *m*

$T'(p)$ saturation temperature at system pressure *p*, *K*

T_2 liquid temperature, *K*

T_w wall temperature, *K*

V_1 surface average and time average velocity of velocity field *l*, *m/s*

V_{1Ku} free rising bubble velocity, *Kutateladze* velocity, *m/s*

$\overline{V'_2}$ volume-averaged fluctuation velocity, *m/s*

$\overline{\overline{V'_2}}$ volume-averaged fluctuation velocity time averaged over $\Delta\tau_w + \Delta\tau_d$, *m/s*

Greek

α_1 volume concentration of velocity field *l*, *dimensionless*

α_{1d} the steam volumetric fraction in the boundary layer having thickness of D_{ld}, *dimensionless*

γ_v volume porosity, *dimensionless*

γ surface permeability, *dimensionless*

$\Delta\rho_{21}$ $= \rho_2 - \rho_1$, *kg/s*

$\Delta\tau_d$ time elapsed from the origination of bubble with the critical size to the bubble departure from the wall, *s*

$\Delta\tau_w$ delay time, *s*

$\Delta \tau_{1d}$	bubble departure time, s
η_2	dynamic viscosity of liquid, $kg/(ms)$
λ_{RT}	$= \left(\dfrac{\sigma}{g \Delta \rho_{21}} \right)^{1/2}$, Rayleigh - Taylor instability wavelength
ℓ_2^t	turbulent length scale, m
ν_2	kinematic viscosity of liquid, m^2/s
ρ_l	density of velocity field l, kg/m^3
ρ_1	gas density, kg/m^3
ρ_2	liquid density, kg/m^3
ρ''	saturated steam density, kg/m^3
ρ'	saturated liquid density, kg/m^3
μ_l	mass introduced into the velocity field l per unit mixture volume and unit time, local volume and time average, $kg/(m^3 s)$
$\mu_{21,gen}$	steam generation per unit mixture volume and unit time due to bubble generation at the wall, local volume and time average, $kg/(m^3 s)$
$\mu_{21,bulk}$	mass evaporated in the bulk per unit mixture volume and unit time, $kg/(m^3 s)$
σ	surface tension, N/m
τ	time, s
φ	contact angle between "liquid drop" and the wall, rad

References

1. Abuaf N, Wu BJC, Zimmer GA and Saha P (June 1981) A study of nonequilibrium flashing of water in a converging diverging nozzle, Vol.1 Experimental, vol 2 Modeling, NUREG/CR-1864, BNL-NUREG-51317,
2. Ahmad SY (1970) Axial distribution of bulk temperature and void fraction in heated channel with inlet subcooling, J. Heat Transfer, vol 92 p 595
3. Blinkov VN, Jones OC and BI Nigmatulin, Nucleation and Flashing in Nozzles - 2. Comparison with Experiments Using a Five - Equation Model for Vapor Void Development, Int. J. Multiphase Flow, vol 19 no 6 pp. 965-986
4. Edwards AR, O'Brien TP (1970) Studies of phenomena connected with the depressurization of water reactors, The Journal of The British Nuclear Energy Society, vol 9 nos 1-4 pp 125-135
5. Kaishev R and Stranski IN (1934) Z. Phys. Chem., vol 26 p 317
6. Kellner H, Gissler (07.02.1984) Programsystem SAPHYR: Anwendungsbeispiel II, Notiz Nr.70.02748.4, Interatom GmbH
7. Kolev NI (1994) The influence of mutual bubble interaction on the bubble departure diameter, Experimental Thermal and Fluid Science, vol 8 no 2 pp 167-174

8. Kolev NI (1995) The Code IVA4: Nucleation and flashing model, Kerntechnik, vol 60, no 6 pp 157-164. Also in: (Apr.3-7, 1995) Proc. Second Int. Conf. On Multiphase Flow, Kyoto; (Aug.13-18, 1995) ASME & JSME Fluid Engineering Conference International Symposium on Validation of System Transient Analysis Codes, Hilton Head (SC) USA; (October 9-11, 1995) Int. Symposium on Two-Phase Flow Modeling and Experimentation, ERGIFE Place Hotel, Rome, Italy
9. Mikic BB, Rohsenow WM and Griffith P (1970) On bubble growth rates, Int. J. Heat Mass Transfer, vol 13 pp 657-666
10. Shin TS and Jones OC (1993) Nucleation and flashing in nozzles - 1, A distributed nucleation model, Int. J. Multiphase Flow, vol 19 no 6 pp 943-964
11. Skripov VP et al. (1980) Thermophysical properties of liquids in metastable state, Moskva, Atomisdat, in Russian
12. Volmer M (1939) Kinetik der Phasenbildung, Dresden und Leipzig, Steinkopf,

18. Boiling of subcooled liquid

18.1 Introduction

The description of the forced convection boiling of subcooled liquid started with using the equilibrium thermodynamics of two-phase flow. In this framework empirical correlations are obtained by generalizing experimental data using the *equilibrium steam mass flow rate concentration* in the flow,

$$X_{eq} = \frac{h - h'(p)}{h''(p) - h'(p)}, \qquad (18.1)$$

where the specific mixture enthalpy is defined as the *mixture enthalpy of the non-inert components*

$$h = \left[\sum_l \left(1 - \sum_n C_{nl}\right) \alpha_l \rho_l w_l h_{Ml}\right] / \left[\sum_l \left(1 - \sum_n C_{nl}\right) \alpha_l \rho_l w_l\right]$$

$$= \left[\sum_l \left(1 - \sum_n C_{nl}\right) X_l h_{Ml}\right] / \left[\sum_l \left(1 - \sum_n C_{nl}\right) X_l\right] \qquad (18.2)$$

and the *mass flow rate* in the axial direction is defined as follows

$$G = \sum_l (\alpha \rho w)_l. \qquad (18.3)$$

18.2 Initiation of visible boiling on the heated surface

Saha and *Zuber* [7] (1974) approximated a number of experimental data for the dependence of the local equilibrium steam mass flow concentration in the flow, defining the *initiation of the visible nucleate boiling*, X^*_{1eq}, as a function of the wall heat flux and the local parameters

$$X_{1eq}^* = -0.0022 \frac{D_h c_{p2}}{\lambda_2} \frac{\dot{q}_w''}{h'' - h'} \qquad Pe_2 \le 70\ 000, \qquad (18.4)$$

or recomputed in terms of liquid temperature at which the subcooled nucleate boiling starts,

$$T_2^* = T'(p) - 0.0022\, \dot{q}_w''\, D_h / \lambda_2 \quad \text{or} \quad Nu =: \frac{\dot{q}_w''}{T'(p) - T_2^*} \frac{D_h}{\lambda_2} = 454, \qquad (18.5)$$

and

$$X_{1eq}^* = -\frac{154}{G} \frac{\dot{q}_w''}{h'' - h'} \qquad Pe_2 > 70\ 000 \qquad (18.6)$$

or

$$T_2^* = T'(p) - 154 \frac{\dot{q}_w''}{c_{p2}} \quad \text{or} \quad St =: \frac{\dot{q}_w''}{c_{p2}\left[T'(p) - T_2^*\right]} = 0.0065 \qquad (18.7)$$

where

$$Pe_2 = GD_{hy}c_{p2}/\lambda_2, \qquad (18.8)$$

and T_2^* is, as already mentioned, the liquid temperature corresponding X_{1eq}^*. The above correlation approximates experimental data in the region: $1 \times 10^5 < p < 138 \times 10^5$ Pa, $95 < G < 2760$ kg/(m²s), $280\ 000 < \dot{q}_w'' < 1\ 890\ 000$ W/m² for channels with rectangular and circular cross sections. For Freon the correlation is valid in $3.2 \times 10^5 < p < 8.5 \times 10^5$ Pa, $101 < G < 2073$ kg/(m²s), $6\ 300 < \dot{q}_w'' < 5\ 360\ 000$ W/m².

Chan [1] analyzed the net vapor generation point in horizontal flow boiling in a tube with $D_h = D_{heat} = 10.7$ mm, and length of 2 m at 4 *bar* pressure and came to the conclusion that the constant in Eq. (18.5) has to be equal to 380. The *Peclet* numbers of his experiments varied between 10^4 and 4×10^4.

18.3 Local evaporation and condensation

18.3.1 Relaxation theory

There are different approaches proposed in the literature to describe the local evaporation and condensation processes. One of the oldest is that by *Levy*. In one of his earlier publications in this field *Levy* [6] (1967) approximates the profile of the experimentally observed steam mass flow concentration in the flow with the following simple expressions

$$X_1 = X_{1eq} - X_{1eq}^* \exp(X_{1eq} / X_{1eq}^* - 1) \qquad (18.9)$$

where X_{eq} is the *equilibrium steam mass flow concentration* in the flow defined by Eq. (18.1) for the case of no inert components

$$\sum X_l h_l = X_{1eq} h'' + (1 - X_{1eq}) h' = X_{1eq} (h'' - h') + h' . \qquad (18.10)$$

Now let as compute the net vapor generation rate defined by the profile described with Eq. (18.9). For one-dimensional steady state flow in a channel with constant cross section we write the mixture mass conservation, the vapor mass conservation and the mixture energy conservation equations as follows

$$G = const, \qquad (18.11)$$

$$\mu_{21} - \mu_{12} = G \frac{dX_1}{dz} \qquad (18.12)$$

$$\frac{d}{dz}\left(\sum X_l h_l\right) = \frac{4}{D_{heat}} \frac{\dot{q}_w''}{G} \qquad (18.13)$$

Having in mind the definition equation (18.10) for X_{1eq}

$$d\left(\sum X_l h_l\right) = d\left[X_{1eq}(h'' - h') + h'\right] = (h'' - h') dX_{1eq} \qquad (18.14)$$

Eq. (18.13) can be rewritten as follows

$$\frac{dX_{1eq}}{dz} = \frac{4}{D_{heat}} \frac{\dot{q}_w''}{G(h'' - h')} \qquad (18.15)$$

Thus from mass conservation equation of the gas phase (18.12) and using Eq. (18.15) we obtain

$$\mu_{21} - \mu_{12} = G \frac{dX_1}{dz} = G \frac{dX_1}{dX_{1eq}} \frac{dX_{1eq}}{dz} = \frac{4}{D_{heat}} \frac{\dot{q}_w''}{h'' - h'} \frac{dX_1}{dX_{1eq}} , \qquad (18.16)$$

or after differentiating the *Levy* approximation, Eq. (18.9),

$$\mu_{21} - \mu_{12} = \frac{4}{D_{heat}} \frac{\dot{q}_w''}{h'' - h'} \left[1 - \exp\left(\frac{X_1}{X_{1eq}^*} - 1\right) \right] \quad \text{for } X_1 > X_{1eq}^* \qquad (18.17)$$

Obviously in accordance with this approach the heat flux at the wall causing all the net evaporation is

$$\dot{q}_{net\_evaporation}'' = \dot{q}_w'' \left[1 - \exp\left(\frac{X_1}{X_{1eq}^*} - 1\right) \right], \qquad (18.18)$$

and the heat flux introduced into the bulk liquid due to recondensation is

$$\dot{q}_2''^{\sigma 1} = \dot{q}_w'' - \dot{q}_w'' \left[1 - \exp\left(\frac{X_1}{X_{1eq}^*} - 1\right) \right] = \dot{q}_w'' \exp\left(\frac{X_1}{X_{1eq}^*} - 1\right). \qquad (18.19)$$

Therefore the effectively condensed mass per unit time and unit mixture volume is

$$\mu_{12} = \frac{4}{D_{heat}} \frac{\dot{q}_2''^{\sigma 1}}{h'' - h'} = \frac{4}{D_{heat}} \frac{\dot{q}_w''}{h'' - h'} \exp\left(\frac{X_1}{X_{1eq}^*} - 1\right). \qquad (18.20)$$

From equation (18.20) and (18.17) we compute the evaporated mass per unit time and mixture volume as

$$\mu_{21} = \frac{4}{D_{heat}} \frac{\dot{q}_w''}{h'' - h'}. \qquad (18.21)$$

This approach does not answer the question about the magnitude of the heat transfer coefficient. One is forced to arbitrarily assume some wall superheat, e.g. 5 K and compute the total heat transfer coefficient by dividing wall heat flux by this value.

An alternative to Eq. (18.9) is the relaxation profile proposed by *Saha and Zuber* [6]

$$X_1 = \frac{X_{1eq} - X_{1eq}^* \exp(X_{1eq}/X_{1eq}^* - 1)}{1 - X_{1eq}^* \exp(X_{1eq}/X_{1eq}^* - 1)}, \qquad (18.22)$$

which results in

$$\dot{q}_2^{"\sigma 1} = \dot{q}_w^{"} \left\{ 1 - \frac{1 - \left(1 - X_{1eq} + X_{1eq}^*\right) \exp(X_{1eq} / X_{1eq}^* - 1)}{\left[1 - X_{1eq}^* \exp(X_{1eq} / X_{1eq}^* - 1)\right]^2} \right\}. \qquad (18.23)$$

Note that there are other proposals for the approximations $X_1 = X_1(X_{1eq}, X_{1eq}^*)$ - see *Kolev* [4] (1986).

Thus, the relaxation theory provided a simple expression for the recondensation process which does not take into account the influence of the local flow pattern and therefore of the interfacial area density. This approach is successfully used for description of steady state boiling in a channel with constant cross sections.

18.3.2 Boundary layer treatment

The treatment of the boundary layers is the more sophisticated method for description of the evaporation and condensation processes. Let us illustrate this approach with the method proposed by *Hughes* et al. [3] (1981), based on one proposal by *Lellouche* [5] (1974) and *Lellouche* and *Zolotar* - see in [3] (1981). The authors consider the heat transfer at the wall as a superposition of the heat flux causing boiling for saturated liquid and a convective heat flux from the wall into the bulk liquid

$$\dot{q}_w^{"} = \dot{q}_{NB}^{"} + \dot{q}_2^{"\sigma w} = h_{NB\_Thom}(T_w - T')^2 + h_{convective}(T_w - T_2). \qquad (18.24)$$

Here in accordance with *Thom* [8] (1966) the saturated boiling heat flux $\dot{q}_{NB}^{"}$ is a quadratic function of the wall superheat with a coefficient being a function only of the system pressure

$$h_{NB\_Thom} = 1942 \exp(p/4.35 \times 10^6) \qquad (18.25)$$

and the convective heat transfer coefficient is that derived by *Dittus* and *Boelter* [2] in 1937 for single-phase liquid flow

$$h_{convective} = 0.023 Re_2^{0.8} Pr_2^{0.4} \frac{\lambda_2}{D_h}, \qquad (18.26)$$

where

$$Re_2 = \rho_2 w_2 D_h / \eta_2, \qquad (18.27)$$

$$Pr_2 = c_{p2} \eta_2 / \lambda_2, \qquad (18.28)$$

are the *Reynolds* and the *Prandtl* numbers respectively. The heat flux released by the recondensation into the bulk flow is in accordance with *Hancox* and *Nicoll* - see in *Hughes* et al. [3], (1981)

$$\dot{q}_2^{"\sigma 1} = h_2^{\sigma 1}(T' - T_2) \quad \text{for} \quad X_1 > X_{1eq}^*, \tag{18.29}$$

where

$$h_2^{\sigma 1} = 0.4 Re_2^{0.662} Pr_2 \frac{\lambda_2}{D_2}. \tag{18.30}$$

The liquid temperature defining the initiation of the visible nucleate boiling, $T_{2,NBI}$, and the corresponding wall temperature are computed from the condition that the evaporation heat flux equals the condensation one

$$h_{NB\_Thom}(T_w - T')^2 = h_2^{\sigma 1}(T' - T_{2,NBI}) \tag{18.31}$$

and Eq. (18.26)

$$\dot{q}_w'' = h_{NB\_Thom}(T_w - T')^2 + h_{convective}(T_w - T_{2,NBI}). \tag{18.32}$$

Solving with respect to the liquid subcooling the authors obtained

$$T' - T_{2,NBI} = \frac{h_{NB\_Thom}}{h_2^{\sigma 1}}(T_w - T')^2$$

$$= \left[\frac{\sqrt{h_{convective}^2 \frac{h_2^{\sigma 1}}{h_{NB\_Thom}} + 4(h_2^{\sigma 1} + h_{convective})\dot{q}'' - h_{convective}\sqrt{\frac{h_2^{\sigma 1}}{h_{NB\_Thom}}}}}{2(h_2^{\sigma 1} + h_{convective})} \right]^2 \tag{18.33}$$

The wall temperature is obtained after solving the slightly rearranged Eq. (18.32)

$$\dot{q}_w'' = h_{NB\_Thom}(T_w - T')^2 + h_{convective}(T' - T_2) \tag{18.34}$$

with respect to the wall superheating

$$T_w - T' = \sqrt{\frac{\dot{q}_w'' - h_{convective}(T' - T_2)}{h_{NB\_Thom}}} \quad \text{for} \quad T_2 > T_{2,NBI} \tag{18.35}$$

The evaporated mass per unit time and unit mixture volume is then

$$\mu_{21} = \frac{4}{D_{heat}} \frac{h_{NB\_Thom}(T_w - T')^2}{h'' - h'}. \tag{18.36}$$

The condensed mass per unit time and unit volume is

$$\mu_{12} = \frac{4}{D_{heat}} \frac{h_2^{\sigma 1}(T' - T_2)}{h'' - h'}. \tag{18.37}$$

The heat introduced into the bulk liquid due to recondensation is

$$\dot{q}_2^{m\sigma 1} = \frac{4}{D_{heat}} h_2^{\sigma 1}(T' - T_2). \tag{18.38}$$

The heat introduced from the wall into the bulk liquid due to forced convection is

$$\dot{q}_2^{m\sigma w} = \frac{4}{D_{heat}} h_{convection}(T_w - T_2). \tag{18.39}$$

Hughes reports very good agreement with experimental data for void fraction in a vertical heated channel using the *Lellouche*'s drift flux correlation. *Kolev* [4] (1985) compared the predictions of this approach with a number of data for void fractions in a vertical boiling channel in the region $38 < G < 2000$ kg/(m²s), $T' - T_2 < 132$ K, $1 < p < 140$ bar, $20.3 < \dot{q}_w'' < 1723$ kW/m², $0.686 < L < 1.835$, $0.0044756 < D_h < 0.0269$ m using the *Chexal* and *Lellouch*'s drift flux correlation, and reported a very good agreement. The agreement was worse for very low mass flow rates.

Nomenclature

Latin

C_{nl} mass concentration of the inert component n inside the gas mixture, dimensionless
c_p specific heat at constant pressure, $J/(kgK)$
D_{hy} hydraulic diameter (4 times flow cross-sectional area / wet perimeter), m
D_{heat} heated diameter (4 times flow cross-sectional area / heated perimeter), m
G $= \sum(\alpha\rho w)_l$, mass flow rate in the axial direction, kg/(m²s)

h	specific enthalpy, J/kg
h_{NB_Thom}	saturated boiling heat transfer coefficient, $W/(m^2K)$
$h_{convective}$	convective heat transfer coefficient, $W/(m^2K)$
$h_2^{\sigma 1}$	heat flux coefficient by the recondensation into the bulk flow based on the heated surface, $W/(m^2K)$
Nu	$\dfrac{\dot{q}_w''}{T'(p)-T_2^*}\dfrac{D_h}{\lambda_2}$, Nusselt number, dimensionless
h	specific mixture enthalpy - mixture enthalpy of the non-inert components, J/kg
L	length, m
p	pressure, Pa
Pe_2	$= GD_{hy}c_{p2}/\lambda_2$, Peclet number, dimensionless
Pr_2	$= c_{p2}\eta_2/\lambda_2$, liquid Prandtl number, dimensionless
Re_2	$= \rho_2 w_2 D_h/\eta_2$, liquid Reynolds number, dimensionless
St	$\dfrac{\dot{q}_w''}{c_{p2}\left[T'(p)-T_2^*\right]}$, Stanton number, dimensionless
T_2^*	liquid temperature corresponding X_{1eq}^*, K
$\dot{q}_w''$	wall heat flux, W/m^2
$\dot{q}_{net\_evaporation}''$	heat flux at the wall causing all the net evaporation, W/m^2
$\dot{q}_{NB}''$	saturated boiling heat flux, W/m^2
$\dot{q}_2''^{\sigma w}$	convective heat flux from the wall into the bulk liquid, W/m^2
$\dot{q}_2''^{\sigma 1}$	heat flux released by the recondensation into the bulk flow, W/m^2
X_l	mass flow rate concentration of the velocity field l inside the multi-phase mixture, dimensionless
X_{1eq}^*	local equilibrium steam mass flow concentration in the flow, defining the initiation of the visible nucleate boiling, dimensionless
X_{eq}	equilibrium steam mass flow rate concentration in the flow, dimensionless
w	axial velocity, m/s

Greek

α	volume fraction, dimensionless
λ	thermal conductivity, $W/(mK)$
ρ	density, kg/m^3
μ_{21}	evaporation mass per unit time and unit mixture volume, $kg/(m^3s)$
μ_{12}	condensation mass per unit time and unit mixture volume, $kg/(m^3s)$

Subscript

'	saturated liquid
"	saturated steam
1	gas
2	liquid
$M1$	non-inert component in the gas mixture
$n1$	inert component in the gas mixture

References

1. Chan AMC (1984) Point on net vapor generation and its movement in transient horizontal flow boiling, Proc. of the 22th Nat. Heat Transfer Conference, Niagara Falls, New York, Dhir VK and Schrock VE (eds)
2. Dittus FW, Boelter LMK (1937) Heat transfer in automobile radiators of the tubular type, University of California Publications in Engineering, vol 2 no 13 p 443
3. Hughes ED, Paulsen MP, Agee LJ (Sept. 1981) A drift-flux model of two-phase flow for RETRAN. Nuclear Technology, vol 54 pp 410-420
4. Kolev NI (August 5-8, 1986) Transiente Zweiphasenstromung, Springer Verlag
5. Lellouche GS (1974) A model for predicting two-phase flow, BNL-18625
6. Levy S (1967) Int. J. Heat Mass Transfer, vol 10 pp 351-365
7. Saha P, Zuber N (1974) Proc. Int. Heat Transfer Conf. Tokyo, Paper 134.7
8. Thom IRS et. al. (1966) Boiling in subcooled water during up heated tubes or annuli. Proc. Instr. Mech. Engs., vol 180 3C pp 1965-1966

19. Natural convection film boiling

19.1 Minimum film boiling temperature

Groeneveld and *Stewart* [3] derived from their measurements the following expression for the minimum film boiling temperature.

For $p \leq 90 \times 10^5 \, Pa$

$$T_{w,MFB} = 557.85 + 0.0441 \times 10^{-3} p - 3.72 \times 10^{-12} p^2$$

$$- \max\left(0, \frac{10^4 X_{1,eq}}{2.82 + 0.00122 \times 10^{-3} p}\right), \qquad (19.1)$$

and for $p > 90 \times 10^5 \, Pa$

$$T_{w,MFB} - T'(p) = 76.722 \frac{p_c - p}{p_{cr} - 90 \times 10^5}. \qquad (19.2)$$

Equation (19.2) guarantees that the minimum film boiling temperature at critical pressure is equal to the critical temperature. This correlation provides the best agreement with data compared to the prediction of seven other correlations reviewed by the authors in [3]. The mass flow rates in the experiments are varied between 110 and 2750 $kg/(m^2 s)$ with no effect on the minimum film boiling temperature. This correlation is valid for pressure between 1 and 220 *bar*, mass flux 50 and 4000 $kg/(m^2 s)$, equilibrium vapor mass flow fraction –19 and 64% for vertical up-flow in circular pipe.

The minimum heat flux is defined as follows

$$q''_{w,MFB} = h_{FB}\left[T_{w,MFB} - T'(p)\right], \qquad (19.3)$$

h_{FB} is the film boiling heat transfer coefficient.

Schroeder-Richter and Bartsch [8] interpreted the minimum film boiling phenomenon as a phase interface normal velocity discontinuity governed by the *Rankine* and *Hugoniot* equation [7, 5],

$$h_2 - h_1 - \frac{1}{2}(p_2 - p_1)\left(\frac{1}{\rho_2} + \frac{1}{\rho_1}\right) = 0 \tag{19.4}$$

assuming that the liquid side boundary layer properties are those at saturation at the minimum film boiling temperature and the vapor side properties are those at saturation at system pressure

$$h''(p) - h'(T_{w,MFB}) - \frac{1}{2}\left[p - p'(T_{w,MFB})\right]\left[\frac{1}{\rho''(p)} + \frac{1}{\rho'(T_{w,MFB})}\right] = 0. \tag{19.5}$$

Henry [4] related the minimum stable film boiling temperature to the homogeneous nucleation temperature and the wall properties

$$T_{w,MFB} = T_{hn} + (T_{hn} - T_2)\sqrt{\frac{(\lambda \rho c_p)_2}{(\lambda \rho c_p)_w}}. \tag{19.6}$$

Henry gives the homogeneous nucleation temperature as a curve fit to those used by *Thurgood* and *Kelley* [9],

$$T_{hn} = \frac{5}{9}\begin{pmatrix} 459.67 + 705.44 - 4.722 \times 10^{-2} p_{psi} \\ +2.3907 \times 10^{-5} p_{psi}^2 - 5.8193 \times 10^{-9} p_{psi}^3 \end{pmatrix}, \tag{19.7}$$

where

$$p_{psi} = 3203.6 - 14.50377 \times 10^{-5} p. \tag{19.8}$$

19.2 Film boiling in horizontal upwards-oriented plates

Film boiling in horizontal upwards-oriented plates in a large pool is possible if the temperature of the wall is larger then the so called minimum film boiling temperature $T_{w,MFB}$,

$$T_w > T_{w,MFB} \tag{19.9}$$

Berenson [1] (1961) obtained the following expression for film boiling on a horizontal plane

$$h_{nc\_FB} = 0.425 \left[\frac{\lambda_1^3 \rho_1 (\rho' - \rho'')(h'' - h')g}{\eta'' \Delta T \lambda_{RT}} \right]^{1/4}, \quad (19.10)$$

where

$$\lambda_{RT} = \left[\frac{\sigma}{g(\rho' - \rho'')} \right]^{1/2}, \quad (19.11)$$

and

$$\Delta T = (T_w - T'). \quad (19.12)$$

The averaged vapor film thickness can be then expressed as

$$\delta_{1F} \approx \frac{\lambda_1}{h_{nc\_FB}} = 2.35 \left[\frac{\eta_1 \lambda_1 \Delta T}{(h'' - h')\rho_1 g(\rho' - \rho'')} \lambda_{RT} \right]^{1/4}, \quad (19.13)$$

where vapor properties are computed as a function of the averaged vapor film temperature $(T_w - T')/2$.

Equating the critical heat flux predicted by the *Zuber*'s correlation [10] to the heat flux obtained for film boiling by the *Berenson* equation an expression for the minimum film boiling temperature can be obtained

$$\Delta T_{w,MFB} = 0.127 \frac{(h'' - h')\rho_{1F}}{\lambda_{1F}} \left[\frac{g(\rho_2 - \rho_1)}{\rho_2 + \rho_1} \right]^{2/3} \left[\frac{\sigma}{g(\rho_2 - \rho_1)} \right]^{1/2} \left[\frac{\eta_2}{\rho_2 - \rho_1} \right]^{1/3} \pm 10\%. \quad (19.14)$$

Leperriere [6] (1983) modified the *Berenson* equation to obtain a correlation for forced convection film boiling in channels with very low mass flux

$$h_{Be,\mathrm{mod}} = h_{Berenson} \left(1 + 25.5 \frac{T_w - T'}{T_2} \right) (1 - \alpha_{1h})^{0.5}, \quad (19.15)$$

where the homogeneous local vapor volume fraction is

$$\alpha_{1h} = \frac{X_{1,eq}/\rho''}{X_{1,eq}/\rho'' + (1 - X_{1,eq})/\rho'}. \quad (19.16)$$

19.3 Horizontal cylinder

For natural convection film boiling on a horizontal cylinder *Bromley* [2] (1950)

$$h_{nc\_FB} = 0.62 \left[\frac{\lambda_{1F}^3 \rho_{1F}(\rho' - \rho'')g(h'' - h')\left[1 + 0.4\dfrac{c_p''(T_w - T')}{h'' - h'}\right]}{\eta'' D_h \Delta T} \right]^{1/4}, \quad (19.17)$$

obtained a very similar result as those by *Berenson* with the exception of the empirical constant. The modification of the latent heat of vaporization takes into account the superheating of the vapor. The analytical derivation of this equation is very similar to those for planes presented in Chapter 21 in details.

19.4 Sphere

For natural convection film boiling on a sphere *Bromley* [2] (1950) computed a different constant in the Eq. (19.17)

$$h_{nc\_FB} = 0.76 \left[\frac{\lambda_{1F}^3 \rho_{1F}(\rho' - \rho'')g(h'' - h')\left[1 + 0.4\dfrac{c_p''(T_w - T')}{h'' - h'}\right]}{\eta'' D_3 \Delta T} \right]^{1/4}. \quad (19.18)$$

We devote a special Section 21.2 to this subject.

Nomenclature

Latin

c_p specific heat at constant pressure, $J/(kgK)$
D_{hy} hydraulic diameter (4 times flow cross-sectional area / wet perimeter), m
g gravitational acceleration, m/s^2
h specific enthalpy, J/kg
h_{FB} film boiling heat transfer coefficient, $W/(m^2K)$
h_{nc_FB} natural convection film boiling heat transfer coefficient, $W/(m^2K)$

p	pressure, *Pa*
p_{psi}	$= 3203.6 - 14.50377 \times 10^{-5} p$, *psi*
p_{cr}	critical pressure, *Pa*
$q''_{w,MFB}$	minimum film boiling heat flux, *W/m²*
$T_{w,MFB}$	minimum film boiling temperature, *K*
T_{hn}	homogeneous nucleation temperature, *K*
X_{eq}	equilibrium steam mass flow rate concentration in the flow, *dimensionless*

Greek

α	volume fraction, *dimensionless*
α_{1h}	$= \dfrac{X_{1,eq}/\rho''}{X_{1,eq}/\rho'' + (1-X_{1,eq})/\rho'}$, homogeneous local vapor volume fraction, *dimensionless*
Δ	finite difference
δ_{1F}	averaged vapor film thickness
λ	thermal conductivity, $W/(mK)$
λ_{RT}	$= \left[\dfrac{\sigma}{g(\rho'-\rho'')}\right]^{1/2}$, *Rayleigh-Taylor* instability wavelength, *m*
ρ	density, *kg/m³*
σ	surface tension, *N/m*
μ_{21}	evaporation mass per unit time and unit mixture volume, *kg/(m³s)*
μ_{12}	condensation mass per unit time and unit mixture volume, *kg/(m³s)*

Subscript

'	saturated liquid
''	saturated steam
1	gas
2	liquid
1F	vapor film

References

1. Berenson PJ (Aug. 1961) Film boiling heat transfer from a horizontal surface, J. Heat Transfer, p 351
2. Bromley LA (1950) Heat transfer in stable film boiling, Chem. Eng. Process, vol 46 no 5 p 221
3. Groeneveld DC, Stewart JC (1982) The minimum boiling temperature for water during film boiling collapse, Proc. 7th International Heat Transfer Conference, Munich, FRG, vol 4 pp 393-398
4. Henry RE (1974) A correlation for minimum film boiling temperature, AIChE Simposium Series, vol 138 pp 81-90
5. Hugoniot PH (1887) Memoire sur la propagation du mouvement dans les corps et specialement dans les gases parfaits, Journal de l'École Polytechnique
6. Leperriere A (1983) An analytical and experimental investigation of forced convective film boiling, M. A. Sc. Thesis, University of Ottawa, Ottawa
7. Rankine WJM (1870) On the thermodynamic theory of waves of finite longitudinal disturbances, Philosophical Transactions of the Royal Society
8. Schroeder-Richter D, Bartsch G (1990) The Leidenfrost phenomenon caused by a thermo-mechanical effect of transition boiling: A revisited problem of non-equilibrium thermodynamics, HTD-Vol.136, Fundamentals of Phase Change: Boiling and Condensation, Witte LC and Avedisian CT (eds) Book No. H00589 – 1990 pp 13-20
9. Thurgood MJ, Kelley JM (December 1979) Battelle Pacific Northwest Laboratories
10. Zuber N (1958) On the stability of boiling heat transfer, Trans. ASME, vol 80 pp 711-720

20. Forced convection boiling

20.1 Convective boiling of saturated liquid

Chen [6, 15] proposed in 1963 to correlate the data for convective heat transfer during forced convection boiling in a tube by superposition of single-phase liquid convection heat flux $h_c(T_w - T_2)$ and of nucleate boiling heat flux $h_{NB}(T_w - T')$ in the following way

$$\dot{q}''_w = h_c(T_w - T_2) + h_{NB}(T_w - T'). \tag{20.1}$$

The mutual influence of both mechanisms is taken into account by modifying the corresponding correlations. The original *Dittus-Boelter* equation

$$h_c = 0.023 \frac{\lambda_1}{D_h} F \times Re_2^{0.8} Pr_2^{0.4}, \tag{20.2}$$

applied as if the liquid is flowing alone in the tube

$$Re = X_2 G D_h / \eta_2, \tag{20.3}$$

$$Pr_2 = c_{p2} \eta_2 / \lambda_2, \tag{20.4}$$

is modified by the multiplier

$$F = 2.35(1/X_{tt} + 0.213)^{0.736}, \qquad 1/X_{tt} > 0.1 \tag{20.5}$$

$$F = 1, \qquad 1/X_{tt} \leq 0.1 \tag{20.6}$$

where the *Martinelli* factor is computed as follows

$$1/X_{tt} = \left(\frac{X_1}{1-X_1}\right)^{0.9} \left(\frac{\rho_2}{\rho_1}\right)^{1/2} \left(\frac{\eta_1}{\eta_2}\right)^{0.1}. \tag{20.7}$$

The nucleate boiling heat transfer coefficient as proposed by *Forster* and *Zuber* [9]

$$h_{NB} = c_1(T_w - T')^{0.24}, \qquad (20.8)$$

where

$$c_1 = c\,\Delta p^{0.75}, \qquad (20.9)$$

$$\Delta p = p'[\min(T_w, 647)] - p,$$

$$c = 0.00122\ S \left(\frac{\lambda' c_p'}{\sigma}\right)^{1/2} Pr_2^{-0.29} \rho'^{1/4} \left[\frac{c_p' \rho'}{\rho''(h'' - h')}\right]^{0.24}, \qquad (20.10)$$

Eq. 17 in [6], is modified by the multiplier S. S is the ratio of the effective superheat to the total superheat of the wall. It is a function of a modified *Reynolds* number computed as follows

$$Re_{Tp} = Re_2 F^{1.25} 10^{-4}, \qquad (20.11)$$

$$S = 1/(1 + 0.12 Re_{Tp}^{1.14}), \quad \text{for } Re_{Tp} \leq 32.5, \qquad (20.12)$$

$$S = 1/(1 + 0.42 Re_{Tp}^{0.78}), \quad \text{for } 32.5 \leq Re_{Tp} \leq 70, \qquad (20.13)$$

$$S = 0.0797 \exp(1 - Re_{Tp}/70), \quad \text{for } 70 > Re_{Tp}, \qquad (20.14)$$

Bjornard and *Grifith* [4]. *Groeneveld* et al. [12] use instead of the last expression simply $S = 0.1$. The correlation was verified in the region $X_1 = 0$ to 0.71, $p = (1.013$ to $69.) \times 10^5$ *Pa*, $G = 54$ to 4070 kg/(m²s), $\dot{q}_w'' = 44$ to $2\,400$ kW/m². Thus the effective heat transfer coefficient is

$$h = \dot{q}_w''/(T_w - T'). \qquad (20.15)$$

The wall temperature by enforced heat flux is

$$T_w = (\dot{q}_w'' + h_c T_2 + h_{NB} T')/(h_c + h_{NB}). \qquad (20.16)$$

Because the nucleate boiling heat transfer coefficient is also function of the wall temperature few iterations are necessary to solve the equation

$$\dot{q}_w'' = h_c (T_w - T_2) + c_1 (T_w - T')^{1.24}, \qquad (20.17)$$

with respect to the wall temperature. The evaporated mass per unit time and mixture volume of the flow is then

$$\mu_{21} = \frac{4}{D_{heat}} \dot{q}_w'' / (h'' - h'). \qquad (20.18)$$

20.2 Forced convection film boiling

20.2.1 Tubes

The forced convection film boiling description started with using the equilibrium thermodynamics of two-phase flow. In this framework the empirical correlations are obtained by generalizing experimental data using the *equilibrium steam mass flow rate concentration* in the flow,

$$X_{eq} = \frac{h - h'(p)}{h''(p) - h'(p)} \qquad (20.19)$$

where the specific mixture enthalpy is defined as the *mixture enthalpy of the non-inert components*

$$h = \left[\sum_l \left(1 - \sum_n C_{nl}\right) \alpha_l \rho_l w_l h_{Ml} \right] / \left[\sum_l \left(1 - \sum_n C_{nl}\right) \alpha_l \rho_l w_l \right] \qquad (20.20)$$

and the *mass flow rate* in the axial direction is defined as follows

$$G = \sum (\alpha \rho w)_l. \qquad (20.21)$$

Dougall and *Rohsenow* [8] proposed one of the oldest correlations in 1963 modifying the well-known single-phase heat transfer correlation of *Dittus* and *Boelter* [7] from 1937

$$h_{fc\_FB} = 0.023 \frac{\lambda_1}{D_h} Re_1^{0.8} Pr_{1w}^{0.4}. \qquad (20.22)$$

Here the *Reynolds* number is defined by using as a characteristic velocity the center of mass velocity

$$w_{eq} = G / \rho_{eq}, \tag{20.23}$$

assuming that the density of the mixture is equal to the equilibrium density

$$\frac{1}{\rho_{eq}} = \frac{1 - X_{eq}}{\rho'} + \frac{X_{eq}}{\rho''}, \tag{20.24}$$

and the properties of the mixture are those of the saturated vapor

$$Re_{1\_mix} = \frac{D_h w_{eq}}{\eta'' / \rho''} = D_h G \frac{\rho''}{\eta''} \left[\frac{1 - X_{eq}}{\rho'} + \frac{X_{eq}}{\rho''} \right]. \tag{20.25}$$

The *Prandtl* number for steam

$$Pr_{1w} = c_{p1} \eta_1 / \lambda_1, \tag{20.26}$$

is computed with the properties at the wall temperature

$$c_{p1}, \eta_1, \lambda_1 = f \left[\max(647, T_w), p \right]. \tag{20.27}$$

Note that the heat flux is then related to the saturation temperature at the local system pressure rather to the real vapor temperature

$$q_w'' = h_{fc\_FB} \left[T_w - T'(p) \right]. \tag{20.28}$$

The heat is then assumed to be completely used for evaporation of the continuous liquid

$$\mu_{21} = \frac{4}{D_{heat\ w}} \frac{q_w''}{h''(p) - h'(p)}, \tag{20.29}$$

$$\mu_{31} = 0, \tag{20.30}$$

if the continuous liquid is available, i.e. if $\alpha_2 > 0$. If there is no continuous liquid $\alpha_2 = 0$ but all the liquid is in dispersed flow $\alpha_3 > 0$ we have

$$\mu_{21} = 0, \tag{20.31}$$

$$\mu_{31} = \frac{4}{D_{heat\ w}} \frac{q_w''}{h''(p) - h'(p)}. \tag{20.32}$$

Polomik [23] proposed in 1967 a modification of the *Dougall and Rohsenow* correlation in order to take into account the liquid subcooling

$$h_{fc\_FB} = 0.023 \frac{\lambda_1}{D_h} Re_{1\_mix}^{0.8} Pr_{1w}^{0.40} \left(\frac{T_2}{T_w}\right)^{0.5}. \tag{20.33}$$

As noted in 1994 by *Andreani* and *Yadigaroglu* [2] p. 15 this correlation provides surprisingly good agreement with data for substantial thermal non-equilibrium.

In the same year as *Dougall* and *Rohsenow*, *Miropolskij* [21] (1963) collected several experimental data in the region of $40 \times 10^5 \leq p \leq 220 \times 10^5$ Pa, $400 \leq G \leq 2000$ kg/(m²s), $0 \leq X_{eq} \leq 1$, $0.008 < D_h < 0.029$ m and successfully reproduced them by the following correlation

$$h_{fc\_FB} = 0.023 \frac{\lambda_1}{D_h} Re_{1\_mix}^{0.8} Pr_{1w}^{0.80} Y. \tag{20.34}$$

Note the difference in the exponent of the *Prandtl* number compared to the *Dougall* and *Rohsenow* correlation and the multiplier

$$Y = 1 - 0.1 \left(\frac{\rho'}{\rho''} - 1\right)^{0.4} \left(1 - X_{eq}\right)^{0.4}. \tag{20.35}$$

Groneveld and *Delorme* [10] 1976 correlated their own measurements by modifying a correlation obtained by *Hadaller* and *Banerjee* [14] for vapor only

$$h_{fc\_FB} = \max\left(3.66, \ 0.008348 \frac{\lambda_1}{D_{hy}} Re_{1\_mix}^{0.8774} Pr_{1w}^{0.6112}\right), \tag{20.36}$$

where the properties for the vapor *Prandtl* number are computed for an averaged temperature as follows

$$c_{p1}, \eta_1, \lambda_1 = f\{\max[647, (T_w + T_1)/2], p\}. \tag{20.37}$$

The authors limited the prediction of this correlation by the value of the modified *Berenson* equation

$$h_{fc\_FB} = \max(h_{nc\_FB\_Berenson}, \ h_{nc\_FB}). \tag{20.38}$$

Groneveld et al. [12] successfully reproduced by this approach in 1989 their experimental data in the following region D_{hy} = 0.0005 to 0.0247 m, p = (1.013 to 204.99)10^5 Pa, G =12.1 to 6874 kg/(m²s), X_{eq} = −0.12 to 3.094. The predicted heat fluxes are in the region $\dot{q}_w''$ = 2660 to 2 700 200 W/m².

20.2.2 Annular channel

For prediction of the forced convection film boiling heat transfer coefficient in an *annular channel* in 1977 *Groeneveld* [11] uses the *Miropolskij* multiplier in the following equation

$$h_{fc\_FB} = 0.052 \frac{\lambda_1}{D_h} Re_1^{0.688} Pr_{1w}^{1.26} / Y^{1.06}. \qquad (20.39)$$

The authors reproduced successfully data in the following region: $450 \times 10^3 \leq \dot{q}_w'' \leq 2250 \times 10^3$ W/m², $800 \leq G \leq 4100$ kg/(m²s), $34 \times 10^5 < p < 100 \times 10^5$ Pa, $0.1 < X_{eq} < 0.9$. This equation is known as *Groeneveld*'s Eq. 5-7 in the literature. *Bjornard* and *Grifith* [4] recommended in 1977 the following modification of the correlation

$$h_{fc\_FB} = 0.052 \frac{\lambda_1}{D_h} Re_1^{0.688} Pr_{1w}^{1.26} / Y^{1.06} \quad \text{for} \quad p > 13.8 \; bar, \qquad (20.40)$$

$$h_{fc\_FB} = 0.052 \frac{\lambda_1}{D_h} Re_1^{0.688} Pr_{1w}^{1.26} \quad \text{for} \quad p < 13.8 \; bar, \qquad (20.41)$$

where

$$Re_1 = w_1 D_1 / v_1, \qquad (20.42)$$

which removes the singularity of the Y factor and is appropriate for direct application in the framework of the non-equilibrium models. Further the authors proposed to combine the film boiling correlation from the type given above valid for high void fractions with natural circulation film boiling in liquid pool by weighting with the void fraction.

Condie-Bengston - see in [2] (1984), reported the correlation

$$h_{fc\_FB} = 0.0524 \frac{\lambda_1}{D_h} \frac{D_h^{0.1905}}{\lambda_1^{0.5407}} \frac{Pr_{1w}^{2.2598}}{\left(1+X_{eq}\right)^{2.0514}} Re_1^{\left[0.6249+0.2043\ln(X_{eq}+1)\right]} \qquad (20.43)$$

valid for the following region: $34 \times 10^3 \le \dot{q}_w'' \le 2\,074 \times 10^3$ W/m², $16.5 \le G \le 5\,234$ kg/(m²s), $4.2 \times 10^5 < p < 215 \times 10^5$ Pa, $-0.12 < X_{eq} < 1.73$.

20.2.3 Tubes and annular channels

Groeneveld generalized in the same work [11] the equation for both *tubes* and *annular channels* as follows

$$h_{fc\_FB} = 0.00327 \frac{\lambda_1}{D_h} Re^{0.901} Pr_{1w}^{1.32} / Y^{1.50} \qquad (20.44)$$

This equation is known as *Groeneveld*'s Eq. 5-9 in the literature. It is reported to be valid in the following region: $120 \times 10^3 \le \dot{q}_w'' \le 2\,250 \times 10^3$ W/m², $700 \le G \le 5300$ kg/(m²s), $34 \times 10^5 < p < 215 \times 10^5$ Pa, $< X_{eq} < 0.9$.

20.2.4 Vertical flow around rod bundles

For vertical flow around rod bundles *Loomis* and *Shumway* [20] (1983) reported the following correlation

$$h_{fc\_FB} = const \frac{\lambda_1}{D_h}(1-0.76\alpha_1)\left[\frac{h''-h'}{(T_w-T_1)c_{p1}}\right]^{0.25} \frac{\lambda_{RT} G}{\eta_2} \frac{\rho_2 - \rho_1}{\rho_1} \qquad (20.45)$$

where

$$\lambda_{RT} = \left[\frac{1}{g}\frac{\sigma}{(\rho_2-\rho_1)}\right]^{1/2} \qquad (20.46)$$

and the constant is

$$const = 18(\pm 0.4)10^{-4}. \qquad (20.47)$$

The correlation is valid in the following region: $p \cong 1.27 \times 10^5$ Pa, $31.55 \le G \le 70.9$ kg/(m²s), $0.1 \le \alpha_1 \le 1$. Note that the correlation is applied in the framework of non-equilibrium thermodynamics, which means that

$$q_1''^{\sigma w} = h_{fc\_FB}(T_w - T_1). \qquad (20.48)$$

20.3 Transition boiling

Ramu and *Weisman* [24] (1974) proposed the following correlation for transition boiling in tubes

$$h_{TB} = \frac{1}{2} S\, h_{CHF\_Addoms} \left\{ \begin{array}{l} \exp\left[-0.01404\left(\Delta T - \Delta T_{CHF\_Addoms}\right)\right] \\ + \exp\left[-0.12564\left(\Delta T - \Delta T_{CHF\_Addoms}\right)\right] \end{array} \right\} + h_{fc\_FB}, \quad (20.49)$$

where

$$\Delta T = T_w - T'(p). \quad (20.50)$$

The component of the forced convection film boiling is computed using the *Dougall* and *Rohsenow* [8] proposal

$$h_{fc\_FB} = 0.023 \frac{\lambda_1}{D_h} Re^{0.800} Pr_{1w}^{0.40}, \quad (20.51)$$

$$Re = \frac{D_h w_{eq}}{\eta''/\rho''} = D_h G \frac{\rho''}{\eta''}\left[\frac{1-X_{eq}}{\rho'} + \frac{X_{eq}}{\rho''}\right]. \quad (20.52)$$

The *Prandtl* number for steam

$$Pr_{1w} = c_{p1}\eta_1/\lambda_1, \quad (20.53)$$

is computed with the properties at the wall temperature

$$c_{p1}, \eta_1, \lambda_1 = f\left[\max\left(647,\, T_w\right),\, p\right]. \quad (20.54)$$

S is the ratio of the effective superheat to the total superheat of the wall from the *Chen* correlation, Eqs.(20.11) to (20.14), [6].

As noted by *Groeneveld* et al. [12] transition-boiling correlations generally provide incorrect asymptotic trends outside their region of validity. An alternative approach uses simply a linear interpolation between the maximum and the minimum points of a log-log plot of $\dot{q}_w''$ vs. ΔT_w, i.e.

$$\frac{\dot{q}_{TB}''}{\dot{q}_{min\_FB}''} = \left(\frac{\dot{q}_{CHF}''}{\dot{q}_{min\_FB}''}\right)^m, \quad \text{where} \quad m = \frac{\ln\left(\Delta T_{min\_FB}/\Delta T_w\right)}{\ln\left(\Delta T_{min\_FB}/\Delta T_{CHF}\right)} \quad (20.55)$$

or

Table 20.1. The critical heat flux and the corresponding temperature difference as a function of pressure – *Addoms* data [1]

p	h_{CHF_Addoms}	ΔT_{CHF_Addoms}
2 75800	86 877	18.9
4 13700	91 874	18.9
41 37000	227 139	16.6
68 95000	397 478	13.9
96 53000	438 930	12.2
169 96175	408 835	6.1
184 78600	384 986	5.0
195 12850	369 087	4.4
204 78150	348 077	3.9
210 02170	323 661	3.3

The correlation was verified in the region $1.72 \times 10^5 < p < 2.06 \times 10^5$ Pa, $20 < G < 50$ kg/(m^{2}s), $30 < \dot{q}''_w < 260$ W/m^2, $0 < X_{eq} < 0.5$.

$$\frac{\Delta T_w}{\Delta T_{min\_FB}} = \left(\frac{\Delta T_{CHF}}{\Delta T_{min\_FB}}\right)^n, \quad \text{where} \quad n = \frac{\ln\left(\dot{q}''_{TB}/\dot{q}''_{min\_FB}\right)}{\ln\left(\dot{q}''_{CHF}/\dot{q}''_{min\_FB}\right)}. \quad (20.56)$$

Film boiling in a large pool exists if $T_w > \Delta T_{min\_FB}$, where ΔT_{min_FB} is the minimum film boiling temperature.

Note that once the value of the critical heat flux is known, the corresponding temperature is obtained from the definition equation of the heat flux for forced convection nucleate boiling as follows

$$\Delta T_{CHF} = T_{CHF} - T'(p) = \dot{q}''_{CHF} / h_{fc\_FB}. \quad (20.57)$$

20.4 Critical heat flux

Natural or forced convection nucleate boiling is a very effective cooling mechanism. Increasing the local heat flux or decreasing the cooling mass flow rate may lead in technical facilities to film boiling in which the heat is transferred to the liquid through a vapor film. This is not an effective heat transfer mechanism. The effect on the material interface is manifested by a temperature jump. At high mass flow rates the jump is not as strong as at low mass flow rates. We distinguish two different mechanisms of occurrence of film boiling. The first one happens at low void fractions and is mainly caused by increasing the bubble population at the heated surface so that they merge together and build a film. The mechanical flow pattern is continuous liquid and dispersed void. This regime is known as departure from nucleate boiling or abbreviated DNB. Different is the mechanism at high void fractions. The mechanical flow pattern is continuous liquid film at the wall and dispersed droplet cared by the void. In this case simply the film dries out and the wall experiences an intimate contact with the steam. This regime is known as the *dry out* regime.

20.4.1 The hydrodynamic stability theory of free convection DNB

Kutateladze [17] established in 1951 the hydrodynamic stability theory. The main idea on which this theory relies is as follows: The vapor film with thickness above the heated surface governed by the capillary constant

$$\delta_{1F} \approx \left(\frac{\sigma}{g\Delta\rho_{21}} \right)^{1/2},$$

experiences a buoyancy force per unit wall surface equal to $g\Delta\rho_{21}\delta_{1F}$. The equilibrium between the dynamic pressure forces $\rho_1 V_{1ev}^2$, and the buoyancy force is distorted if the film is not stable. The transition point is controlled by the conditions

$$\rho_1 V_{1ev}^2 / \left[g\Delta\rho_{21} \left(\frac{\sigma}{g\Delta\rho_{21}} \right)^{1/2} \right] \approx const,$$

or

$$V_{1ev} \approx const \left[\frac{g\sigma(\rho_2 - \rho_1)}{\rho_1^2} \right]^{1/4}.$$

Kutateladze set the critical mass flow rate proportional to the critical velocity

$$\dot{q}''_{CHF} / \Delta h \approx const \left[\frac{g\sigma(\rho_2 - \rho_1)}{\rho_1^2} \right]^{1/4}$$

and found from experiments $const \approx 0.16$. Later the constant was replaced by the dimensionless group

$$const \approx 0.05 \left[10^4 \left(g\lambda_{RT}\rho_1 / p \right)^{1/2} \right]^{2/3}.$$

The obtained result is valid for saturated liquid. Subcooling increases the CHF. *Kutateladse* [18] found in 1962 the following relationship

$$\frac{\dot{q}''_{CHF,sub}}{\dot{q}''_{CHF,sat}} = 1 + 0.065 \left(\frac{\rho_2}{\rho_1}\right)^{4/5} \frac{c_{p2}(T'-T_2)}{h''-h'}.$$

This relationship was verified for $p = (1 \text{ to } 20) \times 10^5 \, Pa$, $\frac{c_{p2}(T'-T_2)}{h''-h'} \leq 0.6$ and $\rho_2/\rho_1 = 45$ to 1650. A very similar expression was obtained in the same year by *Irvey* and *Morris* [16] (1962),

$$\frac{\dot{q}''_{CHF,sub}}{\dot{q}''_{CHF,sat}} = 1 + 0.1 \left(\frac{\rho_2}{\rho_1}\right)^{3/4} \frac{c_{p2}(T'-T_2)}{h''-h'}.$$

The main ideas behind the hydrodynamic theory of boiling by *Zuber* [26] (1958) and *Zuber* and *Tribus* [27] (1958) is slightly different from those of *Kutateladse*. *Zuber* considered the *Rayleigh - Taylor* instability as responsible for the film instability. If the vapor film possesses a wavy vapor-liquid interface the characteristic wavelength is

$$\lambda^*_{RT} \approx 2\pi \left(\frac{\sigma}{g\Delta\rho_{21}}\right)^{1/2}.$$

Vapor jets are then formed with a distance between the axis equal to λ^*_{RT}. The jets are quadratically arranged and take a cross section α_{w1} part of the total surface. The vertical disturbed surface of the jets is unstable - *Helmholtz* instability. The waves are traveling with velocity

$$c^2 = \frac{\sigma n_c}{\rho_2 + \rho_1} - \frac{\rho_2 \rho_1}{(\rho_2 + \rho_1)^2}(V_1 - V_2)^2,$$

where n_c is the wave number. In critical conditions $c \approx 0$ and $n_c \approx 2\pi/\lambda^*_{RT}$ therefore

$$V_1 - V_2 = \left(\frac{\rho_2 + \rho_1}{\rho_2}\right)^{1/2} \left[\frac{g\sigma(\rho_2 - \rho_1)}{\rho_1^2}\right]^{1/4}$$

The rising vapor mass flux is equal to the liquid flux (continuity) and therefore

$$\alpha_{w1}\rho_1 V_1 = -\alpha_{w2}\rho_2 V_2$$

or

$$V_1 - V_2 = V_1 \frac{\alpha_{w2}\rho_2 + \alpha_{w1}\rho_1}{\alpha_{w2}\rho_{w2}}.$$

Thus the critical velocity of the rising vapor jets is

$$V_1 = \frac{\alpha_{w2}\rho_{w2}}{\alpha_{w2}\rho_{w2} + \alpha_{w1}\rho_{w1}} \left(\frac{\rho_2 + \rho_1}{\rho_2}\right)^{1/2} \left[\frac{g\sigma(\rho_2 - \rho_1)}{\rho_1^2}\right]^{1/4}.$$

Thus the critical heat flux is finally

$$\frac{\dot{q}''_{CHF}}{\rho_1(h''-h')} \approx j_1 = \alpha_{1d} V_1 = \alpha_{1d} \frac{\alpha_{w2}\rho_{w2}}{\alpha_{w2}\rho_{w2} + \alpha_{w1}\rho_{w1}} \left(\frac{\rho_2 + \rho_1}{\rho_2}\right)^{1/2} \left[\frac{g\sigma(\rho_2 - \rho_1)}{\rho_1^2}\right]^{1/4}$$

$$\approx \frac{\pi}{24}\left(\frac{\rho_2}{\rho_2 + \rho_1}\right)^{1/2} \left[\frac{g\sigma(\rho_2 + \rho_1)}{\rho_1^2}\right]^{1/4}$$

which is in fact the *Kutateladse* equation with multiplier $\frac{\pi}{24}\left(\frac{\rho_2}{\rho_2 + \rho_1}\right)^{1/2}$ instead of the constant. For low flow, $G \leq 100$ kg/(m²s), Zuber et al. [28] (1961) modified later this correlation to

$$\frac{\dot{q}''_{CHF}}{\rho_1(h''-h')} \approx 0.23164\ (0.96 - \alpha_1)\left(\frac{\rho_2}{\rho_2 + \rho_1}\right)^{1/2} \left[\frac{g\sigma(\rho_2 - \rho_1)}{\rho_1^2}\right]^{1/4}.$$

Note the two completely different approaches to the boiling crises by comparing this theory with the theory presented in Chapter 16. We showed in Chapter 16 that the micro-scale of the nucleation processes, bubble growth and departure, and the turbulence generated by this process naturally lead to self-saturation of the boundary layer with bubbles and dramatically change the mechanism of the heat transfer. In contrast, *Kutateldse* and *Zuber* analyzed the macro-scale above the layer where the micro-physics happens. That is the reason why their analysis gives an absolute upper limit which is about 5 times larger then the observed critical heat flux. This is reflected in the necessity of introducing an empirical constant of about 1/5.

20.4.2 Forced convection DNB and DO correlations

For flow in heated pipes and channels the mass flow rate influences the CHF. There are more than 500 empirical correlations for description of this process demonstrating that the final understanding of this phenomena is not yet reached. We give here three of the successful empirical correlations. *Smolin* et al. [25] considered separately the different flow pattern and for each particular flow pattern proposed a correlation for prediction of the CHF. For equilibrium mass flow fraction less then the one defined by

$$X_0 = 1.5 \frac{\rho_1}{\rho_2 + \rho_1} - 0.1 \text{ (bubble to film flow transition)}$$

the flow is bubbly. In this case *Smolin* et al. propose to use the *Kutateladze* [6] correlation in the form

$$\frac{\dot{q}''_{CHF,0}}{\rho_1(h''-h')} = 0.18 \left(\sqrt{\frac{v''}{v'}} - 1 \right) \left[\frac{g\sigma(\rho_2 - \rho_1)}{\rho_1^2} \right]^{1/4}.$$

For

$$X_{eq} \geq X_0$$

we have inverted annular flow called film boiling. In this case the pipe geometry, the mass flow rate and the equilibrium mass flow concentration influence the CHF

$$\dot{q}''_{CHF,2} = \dot{q}'''_{CHF,0} / \exp\left[0.2 \left(\frac{D_h}{\rho'\sigma} \right)^{1/3} G^{2/3} \left(X_{eq} - X_0 \right) \right].$$

For

$$X_{eq} < 0,$$

$$\dot{q}''_{CHF} = \dot{q}''_{CHF,2}.$$

Saturated liquid boiling: For saturated liquid boiling

$$X_{eq} \geq 0,$$

$$\dot{q}'''_{CHF} = \min\left(\dot{q}''_{CHF,2}, \dot{q}''^{*}_{CHF,3}, \dot{q}''_{CHF,3} \right),$$

the minimum of the CHF in the following regimes is taken:

Fine dispersed flow,

$$\dot{q}''_{CHF,3} = 0.22(h''-h')\frac{\eta''}{\eta'}\left[\frac{\eta''\rho''\rho'}{D_h(\rho'-\rho'')}\right]^{1/3} G^{2/3}(1-X_{eq})^2 / \left[X_{eq}(\rho'-\rho'')+\rho''\right]^{1/3};$$

Drying of the heated wall,

$$\dot{q}''^*_{CHF,2} = \frac{(h''-h')(\rho'-\rho'')}{45}\left(\frac{\sigma g}{\rho''}\right)^{1/4} \ln\left[\frac{0.9(\rho''\sigma)^{1/2}\left(\frac{v''}{v'}\right)^{2/3}(1-X_{eq})^{1/3}}{\eta'^{2/3}G^{2/3}D_h^{1/6}X_{eq}}\right];$$

Continuous vapor: For bubbly flow

$$X_{eq} < X_0,$$

and

$$p \leq 98 \times 10^5$$

we have boiling of subcooled liquid

$$\dot{q}''_{CHF} = \dot{q}''_{CHF,0}$$

$$+8.4\times10^{-3}(h''-h')\sqrt{\rho''}\left\{[\sigma g(\rho'-\rho'')]^{1/4}\left(\frac{\eta'}{\eta''}\right)\right\}^{1.25}\left[\frac{G}{\rho'}\left(\frac{\rho'-\rho''}{\sigma g}\right)^{1/4}\right]^{2/3}(X_0-X_{eq}) \quad (1)$$

For

$$p > 98 \times 10^5$$

$$\dot{q}''_{CHF,2\ at\ X_{eq}=X_0} = \dot{q}''_{CHF,0}$$

film to fine dispersed flow transition (hydraulic theory)

$$\dot{q}''_{CHF,3\ at\ X_{eq}=X_0} = 0.22(h''-h')\frac{\eta''}{\eta'}\left[\frac{\eta''\rho''\rho'}{D_h(\rho'-\rho'')}\right]^{1/3} G^{2/3}\frac{(1-X_0)^2}{\left[X_0(\rho'-\rho'')+\rho''\right]^{1/3}}$$

If

$$\dot{q}''_{CHF,2\ at\ X_{eq}=X_0} \geq \dot{q}''_{CHF,3\ at\ X_{eq}=X_0},$$

then

$\dot{q}''_{CHF}$ = Eq. (1), boiling of subcooled or saturated liquid. Otherwise,

$$\dot{q}''_{CHF,2 \text{ at } X_{eq}=X_0} < \dot{q}''_{CHF,3 \text{ at } X_{eq}=X_0},$$

the CHF is

$$\dot{q}''_{CHF} = \dot{q}''_{CHF,3 \text{ at } X_{eq}=X_0}$$

$$+ 0.01(h''-h')\left(\frac{\rho'}{\rho''}\right)^{1/2}[\sigma g(\rho'-\rho'')]^{1/4}\left[G\left(\frac{\rho'-\rho''}{\sigma g}\right)^{1/4}\right]^{1/2}(X_0-X_{eq})$$

Smolin et al. [25] (1977) verified the above set of correlations for $30 \times 10^5 \le p \le 200 \times 10^5$ Pa, $G \le 7500$ kg/(m²s), $T'-T_2 < 75$ K, $0.004 < D_h < 0.025$ m, 1 m $< L$.

Data for film boiling heat transfer in rod bundles are reported by *Morris* et al. [22].

Biasi et al. [3] proposed in 1967 the following correlation valid for tubes and $100 < G < 6000$ kg/(m²s), $0.003 < D_h < 0.0375$m, $0.2 < L < 6$m, $2.7 \times 10^5 < p < 140 \times 10^5$ Pa, $1/(1+\rho_2/\rho_1) < X_{eq} < 1$:

$$\dot{q}''_{CHF} = \max\left\{1000, \frac{1.883 \times 10^7}{D_h^{*n} G^{*1/6}}(FP/G^{*1/6} - X_{eq}), \frac{3.78 \times 10^7}{D_h^{*n} G^{*0.6}} HP(1-X_{eq})\right\},$$

where

$G^* = G/10$, $D_h^* = D_h/100$, $p^* = p/10^5$, $D_h > 0.01$, $n = 0.4$, $D_h \le 0.01$, $n = 0.6$,

$FP = 0.7249 + 0.099\, p^*/\exp(0.032\, p^*)$,

$HP = -1.159 + 0.149\, p^*/\exp(0.012 p^*) + 8.99\, p^*/(10+p^{*2})$.

Bowring [5] proposed the following correlation validated by comparison with 3800 experimental data points in the region: $2 \times 10^5 < p < 190 \times 10^5$ Pa, $136 < G < 18\,600$ kg/(m²s), $0.002 < D_h < 0.045$ m, $0.15 < L < 3.7$ m. The reported rms error is 7%.

$$\dot{q}''_{CHF} = \left[A - B(h''-h')X_{eq}\right]/C,$$

$$A = 2.317(h'' - h')B\, F_1 /(1+0.0143\, F_2 D_h^{0.5} G),$$

$$B = 0.25 D_h G,$$

$$C = 0.077\, F_3 D_h G / \left[1 + 0.347\, F_4\, (G/1356)^{SN}\right],$$

$$SN = 2 - 0.5\bar{p},$$

$$\bar{p} = 0.14510^{-6}\, p.$$

$\bar{p} \le 1$

$$F_1 = \left\{\bar{p}^{18.942} \exp\left[20.89(1-\bar{p})\right] + 0.917\right\}/1.917,$$

$$F_1 F_2 = \left\{\bar{p}^{1.3610} \exp\left[2.444(1-\bar{p})\right] + 0.309\right\}/1.309,$$

$$F_1 = F_1 /(F_1 F_2),$$

$$F_3 = \left\{\bar{p}^{17.023} \exp\left[16.658(1-\bar{p})\right] + 0.667\right\}/1.667,$$

$$F_4 F_3 = \bar{p}^{1.649},$$

$$F_4 = (F_4 F_3) F_3.$$

$\bar{p} > 1$

$$F_1 = \left\{\exp\left[0.648(1-\bar{p})\right]\right\}/\bar{p}^{0.368},$$

$$F_2 = \left\{F_1 \exp\left[0.245(1-\bar{p})\right]\right\}/\bar{p}^{0.448},$$

$$F_3 = \bar{p}^{0.219},$$

$$F_4 F_3 = \bar{p}^{1.649},$$

$$F_4 = (F_4 F_3) F_3$$

20.4.3 The 1995 look-up table

A look-up table for critical heat flux (CHF) has been developed jointly by AECL Research (Canada) and IPPE (Obninsk, Russia) [13]

$$q''_{CHF} = q''_{CHF}(p, G, X_{1,eq}, D_{hy}).$$

It is based on an extensive data base of CHF values obtained in tubes with a vertical upward flow of steam-water mixture. While the data base covers a wide range of flow conditions, the look-up table is designed to provide CHF values for 8 *mm* tubes at discrete values of pressure, mass flux and dry-out quality covering the ranges 0.1 to 20.0 *MPa*, 0.0 to 8000 $kg/(m^2s)$ and -0.5 to $+1.0$ respectively. Linear interpolation is used to determine the CHF for conditions between the tabulated values, and an empirical correction factor is introduced to extend this CHF table to tubes of diameter values other than 8 *mm*,

$$\dot{q}''_{CHF} = \dot{q}''_{CHF\_8mm}\left(\frac{1000 D_{hy}}{8}\right)^{-1/2}.$$

Compared against the combined AECL-IPPE world data bank (consisting of 22946 data points after excluding duplicate data and obviously erroneous data), the 1995 look-up table predicts the data with overall average and root-mean-square errors of 0.69% and 7.82%, respectively. An assessment of various CHF tables and several empirical correlations shows that the 1995 table consistently provides the best prediction accuracy and is applicable to the widest range of conditions.

Nomenclature

Latin

C_{nl} mass concentration of the inert component *n* inside the gas mixture, dimensionless
c_p specific heat at constant pressure, $J/(kgK)$
c wave travel velocity, *m/s*
D_{hy} hydraulic diameter (4 times flow cross-sectional area / wet perimeter), *m*
D_{heat} heated diameter (4 times flow cross-sectional area / heated perimeter), *m*
G $= \sum(\alpha \rho w)_l$, mass flow rate in the axial direction, $kg/(m^2s)$
g gravitational acceleration, m/s^2
h specific enthalpy, *J/kg*

h	specific mixture enthalpy - mixture enthalpy of the non-inert components, J/kg
h_c	convective heat transfer coefficient, $W/(m^2K)$
h_{NB}	nucleate boiling heat transfer coefficient, $W/(m^2K)$
h_{nc_FB}	natural convection film boiling heat transfer coefficient, $W/(m^2K)$
$h_{nc\_FB\_Berenson}$	natural convection film boiling heat transfer coefficient by Berenson, $W/(m^2K)$
h_{fc_FB}	forced convection film boiling coefficient, $W/(m^2K)$
h_{CHF}	critical heat transfer coefficient, $W/(m^2K)$
p	pressure, Pa
$\dot{q}''_w$	wall heat flux, W/m^2
$\dot{q}''_{TB}$	transition boiling heat flux, W/m^2
$\dot{q}''_{CHF}$	critical heat flux, W/m^2
$\dot{q}''_{CHF,sub}$	critical heat flux for subcooled liquid, W/m^2
$\dot{q}''_{CHF,sat}$	critical heat flux for saturated liquid, W/m^2
$\dot{q}''_{\min\_FB}$	minimum film boiling heat flux, W/m^2
n_c	wave number, -
Pr_{1w}	$= c_{p1}\eta_1/\lambda_1$, gas Prandtl number at wall temperature, dimensionless
Pr_2	$= c_{p2}\eta_2/\lambda_2$, liquid Prandtl number, dimensionless
Re	$= X_2 G D_{hy}/\eta_2$, liquid Reynolds number, dimensionless
Re_1	$= w_1 D_1/\nu_1$, steam Reynolds number, dimensionless
Re_{1_mix}	$= \dfrac{D_h w_{eq}}{\eta''/\rho''}$, mixture Reynolds number based on vapor viscosity, dimensionless
Re_{Tp}	modified Reynolds number, dimensionless
S	ratio of the effective superheat to the total superheat of the wall, dimensionless
T	temperature, K
$T_{w,MFB}$	minimum film boiling temperature, K
X_l	mass flow rate concentration in the flow of the velocity field l, dimensionless
X_{tt}	Martinelli factor, dimensionless
X_{eq}	equilibrium steam mass flow rate concentration in the flow, dimensionless
w	axial velocity, m/s
w_{eq}	$= G/\rho_{eq}$ center of mass velocity

Greek

α	volume fraction, *dimensionless*
ΔT_{CHF}	temperature difference corresponding to the critical heat flux, K
ΔT_{min_FB}	minimum film boiling temperature, K
δ_{1F}	$\approx \left(\dfrac{\sigma}{g\Delta\rho_{21}}\right)^{1/2}$, capillary constant, m
η	dynamic viscosity of liquid, kg/(ms)
λ	thermal conductivity, $W/(mK)$
λ_{RT}	$=\left[\dfrac{\sigma}{g(\rho'-\rho'')}\right]^{1/2}$, *Rayleigh-Taylor* instability wavelength, m
ρ	density, kg/m³
ρ_{eq}	homogeneous mixture equilibrium density, kg/m³
σ	surface tension, N/m
μ_{21}	evaporation mass from the liquid film per unit time and unit mixture volume, kg/(m³s)
μ_{31}	evaporation mass from the droplets per unit time and unit mixture volume, kg/(m³s)
μ_{12}	condensation mass per unit time and unit mixture volume, kg/(m³s)

Subscript

'	saturated steam
"	saturated liquid
1	gas
2	liquid
w	wall
M1	non-inert component in the gas mixture
n1	inert component in the gas mixture
1F	vapor film

References

1. Addoms JN (1948) D. Sc. Thesis, M. I. T.
2. Andreani A, Yadigaroglu G (1994) Prediction methods for dispersed flow film boiling, Int. J. Multiphase Flow, vol 20 pp 1-51
3. Biasi L, Clerici GC, Garribba S, Sala R, Tozzi A (1967) Studies on burnout. Part 3: A new correlation for round ducts and uniform heating and its comparison with world data, Energ. Nucl., vol 14 pp 530-536

4. Bjornard TA, Grifith P (1977) PWR blow-down heat transfer, Thermal and Hydraulic Aspects of Nuclear Reactor Safety, Americal Society of Mechanical Engineers, New York, vol 1 pp 17-41
5. Bowring RW (1972) Simple but accurate round tube, uniform heat flux, dryout correlation over the pressure range 0.7 to 17 MN/m^2 (100 to 2500 psia). AEEW-R-789
6. Chen JC (1963) A correlation for film boiling heat transfer to saturated fluids in convective flow, ASME Publication-63-HT-34, p 2-6
7. Dittus FW, Boelter LMK (1937) Heat transfer in automobile radiators of the tubular type, University of California Publications in Engineering, vol 2 no 13 p 443
8. Dougall RS, Rohsenow WM (1963) Film boiling on the inside of vertical tubes with upward flow of the fluid of low qualities, MIT-TR-9079-26, Massachusetts Institute of Technology
9. Forster HK, Zuber N (1955) Dynamic of the vapor bubbles and boiling heat transfer, AIChE Journal, vol 1 no 4 pp 531-535
10. Groeneveld DC, Delorme GGJ (1976) Prediction of thermal non-equilibrium in the post-dryout regime, NED, vol 36 pp 7-26
11. Groeneveld DC (Mar. 26-28, 1977) Post-dryout heat transfer at reactor operating conditions. Nat. Topical Meet. Water Reactor Safety, Salt Lake City, Utah, American Nuclear Society, Conf. 730304, Rept. AECL-4513, Atomic Energy of Canada Ltd
12. Groeneveld DC, Chen SC, Leung LKH, Nguyen C (1989) Computation of single and two-phase heat transfer rates suitable for water-cooled tubes and subchannels, Nuclear Engineering and Design, vol 114 pp 61-77
13. Groeneveld DC, Leunga LKH, Kirillov PL, Bobkov VP, Smogalev IP, Vinogradov VN, Huangc XC, Royer E (1996) The 1995 look-up table for critical heat flux in tubes, Nuclear Engineering and Design, vol 163 pp 1 -23
14. Hadaller G, Banerjee S (1969) Heat transfer to superheated steam in round tubes, AECL Unpublished Report WDI-147
15. Hetstrony G (ed) (1982) Handbook of multiphase systems, Hemishere Publ. Corp., Washington etc., McGraw-Hill Book Company, New York
16. Irvey HJ, Morris DJ (1962) On the relevance of the vapor - liquid exchange mechanism for subcooled boiling Heat Transfer at High Pressure, AEEW-R127
17. Kutateladze SS (1951) A hydrodynamic theory of changes in the boiling process under free convection conditions. Izv. Akad. Nauk SSSR, Otd. Tech. Nauk, vol 4 pp 529-536; AEC-tr-1991 (1954)
18. Kutateladze SS (1962) Osnovy teorii teploobmena, Moskva, Mashgis, p 456
19. Leung LKH, Hammouda N, Groeneveld DC (1997) A look-up table for film-boiling heat transfer coefficients in tubes with vertical upward flow, Proceedings of the Eighth International Topical Meeting on Nuclear Reactor Thermal-Hydraulics, Kyoto, Japan, vol 2 pp 671-678
20. Loomis GG, Schumway RW (Oct. 1983) Low-pressure transient flow film boiling in vertically oriented rod bundles, Nucl. Technology, vol 63 pp 151-163
21. Miropolskij ZL (1963) Heat transfer in film boiling of steam-water mixture in steam generating tubes, Teploenergetika, vol 10 no 5 pp 49-53
22. Morris DG, Mullins CB, Yoder GL (Apr. 1985) An experimental study of rod bundle dispersed-flow film boiling with high pressure water, Nucl. Tech., vol 69 pp 82-93
23. Polomik EE (October 1967) Transient boiling heat transfer program. Final summary report on program for february 1963- october 1967, GEAP–5563

24. Ramu K, Weisman J (1974) A method for the correlation of transition boiling heat transfer data, Heat Transfer 1974, 5th Int. Heat Transfer Conf., Tokyo, vol 4 pp 160-164
25. Smolin VN, Shpanskii SV, Esikov VI, Sedova TK (1977) Method of calculating burn-out in tubular fuel rods when cooled by water and a water-steam mixture. Teploenergetika, vol 24 no 12 pp 30-35
26. Zuber N (1958) On the stability of boiling heat transfer, Trans. ASME, vol 80 pp 711-720
27. Zuber N, Tribus M (1958) Further remarks on the stability of boiling heat transfer, Report 58-5, Department of Engineering, University of California, Loss Angeles
28. Zuber N et al. (1961) The hydrodynamic crisis in pool boiling of saturated and sub-cooled liquids, Int. Heat Transfer Conf., Boulder, Colorado, International Developments in Heat Transfer, Part 2 no 27 pp 230-236

21. Film boiling on vertical plates and spheres

Section 21.1 of this Chapter presents a closed analytical solution for mixed-convection film boiling on vertical walls. Heat transfer coefficients predicted by the proposed model and experimental data obtained at the Royal Institute of Technology in Sweden by Okkonen et al. are compared. All data predicted are inside the ±10% error band, with mean averaged error being below 4% (using slightly modified analytical solution). The solution obtained is recommended for practical applications. The method presented here is used in Section 22.2 as a guideline for developing a model for film boiling on spheres. The new semi-empirical film boiling model for spheres used in IVA computer code is compared with the experimental data base obtained by Liu and Theofanous. The data are predicted within ±30% error band. This Chapter is a slightly changed variant of the work previously published in [19, 21].

21.1 Plate

21.1.1 Introduction

In nature, film boiling is the controlling heat transfer mechanism during lava-water interaction. In the cryogenic technology rocket engines, it is the inevitable heat transfer mechanism in many cases. Film boiling heat transfer is of particular interest for analysis of accident processes in industry. The theory of film boiling heat transfer has become a classical part of heat and mass transfer textbooks but is often presented in the context of an unwanted heat transfer regime which has to be avoided by design. The close scrutiny paid to accident processes in the industry requires critical re-evaluation of film boiling theory in order to yield a reliable prediction tool for engineering purposes. In particular, calculation of the components of the heat removed by film boiling that leads to vapor heating, liquid heating, evaporation mass flow rate, average vapor volume attached to the surface is very important for increasing the accuracy of the overall flow field prediction of melt interacting with water. As an example of the necessity for improvement of theory, it should be recalled that there is still a need for a closed analytical solution for mixed natural and forced-convection film boiling. In this Chapter a complete analytical solution for film boiling on a vertical wall will first be presented. After this, the attempt will be made to extend the model for spheres. Comparison with experimental data will indicate the success and the limitations of the theoretical results. Where necessary, empirical correction factors are introduced to give more accurate prediction of existing experimental data.

21.1.2 State of the art

Film boiling literature was reviewed by *Sakurai* [32] in 1990, and more recently by *Liu* and *Theofanous* [23] in 1995. There is no need to repeat this here.

In his pioneering work of 1916 *Nusselt* [29] presented a closed model of laminar film condensation on vertical surfaces. The idea used was that the condensation is controlled by the heat conduction through the falling liquid film. This model is so easily applied to film boiling under natural convection on vertical plates, postulating that the heat transfer is controlled by the heat conduction through the vapor film that it deserves to be called the *Nusselt theory for laminar condensation and film boiling on vertical walls due to natural convection*.

Natural-convection film boiling was first investigated by *Bromley* [3] in 1950. He applied *Nusselt*'s idea from 1916. *Frederking* et al. [14] demonstrated in 1965 that the treatment of vertical plates, spheres and horizontal cylinders results in the same expression multiplied by a numerical constant to characterize the appropriate one of the above geometries. Subsequent to this, many theoretical studies related to this subject have been published. The main physical ideas on which the published models rely are similar to those published by *Bromley* et al. [3, 5] for natural and mixed convection, and by *Witte* [36] in 1968 for forced convection.

A rigorous formulation of the steady-state problem for *natural convection* was given by *Nishikawa* et al. [27, 28]. Several numerical solutions of the problem defined with some simplifying assumptions are available, e.g. *Dhir* and *Purohit* [8]. An accurate numerical solution is compared with an extensive set of data for cylinders and spheres by *Sakurai* et al. in [32, 31]. *Sakurai* et al. proposed the most accurate correlation for cylinder geometry known to the author. This forms a good approximation of the accurate numerical solution and compares favorably with experimental data for saturated and subcooled water at atmospheric and elevated pressure, with radiation effects also considered. The same correlation has been recently recommended by *Liu* and *Theofanous* [23] after comparison with an extensive experimental database.

Subsequent to *Witte*'s analytical solution [36] for simplified *forced-convection* steady-state problems in 1968, there have only been numerical solutions provided for this problem in the literature, e.g. *Fodemski* et al. [10], *Liu* and *Theofanous* [23].

A third group within the literature covers the conditions of the *stagnation point*. Assuming that the stagnation point condition is valid in the front half of the sphere, *Epstein* and *Heuser* derived in [9] a correlation which additionally takes into account the liquid subcooling. A numerical solution to this problem is proposed by *Fodemski* [12, 13, 16].

The effect of radiation and subcooling or superheating on the film conduction heat flux is taken into account in different ways. Some authors superposed the separate effects, recommending some reduction of the radiative component [32]. The natural way to take these effects into account is to consider them within the formalism describing the local film thickness. The resulting system of differential equations is solved numerically, e.g. by *Sparrow* [33], *Fodemski* [10], *Liu* and

Theofanous [23]. There is no simple analytical solution to this problem available in the literature.

A numerical solution of the combined natural and forced-convection film boiling problem for spheres in saturated, subcooled or superheated water with radiation is possible as demonstrated in the literature, but requires a very large amount of computation effort if implemented in system computer codes. To analyze the interaction of fragmenting liquid metal in water, a simple method is needed which has the capability (a) to reproduce the available data for the heat transfer coefficient for natural and forced- convection, (b) to separate the effect of radiation, subcooling or superheating in order to compute the heat flux component consumed to cool the hot structure, to heat or cool the liquid, and to produce steam. Dynamic system characteristics can then be predicted by using such a model which provides heat and mass source terms for the macroscopic differential equation governing the flow. Some of the existing models for this case satisfy one or two of the above conditions, but not all of them. It was for this reason that the author developed a new simplified approach which permits separation of the different effects of radiation, subcooling or superheating and computes the particular heat flux components.

21.1.3 Problem definition

Consider the geometry presented in Fig. 21.1. Liquid of temperature T_2 flows along a vertical heated wall. The surface temperature of the wall $T_{3\sigma}$ is so high that a stable vapor film is formed. The evaporation at the liquid-vapor interface forces the interface temperature to be equal to the saturation temperature at the system pressure $T'(p)$. The thickness of the vapor film $\delta_1(z)$ varies with the height z. We introduce two coordinate systems having the same axial coordinate z. The horizontal distance from the wall to the vapor film is r_1. The horizontal distance from the vapor-liquid interface with the liquid is r_2. The vapor velocity has only a vertical component $w_1(r_1)$. The liquid in the boundary layer has a vertical component of $w_2(r_2)$. Various different heat fluxes can immediately be identified first. The radiation heat flux can be split into three different components: $\dot{q}''_{r1}$ is the radiation heat flux from the interface to the vapor. $\dot{q}''_{r2}$ is the radiation heat flux from the wall to the bulk liquid. $\dot{q}''_{r\sigma 2}$ is the radiation heat flux absorbed from the vapor-liquid interface. The partitioning depends mainly on the temperature of the emitting wall. For temperatures below 1000 K almost all radiation energy is dissipated at the liquid interface. This is not the case for the short wavelength light corresponding to higher temperatures. At the present time, it is not clear whether the liquid-vapor interface instabilities create a boundary layer transparent to light or not. The heat flux $\dot{q}''_{FB}$ is due to conduction through the vapor film. The heat flux $\dot{q}''_{3\sigma 1}$ is from the wall into the vapor, due the convective heating of the vapor from the saturation interface temperature to some averaged vapor film temperature. This contribution is frequently neglected in the film boiling literature. In the case of subcooled liquid there is heat flux $\dot{q}''_{\sigma 2}$ from the liquid interface into the liquid due to conduction and convection.

The problem to be solved is to compute all components of the wall heat flux.

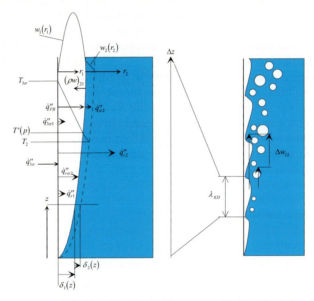

Fig. 21.1. a) Geometry definition for film boiling on vertical wall. b) Instability of film boiling

21.1.4 Simplifying assumptions

The following simplifying assumptions are introduced:

1) The vapor and the liquid are incompressible.

2) The following simplified momentum equations describe flows in the vapor and liquid boundary layers respectively:

$$\eta_1 \frac{\partial^2 w_1}{\partial r_1^2} = \frac{\partial p}{\partial z} + \rho_1 g , \qquad (21.1)$$

$$\eta_2 \frac{\partial^2 w_2'}{\partial r_2^2} = \frac{\partial p}{\partial z} + \rho_2 g \approx 0. \tag{21.2}$$

Since the last equation simplifies to $\frac{\partial p}{\partial z} \approx -\rho_2 g$ one can write

$$\eta_1 \frac{\partial^2 w_1}{\partial r_1^2} = -(\rho_2 - \rho_1)g, \tag{21.3}$$

$$\eta_2 \frac{\partial^2 w_2'}{\partial r_2^2} = 0. \tag{21.4}$$

The above two equations can be integrated analytically to give

$$w_1(r_1) = -\frac{1}{2\eta_1}(\rho_2 - \rho_1)g r_1^2 + C_1 r_1 + C_3 \tag{21.5}$$

$$w_2'(r_2) = C_2 r_2 + C_4. \tag{21.6}$$

The constants are determined from the following boundary conditions

1) at $r_1 = 0$, $w_1 = 0$, \hfill (21.7)

2) at $r_1 = \delta_1$ and $r_2 = 0$, $w_1 = w_2'$, \hfill (21.8)

3) at $r_1 = \delta_1$ and $r_2 = 0$, $\eta_1 \frac{\partial w_1}{\partial r_1} = \eta_2 \frac{\partial w_2'}{\partial r_2}$, \hfill (21.9)

4) at $r_2 = \delta_2$, $w_2' = w_2$. \hfill (21.10)

The solutions to Eqs. (21.5) and (21.6) are

$$w_1 = \frac{\Delta \rho g \delta_1^2}{2\eta_1}\left[\frac{r_1}{\delta_1}\frac{f+2\xi}{f+\xi} - \left(\frac{r_1}{\delta_1}\right)^2\right] + \frac{r_1}{\delta_1}\frac{f}{f+\xi}w_2, \tag{21.11}$$

$$w_2' = \left(1 - \frac{r_2}{\delta_2}\right)\left[\frac{\Delta \rho g \delta_1^2}{2\eta_2}\frac{f\xi}{f+\xi} + \frac{f}{f+\xi}w_2\right] + \frac{r_2}{\delta_2}w_2 \tag{21.12}$$

where

$$f = \eta_2 / \eta_1, \qquad (21.13)$$

$$\xi = \delta_2 / \delta_1. \qquad (21.14)$$

For the case of zero liquid velocity Eqs. (21.11) and (21.12) reduce to those reported by *Liu* and *Theofanous* [23].

In this Chapter the assumption is made that the layers thickness ratio is not a function of the axial coordinate.

The cross-section averaged vapor velocity profile is

$$\langle w_1 \rangle_1 = \frac{1}{\delta_1} \int_0^{\delta_1} w_1 dr_1 = \int_0^1 \left\{ \frac{\Delta \rho g \delta_1^2}{2\eta_1} \left[\frac{r_1}{\delta_1} \frac{f + 2\xi}{f + \xi} - \left(\frac{r_1}{\delta_1} \right)^2 \right] + \frac{r_1}{\delta_1} \frac{f}{f + \xi} w_2 \right\} d \frac{r_1}{\delta_1} \qquad (21.15)$$

or

$$\langle w_1 \rangle_1 = \frac{\Delta \rho g \delta_1^2}{12 \eta_1} \frac{f + 4\xi}{f + \xi} + \frac{1}{2} \frac{f}{f + \xi} w_2. \qquad (21.16)$$

The temperature profile in the vapor layer is assumed to be linear,

$$T_1 - T' = \Delta T_{sp} \left(1 - \frac{r_1}{\delta_1} \right). \qquad (21.17)$$

This assumption results from the observation that the cyclic instabilities do not allow thick films to exist. The order of magnitude is tenths of a *mm*. To permit further treatment, the following average is required

$$\langle w_1 (T_1 - T') \rangle_1 = \frac{1}{\delta_1} \int_0^{\delta_1} w_1 (T_1 - T') dr_1$$

$$= \Delta T_{sp} \left[\langle w_1 \rangle_1 - \int_0^1 \left\{ \frac{\Delta \rho g \delta_1^2}{2\eta_1} \left[\left(\frac{r_1}{\delta_1} \right)^2 \frac{f + 2\xi}{f + \xi} - \left(\frac{r_1}{\delta_1} \right)^3 \right] + \left(\frac{r_1}{\delta_1} \right)^2 \frac{f}{f + \xi} w_2 \right\} d \frac{r_1}{\delta_1} \right]. \qquad (21.18)$$

The last equation can, after integration, be written in the form:

$$\langle w_1 (T_1 - T') \rangle_1 = \frac{1}{2} \Delta T_{sp} \frac{\Delta \rho g \delta_1^2}{12 \eta_1} \frac{f + 3\xi}{f + \xi} + \frac{1}{6} \Delta T_{sp} \frac{f}{f + \xi} w_2. \qquad (21.19)$$

The Reynolds similarity hypothesis for the liquid boundary layer is used:

$$T_2^* - T_2 = \Delta T_{sc}\left(1 - \frac{r_2}{\delta_2}\right). \tag{21.20}$$

This means that the velocity boundary layer is similar to the temperature boundary layer. This hypothesis holds for liquids with $\Pr \approx 1$. For the energy conservation of the liquid boundary layer the following average is required

$$\langle w_2(T_2' - T_2)\rangle_2 = \frac{1}{\delta_2}\int_0^{\delta_2} w_2(T_2' - T_2)\,dr_2$$

$$= \Delta T_{sc}\int_0^1 \left\{\left(1 - \frac{r_2}{\delta_2}\right)\left[\frac{\Delta\rho g \delta_1^2}{2\eta_2}\frac{f\xi}{f+\xi} + \frac{f}{f+\xi}w_2\right] + \frac{r_2}{\delta_2}w_2\right\}d\frac{r_2}{\delta_2}$$

$$- \Delta T_{sc}\int_0^1 \left\{\left(\frac{r_2}{\delta_2} - \left(\frac{r_2}{\delta_2}\right)^2\right)\left[\frac{\Delta\rho g \delta_1^2}{2\eta_2}\frac{f\xi}{f+\xi} + \frac{f}{f+\xi}w_2\right] + \left(\frac{r_2}{\delta_2}\right)^2 w_2\right\}d\frac{r_2}{\delta_2}$$

$$= \Delta T_{sc}\left(\frac{2\Delta\rho g \delta_1^2}{12\eta_2}\frac{f\xi}{f+\xi} + \frac{1}{6}\frac{3f+\xi}{f+\xi}w_2\right). \tag{21.21}$$

21.1.5 Energy balance at the vapor-liquid interface, vapor film thickness, average heat transfer coefficient

In this section the layer thickness ratio ξ is taken as known. A separate section deals with calculation of the layer thickness ratio.

We assume that no energy can be accumulated at the vapor-liquid interface. As a consequence, the energy entering the interface due to conduction from the wall through the vapor, minus the energy removed convectively for heating of the vapor plus the radiation energy dissipated into the surface, minus the energy removed from the interface into the liquid will then evaporate a certain amount of steam at the surface:

$$\left(\lambda_1 \Delta T_{sp}/\delta_1 + \dot{q}''_{r\sigma 2} - \lambda_2 \Delta T_{sc}/\delta_2\right)dA_3 - d\left[\rho_1 c_{p1}\langle w_1[T_1(r_1) - T']\rangle_1 A_1\right]$$

$$= d\left(\rho_1 \Delta h \langle w_1\rangle_1 A_1\right). \tag{21.22}$$

The film cross-section at the position defined by the coordinate z is expressed by

$$A_1 = \delta_1 L, \qquad (21.23)$$

where L is the transversal dimension of the plate. The infinitesimal wall surface is expressed by

$$dA_3 = L dz. \qquad (21.24)$$

By appropriate substitution within the energy jump condition the following is then obtained

$$\lambda_1 \frac{\Delta T_{sp}}{\delta_1}\left(1+\frac{\dot{q}''_{r\sigma 2}}{\lambda_1 \Delta T_{sp}/\delta_1}-\frac{\lambda_2 \Delta T_{sc}/\delta_2}{\lambda_1 \Delta T_{sp}/\delta_1}\right)dz$$

$$= \rho_1 \Delta h d\left\{\frac{\langle w_1\rangle_1 + c_{p1}\langle w_1[T_1(r_1)-T']\rangle_1}{\Delta h}\delta_1\right\}. \qquad (21.25)$$

Introducing the following abbreviations

$$Sp = c_{p1}\Delta T_{sp}/\Delta h, \qquad (21.26)$$

$$r_r = \frac{\dot{q}''_{r\sigma 2}}{\lambda_1 \Delta T_{sp}/\delta_1} \approx const, \qquad (21.27)$$

$$C_2 = \frac{\lambda_2 \Delta T_{sc}}{\lambda_1 \Delta T_{sp}}, \qquad (21.28)$$

$$\xi = \delta_2/\delta_1, \qquad (21.29)$$

$$f_{nc} = \frac{f+4\xi}{f+\xi}\frac{1+\dfrac{1}{2}Sp\dfrac{f+3\xi}{f+4\xi}}{1+r_r-C_2/\xi}, \qquad (21.30)$$

$$f_{fc} = \frac{f}{f+\xi}\frac{1+\dfrac{1}{3}Sp}{1+r_r-C_2/\xi}, \qquad (21.31)$$

Eq. (21.25) is finally obtained in the form:

$$\frac{\lambda_1 \Delta T_{sp}}{\rho_1 \Delta h} dz = \frac{\Delta \rho g}{12\eta_1} f_{nc} \frac{3}{4} d\delta_1^4 + \frac{1}{4} f_{fc} w_2 d\delta_1^2 . \tag{21.32}$$

Before integrating the last equation it is assumed that the ratio of the radiative flux and the local film conduction heat flux is constant, Eq. (21.27). At first glance this assumption would seem arbitrary but as will be discussed later, this turns out to be a good assumption. Integrating with the initial condition $z = 0$, $\delta_1 = 0$ the local vapor film thickness as a function of the axial coordinate is then obtained

$$z = \frac{\rho_1 \Delta h \Delta \rho g}{\lambda_1 \Delta T_{sp} \eta_1} f_{nc} \left(\frac{\delta_1}{2}\right)^4 + \frac{\rho_1 \Delta h w_2}{\lambda_1 \Delta T_{sp}} f_{fc} \left(\frac{\delta_1}{2}\right)^2 . \tag{21.33}$$

Comparing the above equation to Eq. (21.6) by *Bui* and *Dhir* [6] it is evident that they are equivalent only when the conditions $f_{nc} = 1$ and $f_{fc} = 1$ hold, with the reason behind this not directly evident.

Calculation of the average heat transfer coefficients for the limiting case of pure natural convection is given in Appendix 21.1. An estimate of the average heat transfer coefficients for the limiting case of predominant forced convection is given in Appendix 21.2. In order to simplify Eq. (21.33) the resulting average heat transfer coefficients are introduced

$$h_{nc} = c_{nc} \left[\frac{\lambda_1^3 \rho_1 \Delta h \Delta \rho g}{\Delta T_{sp} \eta_1 \Delta z} f_{nc}\right]^{1/4}, \tag{21.34}$$

$$h_{fc} = c_{fc} \left[\frac{\rho_1 \lambda_1 \Delta h w_2}{\Delta T_{sp} \Delta z} f_{fc}\right]^{1/2}, \tag{21.35}$$

where

$$c_{nc} = 2/3, \tag{21.36}$$

and

$$c_{fc} = 1. \tag{21.37}$$

Eqs. (21.34) and (21.35) can be written in terms of dimensionless numbers as follows

$$Nu_{nc} = c_{nc} \left(Gr_1 \frac{Pr_1}{Sp} f_{nc} \right)^{1/4}, \qquad (21.38)$$

$$Nu_{fc} = c_{fc} \left(Re_1 \frac{Pr_1}{Sp} f_{fc} \right)^{1/2}. \qquad (21.39)$$

As it will be demonstrated later h_{nc} is the heat transfer coefficient for the case of zero liquid velocity, that is natural convection only, and h_{fc} is the heat transfer coefficient for predominant forced convection. Equation (21.33) can thus be rewritten in the following form

$$\left(\frac{\delta_1 h_{nc}}{\lambda_1 2 c_{nc}} \right)^4 + 2r^* \left(\frac{\delta_1 h_{nc}}{\lambda_1 2 c_{nc}} \right)^2 - \frac{z}{\Delta z} = 0 \qquad (21.40)$$

where

$$r^* = \frac{1}{2} \left(\frac{h_{fc} c_{nc}}{h_{nc} c_{fc}} \right)^2 = \frac{1}{2} \left(\frac{\rho_1}{\Delta \rho} \frac{Pr_1}{Sp} \frac{f_{fc}^2}{f_{nc}} \right)^{1/2} Fr_2, \qquad (21.41)$$

and the *Froude* number is defined as

$$Fr_2 = \frac{w_2}{\sqrt{\Delta z g}}. \qquad (21.42)$$

The film thickness as a function of the axial coordinate is therefore

$$\frac{\delta_1 h_{nc}}{\lambda_1 2 c_{nc}} = \left[\left(r^{*2} + \frac{z}{\Delta z} \right)^{1/2} - r^* \right]^{1/2}. \qquad (21.43)$$

At location $z = \Delta z$, vapor thickness is at its maximum, i.e.

$$\frac{\delta_{1,\max} h_{nc}}{\lambda_1 2 c_{nc}} = \left[\left(r^{*2} + 1 \right)^{1/2} - r^* \right]^{1/2}. \qquad (21.44)$$

The averaged film boiling heat transfer coefficient is therefore

$$h = \frac{\dot{q}''_{FB}}{\Delta T_{sp}} = \frac{1}{\Delta z}\int_0^{\Delta z}\frac{\lambda_1}{\delta_1}dz = \frac{h_{nc}}{2c_{nc}}\int_0^1\left[\left(r^{*2}+\frac{z}{\Delta z}\right)^{1/2}-r^*\right]^{-1/2}d\frac{z}{\Delta z}$$

$$=\frac{h_{nc}}{2c_{nc}}\frac{4}{3}\left[\left(r^{*2}+1\right)^{1/2}+2r^*\right]\left[\left(r^{*2}+1\right)^{1/2}-r^*\right]^{1/2}$$

$$= h_{nc}\left[\left(r^{*2}+1\right)^{1/2}+2r^*\right]\left[\left(r^{*2}+1\right)^{1/2}-r^*\right]^{1/2}. \tag{21.45}$$

When liquid velocity is equal to zero, $r^*=0$, the expected solution $h = h_{nc}$ is obtained. Equation (21.45) is an important new result. This represents the average heat transfer coefficient for mixed convection as a function of the average flow properties, the heat transfer coefficient for natural convection and the *Froude number*. Note that this equation still contains one unknown variable: the layer thickness ratio ξ. The second equation is obtained in the next section by using the energy balance of the liquid boundary layer.

21.1.6 Energy balance of the liquid boundary layer, layer thickness ratio

The energy balance for the liquid boundary layer has the following form

$$\left(\lambda_2\Delta T_{sc}/\delta_2\right)dA_3 = d\left[\rho_2 c_{p2}\left\langle w_2\left[T_2'(r_2)-T_2\right]\right\rangle_2 A_2\right]. \tag{21.46}$$

As has already been mentioned, it is not clear at the present time whether the liquid-vapor interface instabilities create a boundary layer transparent to light or not. Observations with the naked eye indicate that the interface is rather "milky" and therefore probably not so transparent as a water-steam interface at rest. This point deserves special attention in the future.

The liquid boundary layer cross-section at the position defined by the coordinate z is

$$A_2 = \delta_2 L, \tag{21.47}$$

Substituting Eq. (21.21) for the average used in the last equation, the following is then obtained

$$dz = \frac{\rho_2 c_{p2}}{\lambda_2}\left(\frac{2\Delta\rho g}{12\eta_2}\frac{f\xi^3}{f+\xi}\delta_1 d\delta_1^3 + \frac{1}{6}\frac{3f+\xi}{f+\xi}\xi^2 w_2\delta_1 d\delta_1\right). \tag{21.48}$$

After integrating, this yields:

$$z = \frac{\rho_2 c_{p2}}{\lambda_2}\frac{\Delta\rho g}{\eta_2}\frac{f\xi^3}{f+\xi} 2\left(\frac{\delta_1}{2}\right)^4 + \frac{1}{3}\frac{3f+\xi}{f+\xi}\xi^2 w_2 \frac{\rho_2 c_{p2}}{\lambda_2}\left(\frac{\delta_1}{2}\right)^2. \qquad (21.49)$$

Since the form of Eq. (21.49) is the same as Eq. (21.33), the same transformation can be applied to obtain another expression for the average heat transfer coefficient. Equating the right hand sides of the Eqs. (21.33) and (21.49), the definition equation for computation of the layer thickness ratio is then obtained. The heat transfer coefficients as function of the liquid side properties will next be computed in a similar way to the previous section.

First, the following definition is made

$$h_{2,nc} = c_{2,nc}\left(\lambda_1^4 \frac{\rho_2 c_{p2}}{\lambda_2}\frac{\Delta\rho g}{\eta_2 \Delta z} 2\frac{f\xi^3}{f+\xi}\right)^{1/4}, \qquad (21.50)$$

$$h_{2,fc} = c_{2,fc}\left(\lambda_1^2 \frac{1}{3}\frac{3f+\xi}{f+\xi}\xi^2 w_2 \frac{1}{\Delta z}\frac{\rho_2 c_{p2}}{\lambda_2}\right)^{1/2}, \qquad (21.51)$$

where

$$c_{2,nc} = 1, \qquad (21.52)$$

and

$$c_{2,fc} = 1. \qquad (21.53)$$

With this notation Eq. (21.49) divided by Δz gives

$$\left(\frac{\delta_1 h_{2,nc}}{\lambda_1 2 c_{2,nc}}\right)^4 + 2r_2^* \left(\frac{\delta_1 h_{2,nc}}{\lambda_1 2 c_{2,nc}}\right)^2 - \frac{z}{\Delta z} = 0, \qquad (21.54)$$

where

$$r_2^* = \frac{1}{2}\left(\frac{h_{2,fc} c_{2,nc}}{h_{2,nc} c_{2,fc}}\right)^2 = \frac{1}{2}\left[\frac{1}{18}\frac{\xi(3f+\xi)^2}{f(f+\xi)}\frac{\rho_2}{\Delta\rho} Pr_2\right]^{1/2} Fr_2. \qquad (21.55)$$

Solving for the film thickness yields

$$\frac{\delta_1 h_{2,nc}}{\lambda_1 2 c_{2,nc}} = \left[\left(r_2^{*2} + \frac{z}{\Delta z} \right)^{1/2} - r_2^* \right]^{1/2} \tag{21.56}$$

Using Eq. (21.54) the following is obtained for the average heat transfer coefficient

$$h = h_{2,nc} \left[\left(r_2^{*2} + 1 \right)^{1/2} + 2 r_2^* \right] \left[\left(r_2^{*2} + 1 \right)^{1/2} - r_2^* \right]^{1/2}. \tag{21.57}$$

By comparing Eqs. (21.57) and (21.45) the definition equation for the film thickness ratio is obtained as follows:

$$h_{2,nc} \left[\left(r_2^{*2} + 1 \right)^{1/2} + 2 r_2^* \right] \left[\left(r_2^{*2} + 1 \right)^{1/2} - r_2^* \right]^{1/2}$$

$$= h_{nc} \left[\left(r^{*2} + 1 \right)^{1/2} + 2 r^* \right] \left[\left(r^{*2} + 1 \right)^{1/2} - r^* \right]^{1/2}. \tag{21.58}$$

This is a transcendental equation with respect to the layer thickness ratio and must be solved by iteration. An analytical solution for natural circulation is available as an initial value. Setting the liquid velocity equal to zero yields $h_{2,nc} = h_{nc}$, which results in a cubic equation with respect to the unknown ξ

$$(1 + r_r) C_1 f \xi^3 - f C_1 C_2 \xi^2 - \frac{8}{81} \left(4 + \frac{3}{2} Sp \right) \xi - \frac{8}{81} \left(1 + \frac{1}{2} Sp \right) f = 0. \tag{21.59}$$

In actual fact, Eq. (21.59) is obtained by combining Eqs. (21.34) and (21.50). Here the following new dimensionless parameters inevitably result

$$C_1 = \frac{\lambda_1}{\lambda_2} \frac{v_1}{v_2} \frac{c_{p2} \Delta T_{sp}}{\Delta h} = \frac{1}{f} \frac{a_1}{a_2} Sp. \tag{21.60}$$

An analytical solution is available as follows. Substitute

$$q = -\frac{1}{3} \left\{ \frac{1}{9} \left[\frac{C_2}{1 + r_r} \right]^3 + \frac{4}{81} C_2 \frac{4 + \frac{3}{2} Sp}{(1 + r_r)^2 f C_1} + \frac{4}{27} \frac{1 + \frac{1}{2} Sp}{(1 + r_r) C_1} \right\}, \tag{21.61}$$

$$p = -\frac{1}{3}\left\{\frac{1}{81}\frac{8\left(4+\frac{3}{2}Sp\right)}{\left(1+r_r\right)fC_1} + \frac{1}{3}\left[\frac{C_2}{\left(1+r_r\right)}\right]^2\right\}, \qquad (21.62)$$

$$D = q^2 + p^3. \qquad (21.63)$$

For $D > 0$ the real solution is

$$\xi = \left(-q + D^{1/2}\right)^{1/3} + \left(-q - D^{1/2}\right)^{1/3} + \frac{C_2}{3(1+r_r)}. \qquad (21.64)$$

21.1.7 Averaged heat fluxes

The film boiling heat flux is given by Eq. (21.45)

$$\dot{q}''_{FB} = h\Delta T_{sp}. \qquad (21.65)$$

The average heat flux from the vapor-liquid interface into the liquid is

$$\dot{q}''_{\sigma2} = \frac{1}{\Delta z}\int_0^{\Delta z}\lambda_2\frac{\Delta T_{sc}}{\delta_2}dz = \frac{\lambda_2 \Delta T_{sc}}{\lambda_1 \Delta T_{sp}}\frac{1}{\xi}\frac{1}{\Delta z}\int_0^{\Delta z}\lambda_1\frac{\Delta T_{sp}}{\delta_1}dz = \frac{C_2}{\xi}\dot{q}''_{FB}. \qquad (21.66)$$

The heat flux required for heating saturated steam to a certain average temperature corresponding to the linear temperature profile is

$$\dot{q}''_{3\sigma1}(z)dA_3 = d\Big[\rho_1 c_{p1}\langle w_1[T_1(r_1) - T']\rangle_1 A_1\Big], \qquad (21.67)$$

or after rearrangement

$$\dot{q}''_{3\sigma1}(z) = \frac{1}{2}\rho_1 c_{p1}\Delta T_{sp}\frac{\Delta\rho g}{12\eta_1}\frac{f+3\xi}{f+\xi}\frac{d\delta_1^3}{dz} + \frac{1}{6}\rho_1 c_{p1}\Delta T_{sp}w_2\frac{f}{f+\xi}\frac{d\delta_1}{dz}. \qquad (21.68)$$

The average value can be finally expressed by

$$\dot{q}''_{3\sigma1} = \frac{1}{\Delta z}\int_0^{\Delta z}\dot{q}''_{3\sigma1}(z)dz$$

$$= \frac{1}{2} \rho_1 c_{p1} \Delta T_{sp} \frac{\Delta \rho g}{12 \eta_1 \Delta z} \frac{f+3\xi}{f+\xi} \delta_{1,max}^3 + \frac{1}{6} \frac{\rho_1 c_{p1} \Delta T_{sp} w_2}{\Delta z} \frac{f}{f+\xi} \delta_{1,max}$$

$$= \frac{1}{2} Sp \left[\frac{1}{f_{nc}} \frac{f+3\xi}{f+\xi} \left(\frac{h_{nc}\delta_{1,max}}{2 c_{nc} \lambda_1} \right)^3 \frac{h_{nc}}{h} + \frac{4}{6} \frac{1}{f_{fc}} \frac{f}{f+\xi} \left(\frac{h_{fc}\delta_{1,max}}{2 \lambda_1 c_{fc}} \right) \frac{h_{fc}}{h} \right] \dot{q}''_{FB}. \quad (21.69)$$

The film thickness at the upper end of the integration region is determined by Eq. (21.44).

For the case of the natural convection Eq. (21.69) reduces to

$$\dot{q}''_{3\sigma1} = \frac{1}{2} Sp \frac{1}{f_{nc}} \frac{f+3\xi}{f+\xi} \dot{q}''_{FB}, \quad (21.70)$$

and for the case of predominant forced convection to

$$\dot{q}''_{3\sigma1} = \frac{1}{3} Sp \frac{1}{f_{fc}} \frac{f}{f+\xi} \dot{q}''_{FB}. \quad (21.71)$$

It is evident that the energy transferred to heat up the vapor film is proportional to the energy transferred from the wall to the liquid interface by conduction.

The evaporation mass flow rate is expressed by

$$(\rho w)_{21} = (\dot{q}''_{FB} + \dot{q}''_{r\sigma2} - \dot{q}''_{\sigma2} - \dot{q}''_{3\sigma1})/\Delta h. \quad (21.72)$$

For natural convection the following is obtained

$$(\rho w)_{21} = \left(1 + r_r - \frac{C_2}{\xi} - \frac{1}{2} Sp \frac{1}{f_{nc}} \frac{f+3\xi}{f+\xi}\right) \frac{\dot{q}''_{FB}}{\Delta h}. \quad (21.73)$$

This equation contains an interesting item of information. The condition $(\rho w)_{21} = 0$ leads to a transcendental equation with respect to superheat, as r_r and ξ are also functions of ΔT_{sp}.

The assumption that there is a linear temperature profile is not valid for higher degrees of superheat, because heating of the vapor will first bring the liquid interface to a certain average temperature but will then require more energy than could be supplied at the interface. Under such conditions only radiation heat flux has any influence on film boiling.

Thus, a complete analytical description of the heat transfer mechanism for film boiling is now available.

21.1.8 Effect of the interfacial disturbances

The interfacial disturbances enhance the heat transfer from the interface to the liquid. This effect was recognized by *Coury* and *Dukler* [7] (1970), *Suryanarayana* and *Merte* [34] (1972) and discussed by *Analytis* and *Yadigaroglu* [2] (1985) for vertical surfaces. More recent works by *Bui* and *Dhir* [6], *Nigmatulin* et al. [25], *Nishio* and *Ohtake* [26] substantiate this finding. The effect is:

(a) As it is evident from Fig. 21.1 b, the interfacial instabilities do not allow the film to become very thick. The series of crests indicates that the characteristic length of the process is proportional to the instability wavelength λ_{KH}, rather than to the height of the heated surface so long as $\Delta z > \lambda_{KH}/2$. The classical *Kelvin-Helmholtz* instability analysis applied to this problem defines this instability wavelength by the formula

$$\lambda_{KH} = 2\pi \sqrt{\frac{2\sigma \delta_{1,z=\lambda_{KH}}}{\rho_1 (\langle w_1 \rangle_1 - w_2)^2}} = 2\pi \sqrt{\frac{2\sigma \delta_{1,z=\lambda_{KH}}}{\rho_1 \left(\frac{\Delta \rho g \delta_{1,z=\lambda_{KH}}^2}{12\eta_1} \frac{f+4\xi}{f+\xi} - \frac{1}{2} \frac{f+2\xi}{f+\xi} w_2 \right)^2}}$$

(21.74)

which has to be solved together with Eq. (21.43) for

$$z = \Delta z = \lambda_{KH}/2.$$

(21.75)

(b) We expect that the theory as presented up to this point will underpredict the influence of the subcooling. The instabilities at the surface produce periodical bubble entrainment which causes turbulent mixing in the liquid boundary layer and therefore additional enhancement of heat transfer into the liquid.

This effect can be included in the model by modifying the liquid thermal boundary layer thickness as follows. In place of C_2 the following is used

$$C_2^* = C_2 \left(1 + \frac{2}{5} Sp\right).$$

(21.76)

This is valid for

$$(1+r_r)\xi/C_2 > 1,$$

(21.77)

with the boundary layer thickness left unchanged for all other conditions. For small degrees of superheating this modification is not important; it only starts to influence the computed heat transfer into the liquid in the case of higher degrees of superheat. The data comparison given in the next section provides the justification for this modification.

21.1.9 Comparison of the theory with the results of other authors

Equation (21.58) provides a new general method for layer thickness ratio computation. Equation (21.45) provides a new general method for computation of the averaged film boiling heat transfer coefficient for an arbitrary state of the liquid.

It is worth noting that for the case of natural convection the cubic Eq. (21.59) is obtained for the layer thickness ratio just as do *Sakurai* [32], and *Liu* and *Theofanous* [23], but with different coefficients.

Comparing Eq. (21.28) with Eqs. (21.68) and (21.64) the following is obtained

$$f_{nc} = 1 \Bigg/ \left[\left(1 + \frac{\dot{q}''_{r\sigma2} - \dot{q}''_{\sigma2}}{\dot{q}''_{FB}}\right) \frac{f + \xi}{f + 4\xi} - \frac{\dot{q}''_{3\sigma1}}{\dot{q}''_{FB}} \right] \tag{21.78}$$

or

$$f_{nc} \approx 1 \Bigg/ \left(1 + \frac{\dot{q}''_{r\sigma2} - \dot{q}''_{\sigma2} - \dot{q}''_{3\sigma1}}{\dot{q}''_{FB}}\right). \tag{21.79}$$

This is a very significant equation indicating a simple way to take into account the effects of vapor heating, liquid subcooling and radiation. As a result, Eq. (21.40) can be written in the form

$$\frac{\dot{q}''_{nc}}{\dot{q}''_{nc,0}} \approx \left(1 + \frac{\dot{q}''_{r\sigma2} - \dot{q}''_{\sigma2} - \dot{q}''_{3\sigma1}}{\dot{q}''_{nc}}\right)^{-1/4} \tag{21.80}$$

where

$$\dot{q}''_{nc,0} = c_{nc} \left[\frac{\lambda_1^3 \rho_1 \Delta h \Delta \rho g}{\Delta T_{sp} \eta_1 \Delta z}\right]^{1/4} \Delta T_{sp}. \tag{21.81}$$

Similarly for the case of forced convection the following is obtained

$$\frac{\dot{q}''_{fc}}{\dot{q}''_{fc,0}} \approx \left(1 + \frac{\dot{q}''_{r\sigma2} - \dot{q}''_{\sigma2} - \dot{q}''_{3\sigma1}}{\dot{q}''_{fc}}\right)^{-1/2} \tag{21.82}$$

where

$$\dot{q}''_{fc,0} = c_{fc} \left[\frac{\rho_1 \lambda_1 \Delta h w_2}{\Delta T_{sp} \Delta z}\right]^{1/2} \Delta T_{sp}, \tag{21.83}$$

is the heat flux for predominant forced convection and saturated liquid.

The notation can now be generalized as follows

$$\frac{\dot{q}''_{FB}}{\dot{q}''_{FB,0}} \approx \left(1 + \frac{\dot{q}''_{r\sigma2} - \dot{q}''_{\sigma2} - \dot{q}''_{3\sigma1}}{\dot{q}''_{FB,0}} \frac{\dot{q}''_{FB,0}}{\dot{q}''_{FB}}\right)^{-1/m} . \tag{21.84}$$

The value $m = 2$ applies for predominant forced convection, and $m = 4$ for natural convection. Note that for $(\dot{q}''_{r\sigma2} - \dot{q}''_{\sigma2} - \dot{q}''_{3\sigma1})/\dot{q}''_{FB} \gg 1$ we have $\dot{q}''_{FB}/\dot{q}''_{FB,0} \approx \left[(\dot{q}''_{r\sigma2} - \dot{q}''_{\sigma2} - \dot{q}''_{3\sigma1})/\dot{q}''_{FB,0}\right]^{-1/(m-1)}$. This describes the way in which heat transfer from the surface to the vapor and from the interface to the liquid enhances the film boiling heat transfer component.

It is common practice to incorporate the surface-vapor term into a modified term for latent heat of vaporization. The natural-convection film boiling heat transfer coefficient can thus be expressed with the modified Eq. (21.34a) as follows

$$h_{nc} = c_{nc}\left[\frac{\lambda_1^3 \rho_1 \Delta h^* \Delta\rho g}{\Delta T_{sp} \eta_1 \Delta z} f_{nc}\right]^{1/4} , \tag{21.34a}$$

where

$$\Delta h^* = \Delta h\left\{(f + 4\xi)/(f + \xi) + 0.5\left[(f + 3\xi)/(f + \xi)\right]Sp\right\} \approx \Delta h(1 + 0.5Sp) , \tag{21.85}$$

It is evident that this practice yields a good approximation of the general result as the ratio $\dot{q}''_{3\sigma1}/\dot{q}''_{FB}$ is almost constant for given thermal conditions. The constant 0.5 has been also used by *Sakurai* [32]. Some authors have used other values instead of 0.5: 0.4 - *Bromley* [4], 0.6 - *Frederking* et al. [14] or $0.84/\mathrm{Pr}_1$ - *Sparrow* [33]. If these values are compared with Eq. (21.85) it is evident that the correct value for the linear temperature profile in the vapor is very close to a constant value of 0.5. For degrees of superheat exceeding 700 K at atmospheric pressure the term $\dot{q}''_{3\sigma1}/\dot{q}''_{FB}$ is larger than 0.4, and cannot be neglected. The modification of the latent heat of vaporization as discussed here simplifies the general equation to

$$\frac{\dot{q}''_{FB}}{\dot{q}''_{FB,0}} \approx 1 \Big/ \left(1 + \frac{\dot{q}''_{r\sigma2} - \dot{q}''_{\sigma2}}{\dot{q}''_{FB,0}} \frac{\dot{q}''_{FB,0}}{\dot{q}''_{FB}}\right)^{1/m} . \tag{21.86}$$

For predominant forced convection ($m = 2$ in Eq. (21.84)) the following is then obtained

$$\frac{\dot{q}''_{FB}}{\dot{q}''_{FB,0}} = \left[1 + \frac{1}{4}\left(\frac{\dot{q}''_{r\sigma 2} - \dot{q}''_{\sigma 2}}{\dot{q}''_{FB,0}}\right)^2\right]^{1/2} - \frac{1}{2}\frac{\dot{q}''_{r\sigma 2} - \dot{q}''_{\sigma 2}}{\dot{q}''_{FB,0}}.$$ (21.87)

This equation can be solved numerically for natural convection. As already mentioned, the effect of subcooling or superheating is to either reduce or enhance the effect of radiation.

For the case of natural circulation, film boiling Eq. (21.86) reduces to an equation that was in fact already intuitively introduced by *Bromley*, see Eq. (9) in [3], who considered it as "approximate". It was shown by Sparrow [33, p 234] that for natural-convection film boiling use of Eq. (21.85) permits the effect of the radiation on the conductive heat flux to be predicted to within a few per cent of the accuracy of the exact method. This means that Eq. (21.86) is a very practical approach for the general case of film boiling with radiation in subcooled or superheated liquid as demonstrated in [16].

21.1.10 Verification using the experimental data

Experimental data for averaged heat transfer coefficients were obtained at the Royal Institute of Technology in Sweden by *Okkonen* et al. [30]. The data are for vertical heated walls with $\Delta z = 1.5$ m. Water at atmospheric pressure and subcooling of ΔT_{sc} =3 to 42K, was used as a coolant. The wall superheat was considerable: ΔT_{sp} =487 to 1236K. The heat transfer coefficient was defined by *Okkonen* with respect to the wall-liquid temperature difference. This is recomputed here with respect to wall-saturation temperature difference for comparison purposes. Prediction is performed using Eqs. (21.75), (21.64), (21.30 and (21.34a). A comparison of the predicted data with the 43 data points gained by experiment is given in Fig. 21.2. It is evident that all data are predicted inside the error band of $\pm 10\%$, with a mean averaged error of less than $\pm 4\%$ when the modified theory expressed by Eq. (21.76) is used.

It is interesting to note that for this data we have $f \approx 25$ to 39.6, $\xi \approx 0.35$ to 2.3, $f_{nc} \approx 4.2$ to 205.

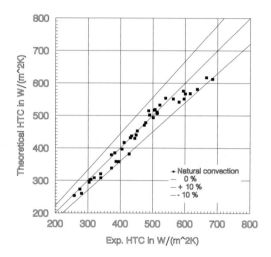

Fig. 21.2. Comparison between the film boiling model and the data base of the Royal Institute of Technology (KTH - Sweden) - Okkonen et al. Vaeth's radiation model (FzK - Karlsruhe) incorporated. $\Delta z = 1.5m$ vertical wall. Water at atmospheric pressure and subcooling $\Delta T_{sc} = 3 - 42K$. Wall superheat $\Delta T_{sp} = 487 - 1236K$

21.1.11 Conclusions

The approximate mathematical description proposed in this work relates to mixed-convection film boiling on a vertical heated plate, in saturated or subcooled liquid. It takes into account the effect of radiation and leads to a new analytical solution. This solution compares well with the experimental data for natural-convection film boiling. The final result is recommended for practical application.

21.1.12 Practical significance

An analytical solution of the mass, momentum and energy conservation equations describing mixed-convection film boiling on vertical walls is now available. From this solution expressions are derived for the particular components of the removed heat that cause vapor heating, liquid heating, for evaporation mass flow rate, and for average vapor volume attached to the surface. This is a step forward in increasing the accuracy of the overall flow field prediction for melt interacting with water in severe accident analysis in industry. Design calculations for vertical surfaces cooled by film boiling can now be implemented with an accuracy of about 10 %.

21.2 Sphere

21.2.1 Introduction

In Section 21.1 of this Chapter the method was described that led to a successful analytical description of film boiling on vertical plates. Here this method is extended to a method for description of mixed-convection film boiling at spheres.

21.2.2 Problem definition

Consider the geometry presented on Fig. 21.3. A hot sphere is flowed by liquid similarly to the case presented in Fig. 21.1. All geometry definitions are similar to those of the plate.

The problem to be solved is: compute all components of the heat flux emitted from the wall as indicated on Fig. 21.3.

21.2.3 Solution method

Encouraged by the successful model development for plane geometry we have repeated the same procedure for the spherical geometry. In addition to the assumption of a linear temperature profile in the liquid we considered also the case for a quadratic profile. The results are analytical solutions for natural and forced convection, respectively. A closed analytical solution for mixed convection could not be obtained. Therefore interpolation between both solutions was necessary to describe the mixed convection. The final result was more complicated than the plane solution. Data comparison shows that several empirical corrections are necessary to bring the model prediction within a 30% error band compared to experiments. An ad hoc application of the plane solution to spherical geometry was much better. Therefore we decided to modify the plane solution appropriately and to verify the solution on the data available. The motivation for this study comes from the *Meyer* and *Schumacher* experiment [24].

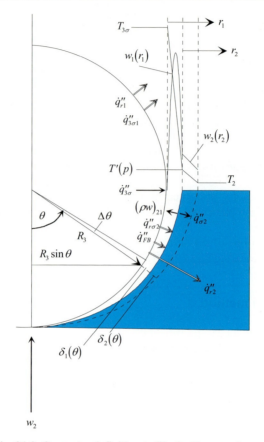

Fig. 21.3. Geometry definition to film boiling at sphere

21.2.4 Model

The interfacial area per unit volume of the flow

$$\frac{F_3}{Vol} = \frac{6\alpha_3}{D_3} \tag{21.88}$$

is necessary to compute power densities from heat fluxes at the surface. At the beginning of the computation we assume that the surface temperature is equal to the particle-averaged temperature $T_{\sigma 3} = T_3$. This assumption can be improved by repeating the procedure described below.

21.2.4.1 Heat flux from the liquid interface into the liquid

In case of subcooling there is a heat transfer from the water interface into the bulk of the water. We compute the *Nusselt* number for the natural convection only using the *Achenbach* correlation [1] verified for $0 < Gr_2 Pr_2 < 5 \times 10^6$

$$Nu_{\sigma 2,nc} = 3.71 + 0.402 (Gr_2 Pr_2)^{1/2}, \qquad (21.89)$$

where the *Grashoff* number for a sphere is approximated as given in the Nomenclature. Interface turbulence due to film boiling is taken into account

$$Nu_{\sigma 2,nc} = 2.1(1 + 15Sp^{*3}) \left[3.71 + 0.402 (Gr_2 Pr_2)^{1/2} \right] \qquad (21.90)$$

by introducing a multiplier depending on the superheating. The forced convection *Nusselt* number is computed by slightly modifying the method of *Gnielinski* [17] (1978)

$$Nu_{\sigma 2, fc} = \max \left(Nu_{\sigma 2, fcl}, Nu_{\sigma 2, fct} \right), \qquad (21.91)$$

where

$$Nu_{\sigma 2, fcl} = 0.664 Re_2^{1/2} Pr_2^{1/3}, \qquad (21.92)$$

$$Nu_{\sigma 2, fct} = 0.037 \frac{Re_2^{0.8} Pr_2}{1 + \frac{2.443}{Re_2^{0.1}} \left(Pr_2^{2/3} - 1 \right)}. \qquad (21.93)$$

This method is verified for $Re_2 < 2 \times 10^4$, $0.7 < Pr_2 < 10^4$. *Achenbach* [1] reported data which demonstrate the validity of the above approach up to $Re_2 < 7.7 \times 10^5$ and pressure up to 40 *bar*. Interface turbulence due to film boiling is taken into account by introducing a multiplier depending on the superheating

$$Nu_{\sigma 2, fc} = 1.5(1 + 5Sp^{*3}) Nu_{\sigma 2, fc}. \qquad (21.94)$$

Again, as in the case of natural convection, the interface turbulence due to film boiling is taken into account by introducing an empirical multiplier $1.5(1 + 5Sp^{*3})$ depending on the superheating. The coefficients are selected by comparison with experimental data as will be discussed later on.

For $Re_2 < 0.001$

$$Nu_{\sigma 2} = Nu_{\sigma 2, nc} \qquad (21.95)$$

otherwise

$$Nu_{\sigma 2} = \max \left(Nu_{\sigma 2, nc}, Nu_{\sigma 2, fc} \right). \qquad (21.96)$$

The thermal power transferred from the interface into the bulk per unit flow volume is therefore

$$\dot{q}'''_{\sigma 2} = \omega_{\sigma 2} \left(T' - T_2 \right), \qquad (21.97)$$

$$\omega_{\sigma 2} = C_{gas} \frac{F_3}{Vol} \frac{\lambda_2}{D_3} Nu_{\sigma 2}. \qquad (21.98)$$

Here the multiplier

$$C_{gas} = \left(\frac{\alpha_2}{\alpha_1 + \alpha_2} \right)^{1/3} \qquad (21.99)$$

takes into account the presence of vapor in the flow outside the film. The power 1/3 rather the proposal by *Liu* and *Theofanous* [23], ¼, is used in Eq. (21.99) for better reproduction of the experimental data for film boiling in two-phase flow with this particular model.

21.2.4.2 Radiation from the sphere

For very high temperatures the radiation cannot be neglected. This is a very important topic which deserves separate and careful discussion. We will confine ourselves to referring to the formalism provided by *Lanzenberger* [22], which gives three radiation components from the particle interface being functions of the volumetric fractions, partial vapor pressure and the temperatures of the emitter and receivers

$$\left[\dot{q}'''_{r1}, \dot{q}'''_{r2}, \dot{q}'''_{r\sigma 2} \right]_L = f\left(\alpha_1, \alpha_2, \alpha_3, p_M, T_1, T_2, T_{3\sigma} \right). \qquad (21.100)$$

For convenience we introduce a splitting factor for correction of the radiation heat redistribution in water due to surface instability based on the increase of the surface absorption

$$\dot{q}'''_{r\sigma 2} = \left[\dot{q}'''_{r\sigma 2} \right]_L + c_{split} \left[\dot{q}'''_{r2} \right]_L, \qquad (21.101)$$

$$\dot{q}_{r2}''' = (1 - c_{split})[\dot{q}_{r2}''']_L, \qquad (21.102)$$

where

$$c_{split} = 0. \qquad (21.103)$$

For superheated liquid

$$T_2 > T' \qquad (21.104)$$

all the radiation reaching the liquid is deposited at the liquid interface facilitating an additional spontaneous flashing condition

$$\dot{q}_{r\sigma2}''' = [\dot{q}_{r\sigma2}''']_L + [\dot{q}_{r2}''']_L \qquad (21.105)$$

$$\dot{q}_{r2}''' = 0. \qquad (21.106)$$

For practical application *Vaeth*'s model [35] in IVA4 was replaced by the new model developed by *Lanzenberger* [22]. The new model solves correctly the radiation transport equation for particular geometries in the multi-phase flows without making use of some simplifying assumptions made by *Vaeth*. One of the results of the new model is, for instance, less radiation energy adsorbed in the steam. This is not significant for the discussed case, but in large-scale melt-water interaction is of importance. The comparisons with experimental data reported here are performed with the model of *Lanzenberger*.

21.2.4.3 Film boiling without radiation in saturated liquid

Here we consider film boiling without radiation in saturated liquid. In Section 21.2.4.3.4 the prediction of the model will be extended to take into account radiation and subcooling.

21.2.4.3.1 Natural convection only

The heat transfer coefficient for natural-convection film boiling is computed by using a somewhat modified plane solution

$$h_{nc,1} = \frac{\lambda_1}{D_3} Nu_{nc,1} \qquad (21.107)$$

where

$$Nu_{nc,1} = 0.7 f(\bar{D}_3)(f_{nc} Gr_1 Pr_1 / Sp)^{1/4}, \qquad (21.108)$$

$$f_{nc} = \frac{f+4\xi}{f+\xi}\left(1+\frac{1}{2}Sp\frac{f+3\xi}{f+4\xi}\right), \qquad (21.109)$$

$$\xi = (-q+\Delta)^{1/3} + (-q-\Delta)^{1/3}, \quad \xi > 0, \qquad (21.110)$$

$$\Delta = \sqrt{D} \text{ for } D \geq 0, \qquad (21.111)$$

$$D = q^2 + p^3, \qquad (21.112)$$

$$q = -\frac{1}{3}\frac{4}{27}\frac{1+\frac{1}{2}Sp}{C_1}, \qquad (21.113)$$

$$p = -\frac{1}{3}\frac{8}{81}\frac{4+\frac{3}{2}Sp}{fC_1}. \qquad (21.114)$$

From the comparison of the data by *Liu* and *Theofanous* [23] for saturated liquid and different particle sizes we obtain correction function values around one.

$\bar{D}_3$	$f(\bar{D}_3)$
2.535	0.995
3.806	0.952
5.070	0.939
7.624	1.031

The natural instability length scale, the *Rayleigh-Taylor* wavelength, is used to normalize the particle size

$$\bar{D}_3 = D_3 / \lambda_{RT}. \qquad (21.115)$$

For smaller values of $\bar{D}_3$ of 2.535 we use $\bar{D}_3 = 0.995$. Finally the thermal power transferred by conduction through the vapor film per unit flow volume is

$$\dot{q}'''_{FB,nc} = \frac{F_3}{Vol} h_{nc,1}(T_{3\sigma} - T'). \qquad (21.116)$$

Assuming that there is a vapor temperature profile automatically leads to the conclusion that there is some thermal power transfer into the vapor. We approximate this part

$$\dot{q}'''_{31,nc} = 0.3 Sp \frac{f + 3\xi}{f_{nc}(f+\xi)} \dot{q}'''_{FB,nc}, \qquad (21.117)$$

For $q^2 + p^3 < 0$ the layers thickness ratio is theoretically not defined. If using minimum film boiling theory strictly consistent with this work no further discussion is necessary. If not, e.g. empirical correlation for minimum film boiling is used, then we set

$$\dot{q}'''_{31,nc} = 0.3\, Sp\, \dot{q}'''_{FB,nc}, \qquad (21.118)$$

and $f_{nc} = 1 + \frac{1}{2} Sp$.

21.2.4.3.2 Predominant forced convection

In earlier development [35] we used for this case the *Bromley-Witte* theory [5, 36],

$$\dot{q}'''_{FB,fc} = \frac{F_3}{Vol} h_{fc,1}(T_{3\sigma} - T'), \qquad (21.119)$$

where

$$h_{fc,1} = \frac{\lambda_1}{D_3} 2.98 \left(\frac{Re_1}{Sp}\right)^{1/2}. \qquad (21.120)$$

The comparison with the forced-convection film boiling data by *Liu* and *Theofanous* [23] resulted in replacing this model by the model of *Epstein-Heuser* [9]:

$$h_{fc,1} = \frac{\lambda_1}{D_3} 0.60 \left(\frac{\rho_2}{\rho_1}\right)^{1/4} Re_1^{1/2} / Sp^{*1/4} \qquad (21.121)$$

in which the constant is changed from 0.65 to 0.6. Again thermal power transfer into the vapor is approximated for $D < 0$ by

$$\dot{q}'''_{31,fc} = 0.05\, Sp\, \dot{q}'''_{FB,fc} \qquad (21.122)$$

otherwise

$$\dot{q}'''_{31,fc} = 0.05\, Sp\, \frac{f}{f_{fc}(f+\xi)} \dot{q}'''_{FB,fc}. \tag{21.123}$$

Here

$$f_{fc} = \frac{f}{f+\xi}\left(1+\frac{1}{3}Sp\right) \tag{21.124}$$

is computed as Eq. (21.31) setting, $C_2 = 0$, $r_r = 0$, and ξ is computed as for natural convection.

21.2.4.3.3 Mixed convection

Mixed convection film boiling is known to exist in the region $Fr \geq 0.2$. For $Fr < 0.2$ the natural-convection only solution

$$\dot{q}'''_{FB,0} = \dot{q}'''_{FB,nc}, \tag{21.125}$$

$$\dot{q}'''_{31} = \dot{q}'''_{31,nc}, \tag{21.126}$$

is the appropriate one. Otherwise, the approximation between natural- and forced-convection solutions is performed following *Liu* and *Theofanous* [23]

$$\dot{q}'''_{FB,0} = \dot{q}'''_{FB,nc}\left[1+\left(\frac{\dot{q}'''_{FB,fc}}{\dot{q}'''_{FB,nc}}\right)^5\right]^{1/5}, \tag{21.127}$$

$$\dot{q}'''_{31} = \dot{q}'''_{31,nc}\left[1+\left(\frac{\dot{q}'''_{31,fc}}{\dot{q}'''_{31,nc}}\right)^5\right]^{1/5}. \tag{21.128}$$

Note that in the previous code development, [35], we used

$$\dot{q}'''_{FB,0} = \max\left(\dot{q}'''_{FB,nc}, \dot{q}'''_{FB,fc}\right).$$

21.2.4.3.4 Feedback of the radiation, subcooling and vapor heating on the vapor film conduction heat transfer

The experience with the plate solution is used here as a guideline. First we define the ratios

$$r_r = \frac{|\dot{q}_{r\sigma2}'''| - \dot{q}_{\sigma2}''' - \dot{q}_{31}'''}{\dot{q}_{FB,0}'''}, \qquad (21.129)$$

$$r_r^* = \frac{1}{2}r_r, \qquad (21.130)$$

influencing the film conduction through influencing the film thickness. Analytical solution of Eq.(21.84) is available for forced convection only,

$$x_{fc} = \sqrt{1 + r_r^{*2}} - r_r^*. \qquad (21.131)$$

For natural convection Eq. (21.84) rewritten in the following form

$$x_{nc}^4 + r x_{nc}^3 + 1 = 0 \qquad (21.132)$$

has to be solving by iteration. We start with the value for forced convection and iterate up to satisfying the equation with residuals less than 0.0001. Finally, we take the maximum of both values.

$$x = \max\left(x_{fc}, x_{nc}\right), \qquad (21.133)$$

$$\dot{q}_{FB}''' = C_{gas} x \dot{q}_{FB,0}'''. \qquad (21.134)$$

The total thermal energy emitted by the interface of the particle is therefore

$$\dot{q}_{\sigma3}''' = -\dot{q}_{FB}''' - \dot{q}_{31}''' - \dot{q}_{r1}''' - \dot{q}_{r2}''' - \dot{q}_{r\sigma2}'''. \qquad (21.135)$$

For implementation in computer codes it is important to check whether there is enough energy which has to be exhausted from the particles. The maximum of this energy can be defined as the energy required to heat up the particle from the liquid temperature to the actual particle temperature within the computational time step,

$$\dot{q}_{\sigma3,\max}''' = \alpha_3 \rho_3 c_{p3} \left(T_{3\sigma} - T_2\right)/\Delta\tau. \qquad (21.136)$$

Should the ratio $f_{\max} = \dot{q}_{\sigma3,\max}''' / \dot{q}_{\sigma3}'''$ be larger than 1 the time step has to be corrected. If the time step is small enough we correct all components as follows

$$\dot{q}_{fb}''' = \dot{q}_{fb}''' f_{\max}, \ \dot{q}_{31}''' = \dot{q}_{31}''' f_{\max}, \ \dot{q}_{r1}''' = \dot{q}_{r1}''' f_{\max}, \ \dot{q}_{r2}''' = \dot{q}_{r2}''' f_{\max}, \ \dot{q}_{r\sigma2}''' = \dot{q}_{r\sigma2}''' f_{\max}.$$

Finally the total film boiling heat transfer coefficient is

$$h_{FB} = -\dot{q}'''_{\sigma 3} / \left[\frac{F_3}{Vol}(T_3 - T') \right]. \qquad (21.137)$$

This definition is used for comparison with the experimental data. The energy crossing the liquid interface and entering the water bulk is composed of radiation and convective components

$$\dot{q}'''_{\sigma 2, total} = \dot{q}'''_{r2} + \dot{q}'''_{\sigma 2}. \qquad (21.138)$$

Similarly the energy transferred from the particle into the gas film is composed by radiation and convective parts

$$\dot{q}'''_{3\sigma 1} = \dot{q}'''_{r1} + \dot{q}'''_{31}. \qquad (21.139)$$

The energy jump condition at the liquid interface gives for the evaporated mass per unit time and unit flow volume the following expression

$$\mu_{21} = \frac{\dot{q}'''_{r\sigma 2} + \dot{q}'''_{FB} - \dot{q}'''_{\sigma 2}}{h'' - h'}. \qquad (21.140)$$

21.2.4.3.5 The interface temperature

The sphere interface temperature during film boiling may differ from the particle-averaged temperature, especially for large particles of poor heat conductors. The limitation come from the fact that only a limited amount can be supplied from the particle bulk to the particle surface. The analytical solution of the *Fourier* equation is the tool used here to estimate the interface temperature. The temperature profile is then averaged to estimate the averaged temperature and therefore the averaged temperature change during the time step. The task is to estimate the possible energy transfer due to the conditions during the time step considered and to compare it with that computed above. The final result is

$$T_{3\sigma} = T_3 + \dot{q}'''_{\sigma 3} / \omega_3 \qquad (21.141)$$

where

$$\omega_3 = \alpha_3 \rho_3 c_{p3} \frac{\Delta \overline{T}_{3\sigma}}{\Delta \tau} \qquad (21.142)$$

The averaged temperature difference is

$$\Delta \overline{T}_{3\sigma} = 1 - \frac{6}{\pi^2} \sum_{n=1}^{\infty} \frac{1}{n^2} / \exp\left(n^2 \frac{\Delta \tau}{\Delta \tau_3} \right), \qquad (21.143)$$

where

$$\Delta \tau_3 = \frac{D_3^2}{4\pi^2 a_3} \tag{21.144}$$

is the time constant of the heat conduction into the spherical particle. The converging *Fourier* series is computed using n as big as necessary (but no larger then 100) to obtain improvement in the summation less than 0.00001. Because $\dot{q}'''_{\sigma 3}$ is a function of the interface temperature itself some iterations may be required to finally obtain the released power density.

21.2.4.3.6 Averaged film thickness in the multiphase flows

To demonstrate the importance of the averaged film thickness in the multi-phase analysis, consider the question how much vapor is associated in the film and how much is in bubble form. The heat and mass transfer processes at the particle interface are completely described in the case of film boiling by the method presented already. The remaining part in bubble form is subject to additional water/vapor heat and mass transfer which will not be considered here. Analysis of the QUEOS [18] experiments demonstrates that at the initial phase of penetration of hot particles in water almost the entire generated vapor is in the film. That means no additional vapor/water heat and mass transfer has to be modeled. For the later phase of interaction it starts to be active.

We approximate the averaged film thickness as follow. For $\Delta V_{23} < 0.001$ we have natural-circulation film boiling leading to

$$\delta_{1F} = \delta_{1F,nc} = \left[\frac{2 D_3^2 \mu_{21} \eta_1}{\alpha_3 g \rho_1 (\rho_2 - \rho_1)} \right]^{1/3} \tag{21.145}$$

otherwise

$$\delta_{1F} = \delta_{1F,fc} = \frac{2}{9} \frac{D_3^2 \mu_{21}}{\alpha_3 \rho_1 \Delta V_{23}} \tag{21.146}$$

with the limitation

$$\delta_{1F} = \min\left(\delta_{1F,nc}, \delta_{1F,fc}\right). \tag{21.147}$$

21.2.5 Data comparison

The sphere film boiling model presented here is implemented in the computer code IVA5. More information about the history of this development and the recent state is available in [17] and [18]. This model is used for the data comparisons presented below. The new experimental data base provided by *Liu* and *Theofanous* [23] in 1995 consists of single- and two-phase flow data. The single-phase data covers subcooling from 0 to 40 *K*, liquid velocity from 0 to 2.3 *m/s*, sphere superheat from 200 to 900 *K* and sphere diameter from 6 to 19 *mm*. The two-phase data are obtained for (a) upward flow having void fraction from 0.2 to 0.65, water velocity from 0.6 to 3.2 *m/s*, and steam velocity from 3 to 9 *m/s*, and (b) for downward flow having void fraction from 0.7 to 0.95, water velocity from 1.9 to 6.5 *m/s*, and steam velocity from 1.1 to 9 *m/s*.

Figure 21.4 presents the comparison between the predicted and calculated heat transfer coefficients for all data. The error band is found to be ±30%. Figure 21.5 presents the comparison between the predicted and calculated heat transfer coefficients for single-phase saturated water and four different sphere diameters. The error band for this subset of data is found to be +20 to -15%. Figure 21.6 presents the comparison between the predicted and calculated heat transfer coefficients for subcooled water and 12.7 *mm* sphere diameter. The error band for this subset of data is found to be +20 to -30%. Figure 21.7 presents the comparison between the predicted and calculated heat transfer coefficients for saturated upward two-phase flow, 12.7 *mm* sphere diameter, void fraction from 0.2 to 0.65, water velocity from 0.6 to 3.2 *m/s*, and steam velocity from 3 to 9 *m/s*. The error band for this subset of data is found to be +30 to −20%.

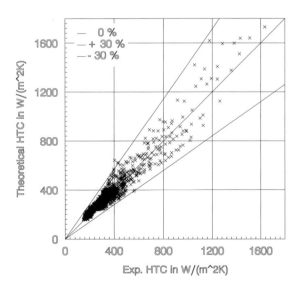

Fig. 21.4. Comparison between predicted and calculated heat transfer coefficients for all data. Error band: ±30%

21.2 Sphere 559

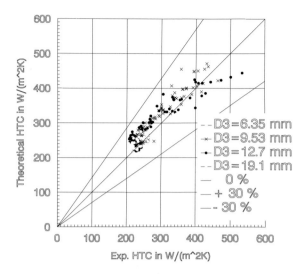

Fig. 21.5. Comparison between predicted and calculated heat transfer coefficients for single-phase saturated water and four different sphere diameters. Error band : +20 to − 15%

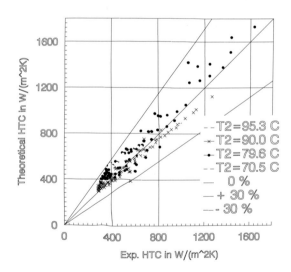

Fig. 21.6. Comparison between predicted and calculated heat transfer coefficients for subcooled water and 12.7 *mm* sphere diameter. Error band: +20 to − 30%.

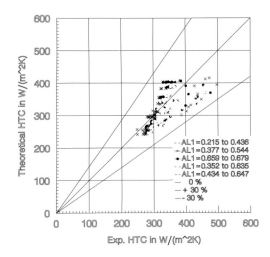

Fig. 21.7. Comparison between the predicted and calculated heat transfer coefficients for saturated upward two phase flow, 12.7 *mm* sphere diameter, void fraction from 0.2 to 0.65, water velocity from 0.6 to 3.2 *m/s*, and steam velocity from 3 to 9 *m/s*. Error band: +30 to − 20%

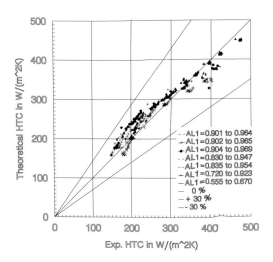

Fig. 21.8. Comparison between the predicted and calculated heat transfer coefficients for saturated downward two phase flow, 12.7 *mm* sphere diameter, void fraction from 0.7 to 0.95, water velocity from 1.9 to 6.5 *m/s*, and steam velocity from 1.1 to 9 *m/s*. Error band: +20 to − 20%

Figure 21.8 presents the comparison between the predicted and calculated heat transfer coefficients for saturated downward two-phase flow, 12.7 *mm* sphere diameter, void fraction from 0.7 to 0.95, water velocity from 1.9 to 6.5 *m/s*, and steam velocity from 1.1 to 9 *m/s*. The error band for this subset of data is found to be +20 to -20%.

21.2.6 Conclusions

The semi-empirical film boiling model used in IVA5 computer code predicts the data base provided by *Liu* and *Theofanous* within ±30% .

Nomenclature

a_1 $\quad = \dfrac{\lambda_1}{\rho_1 c_{p1}}$, vapor temperature diffusivity, m^2/s

a_2 $\quad = \dfrac{\lambda_2}{\rho_2 c_{p2}}$, liquid temperature diffusivity m^2/s

a_3 $\quad = \dfrac{\lambda_3}{\rho_3 c_{p3}}$, particle temperature diffusivity, m^2/s

C_1 $\quad = \dfrac{\lambda_1}{\lambda_2} \dfrac{v_1}{v_2} \dfrac{c_{p2} \Delta T_{sp}}{\Delta h}$, superheat property number, -

C_2 $\quad = \dfrac{\lambda_2 \Delta T_{sc}}{\lambda_1 \Delta T_{sp}}$, superheat to subcooling ratio, -

C_3 $\quad = \dfrac{\rho_2 \lambda_1}{\rho_1 \lambda_2} \dfrac{c_{p2} \Delta T_{sp}}{\Delta h} = \dfrac{a_1}{a_2} Sp$, superheat property number, -

c_{p1} $\quad$ vapor specific heat at constant pressure, $J/(kgK)$

c_{p2} $\quad$ liquid specific heat at constant pressure, $J/(kgK)$

c_{nc} $\quad$ numerical constant for heat transfer coefficient for pure natural convection , -

c_{fc} $\quad$ numerical constant for heat transfer coefficient for pure forced convection,

Fr $\quad = \dfrac{|\Delta V_{23}|}{\sqrt{gD_3}}$, *Froude* number for sphere, -

Fr_2 $\quad = \dfrac{w_2}{\sqrt{2R_3 g}}$, *Froude* number for sphere and cylinder, -

Fr_2 $= \dfrac{w_2}{\sqrt{\Delta z g}}$, *Froude* number for vertical plane, -

f $= \eta_2/\eta_1$, dynamic viscosity ratio, -

Gr_1 $= g(2R_3)^3 \rho_1(\rho_2-\rho_1)/\eta_1^2$, *Grashoff* number for sphere and cylinder, -

Gr_1 $= g\Delta z^3 \rho_1(\rho_2-\rho_1)/\eta_1^2$, *Grashoff* number for vertical plane, -

Gr_2 $= gD_3^3 \left|\dfrac{T'-T_2}{T_2}\right| \left(\dfrac{\rho_2}{\eta_2}\right)^2$, *Grashoff* number for sphere, -

g acceleration due to gravity, m/s^2

h heat transfer coefficient, $W/(m^2 K)$

h_{nc} heat transfer coefficient for pure natural convection, $W/(m^2 K)$

h_{fc} heat transfer coefficient for pure forced convection, $W/(m^2 K)$

$\dot{q}''_{FB}$ heat flux due to conduction through the vapor film, W/m^2

$\dot{q}''_{3\sigma 1}$ heat flux from the wall into the vapor due to convective heating of the vapor from the saturation interface temperature to an averaged vapor film temperature, W/m^2

$\dot{q}''_{\sigma 2}$ heat flux from the liquid interface into the liquid due to convection, W/m^2

$\dot{q}''_{r2}$ radiation heat flux from the wall into the bulk liquid, W/m^2

$\dot{q}''_{3\sigma}$ total heat flux from the wall, W/m^2

$\dot{q}''_{r\sigma 2}$ radiation heat flux absorbed from the vapor-liquid interface, W/m^2

$\dot{q}''_{r1}$ radiation heat flux from the interface into the vapor, W/m^2

Pr_1 $= \dfrac{\eta_1 c_{p1}}{\lambda_1}$, vapor *Prandtl* number, -

Pr_2 $= \dfrac{c_{p2}\eta_2}{\lambda_2}$, liquid *Prandtl* number, -

R_3 radius of sphere or cylinder, m

Re_1 $= \rho_1 \Delta w_{23} 2R_3/\eta_1$, *Reynolds* number, -

Re_2 $= |\Delta V_{23}| D_3 \rho_2/\eta_2$, liquid *Reynolds* numbers, -

r^* $= \left(\dfrac{h_{fc} c_{nc}}{h_{nc} c_{fc}}\right)^2 / 2 = \dfrac{2}{9} r^2$, -

r $= h_{fc}/h_{nc}$, -

r_r $= \dfrac{\dot{q}''_{r\sigma 2}}{\lambda_1 \Delta T_{sp}/\delta_1}$, ratio of the radiation heat flux to the conduction heat flux for plates, -

r_r $= \dfrac{|\dot{q}'''_{r\sigma 2}| - \dot{q}'''_{\sigma 2} - \dot{q}'''_{31}}{\dot{q}'''_{FB,0}}$, for spheres, -

r_r^*	$= \frac{1}{2} r_r$, for spheres, -
r_1	distance from the wall, m
r_2	distance from the interface, m
Nu	$= h 2 R_3 / \lambda_1$, Nusselt number for sphere and cylinder,
Nu	$= h \Delta z / \lambda_1$, Nusselt number for vertical plane,
p	pressure, Pa
p_M	partial vapor pressure in a mixture of vapor and inert components, Pa
Sp	$= c_{p1} \Delta T_{sp} / \Delta h$, superheat number, -
Sp^*	$= \dfrac{c_{p1}(T_{\sigma 3} - T')}{\Delta h^* \mathrm{Pr}_1}$, modified superheat number, -
T_2	liquid temperature, K
$T_{3\sigma}$	temperature at the wall surface, K
$T'(p)$	vapor-liquid interface temperature equal to the saturation temperature of the liquid at the system pressure, K
$w_1(r_1)$	vapor film velocity as a function of the distance from the wall, m/s
$w_2(r_2)$	liquid boundary layer velocity as function of the distance from the interface, m/s
z	axial coordinate, m
$\langle \ \rangle_1$	$= \dfrac{1}{\delta_1} \int_0^{\delta_1} dr_1$, cross-section average, -
α_1	gas volume fraction in three-phase flow, -
α_2	water volume fraction in three-phase flow, -
α_3	volume fraction in three-phase flow, -
ΔT_{sp}	$T_{3\sigma} - T'$, wall superheat, K
ΔT_{sc}	$T' - T_2$, liquid subcooling, K
Δh	latent heat of evaporation, J/kg
Δh^*	$= (h'' - h')\left(1 + \dfrac{1}{2} Sp\right)$, modified latent heat of evaporation, J/kg
Δz	total height of the wall, m
Δw_{12}	vapor film - liquid velocity difference, m
$\Delta \theta$	infinitesimal change of the angular coordinate, rad
$\delta_1(z)$	vapor layer thickness as a function of the axial coordinate, m
$\delta_2(z)$	liquid boundary layer thickness as a function of the axial coordinate, m
$\delta_1(\theta)$	vapor layer thickness as a function of the angular coordinate, m
$\delta_2(\theta)$	liquid boundary layer thickness as a function of the angular coordinate, m

η_1	dynamic viscosity of vapor, $Pa\,s$
η_2	dynamic viscosity of liquid, $Pa\,s$
ρ_1	density of vapor, kg/m^3
ρ_2	density of liquid, kg/m^3
$(\rho w)_{21}$	evaporation mass flow rate, kg/m^2
σ	surface tension, N/m
ω_2	heat transport coefficient liquid bulk/liquid surface, $W/(Km^3)$
ω_3	heat transport coefficient particle bulk/particle surface, $W/(Km^3)$
θ	angular coordinate, rad
λ_{KH}	Kelvin-Helmholtz instability wavelength, m
λ_{RT}	$= \sqrt{\dfrac{\sigma_2}{g(\rho_2-\rho_1)}}$, Rayleigh-Taylor wavelength, m
ξ	$= \delta_2/\delta_1$, film thickness ratio, -

Subscripts

nc	natural convection
fc	forced convection
l	laminar
t	turbulent
1	gas
2	water
3	particles
mn	from field m to field n
FB	film boiling
σ	surface

References

1. Achenbach E (1993) Heat and flow characteristics of packed beds, Experimental Heat Transfer, Fluid Mechanics and Thermodynamics, Kelleher MD et al. (eds.) Elsevier, pp 287-293
2. Analytis GT, Yadiragoglu G (Oct. 15-18, 1985) Analytical modeling of IAFB, Proc. of the Third Int. Topical Meeting on Reactor Thermal Hydraulics, Newport, Rhode Island, USA, Chiu C and Brown G (eds) vol 1 paper no 2.B
3. Bromley LA (1950) Heat transfer in stable film boiling, Chem. Eng. Prog., vol 46 no 5 pp 221-227
4. Bromley LA (Dec. 1952) Effect of heat capacity of condensate, Industrial and Engineering Chemistry, vol 44 no12 pp 2966-2969
5. Bromley LA, LeRoy NR, Rollers JA (Dec. 1953) Heat transfer in forced convection film boiling, Industrial and Engineering Chemistry, vol 45 pp 2639-2646

6. Bui TD, and Dhir VK (November 1985) Film boiling heat transfer on isothermal vertical surface. ASME Journal of Heat Transfer, vol 107 pp 764-771
7. Coury GE, Dukler AE (1970) Turbulent film boiling on vertical surfaces - A study including the interface waves, 4th International Heat Transfer Conference, Paper no B.3.6
8. Dhir VK, Purohit GP (1978) Subcooled film-boiling heat transfer from spheres, Nuclear Engineering and Design, vol 47 pp 49 - 66
9. Epstein M, Heuser GM (1980) Subcooled forced-convection film boiling in the forward stagnation region of a sphere or cylinder, Int. J. Heat Mass Transfer, vol 23 pp 179-189
10. Fodemski TR, Hall WB (1982) Forced convection film boiling on a sphere immersed in (a) sub-cooled or (b) superheated liquid, Proc. 7th Int. Heat Transfer Conference, Muenchen, vol 4 pp 375 - 739
11. Fodemski TR (1985) The influence of liquid viscosity and system pressure on stagnation point vapor thickness during forced convection film boiling, Int. J. Heat Mass Transfer, vol 28 pp 69 - 80
12. Fodemski TR, Stagnation point oscillation and stability of forced convection film boiling on a sphere immersed in (a) sub-cooled or (b) superheated liquid, Proc. 7th Int. Heat Transfer Conference, Muenchen, vol 4 pp 369 - 773
13. Fodemski TR (1992) Forced convection film boiling in the stagnation region of molten drop and its application to vapor explosions, Int. J. Heat Mass Transfer, vol 35 pp 2005 - 2016
14. Frederking THK, Chapman RC, Wang S (1965) Heat transport and fluid motion during cooldown of single bodies to low temperatures, International Advances in Cryogenic Engineering, Timmerhaus KD (ed) Plenum Press, New York, pp 353-360
15. Gnielinski V (1978) Gleichungen zur Berechnung des Waerme und Stoffaustausches in durchstroemten ruhenden Kugelschuettungen bei mittleren und grossen Pecletzahlen. Verfahrenstechnik, vol 12 pp 363-366
16. Kolev NI (April 3-7 1995) IVA4 Computer code: The model for film boiling on a sphere in subcooled, satutated and superheated water, Proc. Second Int. Conference On Multiphase Flow, Kyoto, Japan. Presented also in (14-15 Nov. 1994) „Workshop zur Kühlmittel/Schmelze - Wechselwirkung", Köln, Germany
17. Kolev NI (27-29 September 1995) IVA4 Computer code: An universal analyzer for multi-phase flows and its applicability to melt water interaction, Proc. of the Technical committee meeting on advances in and experiences with accidents consequences analysis, held at Vienna, Austria, IAEA-TC-961, Vienna, Austria
18. Kolev NI (October 15-16, 1996) Three fluid modeling with dynamic fragmentation and coalescence fiction or daily practice? 7th FARO Experts Group Meeting Ispra. (5th-8th November 1996) Proceedings of OECD/CSNI Workshop on Transient thermal-hydraulic and neutronic codes requirements, Annapolis, MD, USA. (June 2-6, 1997) 4th World Conference on Experimental Heat Transfer, Fluid Mechanics and Thermo-dynamics, ExHFT 4, Brussels. (June 22-26, 1997) ASME Fluids Engineering Conference & Exhibition, The Hyatt Regency Vancouver, Vancouver, British Columbia, CANADA, Invited Paper
19. Kolev NI (19th-21st May 1997) Verification of the IVA4 film boiling model with the data base of Liu and Theofanous, Proceedings of OECD/CSNI Specialists Meeting on Fuel-Coolant Interactions (FCI), JAERI-Tokai Research Establishment, Japan

20. Kolev NI (June 2-6, 1997) Film boiling: vertical plates, Proceedings of 4th World Conference on Experimental Heat Transfer, Fluid Mechanics and Thermodynamics EXHFT 4, Brussels, Belgium
21. Kolev NI (1998) Film boiling on vertical plates and spheres, Experimental Thermal and Fluid Science, vol 18 (1998) pp 97-115
22. Lanzenberger K (1997) Radiation models for multi-phase flows. Part 1: Film boiling, Siemens report 1997, KWU NA-M/97/E024
23. Liu C, Theofanous TG (August 1995) Film boiling on spheres in single- and two-phase flows Part 1: Experimental Studies ANS Proceedings, Part 2: A Theoretical Study, National Heat Transfer Conference, Portland
24. Meyer L, Schumacher G (April 1996) QUEOS a simulation-experiment of the premixing phase of steam explosion with hot spheres in water base case experiments, Forschungs-zentrum Karlsruhe Technik und Umwelt, Karlsruhe, Wissenschaftliche Berichte FZKA 5612
25. Nigmatulin NI et al. (1994) Interface oscilations and heat transfer mechanism at film boiling, Proc. of the Tenth International Heat Transfer Conference, Brighton, UK
26. Nishio S, Ohtake H (1993) Vapor film unit model and heat transfer correlation for natural convection film boiling with wave motion under subcooled conditions. Int. J. Heat Mass Transfer, vol 36 no 10 pp 2541-2552
27. Nishikawa K, Ito T (1966) Two-phase boundary-layer treatment of free convection film boiling, Int. J. Heat Mass Transfer, vol 9 pp 103-115
28. Nishikawa K, Ito T, Matsumoto K (1976) Investigation of variable thermophysical Property problem concerning pool film boiling from vertical plate with prescribed uniform temperature, Int. J. Heat Mass Transfer, vol 19 pp 1173-1182
29. Nusselt WZ (1916) Z. Ver. deut. Ing., vol 60 pp 541-569
30. Okkonen T et al (August 3-6, 1996) Film boiling on a long vertical surface under high heat flux and water subcooling conditions, Proc. of the 31st Nat. Heat Transfer Conference, Houston, Texas
31. Sakurai A, Shiotsu M, Hata K (1986) Effect of subcooling on film boiling heat transfer from horizontal cylinder in a pool water, Proc. 8th Int. Heat Transfer Conf., vol 4 pp 2043 - 2048
32. Sakurai A (1990) Film boiling heat transfer, Proc. of the Ninth Intrn. Heat Transfer Conference, Jerusalem, Israel, Hetstroni G (ed), vol 1 pp 157 - 187
33. Sparrow EM (1964) The effect of radiation on film boiling heat transfer, Int. J. Heat and Mass Transfer, vol 7 pp 229 - 238
34. Suryanarayana NV, Merte H (1972) Film boiling on vertical surfaces, J. Heat Transfer, vol 94 pp 337-383
35. Vaeth L (March 1995) Radiative heat transfer for the transient three phase three component flow model IVA3, Kernforschungszentrum Karlsruhe GmbH, INR, Interner Bericht 32.21.02/10A, INR-1914, PSF-3212
36. Witte LC (August 1968) Film boiling from a sphere, I&EC Fundamentals, vol 7 no 3 pp 517-518

Appendix 21.1 Natural convection at vertical plate

From Eq. (21.31) the local vapor film thickness as a function of the axial coordinate for natural-convection film boiling at a vertical plate is then obtained

$$\frac{1}{\delta_1} = \frac{1}{2}\left[\frac{\rho_1 \Delta h \Delta \rho g}{\lambda_1 \Delta T_{sp} \eta_1 z} f_{nc}\right]^{1/4}. \tag{A21.1-1}$$

The film boiling heat flux averaged over the height Δz is therefore

$$\dot{q}''_{FB} = \frac{1}{\Delta z}\int_0^{\Delta z} \lambda_1 \frac{\Delta T_{sp}}{\delta_1} dz = \Delta T_{sp} \frac{1}{2}\left[\frac{\lambda_1^3 \rho_1 \Delta h \Delta \rho g}{\Delta T_{sp}\eta_1} f_{nc}\right]^{1/4} \frac{1}{\Delta z}\int_0^{\Delta z} dz/z^{1/4}$$

$$= \Delta T_{sp} \frac{2}{3}\left[\frac{\lambda_1^3 \rho_1 \Delta h \Delta \rho g}{\Delta T_{sp}\eta_1 \Delta z} f_{nc}\right]^{1/4}. \tag{A21.1-2}$$

The heat transfer coefficient is

$$h_{nc} = c_{nc}\left[\frac{\lambda_1^3 \rho_1 \Delta h \Delta \rho g}{\Delta T_{sp}\eta_1 \Delta z} f_{nc}\right]^{1/4} \tag{A21.1-3}$$

where

$$c_{nc} = 2/3. \tag{A21.1-4}$$

Equation (A21.1-3) for $f_{nc} = 1$ reduces to the *Nusselt* equation for laminar film condensation at a cold wall [29] with correspondingly inverted subscripts.

Appendix 21.2 Predominant forced convection only at vertical plate

Consider the predominant forced-convection film boiling at a vertical plate. Neglecting the natural convection terms as small against the predominant forced convection terms in Eq. (21.31), the following is then obtained

$$\frac{1}{\delta_1} = \frac{1}{2}\left[\frac{\rho_1 \Delta h w_2}{\lambda_1 \Delta T_{sp} z} f_{fc}\right]^{1/2}. \tag{A21.2-1}$$

The film boiling heat flux averaged over the height Δz is therefore

$$\dot{q}''_{FB} = \frac{1}{\Delta z}\int_0^{\Delta z} \lambda_1 \frac{\Delta T_{sp}}{\delta_1} dz = \Delta T_{sp} \frac{1}{2}\left[\frac{\rho_1 \lambda_1 \Delta h w_2}{\Delta T_{sp}} f_{fc}\right]^{1/2} \frac{1}{\Delta z}\int_0^{\Delta z} dz/z^{1/2}$$

$$= \Delta T_{sp}\left[\frac{\rho_1 \lambda_1 \Delta h w_2}{\Delta T_{sp} \Delta z} f_{fc}\right]^{1/2} \qquad (A21.2\text{-}2)$$

The heat transfer coefficient is

$$h_{fc} = c_{fc}\left[\frac{\rho_1 \lambda_1 \Delta h w_2}{\Delta T_{sp} \Delta z} f_{fc}\right]^{1/2} \qquad (A21.2\text{-}3)$$

where

$$c_{fc} = 1. \qquad (A21.2\text{-}4)$$

22. Liquid droplets

22.1 Spontaneous condensation of pure subcooled steam – nucleation

Consider steam at local temperature T_1 and local static pressure p. The saturation temperature at the system pressure is

$$T' = T'(p). \tag{22.1}$$

The saturation pressure as a function of the steam temperature is

$$p' = p'(T_1). \tag{22.2}$$

The subcooling is defined as a temperature difference

$$\Delta T_{1,sub} = T'(p) - T_1. \tag{22.3}$$

For steam temperature larger than the saturation temperature there is no condensation. One expects intuitively condensation if the local steam temperature is lower than the saturation temperature at the system pressure, that is if

$$\Delta T_{1,sub} > 0. \tag{22.4}$$

The originating nucleus releases energy at saturation temperature $T'(p)$. The released energy is then transferred into the subcooled vapor acting towards reducing subcooling and towards stopping condensation. Therefore, in nature subcooling is a necessary condition to observe condensation. The condensation itself starts around a preferred number of points in the unit volume n_3. Around these points the predominantly low energy molecules form microscopic drops called activated clusters - nucleus with size R_3. Very small clusters are not stable and degrade spontaneously. Clusters with larger sizes may become stable. Intuitively we expect increasing origination frequency of clusters if the subcooling of the vapor is larger. We also intuitively cannot imagine that arbitrary large meta-stable subcooling is sustainable in nature.

In order to describe nucleation it is necessary to answer the following questions:

a) what is the nucleation size separating the stable from the unstable clusters,
b) how does nucleation site density depend on the subcooling, and
c) what is the maximum allowable subcooling in the steam.

The purpose of this Section is to give answers to these questions for pure vapors.

22.1.1 Critical nucleation size

The considerations leading to estimate of the critical nucleation size, probably first formulated by *Gibbs* [17] in 1878, will be next summarized. Consider a subcooled non-stable spherical steam volume, which is just about forming a nucleus, having an initial radius R_{30}, and volume $\frac{4}{3}\pi R_{30}^3$, pressure p, temperature T_1, and density $\rho_1 = \rho_1(p,T_1)$. Note that this density is computed for unstable steam. For practical application attention should be paid to this point because usually the state equation is obtained for stable fluids and an extrapolation into the unstable region is not straightforward for all approximations of the equation of state. During the origination of the nucleus the sphere collapse from the initial radius to R_3, that is to the volume $\frac{4}{3}\pi R_3^3$. The final state is characterized by the same temperature T_1 but the state is stable, that is, the local pressure inside the nucleus is $p' = p'(T_1)$. Because of the small nucleus size the pressure inside is considered to be uniform. The density is therefore $\rho' = \rho'(T_1)$. During the origination of the nucleus the initial and the end mass are the same. Therefore

$$\left(\frac{R_{30}}{R_3}\right)^3 = \frac{\rho'(T_1)}{\rho_1}. \tag{22.5}$$

Thus the initial volume of the sphere is changed by

$$\frac{4}{3}\pi\left(R_3^3 - R_{30}^3\right) = \frac{4}{3}\pi R_3^3\left[1 - \frac{\rho'(T_1)}{\rho_1}\right]. \tag{22.6}$$

Volume reduction requires mechanical work. The amount of this work

$$4\pi \int_{R_{30}}^{R_3} \left[p'(T_1) - p \right] r^2 dr \approx \frac{4}{3} \pi \left(R_3^3 - R_{30}^3 \right) \left[p'(T_1) - p \right]$$

$$= \frac{4}{3} \pi R_3^3 \left[1 - \frac{\rho'(T_1)}{\rho_1} \right] \left[p'(T_1) - p \right] \quad (22.7)$$

is reducing the internal energy of sphere. The initial sphere interface was not a real one. The final sphere interface is a real one. The interface is a prominent characteristic of the origination of new liquid phase inside the steam. This well-defined surface requires additional mechanical work

$$4\pi \int_0^{R_3} \frac{4\sigma}{r} r^2 dr = 4\pi R_3^2 \sigma \quad (22.8)$$

to be created. The surface tension σ is a function of the local temperature which may strongly vary during transient processes. It is usually measured at macroscopic droplets. Whether the so obtained information is still valid for a very small nucleus is not clear. We will assume that the so measured surface tension holds also for very small droplets. One example of the dependence of the surface tension on the temperature is given here for water surrounded by its steam,

$$\sigma = 0.2358 \left(1 - T_3 / T_c \right)^{1.256} \left[1 - 0.625 \left(1 - T_3 / T_c \right) \right], \quad (22.9)$$

Lienhard [24]. We see that if the droplet surface approaches the critical temperature for water, $T_{c,H_2O} = 647.2 K$, the surface tension diminishes. Within the region 366 to 566 K the above equation can be approximated by

$$\sigma = 0.14783 \left(1 - T_3 / T_c \right)^{1.053} \quad (22.10)$$

with an error of $\pm 1 \%$.

Thus the work required to create a single nucleus

$$\Delta E_3 = 4\pi R_3^2 \sigma - \frac{4}{3} \pi R_3^3 \left(1 - \frac{\rho'(T_1)}{\rho_1} \right) \left[p'(T_1) - p \right] = 4\pi \sigma \left(R_3^2 - \frac{2}{3} \frac{R_3^3}{R_{3c}} \right),$$
$$(22.11)$$

where

$$R_{3c} = 2\sigma \bigg/ \left\{ \left[1 - \frac{\rho'(T_1)}{\rho_1} \right] \left[p'(T_1) - p \right] \right\}, \quad (22.12)$$

possesses a maximum at

$$\Delta E_{3c} = \frac{4}{3}\pi\sigma R_{3c}^2,\tag{22.13}$$

"...which does not involve any geometrical magnitudes", see *Gibbs* [17] 1878. Equation (22.12) is known in the literature as the *Laplace and Kelvin* equation. Clusters with sizes smaller than the critical one are unstable and disappear quickly after creation absorbing the excess energy from the surrounding vapor. Droplets with sizes larger than the critical one remain stable and release heat into the surrounding vapor. Obviously to create a nucleus a deficiency of internal energy of the steam is necessary which has to be greater than the critical energy. The energy required to create a stable nucleus with critical size divided by the vapor temperature defines the entropy change of the system between the initial and final state.

$$\Delta S_{13c} = \Delta E_{3c}/T_1.\tag{22.14}$$

The entropy change is made dimensionless by dividing by the *Boltzmann* constant $k = 13.805 \times 10^{-24}$. The resulting dimensionless number

$$Gb_1 = \frac{\Delta E_{3c}}{kT_1},\tag{22.15}$$

is called the *Gibbs* number and is frequently used in the literature as a measure of the subcooling. We see that the larger the subcooling the smaller the *Gibbs* number. Two consequences of Eqs. (22.12 to 22.15) are of practical importance.

a) If the steam temperature approaches the critical temperature the surface tension diminishes and therefore the energy required for creation of a nucleus is zero, which means that all molecules are nuclei for themselves and the origination of liquid inside the steam is impossible. The *Gibbs* number in this case is also equal to zero.

b) Rapid decompression can lead to large subcooling. The maximum possible subcoolings experimentally observed are in the range of *Gibbs* numbers from 6.8 to 9.6, *Skripov* et al. [37], p. 136. It follows from the Eqs. (22.12 to 22.14)

$$p'(T_{1,spin}) - p = (1.57\_to\_1.32)\frac{\sigma^{3/2}}{(kT_{1,spin})^{1/2}\left(1 - \frac{p'(T_{1,spin})}{\rho_1(p,T_{1,spin})}\right)}.\tag{22.16}$$

This equation is known as the *Furth* equation [16]. It describe a dependence in the p,T-plane called a *spinoidal line*.

22.1.2 Nucleation kinetics, homogeneous nucleation

Origination of nucleation clusters within the steam is called homogeneous nucleation. It has to be distinguished from the so called heterogeneous nucleation. In the heterogeneous nucleation the origination of the liquid drops happens at the surfaces of other materials, e.g. walls, solid particles etc. Statistical thermodynamics provides the mathematical tools for the derivation of an equation defining the generated nucleus per unit volume and unit time, $\dot{n}_{3,kin}$, as a function of the local parameters. If we observe over the time $\Delta \tau$ a point in the space occupied by subcooled vapor, we will see that the stronger the subcooling the more frequently the system departs from the averaged state. With other words over an integral time $\Delta \tau_{13}$ the system is outside of the averaged state. *Volmer* [42] p.81 following *Boltzman* and *Einstein* defined the probability of departure from the averaged state as $\Delta \tau_{13} / \Delta \tau$ and assumed that this probability is proportional to $\exp(-Gb_1)$, that is

$$\Delta \tau_{13} / \Delta \tau = \exp(-Gb_1). \tag{22.17}$$

This mathematical expression says that the smaller the needed entropy change for a departure from the averaged state, the higher the probability of its occurrence. The probability of departure from the averaged state is usually set equal to the probability of origination of a single nucleus, that is

$$\frac{nucleation\_frequency}{molecular\_colisions\_frequency} = \exp(-Gb_1). \tag{22.18}$$

Zeldovich [43] obtained the final expression for the generated nuclei per unit volume and unit time in the form

$$\dot{n}_{3,kin} = c_T Z e_1 \left(\frac{1}{4} N_1 a_1 F_{3c} \right) N_1 \exp(-Gb_1), \tag{22.19}$$

- see also in *Fieder* et al. [13]. The assumptions used to derive this expression are at the same time the limitation of the model as discussed by *Ehrler* in [41]:

a) The clusters are distributed in the vapor and are assumed to have a degree of freedom characteristic for big molecules in the vapor;

b) The nuclei are solid non-oscillating spheres with respect to the center of the mass of the system;

c) Macroscopic properties like density and surface tension are applicable to clusters having sizes of the order of nanometers.

Here the nucleation frequency per unit volume is considered as proportional to the number of the clusters with critical size originating per unit time and unit volume, $N_1 \exp(-Gb_1)$, and the number of the molecules striking the cluster, $\frac{1}{4}N_1 a_1 F_{3c}$, and remaining there $c_c \frac{1}{4}N_1 a_1 F_{3c}$. The proportionality factor was found to be $c_T Ze_1$, where

$$Ze_1 = \sqrt{\frac{Gb_1}{3\pi n_{3c}^2}} \qquad (22.20)$$

is the so called *Zeldovich* number [43] and c_T is the so called temperature correction factor taking into account the heat transfer from the nucleus to the surrounding vapor. For pure water steam $c_T \approx 0.001$. The components of these two expressions are computed as follows. The molecular mass, m_μ, and the *Avogadro* number, $N_A = 6.02 \times 10^{23}$, uniquely define the mass of a single molecule $m_1 = m_\mu / N_A$. The number of the vapor molecules per unit volume is therefore $N_1 = \rho_1 / m_1$. The thermal velocity of the vapor molecules is defined by

$$a_1 = \sqrt{\frac{8}{\pi} R_1 T_1} \,. \qquad (22.21)$$

After its origination the surface F_{3c} of the cluster is struck by the following number of molecules per unit time $\frac{1}{4}N_1 a_1 F_{3c}$. Only a part of them, $c_c \frac{1}{4} N_1 a_1 F_{3c}$ remain there, where $c_c \approx 0.1$ for water-steam plus non-condensable gases, and $c_c \approx 0.5$ to 1, for pure water steam. The number of the molecules in a single nucleus with critical size is

$$n_{3c} = \rho'(T_1) \frac{4}{3} \pi R_{3c}^3 / m_1 \,. \qquad (22.22)$$

For successful prediction of several experiments with steam flows in *Laval* nozzles *Deitsch* and *Philiphoff* [11] p.153 provided the following modification of one expression obtained by *Frenkel* [15]

$$\dot{n}_{3,kin} = \left(\frac{p}{kT_1}\right)^2 \frac{c_T}{\rho'(T_1)} \sqrt{\frac{2\sigma m_1}{\pi}} \exp(-\beta Gb_1) \,. \qquad (22.23)$$

The modification contains the following empirical factor

$$\beta = 0.25 + 6.7\left\{\left[p'(T_1) + \frac{1}{2}\rho_1 w^2\right]/p_c\right\}^{1/3} \qquad (22.24)$$

for $c_T = 1$. The validity of Eq. (22.23) is proven for pressures between 0.25 and 4 MPa.

The mass condensed per unit volume and time due to nucleation is

$$\mu_{13} = \rho'(T_1)\frac{d}{d\tau}(n_3 V_3) = \rho'(T_1)V_{3c}\dot{n}_{3,kin} + \rho'(T_1)n_3\frac{dV_3}{d\tau}. \qquad (22.25)$$

We see that it consists of two parts, one due to nucleation, and the other due to growth of the droplets. The energy released during the condensation into the surrounding vapor is therefore

$$\dot{q}_1'''^{\sigma 3} = \dot{n}_{3,kin}\Delta E_{3c} + \rho'(T_1)n_3\frac{dV_3}{d\tau}[h_1 - h'(T_1)]. \qquad (22.26)$$

22.1.3 Droplet growth

Small droplets: For very small droplets with sizes comparable with the molecular free path length the kinetic gas theory provides the basis for computation of the heat and mass transfer. The net mass flow rate of condensation is then

$$(\rho w)_{13} = \frac{p_{M1}}{\sqrt{2\pi R_{M1}T_1}} - \frac{p'(T_3^{\sigma 1})}{\sqrt{2\pi R_{M1}T_3^{\sigma 1}}}. \qquad (22.27)$$

or having in mind that

$$\dot{q}_3'''^{\sigma 1} = a_{13}h_c(T_1 - T_3^{\sigma 1}) = a_{13}(\rho w)_{13}[h_1 - h'(T_1)],$$

we obtain for the effective heat transfer coefficient the following expression

$$h_c = (\rho w)_{13}\frac{h_1 - h'(T_1)}{T_1 - T_3^{\sigma 1}} = \frac{h_1 - h'(T_1)}{T_1 - T_{3\sigma}}\left(\frac{p_{M1}}{\sqrt{2\pi R_{M1}T_1}} - \frac{p'(T_3^{\sigma 1})}{\sqrt{2\pi R_{M1}T_3^{\sigma 1}}}\right). \qquad (22.28)$$

Mason [27, 28] (1951-1957), Ludvig [26] (1975) proposed the following equation describing the change of the droplet radius under such conditions

$$\rho_3 \frac{dR_3}{d\tau} = \frac{1 - \frac{R_{3c}}{R_3}}{R_3 + 1.59\ell_1} \frac{T'(p_{M1}) - T_1}{\frac{h'' - h'}{\lambda_1} + \frac{p - p_{M1}}{p} \frac{R_{M1} T'(p_{M1}) p_{M1}^2}{\rho_{M1}(h'' - h') D_{M \to \Sigma n}}}, \qquad (22.29)$$

for $R_3 > R_{3c}$ which is equivalent to the equation

$$\rho_3 \frac{dR_3}{d\tau} = \frac{\mu_{13}}{a_{13}} = \frac{\dot{q}_1^{\prime\prime\sigma 3}}{h_{M1} - h'}, \qquad (22.30)$$

where

$$h_c = (h_{M1} - h') \frac{1 - \frac{R_{3c}}{R_3}}{R_3 + 1.59\ell_1} \frac{T'(p_{M1}) - T_1}{\frac{h'' - h'}{\lambda_1} + \frac{p - p_{M1}}{p} \frac{R_{M1} T'(p_{M1}) p_{M1}^2}{\rho_{M1}(h'' - h') D_{M \to \Sigma n}}}, \qquad (22.31)$$

is the heat transfer coefficient. Here

$$\ell_1 = \frac{\eta_1}{p} \sqrt{\frac{\pi R_{M1} T_1}{2 m_{\mu,M1}}} \qquad (22.32)$$

is the *free path of a steam molecule* and $m_{\mu,M1}$ is the *mole weight*. For water we have $m_{\mu,M1} = m_{\mu,H_2O} = 18$. Note that this equation is valid for steam-air mixtures.

Large droplets: For *larger macroscopic droplets* heat conduction and convection govern the heat and mass transfer. In this case the droplet grows due to condensation at the surface which depends on the properties of the environment to absorb the released heat. For a droplet in stagnant steam the results obtained by the modeling of *thermally controlled bubble growth* can be applied. Note that the continuous phase is the gas phase in this case. This means that in the relationship

$$R_3 = f \sqrt{a_1 \tau^*}, \qquad (22.33)$$

where $f = f(..., Ja)$, $a_1 = \lambda_1 / (\rho_1 c_{p1})$, and

$$Ja = \frac{\rho_1 c_{p1}(T' - T_1)}{\rho'(h'' - h')}. \qquad (22.34)$$

is the *Jakob* number. Remember that the denominator of the *Jakob* number is obtained from the boundary condition on the droplet surface, namely the equality of the heat released by the condensation, and of the heat taken away by heat conduction from the environment. The mass corresponding to this heat changes the radius of the droplet. The numerator in the *Jakob* number is obtained from the energy conservation equation for the continuum.

Analogously for convective heat transfer we can compute a heat transfer coefficient for a moving droplet and the corresponding mass and energy source terms.

22.1.4 Self-condensation stop

As already mentioned, the released heat during the condensation increases the gas temperature, which for closed systems may decrease the condensation to zero. For a closed control volume or negligible convection and the diffusion from the neighboring elementary cells, the approximate form of the energy conservation equation is

$$\alpha_1 \rho_1 c_{p1} \frac{dT_1}{d\tau} = \dot{q}_1^{m\sigma 3} = \frac{3\alpha_3}{R_3} h_c \left(T_3^{\sigma 1} - T_1 \right) \tag{22.35}$$

or

$$\frac{dT_1}{d\tau} = \frac{T_3^{\sigma 1} - T_1}{\Delta \tau_1^*}, \tag{22.36}$$

where

$$\Delta \tau_1^* = \frac{\alpha_1 \rho_1 c_{p1} R_3}{3 \alpha_3 h_c} \tag{22.37}$$

approximates the *time constant of the phenomena*. The product of the density and the specific heat at constant pressure for a gas is less than the analogous product for a liquid. That is why the time constant for the analyzed case is smaller compared to the time constant for processes controlled by the heat accumulation in a liquid. This fact is very important and has to be taken into account for the construction of stable solution algorithms especially for condensing flows.

Thus, under these circumstances the difference between the saturation temperature and the gas temperature decreases exponentially

$$T_3^{\sigma 1} - T_1 = \left(T_{30}^{\sigma 1} - T_{10} \right) e^{-\Delta \tau / \Delta \tau_1^*}. \tag{22.38}$$

Therefore, the *average* heat released during the condensation within the time step $\Delta\tau$ per unit mixture volume and unit time is

$$\dot{q}_1^{m\sigma 3} = \frac{3\alpha_3}{R_3} h_c \left(T_3^{\sigma 1} - T_1\right) f, \qquad (22.39)$$

where

$$f = \frac{\Delta\tau_1^*}{\Delta\tau}\left(1 - e^{-\Delta\tau/\Delta\tau_1^*}\right). \qquad (22.40)$$

The corresponding *averaged* mass source term is

$$\mu_{13} = \frac{\dot{q}_1^{m\sigma 3}}{h_{M1} - h'}. \qquad (22.41)$$

22.2 Heat transfer across droplet interface without mass transfer

Consider a family of mono-disperse droplet designated with d, moving in a continuum, designated with c, with relative velocity ΔV_{cd}. We are interested in how much thermal energy is transferred between the surface of a droplet and the surrounding continuum. We will consider first the steady state case, thereafter the transient heat conduction problem within a droplet and finally we will give some approximate solutions for the effective interface temperature and the average energy transferred.

The energy transported between the drop velocity field and the continuum per unit time and unit mixture volume for a steady state case is frequently approximated by

$$\dot{q}_{cd}''' = a_{cd} h_{cd} \left(T_c - T_d\right) = 6\alpha_d \lambda_c Nu_d \left(T_c - T_d\right)/D_d^2. \qquad (22.42)$$

Here $a_{cd} = 6\alpha_d / D_d$ is the interfacial area density, $h_{cd} = \lambda_c Nu_d / D_d$ is the heat transfer coefficient and Nu_d is the *Nusselt* number defined as given in Table 22.1. Table 22.1 contains analytically and experimentally obtained solutions for the *Nusselt* number as a function of the relative velocity and the continuum properties.

22.2 Heat transfer across droplet interface without mass transfer

Table 22.1. Heat transfer coefficient on the surface of moving solid sphere and water droplets

$$Nu_d = \frac{D_d h_c}{\lambda_c}, \quad Pe_c = \frac{D_d |\Delta V_{cd}|}{a_c} = Re_d Pr_c, \quad Re_d = \frac{D_d \rho_c |\Delta V_{cd}|}{\eta_c}, \quad Pr_c = \frac{\eta_c}{\rho_c a_c} = \frac{\eta_c c_{pc}}{\lambda_c},$$

$$a_c = \frac{\lambda_c}{\rho_c c_{pc}}$$

$Re_d \ll 1$		Potential flow
$Pe_c \ll 1$		
$Nu_d = 2 + \frac{1}{2}Pe_c + \frac{1}{6}Pe_c^2 + ...$		Soo [38] (1965)
$Pe_c \gg 1$		
$Nu_d = 0.98 Pe_c^{1/3}$	$\eta_d/\eta_c = \infty$	Nigmatulin [32] (1978)
$Nu_d = \frac{2}{\pi^{1/2}} Pe_c^{1/3}$		Boussinesq [6] (1905), isothermal sphere
$Nu_d = \left(\frac{3}{4\pi}\frac{1}{1+\eta_d/\eta_c} Pe_c\right)^{1/2}$		Levich [44] (1962)
$Nu_d = 0.922 + 0.991 Pe_c^{1/3}$		Acrivos-Goddard [3] (1965)
$Nu_d = 4.73 + 1.156 Pe_c^{1/2}$		Watt [43] (1972) isothermal sphere
$Re_d < 1$,		
$Pe_c < 10^3$		
$Nu_d = 2 + \dfrac{\frac{1}{3} Pe_c^{0.84}}{1+\frac{1}{3} Pe_c^{0.51}}$	$\eta_d/\eta_c = \infty$	Nigmatulin [32] (1978)
$Pe_c < 1$		

$$Nu_d = 2 + \frac{1}{2}Pe_c^{1/3}$$

Acrivos and Taylor [2] (1965)

only the resistance in the gas is taken into account

$1 < Re_d < 7 \times 10^4$, $0.6 < Pr_c < 400$

$$Nu_d = 2 + (0.55 \text{ to } 0.7) Re_d^{1/2} Pr_c^{1/3}$$

Soo [38] (1965)

Droplets and bubbles

Michaelides [45] (2003)

$0 < Re_d < 1$, $Pe_c > 10$

$$Nu_d = \left(\frac{0.651}{1+0.95\eta_d/\eta_c}Pe_c^{1/2} + \frac{0.991\eta_d/\eta_c}{1+\eta_d/\eta_c}Pe_c^{1/3}\right)(1.032+A)$$

$$+\frac{1.651(0.968-A)}{1+0.95\eta_d/\eta_c}+\frac{\eta_d/\eta_c}{1+\eta_d/\eta_c}, \quad A = \frac{0.61Re_d}{21+Re_d}.$$

$10 \leq Pe_c \leq 1000$, $Re_d > 1$

$$Nu_d = \frac{2-\eta_d/\eta_c}{2}Nu_{d,0} + \frac{4\eta_d/\eta_c}{6+\eta_d/\eta_c}Nu_{d,2} \text{ for } 0 \leq \eta_d/\eta_c \leq 2$$

$$Nu_d = \frac{4}{\eta_d/\eta_c+2}Nu_{d,2} + \frac{\eta_d/\eta_c-2}{\eta_d/\eta_c+2}Nu_{d,\infty} \text{ for } 2 < \eta_d/\eta_c < \infty$$

where

$$Nu_{d,0} = 0.651Pe_c^{1/2}(1.032+A)+1.6-A \text{ for } \eta_d/\eta_c = 0,$$

$$Nu_{d,\infty} = 0.852Pe_c^{1/3}\left(1+0.233Re_d^{0.287}\right)+1.3-0.182Re_d^{0.355} \text{ for } \eta_d/\eta_c = \infty,$$

$$Nu_{d,2} = 0.64Pe_c^{0.43}\left(1+0.233Re_d^{0.287}\right)+1.41-0.15Re_d^{0.287} \text{ for } \eta_d/\eta_c = 2.$$

Solid sphere undergoing a step temperature change, Feng and Michaelides [46] (1986)

$$Nu_d = 2\left\{1 + 2Pe_c^2 \ln(2Pe_c) + Pe_c \left[\frac{1}{2}\mathrm{erf}\left(2Pe_c\sqrt{\tau^*}\right) + \frac{\exp(-4Pe_c^2 \tau^*)}{4Pe_c\sqrt{\pi\tau^*}}\right]\right\}$$

Water droplets

$$Nu_d = 2 + 0.6 Re_d^{1/2} Pr_c^{1/3} \qquad\qquad \textit{Ranz} \text{ and } \textit{Marshal} \text{ [33] (1952)}$$

$64 < Re_d < 250$, $0.00023 < D_d < 0.00113 m$, $1.03 \times 10^5 < p < 2.03 \times 10^5 Pa$, $2.8 < T_c - T_d < 36K$, $2.7 < \Delta V_{cd} < 11.7 m/s$

$$Nu_d = 2 + 0.738 Re_d^{1/2} Pr_c^{1/3}. \qquad\qquad \textit{Lee} \text{ and } \textit{Ryley} \text{ [25] (1968)}$$

In nature the temperature of both media changes with the time. That is why we consider next the transient heat conduction problem inside the droplet associated with the heat transfer at the surface.

Consider the case for which the continuum temperature is T_c and the droplet temperature at the beginning of the time $\Delta\tau$ is T_d. The *Fourier* equation

$$\frac{\partial T}{\partial \tau} = \frac{\lambda_d}{\rho_d c_{cd}}\left(\frac{\partial^2 T}{\partial r^2} + \frac{2}{r}\frac{\partial T}{\partial r}\right). \qquad (22.43)$$

describes the temperature profile in the droplet. Usually the equation is written in terms of the dimensionless temperature

$$\overline{T} = \frac{T(r,\tau) - T_d}{T_d^{\infty} - T_d} \qquad (22.44)$$

for $r = 0$, R_d, $\tau = \tau, \tau + \Delta\tau$ in the following form

$$\frac{\partial \overline{T}}{\partial \tau} = a_d\left(\frac{\partial^2 \overline{T}}{\partial r^2} + \frac{2}{r}\frac{\partial \overline{T}}{\partial r}\right), \qquad (22.45)$$

where $a_c = \lambda_c/(\rho_c c_{pc})$ is the droplet thermal diffusivity. Text book analytical solution of the *Fourier* equation is available for the following boundary conditions:

(a) Drop initially at uniform temperature: $\overline{T}(r,0) = 0$;

(b) Drop surface at $r = R_d$ immediately reaches the temperatures $T_d^{\sigma c}$, $\bar{T}(R_d,\tau) = 1$;

(c) Symmetry of the temperature profile: $\left.\dfrac{\partial \bar{T}}{\partial r}\right|_{r=0} = 0$.

The solution is represented by the converging *Fourier* series

$$\bar{T}(r,\tau) = 1 - \frac{2}{\pi} \frac{R_d}{r} \sum_{n=1}^{\infty} \frac{(-1)^n}{n} \sin\left(n\pi \frac{r}{R_d}\right) \exp\left(-n^2 \tau / \Delta \tau_d\right) \qquad (22.46)$$

where $\Delta \tau_d = D_d^2 / (4\pi^2 a_d)$ is the characteristic time constant of the heat conduction process. The volume-averaged non-dimensional temperature is

$$\bar{T}_m = 1 - \frac{6}{\pi^2} \sum_{n=1}^{\infty} \frac{1}{n^2} \exp\left(-n^2 \tau / \Delta \tau_d\right), \qquad (22.47)$$

Convection inside the droplet caused by the interfacial shear due to relative velocity can improve the thermal conductivity of the droplet a_d by a factor $f \geq 1$. Therefore

$$\bar{T}_m = 1 - \frac{6}{\pi^2} \sum_{n=1}^{\infty} \frac{1}{n^2} \exp\left(-fn^2 \tau / \Delta \tau_d\right). \qquad (22.48)$$

Celata et al. [7] (1991) correlated their experimental data for condensation with the following expression for *f*,

$$f = 0.53 Pe_c^{*0.454}, \qquad (22.49)$$

where a special definition of the *Peclet* number is used $Pe_c^* = \dfrac{D_d |\Delta V_{cd}|}{a_c} \dfrac{\eta_c}{\eta_c + \eta_d}$.

For a water droplet with D_d = 0.00002, 0.0002, 0.001m and $T_c - T_d = 20K$ the time necessary to reach $\bar{T}_m$ = 0.99 is 0.00006, 0.00613, 0.612 s respectively. For values of $\bar{T}_m$ greater than 0.95 (long contact times) the first term only of Eq. (22.48) is significant

$$\bar{T}_m \approx 1 - \frac{6}{\pi^2} \exp\left(-fn^2 \tau / \Delta \tau_d\right). \qquad (22.50)$$

22.2 Heat transfer across droplet interface without mass transfer

One alternative solution of Eq. (22.48) can be obtained by the method of *Laplace* transformations which takes the form of a diverging infinite series. For values of $\overline{T}_m$ less than 0.4 (short contact times) the first term only of this series

$$\overline{T}_m = \frac{3\sqrt{\pi}}{D_d \sqrt{a_d \tau}} \qquad (22.51)$$

is significant. This equation applies for short contact times when the temperature gradient of the surface surroundings has not penetrated to the center of the sphere which consequently behaves as a semi-infinite body.

From the energy balance of the droplet velocity field we obtain the time-averaged energy transfer from the surface into the droplet per unit time and unit mixture volume

$$\dot{q}_d^{m\sigma c} = \alpha_d \rho_d c_{pd} \frac{1}{\Delta \tau} \int_0^{\Delta \tau} \frac{dT_{dm}}{d\tau} d\tau = \alpha_d \rho_d c_{pd} \frac{1}{\Delta \tau} (T_d^{\sigma c} - T_d) \left[\overline{T}_{dm}(\Delta \tau) - \overline{T}_{dm}(0) \right]$$

leading to

$$\alpha_d \rho_d c_{pd} \frac{\overline{T}_{dm}(\Delta \tau)}{\Delta \tau} (T_d^{\sigma c} - T_d) = \omega_d (T_d^{\sigma c} - T_d) \qquad (22.52)$$

If the surface temperature differs from the continuum temperature there is a convective component of the energy transported from the continuum to the surface for $T_c > T_d^{\sigma c}$ or from the surface to the continuum for $T_c < T_d^{\sigma c}$, namely

$$\dot{q}_c^{m\sigma d} = 6\alpha_d \lambda_c Nu_d (T_d^{\sigma c} - T_c) / D_d^2 = \omega_c (T_d^{\sigma c} - T_c), \qquad (22.53)$$

where Nu_d is the *Nusselt* number for convective heat transfer between a sphere and the continuum.

For $T_c \approx const$ the effective interface temperature is obtained from the energy jump condition $\dot{q}_d^{m\sigma c} = -\dot{q}_c^{m\sigma d}$ after solving for $T_d^{\sigma c}$

$$T_d^{\sigma c} = \frac{\omega_d T_d + \omega_c T_c}{\omega_d + \omega_c}. \qquad (22.54)$$

For the case when $T_c \neq const$ the heat-transfer causes changes in the continuum temperature too. If we neglect the convection and the diffusion from the neighboring elementary cells, which is valid for a heat transfer time constant considerably smaller than the flow time constant, the continuum energy conservation can be approximated by

$$\alpha_c \rho_c c_{pc} \frac{dT_c}{d\tau} = 6\alpha_d \lambda_c Nu_d \left(T_d^{\sigma c} - T_c\right)/D_d^2 \qquad (22.55)$$

or

$$\frac{dT_c}{d\tau} = \frac{T_d^{\sigma c} - T_c}{\Delta \tau_c^*}, \qquad (22.56)$$

where

$$\Delta \tau_c^* = \frac{\alpha_c \rho_c c_{pc} D_d^2}{6\alpha_d \lambda_c Nu_d} \qquad (22.57)$$

approximates the characterizing time constant of the continuum heating or cooling. Note that the product of the density and the specific heat at constant pressure for gas is less than the analogous product for liquid. That is why the time constant for gases as a continuum should be smaller compared to the time constant for liquids as a continuum. This make the gas temperature very sensitive to the heat and mass transfer process.

Thus for some effective $T_d^{\sigma c}$ the difference between the interface temperature and the continuum temperature decreases exponentially

$$\overline{T}_{cm}(\Delta \tau) = \frac{T_d - T_c}{T_d^{\sigma c} - T_c} = \exp\left(-\frac{\Delta \tau}{\Delta \tau_c^*}\right) \qquad (22.58)$$

during the time interval $\Delta \tau$. Therefore, the time-averaged heat, transferred from the surface into the continuum per unit mixture volume and unit time is

$$\dot{q}_c^{m\sigma d} = 6\alpha_d \lambda_c Nu_d \left(T_d^{\sigma c} - T_c\right) f_c / D_d^2 = \omega_c \left(T_d^{\sigma c} - T_c\right) \qquad (22.59)$$

where

$$f_c = \left[1 - \exp\left(-\frac{\Delta \tau}{\Delta \tau_c^*}\right)\right] \frac{\Delta \tau_c^*}{\Delta \tau}. \qquad (22.60)$$

Note that for $\Delta \tau \to 0$, $f_c \to 1$.

Thus, from the energy jump conditions $\dot{q}_d^{m\sigma c} = -\dot{q}_c^{m\sigma d}$ we obtain the effective interface temperature by Eq. (22.12). Having this temperature we can compute the average heat power density $\dot{q}_c^{m\sigma d}$ by Eq. (22.58).

22.3 Direct contact condensation of pure steam on subcooled droplet

Consider the case where the steam temperature is equal to the saturation temperature by the system pressure

$$T_1 = T'(p), \qquad (22.61)$$

and the droplet temperature at the beginning of the time step considered is less than the saturation temperature

$$T_3 < T'(p). \qquad (22.62)$$

The heat conduction from the surface, which has the saturation temperature, to the bulk of the droplet is the condensation controlling mechanism in this case. Therefore, the *Fourier* equation describes the temperature profile inside the droplet. *Ford* and *Lekic* [14] (1973) considered additionally the following effects

(a) convection inside the droplet;
(b) limitation of the steam supply towards the droplet surface

and introduced the corrector f in Eq. (22.48),

$$\overline{T}_m \approx 1 - \exp(-f\tau/\Delta\tau). \qquad (22.63)$$

From the energy balance of the droplet velocity field we obtain the time-averaged energy transfer from the surface into the droplet per unit time

$$\dot{q}_3^{m\sigma 1} = \alpha_3 \rho_3 c_{p3} \left[T'(p) - T_3 \right] \overline{T}_{3m}(\Delta\tau) / \Delta\tau. \qquad (22.64)$$

If the steam temperature differs from the saturation temperature there is a convective component of the energy transported from the steam to the surface for $T_1 > T'(p)$ or from the surface to the steam for $T_1 < T'(p)$, namely

$$\dot{q}_1^{m\sigma 3} = a_{13} \frac{\lambda_1'}{D_3} Nu_{13}(T' - T_1), \qquad (22.65)$$

where Nu_{13} is the *Nusselt* number for convective heat transfer between sphere and steam. For the case when $T_1 < T'(p)$ part of the released heat during the condensation increases the gas temperature. If we neglect the convection and the diffusion from the neighboring elementary cells, the energy conservation can be approximated by

$$\alpha_1 \rho_1 c_{p1} \frac{dT_1}{d\tau} = 6\alpha_3 \lambda_1 Nu_3 \left(T_3^{\sigma 1} - T_1\right)/D_3^2, \qquad (22.66)$$

or

$$\frac{dT_1}{d\tau} = \frac{T_3^{\sigma 1} - T_1}{\Delta \tau_1^*}, \qquad (22.67)$$

where

$$\Delta \tau_1^* = \frac{\alpha_1 \rho_1 c_{p1} D_3^2}{6\alpha_3 \lambda_1 Nu_3} \qquad (22.68)$$

approximates the time constant of the steam heating. For nearly constant pressure, and therefore constant T', the difference between the saturation temperature and the gas temperature decreases exponentially during the time interval $\Delta \tau$,

$$\overline{T}_{cm}(\Delta \tau) = \frac{T_3^{\sigma 1} - T_1}{T_3^{\sigma 1} - T_{10}} = \exp\left(-\frac{\Delta \tau}{\Delta \tau_1^*}\right). \qquad (22.69)$$

Therefore, the time-averaged heat transferred from the surface into the gas per unit mixture volume and unit time is

$$\dot{q}_1^{''' \sigma 3} = 6\alpha_3 \frac{\lambda_1'}{D_3^2} Nu_{13} \left(T' - T_1\right) \qquad (22.70)$$

where

$$f = \left[1 - \exp\left(-\frac{\Delta \tau}{\Delta \tau_1^*}\right)\right] \frac{\Delta \tau_1^*}{\Delta \tau}. \qquad (22.71)$$

Thus, from the assumption that there is no energy and mass accumulation the interface (energy and mass jump conditions) we obtain for the averaged condensing mass per unit mixture volume and unit time the following expression

$$\mu_{13} = \frac{\dot{q}_3^{m\sigma 1} - \dot{q}_1^{m\sigma 3}}{h'' - h'}. \qquad (22.72)$$

22.4 Spontaneous flashing of superheated droplet

Next we consider a single droplet in a gas environment having pressure p. The averaged droplet temperature is T_3. The surface temperature of the droplet, $T_3^{\sigma 1}$, is *higher* than the saturated temperature $T'(p)$. In the first instant of the process considered here, the spontaneous emission of mass will be *explosive*. In the next moment the surface temperature would drop and reach the saturation temperature, if no heat is transferred from the bulk region of the droplet to the surface. The heat transferred from the bulk region to the surface maintains the surface temperature *higher* than the saturation temperature and causes further evaporation. The *thermal resistance inside the drop* has a delaying effect on the evaporation, because the heat conduction is the slower process compared to the mass emission from the surface.

The net mass flow rate of evaporation from the droplet surface can be calculated by solving numerically the *Boltzmann* kinetic equation for such conditions. This is a quite complicated task. For the practical application the theory first proposed by *Hertz* is used. *Hertz* [18] (1882) computed the evaporation mass flow rate from a surface with temperature $T_3^{\sigma 1}$ and surface pressure $p'(T_3^{\sigma 1})$ multiplying the frequency of the molecules leaving unit surface $N_1 a_1 / 4$ by the mass of a single molecule m_1. Here N_1 is the number of vapor molecules in unit volume. The averaged thermal velocity of the molecules at temperature $T_3^{\sigma 1}$ is

$$a_1 = \sqrt{\frac{8}{\pi} R_{M1} T_3^{\sigma 1}}. \qquad (22.73)$$

The result is

$$(\rho w)_{31}^* = m_1 \frac{1}{4} N_1 a_1. \qquad (22.74)$$

Having in mind that

$$m_1 N_1 = \rho_1 \left[p'(T_3^{\sigma 1}), T_3^{\sigma 1} \right], \qquad (22.75)$$

$$p'(T_3^{\sigma 1}) / \rho_1 \left[p'(T_3^{\sigma 1}), T_3^{\sigma 1} \right] \approx R_{M1} T_3^{\sigma 1}, \qquad (22.76)$$

Hertz obtain finally

$$(\rho w)^*_{31} = \frac{p'(T_3^{\sigma 1})}{\sqrt{2\pi R_{M1} T_3^{\sigma 1}}}. \tag{22.77}$$

Similarly *Hertz* computed the condensing mass flow rate from the steam with temperature T_1 equal to the gas temperature and partial pressure p_{M1} multiplying the frequency of the molecules striking unit surface by the mass of a single molecule. The averaged thermal velocity of the molecules at temperature T_1 across the plane surface in local thermodynamical equilibrium is

$$a_1 = \sqrt{\frac{8}{\pi} R_{M1} T_1}. \tag{22.78}$$

The result is

$$(\rho w)^*_{13} = \frac{p_{M1}}{\sqrt{2\pi R_{M1} T_1}}. \tag{22.79}$$

Assuming (i) reversible process, (ii) $T_1 \approx T_3^{\sigma 1}$, and (iii) condensation and evaporation flows do not affect each other, *Hertz* obtained finally

$$(\rho w)_{31} = (\rho w)^*_{31} - (\rho w)^*_{13} = \frac{1}{\sqrt{2\pi R_{M1} T_1}} \left[\sqrt{\frac{T_1}{T_3^{\sigma 1}}} p'(T_3^{\sigma 1}) - p_{M1} \right]$$

$$\approx c \frac{1}{\sqrt{2\pi R_{M1} T_1}} \left[p'(T_3^{\sigma 1}) - p_{M1} \right], \tag{22.80}$$

where c is the so called accommodation coefficient being of order of

$$c = 0.1 \text{ to } 1 \tag{22.81}$$

as shown by *Volmer*'s review in [42] (1931). Later improvements by *Knudsen* [20] (1915) and *Langmuir* [22] (1913), *Langmuer* et al. [23] (1927) led to the formula for the *net mass flow rate of the spontaneous evaporation*

$$(\rho w)_{31} = (\rho w)^*_{31} - (\rho w)^*_{13} = c_e \frac{p'(T_3^{\sigma 1})}{\sqrt{2\pi R_{M1} T_3^{\sigma 1}}} - c_c \frac{p_{M1}}{\sqrt{2\pi R_{M1} T_1}}. \tag{22.82}$$

Here c_e is the *probability of escape* for a liquid molecule at the interface, and c_c is the *probability of capture* for a vapor molecule at the interface. This relationship is known in the literature as the equation of *Hertz, Knudsen* and *Langmuir* (see in [34] (1985). For no sound physical reason, Eq. (22.81) is generally approximated in the literature by

$$(\rho w)_{31} = c \left(\frac{p'(T_3^{\sigma 1})}{\sqrt{2\pi R_{M1} T_3^{\sigma 1}}} - \frac{p_{M1}}{\sqrt{2\pi R_{M1} T_1}} \right). \tag{22.83}$$

For pressures below 0.1 *bar* measurements reported by *Fedorovich* and *Rohsenow* [12] (1968) for liquid metals tightly clustered about unity. For larger pressure Eq. (22.81) is a good approximation - see the discussion by *Mills* [29] and *Mills* and *Seban* [30]. For water the discussion and the measurements in [30,31,5] give

$$c = 0.35 \text{ to } 1. \tag{22.84}$$

Knowles [19] (1985) derived an analytical expression taking into account the influence of the condensation due to the relative motion of the interface with respect to the steam. The result is

$$(\rho w)_{31} = c \left(\frac{p'(T_3^{\sigma 1})}{\sqrt{2\pi R_{M1} T_3^{\sigma 1}}} - f(\Delta |\mathbf{V}_{21}|) \frac{p_{M1}}{\sqrt{2\pi R_{M1} T_1}} \right), \tag{22.85}$$

where the so called *transport enhancement factor* approximating the analytical solution is

$$f(\Delta \mathbf{V}_{21}) = 1 + 1.7 \frac{\Delta |\mathbf{V}_{21}|}{\sqrt{2R_{M1} T_1}} + 0.7 \left(\frac{\Delta |\mathbf{V}_{21}|}{\sqrt{2R_{M1} T_1}} \right)^2 \tag{22.86}$$

for $\frac{\Delta |\mathbf{V}_{21}|}{\sqrt{2R_{M1} T_1}} \geq -1$ and

$$f(\Delta |\mathbf{V}_{21}|) = 0 \tag{22.87}$$

otherwise. For completeness we give some other modifications given in the literature. *Schrage*, see in [4] (1980), used the following form:

$$(\rho w)_{d1} = \frac{2c}{2-c} \frac{1}{\sqrt{2\pi R_{M1} T_3^{\sigma 1}}} \left[p'(T_3^{\sigma 1}) - p_{M1} - \frac{p'(T_3^{\sigma 1})}{2 T_3^{\sigma 1}} (T_3^{\sigma 1} - T_1) \right], \tag{22.88}$$

Here a modified *accommodation* coefficient is used having values for water $c_{H_2O} = 0.01$ to 1. *Samson* and *Springer* [36] (1969) used the following form for very small droplets, for $\ell_1 / D_d \gg 1$,

$$(\rho w)_{d1} = \frac{1}{\sqrt{2\pi R_{M1} T_1}} \left[p'(T_d^{\sigma c}) - p_{M1} - \frac{p_{M1}}{2T_1}(T_d^{\sigma c} - T_1) \right], \qquad (22.89)$$

and for large droplets, $\ell_1 / D_d \ll 1$

$$(\rho w)_{d1} = \frac{1}{\sqrt{2\pi R_{M1} T_1}} \left[p'(T_d^{\sigma c}) - p_{M1} \right]. \qquad (22.90)$$

Here

$$\ell_1 = \frac{\eta_1}{p}\sqrt{\frac{\pi R_{M1} T_1}{2 M_{M1}}}, \qquad (22.91)$$

is the free path of a steam molecule and $M_{M1} = M_{H_2O} = 18$ is the mole weight. *Labunzov* and *Krjukov* [21] (1977) used the following form:

$$(\rho w)_{d1} = 2\left[p''(T_{di}) - p_{M1}(T_1, p_{M1}) \right] \sqrt{\frac{R_{M1} T_{di} \rho_{M1}(T_1, p_{M1})}{2\pi \rho''(T_{di})}}. \qquad (22.92)$$

The spontaneously evaporating mass from the superheated velocity field per unit mixture volume is

$$\mu_{d1} = (6\alpha_d / D_d)(\rho w)_{d1}. \qquad (22.93)$$

In the case of strong emission of steam from the surface into the surrounding gas the heat transfer from the surface to the surrounding continuum can be neglected and therefore

$$\dot{q}_1^{m\sigma 3} = 0. \qquad (22.94)$$

The energy jump condition for this situation reads

$$\mu_{d1} \left[h''(T_d^{\sigma c}) - h'(T_d^{\sigma c}) \right] = -\dot{q}_d^{m\sigma c}. \qquad (22.95)$$

If we use the approximate solution of the *Fourier* equation we obtain the one equation with respect the unknown surface temperature, namely

$$\frac{6}{D_d}(\rho w)_{d1}\left[h''(T_d^{\sigma c})-h'(T_d^{\sigma c})\right]=\rho_d c_{pd}\left(T_d-T_d^{\sigma c}\right)\frac{1}{\Delta \tau}\left[1-\frac{6}{\pi^2}\exp\left(-\frac{\Delta \tau}{\Delta \tau_d}\right)\right].$$
(22.96)

This equation can be solved by iteration starting with

$$T_d^{\sigma c} \approx T_d.$$
(22.97)

Having the surface temperature it is very easy to compute the mass source term from the Eqs. (22.85) and (22.93), and the corresponding heat power density from Eq. (22.95).

The spontaneous evaporation will cease if the droplet surface temperature reaches the saturation temperature corresponding to the system pressure. If the saturation temperature corresponding to the partial steam pressure $T'(p_{M1})$ is lower than the surface temperature the slow evaporation continues due to molecular diffusion. This process will be considered in Section 22.5.

22.5 Evaporation of saturated droplets in superheated gas

Let us consider the evaporation of saturated droplets in superheated gas. *Lee - Ryley* [25] (1968) found that the heat needed for the evaporation is convectively transported from the gas to the surface, and the emitted mass from the surface does not influence considerably this heat transfer process. In this case we have

$$\mu_{1d}=0,$$
(22.98)

$$\dot{q}_3^{'''\sigma 1}=0.$$
(22.99)

During the droplet evaporation the gas is *cooled* with

$$\dot{q}_1^{'''\sigma d}=-a_{d1}h_c\left[T_1-T'(p)\right]=-(6\alpha_d/D_d)h_c\left[T_1-T'(p)\right].$$
(22.100)

This is the instantaneous power density. From the energy jump condition (no energy accumulation on the droplet interface) we obtain for the instantaneous evaporating mass per unit mixture volume and unit time

$$\mu_{31} = (6\alpha_d / D_d) h_c \left[T_1 - T'(p)\right] / \left[h''(p) - h_{Md}\right]. \tag{22.101}$$

Actually we are interested in the *averaged* values of the mass source term during the time step $\Delta\tau$. For their approximate estimation we use the following way. Because the mass change of the droplet is caused by the evaporated mass per unit time, we have the following form of the mass conservation equation for a single droplet

$$\rho' \frac{dR_d}{d\tau} = -h_c \left[T_1 - T'(p)\right] / \left[h''(p) - h_{Md}\right] \tag{22.102}$$

or

$$\frac{dR_d}{d\tau} = -\frac{h_c}{\rho'} \left[T_1 - T'(p)\right] / \left[h''(p) - h_{Md}\right]$$

$$= -\frac{Nu_{d1} \lambda_1}{\rho' 2 R_d} \left[T_1 - T'(p)\right] / \left[h''(p) - h_{Md}\right] = -c / R_3. \tag{22.103}$$

This is the relationship between the change of the radius, the *Nusselt* number, and the temperature difference driving the evaporation. Integrating the above equation we obtain

$$\frac{R_d}{R_{d0}} = \sqrt{1 - 2c\Delta\tau / R_{d0}^2} = \sqrt{1 - \Delta\tau / \Delta\tau^*}. \tag{22.104}$$

If sufficient internal energy of the gas is available the *evaporation* of the entire droplet $R_d = 0$ happens after the time interval

$$\Delta\tau^* = R_{d0}^2 / (2c). \tag{22.105}$$

After this time interval the Eq. (22.103) is no longer valid.

The third power of the radii ratio

$$\left(\frac{R_d}{R_{d0}}\right)^3 = \left(1 - \frac{\Delta\tau}{\Delta\tau^*}\right)^{3/2}. \tag{22.106}$$

is necessary to compute the averaged value of the evaporated mass per unit mixture volume and unit time within the time interval $\Delta\tau$

$$\mu_{d1} = \rho' n_d \frac{V_{d0} - V_d}{\Delta \tau} = \frac{\rho' n_d V_{d0}}{\Delta \tau}\left(1 - \frac{V_d}{V_{d0}}\right) = \frac{\rho' \alpha_d}{\Delta \tau}\left[1 - \left(\frac{R_d}{R_{d0}}\right)^3\right] \quad (22.107)$$

for $\Delta \tau < \Delta \tau^*$ and

$$\mu_{d1} = \frac{\rho' \alpha_d}{\Delta \tau} \quad (22.108)$$

for $\Delta \tau \geq \Delta \tau^*$. The last equation is practically the equation of the entire evaporation of the available droplets inside the time interval $\Delta \tau$.

Having in mind that during the evaporation

$$\mu_{d1} = 0 \quad (22.109)$$

$$\dot{q}_d^{m\sigma c} = 0, \quad (22.110)$$

we obtain from the energy jump condition the following relationship

$$\dot{q}_1^{m\sigma d} = -\mu_{d1}\left[h''(p) - h'(p)\right]. \quad (22.111)$$

The above analysis was performed under the assumption that the *Nusselt* number does not depend on the radius during the considered time interval. In fact, the radius decreases in accordance with the relationship of *Lee - Ryley* [25] (1968),

$$Nu_{d1} = 2 + 0.738 Re_{d1}^{1/2} Pr_1^{1/3} = 2 + 0.738\left(2|\Delta V_{1d}|/v_1\right)^{1/2} Pr_1^{1/3} D_d^{1/2} = 2 + c_3 R_d^{1/2}, \quad (22.112)$$

and therefore, the *Nusselt* number decreases too. This makes the process slower than that previously described (assuming the *Nusselt* number not depending on the radius). Taking into account this relationship

$$\frac{R_d}{2 + c_3 R_d^{1/2}} \frac{dR_d}{d\tau} = c_2, \quad (22.113)$$

Lee and *Ryley* [25] (1968) obtained an analytical integral of the above equation, namely

$$\Delta \tau = \Delta \tau^* - \frac{1}{c_2}\left[\frac{2}{3}\frac{\sqrt{R_d}}{c_3} - \frac{D_d}{c_3^2} + \frac{8\sqrt{D_d}}{c_3^3} - \frac{16}{c_3^4}\ln\left(\frac{2}{2 + c_3 D_d^{1/2}}\right)\right]. \quad (22.114)$$

For a known time interval $\Delta \tau$ we can compute the corresponding droplet diameter after solving the above transcendental equation iteratively. The time required for the full droplet evaporation (if sufficient excess internal gas energy is available), i.e. $R_d = 0$ is easily obtained

$$\Delta \tau^* = \frac{1}{c_2} \left[\frac{2}{3} \frac{\sqrt{R_{d0}}}{c_3} - \frac{D_{d0}}{c_3^2} + \frac{8\sqrt{D_{d0}}}{c_3^3} - \frac{16}{c_3^4} \ln\left(\frac{2}{2 + c_3 D_{d0}^{1/2}} \right) \right]. \qquad (22.115)$$

Finally for completeness let us mention the empirical correlation proposed by *Saha* [35] (1980), describing the droplet evaporation in superheated steam in dispersed non-adiabatic flow, in dry out condition on the heated wall

$$\alpha_{1d} h_c = 6300 \alpha_d \frac{\lambda_1}{D_h^2} \left(1 - \frac{p}{p_{kr}} \right)^2 \left(\frac{\rho_1 V_1^2 D_h}{\sigma} \right)^{1/2}. \qquad (22.116)$$

Condie et al. [10] (1984) show that this correlation overestimates the evaporation rate compared to the experimental data and recommend *Webb*'s correlation (see [10])

$$\alpha_{1d} h_c = 1.32 \left(1 - \alpha_1^* \right)^{2/3} \frac{\lambda_1}{D_h^2} \left(1 - \frac{p}{p_{kr}} \right)^2 \left(\frac{\rho_1 V_1^2 D_h}{\sigma} \right)^{1/2} \qquad (22.117)$$

approximating better a number of experimental data (α_1^* is the homogeneous void fraction).

22.6 Droplet evaporation in gas mixture

Consider a gas mixture with inert mass concentration $\sum C_{n1}$ and vapor concentration C_{M1}, having total pressure p and gas temperature T_1 see Fig. 22.1. A family of mono-dispersed droplets is flowing through the gas with relative velocity ΔV_{13}. At the beginning of the time interval considered, $\Delta \tau$, the droplets possess temperature T_3 and diameter D_3. Consider the case when the droplet temperature T_3 is greater than the saturation temperature T' corresponding to the partial vapor pressure p_{M1}

$$T_3 > T'(p_{M1}). \qquad (22.118)$$

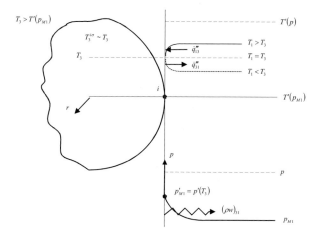

Fig. 22.1. Evaporation of a water droplet in a two-component gas mixture

In this case the interface temperature $T_{3\sigma}$ is somewhere between $T'(p_{M1})$ and T_1

$$T'(p_{M1}) < T_{3\sigma} < T_1 . \tag{22.119}$$

The surface is emitting vapor molecules which are struggling through the gas mixture boundary layer. The molecules of the inert gas present considerable resistance to their movement. The transport through the boundary layer is diffusion controlled. The evaporation mass flow per unit surface and unit time is $(\rho w)_{31}$. The heat needed for the evaporation is absorbed partially from the bulk of the droplet by thermal diffusion. Heat is transferred from or to the surface to or from the surrounding gas due to convection. The purpose of this section is to describe a method for computation of the evaporation mass flow rate.

The vapor pressure of the immediate neighborhood of the interface is equal to the saturation pressure corresponding to the surface temperature

$$p_{M1\sigma} = p'(T_3^{\sigma 1}) . \tag{22.120}$$

Therefore there is a pressure difference $p_{M1\sigma} - p_{M1}$ across the diffusion boundary layer driving the mass flow rate $(\rho w)_{31}$. The simplest way to compute $(\rho w)_{31}$ is to use the assumption for the similarity between temperature and concentration profile around the droplets and therefore the analogy between the heat and mass transfer, namely

$$(\rho w)_{31} = \rho_1 \beta (C_{M1\sigma} - C_{M1}) \tag{22.121}$$

where

$$\beta = -\frac{D_{M \to \Sigma n}}{C_{M1\sigma} - C_{M1}} (\nabla C_{M1})_{\sigma 3} \qquad \text{(definition)} \qquad (22.122)$$

$$C_{M1\sigma} = \frac{\rho_{M1\sigma}}{\sum \rho_{n1\sigma} + \rho_{M1\sigma}} = \frac{1}{1 + \sum \rho_{n1\sigma} / \rho_{M1\sigma}}, \qquad (22.123)$$

$$\rho_{M1\sigma} = \rho''(T_{3\sigma}), \qquad (22.124)$$

$$\sum \rho_{n1\sigma} = \frac{p - p'(T_3^{\sigma 1})}{R_{\Sigma n1} T_3^{\sigma 1}}, \qquad (22.125)$$

where

$$R_{\Sigma n1} = 287.04 \qquad (22.126)$$

is the gas constant for air.

In these case the mass transfer coefficient is computed from an appropriate heat transfer correlation, e.g. the correlation of *Renz* and *Marshall* [33] (1952) replacing the *Nu* and *Pr* numbers with *Sh* and *Sc* numbers respectively

$$Sh = \frac{\beta D_3}{D_{M \to \Sigma n}} = 2 + 0.6 Re_{13}^{1/2} Sc_1^{1/3}, \qquad (22.127)$$

where

$$Re_{13} = \rho_1 D_1 \Delta V_{13} / \eta_1, \qquad (22.128)$$

$$Sc_1 = \eta_1 / (\rho_1 D_{M \to \Sigma n}). \qquad (22.129)$$

An additional simplifying assumption

$$T_{3\sigma} \approx T_3 \qquad (22.130)$$

completes this model.

Having in mind that

$$\mu_{13} = 0, \qquad (22.131)$$

22.6 Droplet evaporation in gas mixture

$$\dot{q}_1^{m\sigma3} = 0, \qquad (22.132)$$

we obtain from the energy jump condition the heat extracted from the droplet during the slow evaporation per unit mixture volume and unit time

$$\dot{q}_3^{m\sigma1} = -\mu_{31}\left[h''(p) - h_{M3}\right]. \qquad (22.133)$$

In order to obtain the averaged value of the mass source term, we use the sum of the mass conservation equations for the inert components in the gas mixture, neglecting convection and diffusion from the neighbor control volumes,

$$\alpha_1 \rho_1 \frac{d}{d\tau}\sum C_{n1} = -\mu_{31}\sum C_{n1} = a_{13}\rho_1 \beta \sum C_{n1}\left(\sum C_{n1\sigma} - \sum C_{n1}\right) \qquad (22.134)$$

or after the substitution

$$\overline{\tau} = \frac{\tau a_{13}\rho_1 \beta \sum C_{n1\sigma}}{\alpha_1}, \qquad (22.135)$$

$$C = \frac{\sum C_{n1}}{\sum C_{n1\sigma}}, \qquad (22.136)$$

$$\frac{dC}{d\overline{\tau}} = C(1-C). \qquad (22.137)$$

The solution of this equation for $\tau = 0$, $C = C_0$ is

$$\frac{1-C}{1-C_0} = \frac{1}{1+C_0\left(e^{\overline{\tau}}-1\right)}. \qquad (22.138)$$

Using the so obtained solution, we compute the *averaged* value of the evaporated mass per unit mixture volume and unit time within the time interval $\Delta\tau$,

$$\mu_{31} = -a_{13}\rho_1 \beta \left(\sum C_{n1\sigma} - \sum C_{n1}\right) f, \qquad (22.139)$$

where

$$f = \frac{1}{1-C_0}\left\{1 - \frac{1}{\Delta\overline{\tau}}\ln\left[1-C_0\left(1-e^{\overline{\tau}}\right)\right]\right\} \le 1, \qquad (22.140)$$

This expression takes into account the *reduction of the driving forces* of the evaporation under the above discussed conditions.

There are other proposals in the literature how to compute $(\rho w)_{31}$. Following *Spalding* [39] (1953) the diffusion mass flow rate at macroscopic surface is

$$(\rho w)_{31} = \rho_1 \beta \ln(1+Sp) = \rho_1 \beta \ln C, \qquad (22.141)$$

where

$$Sp = \frac{C_{M1\sigma} - C_{M1}}{1 - C_{M1\sigma}}, \qquad (22.142)$$

is the *Spalding* mass transfer number and $C_{M1\sigma}$ is the vapor mass concentration at the boundary layer.

Clift et al. [9] (1978) use for the region $0.25 < Sc_1 < 100$ and $Re_{13} < 400$ the following correlation to compute

$$Sh = 1 + \left(1 + Re_{13} Sc_1\right)^{1/3} f(Re_{13}), \qquad (22.143)$$

$$f(Re_{13}) = 1 \qquad \text{for} \qquad Re_{13} < 1, \qquad (22.144)$$

$$f(Re_{13}) = Re_{13}^{0.077} \qquad \text{for} \quad 1 < Re_{13} < 400. \qquad (22.145)$$

Approximating numerical results *Chiang* and *Sirignano* [8] (1990) use the following correlation

$$(\rho w)_{31} = \rho_1 \beta \frac{Sp}{(1+Sp)^{0.7}}, \qquad (22.146)$$

where β is computed by

$$Sh = 2 + 0.46 Re_{13}^{0.6} Sc_1^{1/3}. \qquad (22.147)$$

The validity of this approach is recommended by the authors within the region $0.3 < Sp < 4.5$, $30 < Re_{13} < 250$.

22.6 Droplet evaporation in gas mixture

Tanaka [41] (1980) uses the following equation

$$(\rho w)_{31} = \rho_1 \beta \left(C_{M1\sigma} - C_{M1}\right) 1.39 \frac{\left(\dfrac{\sum C_{n1\sigma}}{\sum C_{n1}}\right)^{0.52}}{\left(1 + \dfrac{\sum C_{n1\sigma}}{\sum C_{n1}}\right)^{0.48} \sum C_{n1}} \quad (22.148)$$

where the mass transfer coefficient β was computed by means of the *Ranz* and *Marchal* [33] correlation.

Abramson and *Sirignano* [1] (1989) used the *Spalding* correlation by computing the mass transfer coefficient taking into account the heat transfer from the droplet bulk to the surface by modifying the *Ranz* and *Marchal* equation for *Sh*-number as follows

$$Sh = 2 + B \frac{Sh_{RM} - 2}{(1 + Sp)^{0.7} \ln(1 + Sp)}, \quad (22.149)$$

where

$$B = \frac{c_{p3}(T_1 - T_3)}{h'' - h'}. \quad (22.150)$$

All properties for Re_{13}, Sc_1, except ρ_1 are computed at reference parameters, C_{ref}, defined as follows

$$C_{ref} = \frac{1}{n}\left[(n-1)C_{M1\sigma} + C_{M1}\right], \quad n = 3. \quad (22.151)$$

Next we try to remove the simplifying assumption $T_3^{\sigma 1} \approx T_3$.

The time-averaged heat transfer from the bulk of the droplet to the surface is

$$\dot{q}_3^{m\sigma 1} = \alpha_3 \rho_3 c_{p3} \left(T_3^{\sigma 1} - T_3\right) \overline{T}_{3m}(\Delta \tau)/\Delta \tau. \quad (22.152)$$

The time-averaged convective heat transfer from the surface to the gas is

$$\dot{q}_1^{m\sigma 3} = 6\alpha_3 \lambda_{1\sigma} Nu_{13}\left(T_3^{\sigma 1} - T_1\right) f_1 / D_3^2. \quad (22.153)$$

From the energy jump condition at the surface we have

$$\dot{q}_3^{m\sigma 1} + \dot{q}_1^{m\sigma 3} = -6\alpha_3 (\rho w)_{31}\left[h''\left(T_3^{\sigma 1}\right) - h'\left(T_3^{\sigma 1}\right)\right]/D_3. \quad (22.154)$$

Solving this equation with respect to the surface temperature we obtain

$$T_3^{\sigma 1} = \frac{\rho_3 c_{p3} T_3 \overline{T}_{3m}(\Delta\tau)/\Delta\tau + 6\lambda_{1\sigma} Nu_{13} T_1 f_1 / D_3^2 - 6(\rho w)_{31}\left[h''(T_3^{\sigma 1}) - h'(T_3^{\sigma 1})\right]/D_3}{\rho_3 c_{p3} \overline{T}_{3m}(\Delta\tau)/\Delta\tau + 6\lambda_{1\sigma} Nu_{13} T_1 f_1 / D_3^2}.$$

(22.155)

$(\rho w)_{31}$, h'' and h' are non-linear functions of $T_3^{\sigma 1}$ and therefore the above equation should be solved by iterations. Having $T_3^{\sigma 1}$ it is a simple matter to compute $(\rho w)_{31}$, $\dot{q}_3^{m\sigma 1}$ and $\dot{q}_1^{m\sigma 3}$.

The method to describe vapor condensation from a gas mixture on a liquid droplet is similar and will not be described separately.

Nomenclature

Latin

a_1 $= \sqrt{\dfrac{8}{\pi} R_1 T_1}$, thermal velocity of the vapor molecules, m/s

c_c ≈ 0.1 for water steam plus non-condensable gases, ≈ 0.5 to 1, for pure water-steam, part of the molecules striking the nucleus and remaining there, *dimensionless*

c_T temperature correction factor taking into account the heat transfer from the nucleus to the surrounding vapor, *dimensionless*

F_{3c} surface of the cluster, m²

Gb_1 $= \dfrac{\Delta E_{3c}}{kT_1}$, Gibbs number, *dimensionless*

k $= 13.805 \times 10^{-24}$ is the *Boltzmann* constant, J/K

m_μ molar mean molecular mass, as many kilograms of a given substance as its molecular mass is, e.g. $m_{\mu, H_2O} = 18$, $m_{\mu, Air} = 28.9$, kg/mol

m_1 $= \dfrac{m_\mu}{N_A}$ mass of a single molecule, kg

N_A $= 6.02 \times 10^{23}$, *Avogadro* number, 1/mol

N_1 $= \rho_1 / m_1$, number of the vapor molecules per unit volume, 1/m³

$\dot{n}_{3,kin}$ generated nuclei per unit volume and unit time, 1/(m³s)

n_3	number of droplets per unit flow, $1/m^3$
n_{3c}	$= \rho'(T_1)\frac{4}{3}\pi R_{3c}^3/m_1$, number of the molecules in a single nucleus with critical size, $1/m^3$
p	local static pressure, Pa
p'	$= p'(T_1)$, saturation pressure as a function of the steam temperature, Pa
p_c	critical pressure, for water 221.2×10^5, Pa
R_3	droplet radius, m
R_{30}	radius of a subcooled non-stable spherical steam volume, m
R_{3c}	critical nucleus radius, m
T_1	steam at local temperature, K
T'	$= T'(p)$, saturation temperature at the system pressure, K
T_c	critical temperature, K
T_{c,H_2O}	$= 647.2$, critical temperature of water, K
$T_{1,spin}$	temperature at the spinoidal line, K
w	local velocity, m/s
Ze_1	$= \sqrt{\dfrac{Gb_1}{3\pi n_{3c}^2}}$, Zeldovich number, *dimensionless*

Greek

β	empirical coefficient, *dimensionless*
ΔE_3	work required to create a single nucleus, J
ΔE_{3c}	work required to create a stable single nucleus, J
ΔS_{13c}	$= \Delta E_{3c}/T_1$, entropy increase required to create a stable single nucleus, J/K
$\Delta T_{1,sub}$	$= T'(p) - T_1$, subcooling, K
ρ_1	$= \rho_1(p, T_1)$, steam density, kg/m^3
σ	surface tension, Nm

References

1. Abramson B, Sirignano WA (1989) Droplet vaporization model for spray combustion calculations, Int. Heat Mass Transfer, vol 32 pp 1605-1618
2. Acrivos A, Taylor TD (1962) Heat and mass transfer from single spheres in stokes flow, The Physics of Fluids, vol 5 no 4 pp 387-394

3. Acrivos A, Goddard J (1965) Asymptotic expansion for laminar forced- convection heat and mass transfer, Part 1, J. of Fluid Mech., vol 23 p 273
4. Bankoff SS (1980) Some Condensation Studies Pertinent to LWR Safety, Int. J. Multiphase Flow, vol 6 pp 51-67
5. Berman LD (1961) Soprotivlenie na granize razdela fas pri plenochnoi kondensazii para nizkogo davleniya, Tr. Vses. N-i, i Konstrukt In-t Khim. Mashinost, vol 36 p 66
6. Boussinsq M (1905) Calcul du pourvoir retroidissant des courant fluids, J. Math. Pures Appl., vol 1 p 285
7. Celata GP, Cumo M, D'Annibale F, Farello GE (1991) Direct contact condensation of steam on droplets, Int. J. Multiphase Flow, vol 17 no 2 pp 191-211
8. Chiang CH, Sirignano WA (Nov. 1990) Numerical analysis of convecting and interacting vaporizing fuel droplets with variable properties, Presented at the 28th AIAA Aerospace Sciences Mtg, Reno
9. Clift R, Ggarce JR, Weber ME (1978) Bubbles, drops, and particles, Academic Press, New York 1978
10. Condie KG et al. (August 5-8, 1984) Comparison of heat and mass transfer correlation with forced convective non equilibrium post-CHF experimental data, Proc. of 22nd Nat. Heat Transfer Conf. & Exhibition, Niagara Falls, New York, in Dhir VK, Schrock VE (ed) Basic Aspects of Two-Phase Flow and Heat Transfer, pp 57-65
11. Deitsch ME, Philiphoff GA (1981) Two phase flow gas dynamics, Moscu, Energoisdat, in Russian
12. Fedorovich ED, Rohsenow WM (1968) The effect of vapor subcooling on film condensation of metals, Int. J. of Heat Mass Transfer, vol 12 pp 1525-1529
13. Fieder J, Russel KC, Lothe J, Pound GM (1966) Homogeneous nucleation and growth of droplets in vapours. Advances in Physics, vol 15 pp 111-178
14. Ford JD, Lekic A (1973) Rate of growth of drops during condensation. Int. J. Heat Mass Transfer, vol 16 pp 61-64
15. Frenkel FI (1973) Selected works in gasdynamics, Moscu, Nauka, in Russian
16. Furth R (1941) Proc. Cambr. Philos. Soc., vol 37 p 252
17. Gibbs JW (1878) Thermodynamische Studien, Leipzig 1982, Amer. J. Sci. and Arts, Vol.XVI, pp 454 - 455
18. Hertz H (1882) Wied. Ann., vol 17 p 193
19. Knowles J B (1985) A mathematical model of vapor film destabilization, Report AEEW-R-1933
20. Knudsen M (1915) Ann. Physik, vol 47 p 697
21. Labunzov DA, Krjukov AP (1977) Processes of intensive flushing, Thermal Engineering, in Russian, vol 24 no 4 pp 8-11
22. Langmuir I (1913) Physik. Z., vol 14 p 1273
23. Langmuir I (1927) Jones HA and Mackay GMJ, Physic., rev vol 30 p 201
24. Lienhard JH   A heat transfer textbook, Prentice-Hall, Inc., Engelwood Cliffts, New Jersey 07632
25. Lee K, Ryley DJ (Nov. 1968) The evaporation of water droplets in superheated steam, J. of Heat Transfer, vol 90
26. Ludvig A (1975) Untersuchungen zur spontaneous Kondensation von Wasserdampf bei stationaerer Ueberschalllstroemung unter Beruecksichtigung des Realgasverhaltens. Dissertation, Universitaet Karlsruhe (TH)
27. Mason BJ (1951) Spontaneous condensation of water vapor in expansion chamber experiments. Proc. Phys. Soc. London, Serie B, vol 64 pp 773-779

28. Mason BJ (1957) The Physics of Clouds. Clarendon Press, Oxford
29. Mills AF The condensation of steam at low pressure, Techn. Report Series No. 6, Issue 39. Space Sciences Laboratory, University of California, Berkeley
30. Mills AF, Seban RA (1967) The condensation coefficient of water, J. of Heat Transfer, vol 10 pp 1815-1827
31. Nabavian K, Bromley LA (1963) Condensation coefficient of water; Chem. Eng. Sc., vol 18 pp 651-660
32. Nigmatulin RI (1978) Basics of the mechanics of the heterogeneous fluids, Moskva, Nauka, in Russian
33. Ranz W, Marschal W Jr (1952) Evaporation from drops, Ch. Eng. Progress, vol 48 pp 141-146
34. Ranz W, Marschal W Jr. (1952) Evaporation from drops, Ch. Eng. Progress, vol 48 pp 141-146
35. Saha P (1980) Int. J. Heat and Mass Transfer, vol 23 p 481
36. Samson RE, Springer GS (1969) Condensation on and evaporation from droplets by a moment method, J. Fluid Mech., vol 36 pp 577-584
37. Skripov WP, Sinizyn EN, Pavlov PA, Ermakov GW, Muratov GN, Bulanov NB, Bajdakov WG (1980) Thermophysical properties of liquids in metha-stable state, Moscu, Atomisdat, in Russia
38. Soo SL (1969) Fluid dynamics of multiphase systems, Massachusetts, Woltham
39. Spalding DB (1953) The combustion of liquid fuels, Proc. 4th Symp. (Int.) on Combustion, Williams & Wilkins, Baltimore MD, pp 847-864
40. Tanaka M (Feb. 1980) Heat transfer of a spray droplet in a nuclear reactor containment. Nuclear Technology, vol 47 p 268
41. VDI-Waermeatlas (1984) 4. Auflage VDI-Verlag
42. Volmer M (1939) Kinetik der Phasenbildung, Dresden und Leipzig, Verlag von Theodor Steinkopff
43. Zeldovich JB (1942) To the theory of origination of the new phase, cavitation, Journal of Experimental and Theoretical Physics, in Russian, vol 12 no 11/12 pp 525-538
44. Levich VG (1962) Physicochemical Hydrodynamics, Prentice-Hall, Englewood Cliffs, NJ
45. Michaelides EE (March 2003) Hydrodynamic force and heat/mass transfer from particles, bubbles and drops – The Freeman Scholar Lecture, ASME Journal of Fluids Engineering, vol 125 pp 209-238
46. Feng Z-G and Michaelides EE (1986) Unsteady heat transfer from a spherical particle at finite Peclet numbers, ASME Journal of Fluids Engineering, vol 118 pp 96-102

23. Heat and mass transfer at the film-gas interface

23.1 Geometrical film-gas characteristics

There are different flow pattern leading to a film-gas interface in multi-phase flows. Stratified and annular flows in channels with different geometry, walls of large pools with different orientations and part of the stratified liquid gas configuration in large pools are the flow patterns of practical interest. The problem of the mathematical modeling of interface heat and mass transfer consists of (a) macroscopic predictions of the geometrical sizes and local averaged temperatures, velocities, concentrations and pressure and (b) microscopic modeling of the interface heat and mass transfer. The purpose of this Chapter is to review the state of the art for modeling of the interface heat and mass transfer for known geometry.

Next we summarize some important geometrical characteristics needed further.

The cross section occupied by gas and liquid in a channel flow is

$$F_1 = \alpha_1 F \tag{23.1}$$

and

$$F_2 = \alpha_2 F \tag{23.2}$$

respectively, where F is the channel cross section. Depending on the gas velocity the film structure can be

(a) symmetric, or
(b) asymmetric.

The symmetric film structure is characterized by uniformly distributed film on the wet perimeter in the plane perpendicular to the main flow direction. For this case the film thickness is

$$\delta_{2F} = D_h(1-\sqrt{1-\alpha_2})/2, \tag{23.3}$$

where D_h is the hydraulic diameter of the channel in a plane perpendicular to the main flow direction.

The liquid gas interface per unit flow volume that is called the interfacial area density is

$$a_{12} = \frac{4}{D_h}\sqrt{1-\alpha_2} \ . \tag{23.4}$$

The hydraulic diameter of the gas is

$$D_{h1} = D_h\sqrt{1-\alpha_2} \ . \tag{23.5}$$

Asymmetric film structure results in the case of dominance of the gravitation force. The flow can be characterized by three perimeters, one of the gas-wall contact, Per_{1w}, one of the gas-liquid contact, Per_{12}, and one of the liquid-wall contact, Per_{2w}. To these three perimeters correspond three shear stresses, one between the gas and wall, τ_{1w}, one between gas and liquid, τ_{12}, and one between the liquid and wall, τ_{2w}. The hydraulic diameters for computation of the friction pressure loss of the both fluids can be defined as

$$D_{h1} = 4\alpha_1 F / (Per_{1w} + Per_{12}) , \tag{23.6}$$

$$D_{h1} = 4\alpha_2 F / (Per_{2w} + Per_{12}) . \tag{23.7}$$

Two characteristic *Reynolds* numbers can be defined by using these length scales

$$Re_1 = \frac{\rho_1 V_1 D_{h1}}{\eta_1} = \frac{\alpha_1 \rho_1 V_1}{\eta_1} \frac{4F}{Per_{1w} + Per_{12}} , \tag{23.8}$$

$$Re_2 = \frac{\rho_2 V_2 D_{h2}}{\eta_2} = \frac{\alpha_2 \rho_2 V_2}{\eta_2} \frac{4F}{Per_{2w} + Per_{12}} . \tag{23.9}$$

In the three-dimensional space it is possible that the liquid in a computational cell is identified to occupy the lower part of the cell. In this case the gas-liquid interfacial area density is

$$a_{12} = 1/\Delta z \ . \tag{23.10}$$

The film thickness in this case is

$$\delta_{2F} = \alpha_2 \Delta z, \qquad (23.11)$$

where the α_2 is the local liquid volume fraction in the computational cell. In the case of film attached at the vertical wall of radius r of a control volume in cylindrical coordinates

$$\delta_{2F} = r\left(1 - \sqrt{1 - \alpha_2\left[1 - (r_{i-1}/r)^2\right]}\right). \qquad (23.12)$$

For the limiting case of $r_{i-1} = 0$ Eq. (23.12) reduces to Eq. (23.3). The interfacial area density in this case is

$$a_{12} = 2\frac{1 - \delta_{2F}/r}{r\left[1 - (r_{i-1}/r)^2\right]} = 2\frac{\sqrt{1 - \alpha_2\left[1 - (r_{i-1}/r)^2\right]}}{r\left[1 - (r_{i-1}/r)^2\right]}. \qquad (23.13)$$

For the case of $r_{i-1} = 0$ Eq. (23.13) reduces to Eq. (23.4). For such flow pattern the following film *Reynolds* number is used

$$Re_{2F} = \frac{\rho_2 V_2 \delta_{2F}}{\eta_2}.$$

23.2 Convective heat transfer

Next we consider a flow without mass transfer. Detailed consideration of the convective heat transfer is useful not only for flows without mass transfer but also for flow with mass transfer at the film-gas interface.

The gas and liquid temperatures are T_1 and T_2, respectively. The interface temperature, which plays an important role in the mathematical description is $T_2^{1\sigma}$. The total heat transfer mechanism can be considered as

(a) heat transfer from the interface into the gas, $\dot{q}_1^{m2\sigma}$, and
(b) heat transfer from the interface into the bulk liquid $\dot{q}_2^{m1\sigma}$.

23.2.1 Gas side heat transfer

To model the heat transfer between interface and gas we can use known correlations for heat transfer from gas to solid surface

$$\dot{q}_1^{m2\sigma} = a_{12} h_1^{2\sigma} \left(T_1^{2\sigma} - T_1 \right). \tag{23.14}$$

Here $h_1^{2\sigma}$ is the heat transfer coefficient.

For laminar gas flow in channels, $Re_1 < 1450$ in accordance with *Hausen*, see in [10], the maximum *Nusselt* number is 3.66 corrected for the effect of the temperature gradients,

$$h_1^{2\sigma} = 3.66 \frac{\lambda_1}{D_{h1}} \left(\frac{T_1}{T_1^{2\sigma}} \right)^{1/4} \tag{23.15}$$

For turbulent gas flow in channels, $Re_1 < 1450$ in accordance with *Mc Eligot*, see [10], the *Nusselt* number is computed by modified similarity to momentum transfer

$$h_1^{2\sigma} = 0.021 \frac{\lambda_1}{D_{h1}} Re_1^{0.8} Pr_1^{0.4} \left(\frac{T_1}{T_1^{2\sigma}} \right)^{1/2} \tag{23.16}$$

Sometimes this correlation is used in the form

$$\beta_1 = \frac{h_{1\sigma}}{\rho_1 c_{p1}} = \frac{Nu_1}{Re_1 Pr_1} V_1 = St_1 V_1 \tag{23.17}$$

or

$$\beta_1 = 0.021 Pr_1^{-0.6} Re_1^{-1/5} \left(\frac{T_1}{T_1^{2\sigma}} \right)^{1/2} V_1. \tag{23.18}$$

Similar is the equation introduced by *Bunker* and *Carey* [5] (1986) for gas side heat transfer namely

$$\beta_1 = 0.037 Pr_1^{-1/3} Re_1^{-1/5} V_1. \tag{23.19}$$

The order of magnitude of the constants in the above equations, 0.021 to 0.037, was derived by comparison with experiments.

In the case of natural circulation of gas at a vertical wall we can use the proposal made by *Rohsenow* and *Choi* [19],

$$h_{1\sigma}\Delta z / \lambda_1 = 0.56(Gr_1 Pr_1)^{1/4} \quad (10^4 < Gr_1 Pr_1 < 10^8 \text{ laminar}), \quad (23.20)$$

$$h_{1\sigma}\Delta z / \lambda_1 = 0.13(Gr_1 Pr_1)^{1/3} \quad (10^8 < Gr_1 Pr_1 < 10^{10} \text{ transition}), \quad (23.21)$$

$$h_{1\sigma}\Delta z / \lambda_1 = 0.021(Gr_1 Pr_1)^{2/5} \quad (10^{10} < Gr_1 Pr_1 \text{ turbulent}), \quad (23.22)$$

where the *Grashof* number

$$Gr_1 = g\left(-\frac{1}{\rho_1}\frac{\partial \rho_1}{\partial T_1}\right)\left|T_1 - T_1^{2\sigma}\right|\Delta z^3 / v_1^2 \approx g\left(\left|T_1 - T_1^{2\sigma}\right|/T_1\right)\Delta z^3 (\rho_1 / \eta_1)^2 \quad (23.23)$$

is defined using the vertical size Δz of the considered wall.

For a horizontal plane the *Jakob* correlation [12] for natural circulation can be used

$$h_1^{2\sigma} = 2.417 \lambda_1 \left[g\left(\left|T_1 - T_1^{2\sigma}\right|^{20/16} / T_1\right)(\rho_1/\eta_1)^2 Pr_1 \right]^{1/4}. \quad (23.24)$$

The heat transfer coefficient for forced convection at a plane surface can be computed using the classical *Reynolds* analogy, assuming similarity of the velocity and temperature profiles and no distortion of the boundary layer due to mass transfer

$$\beta_1 = \frac{\dot{q}_1^{\prime\prime 2\sigma}/(T_1 - T_1^{2\sigma})}{\rho_1 c_{p1}} = \frac{1}{2}\lambda_{w1}V_1 \quad (23.25)$$

or

$$\beta_1 = \frac{h_{1\sigma}}{\rho_1 c_{p1}} = V_1 \lambda_{w1}/2 \quad (23.26a)$$

or

$$Nu_{1z} = Re_{1z} Pr_1 \lambda_{w1}/2 \quad (23.26b)$$

where

$$Nu_{1z} = h_{1\sigma}\Delta z / \lambda_1, \quad (23.27)$$

$$Re_{1z} = V_1 \Delta z / v_1. \quad (23.28)$$

For plane walls the average shear stress along Δz is

$$\tau_{w1} = \lambda_{w1} \tfrac{1}{2}\rho_1 V_1^2, \tag{23.29}$$

where for laminar flow

$$Re_{1z} < (1 \text{ to } 3)\,10^5, \tag{23.30}$$

in accordance with *Prandtl* and *Blasius*

$$\lambda_{w1} = 1.372 / Re_{1z}^{1/2}, \tag{23.31}$$

and for turbulent flow - see in *Albring* [1]

$$\lambda_{w1} = 0.072 / Re_{1z}^{1/5}. \tag{23.32}$$

For the laminar flow Eq. (23.31) substituted in Eq. (23.26) gives

$$Nu_{1z} = 0.686 Re_{1z}^{1/2} Pr_1, \tag{23.33}$$

which is very similar to the widely used *Pohlhausen* equation [17]

$$Nu_{1z} = 0.664 Re_{1z}^{1/2} Pr_1^{1/3} \tag{23.34}$$

recommended by *Gnielinski* [9] after comparison with data for the laminar flow. For turbulent flow Eq. (23.26) was improved by *Petukhov* and *Popov* [18]

$$Nu_{1z} = \left(Re_{1z} Pr_1 \lambda_{w1}/2\right)/\left[1 + 12.7\left(\lambda_{w1}/2\right)^{1/2}\left(Pr_1^{2/3} - 1\right)\right]. \tag{23.35}$$

The influence of the change of the gas properties at the wall can be taken into account in accordance with *Churchill* and *Brier* [6] by multiplying the results by $\left(T_1/T_{1\sigma}\right)^{0.12}$. After comparing with a large number of data for turbulent flows *Gnielinski* recommended Eqs. (23.35) and (23.32). For comparison we give here the recently successfully used modification of the *Petukhov* and *Popov* result, by *Kim* and *Corradini* [15]

$$Nu_{1z} = \left(Re_{1z} Pr_1 \lambda_{w1}/2\right)/Pr_1^t, \tag{23.36}$$

where the turbulent *Prandtl* number is

$$\Pr_1^t = c_1 + c_2 / \Pr_1 \qquad (23.37)$$

$$c_1 = 0.85, \qquad (23.38)$$

$$c_2 = 0.012 \text{ to } 0.05 \text{ for } Re_{1z} = 2.10^4, \qquad (23.39)$$

$$c_2 = 0.005 \text{ to } 0.015 \text{ for } Re_{1z} = 10^5. \qquad (23.40)$$

In pure vapor condensation the condensate film provides the only heat transfer resistance.

23.2.2 Liquid side heat transfer due to conduction

Next we consider the heat transfer from the interface into the bulk liquid. The liquid can be laminar or turbulent. For laminar liquid the heat is transferred from the interface into the liquid due to heat conduction described by the *Fourier* equation, [8] (1822),

$$\frac{\partial T}{\partial \tau} = -a_2 \frac{\partial^2 T}{\partial y^2}, \qquad (23.39)$$

where the positive y direction is defined from the interface into the bulk. The text book solution for the following boundary conditions

$\tau \geq 0, y = 0, T = T$ (interface), $\qquad (23.40)$

$\tau = 0, y > 0, T = T_2$ (bulk liquid), $\qquad (23.41)$

$\tau > 0, y \to 1, T = T_2$ (bulk liquid), $\qquad (23.42)$

is

$$T - T_2 = \left(T_2^{1\sigma} - T_2\right) erfc \frac{y}{2(a_2\tau)^{1/2}}. \qquad (23.43)$$

Using the temperature gradient

$$\frac{\partial T}{\partial \tau} = -\frac{T_2^{1\sigma} - T_2}{(\pi a_2 \tau)^{1/2}} \exp\left[-y^2/(4a_2\tau)\right] \qquad (23.44)$$

at $y = 0$ we compute the heat flux at the interface

$$\dot{q}_2''^{1\sigma} = -\lambda_2 \frac{\partial T}{\partial y}\bigg|_{y=0} = \sqrt{\frac{2a_2}{\pi\tau}} \rho_2 c_{p2} \left(T_2^{1\sigma} - T_2\right) = \frac{B}{\sqrt{\tau}} \rho_2 c_{p2} \left(T_{2\sigma} - T_2\right)$$

$$= B_2 \rho_2 c_{p2} \left(T_2^{1\sigma} - T_2\right). \qquad (23.45)$$

The averaged heat flux over the time period $\Delta\tau$ is

$$\dot{q}_2''^{1\sigma} = B\rho_2 c_{p2} \left(T_2^{1\sigma} - T_2\right) 2 / \sqrt{\Delta\tau} \qquad (23.46)$$

Assuming $T_1 \approx$ const, $T_2 \approx$ const and using the energy jump condition $\dot{q}_1''^{2\sigma} + \dot{q}_2''^{1\sigma} = 0$ in or

$$h_1^{2\sigma} \left(T_1^{2\sigma} - T_1\right) + \frac{2B}{\sqrt{\Delta\tau}} \rho_2 c_{p2} \left(T_2^{1\sigma} - T_2\right) = 0, \qquad (23.47)$$

and assuming $T_2^{1\sigma} = T_1^{2\sigma}$ which is valid for the case of no mass transfer accross the interface we obtain

$$T_2^{1\sigma} = \left(h_1^{2\sigma} T_1 + \frac{2B}{\sqrt{\Delta\tau}} \rho_2 c_{p2} T_2\right) / \left(h_1^{2\sigma} + \frac{2B}{\sqrt{\Delta\tau}} \rho_2 c_{p2}\right). \qquad (23.48)$$

Thus having $T_2^{1\sigma}$ we compute for Eq. (23.46) the heat flux from the interface into the bulk liquid

$$\dot{q}_2''^{1\sigma} = h_{12}^* \left(T_1 - T_2\right) \qquad (23.49)$$

where

$$\frac{1}{h_{12}^*} = \frac{1}{h_1^{2\sigma}} + \frac{1}{\frac{2B}{\sqrt{\Delta\tau}} \rho_2 c_{p2}}. \qquad (23.50)$$

The heat transferred per unit control volume and unit time is

$$\dot{q}_2'''^{1\sigma} = -\dot{q}_1'''^{2\sigma} = a_{12} \dot{q}_2''^{1\sigma}. \qquad (23.51)$$

23.2.3 Liquid side heat conduction due to turbulence

Let $\Delta \tau_2$ be the average time in which a turbulent eddy stays in the neighborhood of the free surface before jumping apart called sometimes the renewal period. During this time the average heat flux from the interface to the eddies by heat conduction is

$$\dot{q}_2''^\sigma = \left(2B/\sqrt{\Delta \tau_2}\right)\rho_2 c_{p2}\left(T_2^{1\sigma} - T_2\right) = \beta_2 \rho_2 c_{p2}\left(T_2^{1\sigma} - T_2\right). \tag{23.52}$$

If the volume-averaged pulsation velocity is V_2' and the turbulent length scale in the film is ℓ_{e2} the time in which the eddy stays at the interface is

$$\Delta \tau_2 = f\left(\ell_{e2}, V_2'\right). \tag{23.53}$$

Thus, the task to model turbulent heat conduction from the interface into the liquid is reduced to the task to model

(i) pulsation velocity V_2' and

(ii) turbulent length scale ℓ_{e2}.

Usually β_2 is written as a function of the dimensionless turbulent *Reynolds* number

$$Re_2' = \rho_2 V_2' \ell_{e2} / \eta_2, \tag{23.54}$$

and *Prandtl* number Pr_2,

$$\beta_2 = \sqrt{\frac{2a_2}{\pi \Delta \tau_2}} = \sqrt{\frac{2}{\pi}} Pr_2^{-1/2} \sqrt{\frac{\eta_2}{\rho_2 \Delta \tau_2}} = \sqrt{\frac{2}{\pi}} Pr_2^{-1/2} Re_2'^{-1/2} \sqrt{\frac{V_2' \ell_{e2}}{\Delta \tau_2}}, \tag{23.55}$$

and is a subject of modeling work and verification with experiments. Once we know β_2 we can compute some effective conductivity by the equation

$$\sqrt{\frac{2\lambda_{2\text{eff}}}{\rho_2 c_{p2} \pi \Delta \tau_2}} \rho_2 c_{p2}\left(T_{2\sigma} - T_2\right) = \beta_2 \rho_2 c_{p2}\left(T_{2\sigma} - T_2\right) \tag{23.56}$$

or

$$\lambda_{2\mathit{eff}} = \frac{\pi}{2} \rho_2 c_{p2} \Delta \tau_2 \beta_2^2, \qquad (23.57)$$

and use Eq. (23.46) for both laminar and turbulent heat transfer.

23.2.3.1 High Reynolds number

One of the possible ways for computation of the renewal period $\Delta \tau_2$ for high turbulent *Reynolds* numbers

$$Re_2' > 500 \qquad (23.58)$$

is the use the hypothesis by *Kolmogoroff* for isotropic turbulence

$$\Delta \tau_2 \approx c_1 \left(\nu_2 \ell_{e2} / V_2'^3 \right)^{1/2}, \qquad (23.59)$$

where

$$\ell_{e2} \approx c_1 \delta_{2F}, \qquad (23.60)$$

where c_1 is a constant. In this region the turbulence energy is concentrated in microscopic eddies (mechanical energy dissipating motion). Substituting $\Delta \tau_2$ from Eq. (23.59) into (23.55) we obtain

$$\beta_2 = \sqrt{\frac{2}{\pi}} Pr_2^{-1/2} Re_2'^{-1/4} V_2'. \qquad (23.61)$$

The qualitative relationship $\beta_2 Pr_2^{1/2} \approx Re_2'^{-1/4}$ was originally proposed by *Banerjee* et al. in [21] (1968) and experimentally confirmed by *Banerjee* et al. [3] (1990). *Lamont* and *Scot*, see in *Lamont* and *Yuen* [16] (1982), describe successfully heat transfer from the film surface into the flowing turbulent film for high Reynolds numbers using Eq. (23.61) with a constant 0.25 instead of $\sqrt{2/\pi}$. Recently *Hobbhahn* [11] (1989) obtained experimental data for condensation on a free surface, which are successfully described by Eq. (23.61) modified as follows

$$\beta_2 = 0.07 Pr_2^{-0.6} Re_2'^{-1/5} V_2' \qquad (23.62)$$

where the discrepancy between the assumption made by the authors

$$\ell_{e2} \approx \delta_{2F} \qquad (23.63)$$

and

$$V_2' \approx |V_2| + \sqrt{\rho_1/\rho_2}\,|V_1 - V_2| \qquad (23.64)$$

and the reality are compensated by the constant 0.07.

23.2.3.2 Low Reynolds number

For low turbulence *Reynolds* number

$$Re_2' < 500 \qquad (23.65)$$

the *Kolmogoroff* time scale is no longer applicable. In this region the turbulence energy is concentrated in macroscopic eddies (mechanical energy containing motion). The choice of the liquid film thickness, as a length scale of turbulence

$$\ell_{e2} \approx \delta_{2F} \qquad (23.66)$$

is reasonable. Therefore the renewal period is

$$\Delta \tau_2 \approx \delta_{2F}/V_2' \qquad (23.67)$$

and Eq. (23.55) reduces to

$$\beta_2 = \sqrt{\frac{2}{\pi}} Pr_2^{-1/2} Re_2^{t-1/2} V_2'. \qquad (23.68)$$

For low *Reynolds* numbers *Fortescue* and *Pearson* [7] (1967) recommended instead of $\sqrt{2/\pi}$ in Eq. (23.68) to use

$$\text{const} = 0.7(1+0.44/\Delta\tau_2^*), \qquad (23.69)$$

where

$$\Delta\tau_2^* = \Delta\tau_2 V_2'/\ell_{e2} \approx 1, \qquad (23.70)$$

which is > 0.85, as recommended by *Brumfield* et al. [4] (1975).

Theofanous et al. [22] (1976) show that most of their experimental data for condensation on channels with free surface can be described by using Eqs. (23.61) and (23.68) with the above discussed corrections introduced by *Lamont* and *Scot*, and *Fortescue* and *Person*, respectively.

23.2.3.3 Pulsation velocity

a) Time scale of turbulence pulsation velocity based on average liquid velocity.

A very rough estimate of the pulsation velocity is

$$V' = c_1 V_2 \approx (0.1 \text{ to } 0.3) V_2, \qquad (23.71)$$

With this approach and

$$\ell_{e2} \approx c_2 \delta_{2F} \qquad (23.72)$$

the two equations for β_2 read

$$Re_{2F} < c_1 500, \qquad (23.73)$$

$$\beta_2 = c_2^{-1/4} c_1^{3/4} \sqrt{\frac{2}{\pi}} Pr_2^{-1/2} Re_{2F}^{-1/4} V_2 = const Pr_2^{-1/2} Re_{2F}^{-1/4} V_2. \qquad (23.74)$$

Comparing this equation with the *McEligot* equation we see that

$$const \approx 0.021 \text{ to } 0.037. \qquad (23.75)$$

For

$$Re_2 < c_1 500, \qquad (23.76)$$

$$\beta_2 = c_1^{1/2} \sqrt{\frac{2}{\pi}} Pr_2^{-1/2} Re_{2F}^{-1/2} V_2. \qquad (23.77)$$

b) The scale of turbulence pulsation velocity based on friction velocity

Improvement of the above theory requires a close look to the reasons for the existence of turbulence in the liquid film. This is either the wall shear stress, τ_{2w}, or the shear stress acting at the gas-liquid interface, τ_{12}, or both simultaneously. The shear stress at the wall for channel flow is

$$\tau_{w2} = \frac{1}{4}\lambda_{w2}\frac{1}{2}\rho_2 V_2^2, \qquad (23.78)$$

where

$$\lambda_{w2} = \lambda_{friction}\left(\frac{\rho_2 V_2 D_{h2w}}{\eta_2}, k_w / D_{h2w}\right).\qquad(23.79)$$

Here

$$D_{h2w} = \frac{4\alpha_2 F}{Per_{2w}}.\qquad(23.80)$$

is the hydraulic diameter of the channel for the liquid, and k_w / D_{h2w} is the relative roughness. The shear stress of the gas liquid interface in channel flow is

$$\tau_{12} = \frac{1}{4}\lambda_{12}\frac{1}{2}\rho_1 \Delta V_{12}^2,\qquad(23.81)$$

where

$$\lambda_{w2} = \lambda_{friction}\left(\frac{\rho_1 V_{12} D_{h1}}{\eta_2}, \frac{\delta_{2F}/4}{D_{h1}}\right).\qquad(23.82)$$

Here

$$D_{h1} = \frac{4\alpha_1 F}{Per_{1w} + Per_{12}},\qquad(23.83)$$

is the hydraulic diameter of the "gas channel" and $(\delta_{2F}/4)/D_{h1}$ is the relative roughness for the "gas channel" taken to be a function of the waviness of the film.

For vertical plane walls the average shear stress at the wall along Δz is

$$\tau_{w2} = \lambda_{w2}\frac{1}{2}\rho_2 V_2^2\qquad(23.84)$$

where for laminar flow

$$Re_{2z} < (1 \text{ to } 3)10^5\qquad(23.85)$$

in accordance to *Prandtl* and *Blasius*

$$\lambda_{w2} = 1.372 / Re_{2z}^{1/2}\qquad(23.86)$$

and

$$Re_{2z} = V_2 \Delta z / v_2\qquad(23.87)$$

and for turbulent flow - see in *Albring* [1]

$$\lambda_{w2} = 0.072 / Re_{2z}^{1/5}. \tag{23.88}$$

Similarly the gas side averaged shear stress is computed using

$$Re_{1z} = \Delta V_{12} \Delta z / v_2. \tag{23.89}$$

Thus the effective shear stress in the film is

$$\tau_{2eff} = \frac{Per_{2w}\tau_{2w} + Per_{12}\tau_{12}}{Per_{2w} + Per_{12}}. \tag{23.90}$$

Now we can estimate the time scale of the turbulence in the shear flow

$$\Delta \tau_2 \approx \left(v_2 \delta_{2F} / V_2'^3\right)^{1/2}, \tag{23.91}$$

using the dynamic friction velocity

$$V_{2eff}^* = \sqrt{\tau_{2eff} / \rho_2}. \tag{23.92}$$

Assuming that the pulsation velocity is of the order of magnitude of the friction velocity

$$V_2' \approx const\ V_{2eff}^*, \tag{23.93}$$

where the

$$const \approx 2.9, \tag{23.94}$$

we obtain for

$$Re_{2F}^* > const\ 500, \tag{23.95}$$

$$\beta_2 = const^{3/4} \sqrt{\frac{2}{\pi}} Pr_2^{-1/2} Re_{2F}^{*-1/4} V_{2eff}^*, \tag{23.96}$$

and for

$$Re_2 < const\ 500, \tag{23.97}$$

$$\beta_2 = const^{1/2} \sqrt{\frac{2}{\pi}} Pr_2^{-1/2} Re_{2F}^{*-1/2} V_{2eff}^*, \tag{23.98}$$

where

$$Re^*_{2F} = \frac{\rho_2 V^*_{2eff} \delta_{2F}}{\eta_2}. \quad (23.99)$$

More careful modeling of the turbulent length scale for the derivation of Eq. (23.98) was done by *Kim* and *Bankoff* [14] (1983). The authors used the assumption

$$V'_2 \approx V^*_{2eff}, \quad (23.100)$$

and modified Eq. (23.98) as follows

$$\beta_2 = 0.061 Pr_2^{-1/2} Re_2^{0.12} V^*_{2eff}, \quad (23.101)$$

where

$$\ell_{e2} = \left[\frac{\sigma}{(\rho_2 - \rho_1)g} \right]^{1/2} 3.03 \times 10^{-8} Re_1^{*1.85} Re_2^{*0.006} Pr_2^{-0.23}, \quad (23.102)$$

for $3000 < Re_1^* < 18000$ and $800 < Re_2^* < 5000$, $\lambda_{12} = 0.0524 + 0.92 \times 10^{-5} Re_2^*$ valid for $Re_2^* > 340$, and $\tau_{12} \gg \tau_{2w}$. Note the special definition of the *Reynolds* numbers as mass flow per unit width of the film

$$Re_1^* = \frac{\alpha_1 \rho_1 V_1}{\eta_1} \frac{F}{Per_{12}}, \quad (23.103)$$

$$Re_2^* = \frac{\alpha_2 \rho_2 V_2}{\eta_2} \frac{F}{Per_{12}}. \quad (23.104)$$

Assuming that the pulsation velocity is of order of the magnitude of the friction velocity as before but the time scale of the turbulence is

$$\Delta \tau_2 \approx (v_2/\varepsilon_2)^{1/2} = v_2/V^{*2}_{2eff}, \quad (23.105)$$

results for

$$Re^*_{2F} > const\ 500, \quad (23.106)$$

in

$$\beta_2 = \sqrt{\frac{2}{\pi}} Pr_2^{-1/2} V_{2\text{eff}}^* \,. \qquad (23.107)$$

This equation is recommended by *Jensen* and *Yuen* [13] (1982) with a constant 0.14 instead $\sqrt{2/\pi}$ for $\tau_{12} \gg \tau_{2w}$. Hughes and Duffey reproduced an excellent agreement with experimental data for steam condensation in horizontal liquid films by using Eq. (23.96) and the assumption

$$\tau_{2\text{eff}} = (\tau_{12} + \tau_{2w})/2 \,. \qquad (23.108)$$

Nevertheless one should bear in mind that Eq. (23.90) is more general.

An example of detailed modeling of the turbulence structure in the film during film condensation from stagnant steam on vertical cooled surfaces is given by *Mitrovich* [19]. For this case

$$\dot{q}_2^{m 1\sigma} = h_2^{1\sigma} \left(T_2^{1\sigma} - T_w \right) \qquad (23.109)$$

where

$$h_2^{1\sigma} \delta_{2F} / \lambda_2 = 1.05 Re_{2F}^{-0.33} \left[1 + C^{1.9} Re_{2F}^{1.267} Pr_2^{1.11} \right]^{0.526}, \qquad (23.110)$$

$$C = 8.8\ 10^{-3}/(1 + 2.29\ 10^{-5} Ka_2^{0.269}), \qquad (23.111)$$

$$Ka_2 = \rho_2 \sigma^3 / (g \eta_2^4) \,. \qquad (23.112)$$

Summarizing the results discussed above we can say the following:

(a) Gas in the two-phase film flow behaves as a gas in a channel. Therefore the gas side heat transfer can be considered as a heat transfer between gas and the "channel wall" taking into account that the "wall roughness" is influenced by the waves at the liquid surface.

(b) The liquid side heat transfer at the interface is due to molecular and turbulent heat conduction. The modeling of the turbulent heat conduction can be performed by modeling the time and length scale of the turbulence taking into account that turbulence is produced mainly

(i) at the wall - liquid interface, and
(ii) at the gas - liquid interface.

(c) Gas side heat transfer in a pool flow can be considered as a heat transfer at plane interface.

(d) The liquid side heat transfer from the interface into the bulk liquid is governed by the solution of the transient *Fourier* equation where in case of the turbulence the use of effective heat conductivity instead of the molecular conductivity is recommended.

23.3 Spontaneous flashing of superheated film

The considerations for the estimation of the spontaneous evaporation of the superheated film are identical to those presented in Chapter 22 Section 5 for the spontaneous evaporation of a superheated droplet. In order to calculate the source terms we simply replace the subscripts 3 with 2 and use the relationships from Chapter 22 Section 5.

Having in mind that

$$\mu_{12} = 0, \tag{23.113}$$

$$\dot{q}_1'''^{2\sigma} = 0, \tag{23.114}$$

we obtain for the energy jump condition

$$(\rho w)_{21} \left[h''\left(T_2^{1\sigma}\right) - h'\left(T_2^{1\sigma}\right) \right] = -\beta_2 \rho_2 c_{p2} \left(T_2^{1\sigma} - T_2\right), \tag{23.115}$$

or

$$T_2^{1\sigma} = T_2 - (\rho w)_{21} \left[h''\left(T_2^{1\sigma}\right) - h'\left(T_2^{1\sigma}\right) \right] / \left(\beta_2 \rho_2 c_{p2}\right). \tag{23.116}$$

From this equation the interface temperature can be computed by iterations starting with $T_2^{1\sigma} \approx T'(p)$. The heat transfer from the interface to the bulk liquid is

$$\dot{q}_2''^{1\sigma} = \beta_2 \rho_2 c_{p2} \left(T_2^{1\sigma} - T_2\right), \tag{23.117}$$

and the evaporated mass per unit time and unit mixture volume is

$$\mu_{21} = a_{12} (\rho w)_{21}. \tag{23.118}$$

23.4 Evaporation of saturated film in superheated gas

If the flow consists of superheated gas, $T_1 > T'(p)$, and saturated liquid, $T_2 = T'(p)$ the heat transferred from the gas to the interface by convection is

$$\dot{q}_2^{m1\sigma} = -a_{12}h_c\left[T_1 - T'(p)\right]. \tag{23.119}$$

The energy jump condition at the surface is

$$\mu_{21} = -\dot{q}_2^{m1\sigma}/\left[h''(p) - h'(p)\right] = a_{12}h_c\left[T_1 - T'(p)\right]/\left[h''(p) - h'(p)\right]. \tag{23.120}$$

The energy balance for the gas at $p \approx const$ and negligible gas diffusion and convection from the neighboring cells is

$$\alpha_1\rho_1 c_{p1}\frac{dT_1}{d\tau} = -\dot{q}_2^{m1\sigma}\frac{h_1 - h''(p)}{h''(p) - h'(p)} \approx -\dot{q}_2^{m1\sigma}\frac{c_{p1}\left[T_1 - T'(p)\right]}{h''(p) - h'(p)}, \tag{23.121}$$

or

$$\frac{dT_1}{d\tau} = \left[T_1 - T'(p)\right]^2/\Delta\tau_1, \tag{23.122}$$

where

$$\Delta\tau_1 = \left[h''(p) - h'(p)\right]\alpha_1\rho_1/a_{12}.$$

We see from the above equation that the colder steam entering the bulk steam has a cooling effect on the bulk steam. For $\tau = 0$, $T_1 = T_{1a}$ the analytical solution gives

$$\frac{T_1 - T'(p)_1}{T_{1a} - T'(p)} = \frac{1}{1 - \left[T_1 - T'(p)\right]\dfrac{\Delta\tau}{\Delta\tau_1}}. \tag{23.123}$$

The averaged temperature difference $T_1 - T'(p)$ along the time $\Delta\tau$ is

$$T_1 - T'(p) = -\frac{\Delta\tau_1}{\Delta\tau}\ln\left\{1 - \left[T_{1a} - T'(p)\right]\frac{\Delta\tau}{\Delta\tau_1}\right\}. \tag{23.124}$$

Thus, for evaporation of a saturated film in superheated gas the instantaneous heat and mass transfer is defined by Eqs. (23.119) and (23.120) and the average heat and mass transfer along the time $\Delta\tau$ is computed as follows:

(i) Compute the averaged heat temperature difference using Eq. (23.124);

(ii) Compute the average heat transfer using Eq. (23.119);

(iii) Compute the average mass transfer using the energy jump condition, Eq. (23.120).

23.5 Condensation of pure steam on subcooled film

Consider steam having temperature T_1 interacting with a liquid film having temperature T_2 where

$$T_2 < T'(p). \tag{23.125}$$

The steam side interface temperature is $T_1^{2\sigma}$. The heat transferred from the interface into the gas per unit time and unit mixture volume is

$$\dot{q}_1'''^{2\sigma} = a_{12} h_c \left(T_1^{2\sigma} - T_1\right) \tag{23.126}$$

where h_c is computed as discussed in Section 23.1. The heat transferred from the interface into the bulk liquid is

$$\dot{q}_2'''^{1\sigma} = a_{12} \beta_2 \rho_2 c_{p2} \left(T_2^{1\sigma} - T_2\right). \tag{23.127}$$

A simplified way to compute the condensing mass per unit time and unit mixture volume is to assume

$$T_2^{1\sigma} = T_1^{2\sigma} \approx T'(p). \tag{23.128}$$

Using the energy jump condition we obtain

$$\mu_{12} = a_{12} \left\{ h_c \left[T'(p) - T_1\right] + \beta_2 \rho_2 c_{p2} \left[T'(p) - T_2\right] \right\} / \left[h''(p) - h'(p)\right]. \tag{23.129}$$

An improvement of the description can be achieved by dropping the assumption (23.128). In this case one must compute the condensing mass flow rate $(\rho w)_{12}$ as a function of the interface temperature using the kinetic theory as discussed in Section 22.1.2. Consequently we have

$$(\rho w)_{12} = f\left(T_2^{1\sigma}, T_1, p\right) \tag{23.130}$$

The energy jump condition

$$(\rho w)_{12} \left[h''(T_{2i}) - h'(T_{2i})\right] = h_c \left(T_1^{2\sigma} - T_1\right) + \beta_2 \rho_2 c_{p2} \left(T_2^{1\sigma} - T_2\right), \tag{23.131}$$

governs the interface temperature

$$T_2^{1\sigma} = \left\{(\rho w)_{12} \left[h''(T_2^{1\sigma}) - h'(T_2^{1\sigma})\right] + h_c T_1 + \beta_2 \rho_2 c_{p2} T_2\right\} / \left(h_c + \beta_2 \rho_2 c_{p2}\right). \tag{23.132}$$

Obviously $T_2^{1\sigma}$ has to be obtained by iterations starting with $T_2^{1\sigma} \approx T'(p)$.

23.6 Evaporation or condensation in presence of non-condensable gases

In pure vapor condensation the condensate film provides the only heat transfer resistance, whereas if a small amount of non-condensable gas is present, then the main resistance to heat transfer is the gas/vapor boundary layer. Air mass fractions as low as 0.5 % can reduce 50 % or more the condensation, see [17, 18, 9, 2, 20, 24].

Next we consider gas-film flow. The gas has temperature T_1 and concentration of steam C_{M1}. The system pressure is p and the steam partial pressure in the bulk of the gas is p_{M1}. The liquid has temperature T_2, that is less than the saturation temperature $T'(p)$ for the system pressure. This means that the liquid is subcooled and spontaneous flashing is not possible. In this case diffusion controlled evaporation or condensation at the interface is possible depending on the interface temperature satisfying the energy jump condition. If the partial pressure in the gas boundary layer $p'(T_2^{1\sigma}) > p_{M1}$ steam is transported from the interface into the gas bulk. The opposite is the case if $p'(T_2^{1\sigma}) < p_{M1}$.

Thus for the analysis the estimation of the interface temperature is important. The way we propose here is:

(1) Assume as a first estimation $T_2^{1\sigma} \approx T_2$;

(2) Compute the steam partial pressure at the interface $p'(T_2^{1\sigma})$;

(3) If $p'(T_2^{1\sigma}) > p_{M1}$ compute the diffusion controlled evaporation mass flow rate

$$(\rho w)_{21} = -(\rho w)_{12}^*,\qquad(23.133)$$

where

$$(\rho w)_{12}^* = h_c \frac{M_{M1}}{M_1} \ln \frac{p - p'(T_2^{1\sigma})}{\sum p_{n1}} / (c_{p1} Le^{2/3}),\qquad(23.134)$$

see in [23] (1984).

(4) If $p'(T_2^{1\sigma}) < p_{M1}$ compute the diffusion controlled condensation mass flow rate

$$(\rho w)_{21} = (\rho w)_{12}^*\qquad(23.135)$$

Here

$$M_1 = \frac{\sum p_{n1}}{p} M_{\Sigma n1} + \frac{p - \sum p_{n1}}{p} M_{M1}\qquad(23.136)$$

is the mixture mole mass and $M_{\Sigma n1}$ and M_{M1} are the mole masses of the inert components and of the steam, respectively, e.g. $M_{\Sigma n1}$ = 28.96 *kg/mol* for air and M_{M1} = 18.96 *kg/mol* for water steam.

$$Le = \lambda_1 / (\rho_1 c_{p1} D_{M \to \Sigma n})\qquad(23.137)$$

is the *Lewis* number, $D_{M \to \Sigma n}$ is the diffusion constant for steam in the gas mixture.

(5) Compute the gas and the liquid heat flux from the interface

$$\dot{q}_1''^{2\sigma} = f\, h_c (T_1^{2\sigma} - T_1),\qquad(23.138)$$

$$\dot{q}_2^{\prime\prime 1\sigma} = \beta_2 \rho_2 c_{p2}\left(T_2^{1\sigma} - T_2\right). \tag{23.139}$$

Here f is a factor reflecting the gas side convective heat transfer. In the case of no mass transfer $f = 1$.

For condensation the following proposal for computation of f was made in [23]:

$$f = |F|/\left(1 - e^{-|F|}\right), \tag{23.140}$$

where

$$F = (\rho w)_{12}\, c_{pM1}/h_c. \tag{23.141}$$

(6) Compute by iterations the interface temperature $T_2^{1\sigma}$ from the energy jump condition

$$T_2^{1\sigma} = \left\{(\rho w)_{12}^*\left[h''\left(T_2^{1\sigma}\right) - h'\left(T_2^{1\sigma}\right)\right] + fh_c T_1 + \beta_2 \rho_2 c_{p2} T_2\right\}/\left(fh_c + \beta_2 \rho_2 c_{p2}\right) \tag{23.142}$$

starting with $T_2^{1\sigma} = T_2$.

(7) Check again the initial hypothesis and compute the interfacial evaporation or condensation mass flow rate, respectively, and the heat fluxes using the final interfacial temperature $T_2^{1\sigma}$.

Nomenclature

Latin

a	$= \lambda/(\rho c_p)$, temperature conductivity, m^2/s
a_{12}	$= 1/\Delta z$ gas liquid interfacial area density in Cartesian coordinates, $1/m$
D_{h1}	hydraulic diameter for the gas, m
D_{h2}	hydraulic diameter for the liquid, m
D_{h12}	hydraulic diameter for computation of the gas friction pressure loss component in a gas-liquid stratified flow, m
F	channel cross section, m

Gr_1 $= g\left(-\dfrac{1}{\rho_1}\dfrac{\partial \rho_1}{\partial T_1}\right)\left|T_1 - T_1^{2\sigma}\right|\Delta z^3 / v_1^2$, gas site *Grashof* number, dimensionless

$h_1^{2\sigma}$ gas site interface heat transfer coefficient, $W/(m^2K)$

k_w wall roughness, m

Le $= \lambda_1 / \left(\rho_1 c_{p1} D_{M \to \sum n}\right)$ *Lewis* number, *dimensionless*

$D_{M \to \sum n}$ diffusion coefficient for steam in the gas mixture, m^2/s

ℓ_{e2} turbulent length scale in the film, m

M_1 mixture mole mass, $kg/1\_mol$

M_{n1} mole masses as the inert components, $kg/1\_mol$

M_{M1} mole masses of the steam, $kg/1\_mol$

Nu_{1z} $= h_{1\sigma}\Delta z / \lambda_1$, gas *Nusselt* number, *dimensionless*

Per_w perimeter of the rectangular channel, m

Per_{1w} wetted perimeters for the gas, m

Per_{2w} wetted perimeters for the liquid, m

Pr_1^t turbulent gas *Prandtl* number, *dimensionless*

Pr_1 gas *Prandtl* number, *dimensionless*

Pr_2 liquid *Prandtl* number, *dimensionless*

p pressure, Pa

$\dot{q}_1^{m2\sigma}$ heat flux density from the interface into the gas, W/m^3

$\dot{q}_2^{m1\sigma}$ heat flux density from the interface into the bulk liquid, W/m^3

Re_1 $= \rho_1 w_1 D_{h1} / \eta_1$, gas *Reynolds* number, *dimensionless*

Re_2 $= \rho_2 w_2 D_{h2} / \eta_2$, liquid *Reynolds* number, *dimensionless*

Re_1^* $= \dfrac{\alpha_1 \rho_1 V_1}{\eta_1}\dfrac{F}{Per_{12}}$, modified gas *Reynolds* number, *dimensionless*

Re_2^* $= \dfrac{\alpha_2 \rho_2 V_2}{\eta_2}\dfrac{F}{Per_{12}}$, modified liquid *Reynolds* number, *dimensionless*

Re_{2F} $= \dfrac{\rho_2 V_2 \delta_{2F}}{\eta_2}$ film *Reynolds* number, *dimensionless*

Re_{1z} $= V_1 \Delta z / v_1$, gas *Reynolds* number, *dimensionless*

Re_2^t $= \rho_2 V_2' \ell_{e2} / \eta_2$, liquid turbulent *Reynolds* number, *dimensionless*

Re_{2F}^* $= \dfrac{\rho_2 V_{2eff}^* \delta_{2F}}{\eta_2}$, liquid turbulent *Reynolds* number based on the friction velocity, *dimensionless*

r radius, m

Sr_1	$= \dfrac{Nu_1}{Re_1 \, Pr_1}$, Struhal number, *dimensionless*	
T	temperature, *K*	
$T_2^{1\sigma}$	liquid side interface temperature, *K*	
V_1	gas velocity, *m/s*	
V_2	liquid velocity, *m/s*	
V_2'	volume-averaged liquid pulsation velocity, *m/s*	
$V_{2\mathit{eff}}^*$	$= \sqrt{\tau_{2\mathit{eff}}/\rho_2}$, dynamic friction velocity, *m/s*	
w_1	axial gas velocity, *m/s*	
w_2	axial liquid velocity, *m/s*	
y	distance, *m*	
z	axial coordinate, *m*	

Greek

α_2	local liquid volume fraction in the computational cell, *dimensionless*
β_1	$= \dfrac{h_{1\sigma}}{\rho_1 c_{p1}} = \dfrac{Nu_1}{Re_1 Pr_1} V_1 = Sr_1 V_1$, weighted heat transfer coefficient, *m/s*
Δ	finite difference, -
$\Delta \tau_2$	time for which the eddy stays at the interface, renewal period, *s*
δ_{2F}	film thickness, *m*
η	dynamic viscosity of liquid, *kg/(ms)*
λ	thermal conductivity, $W/(mK)$
$\lambda_{2\mathit{eff}}$	effective conductivity, $W/(mK)$
λ_{w1}	friction coefficient, *dimensionless*
μ_{21}	steam mass generation per unit mixture volume and unit time due to evaporation, $kg/(sm^3)$
μ_{12}	liquid mass generation per unit mixture volume and unit time due to condensation, $kg/(sm^3)$
ρ	density, kg/m^3
$(\rho w)_{21}$	steam mass generation per unit surface and unit time due to evaporation, $kg/(m^2 s)$
$(\rho w)_{12}$	liquid mass generation per unit surface and unit time due to condensation, $kg/(m^2 s)$
τ_{w1}	$= \lambda_{w1} \tfrac{1}{2} \rho_1 V_1^2$, shear stress, N/m^2
τ_{2w}	liquid-wall shear stress, N/m^2
τ_{12}	gas-liquid interface shear stress, N/m^2

τ time, s

Subscripts

1   gas
2   liquid film
$2F$ liquid film
$M1$ vapor inside the gas mixture
$n1$ inert components inside the gas

References

1. Albring W (1970) Angewandte Stroemungslehre, Verlag Theodor Steinkopf, Dresden, 4. Auflage,
2. Al-Diwani HK and Rose JW (1973) Free convection film condensation of steam in presence of non condensing gases, Int. J. Heat Mass Transfer, vol 16 p 1959
3. Banerjee S (1990) Turbulence structure and transport mechanisms at interfaces, Proc. Ninth Int. Heat Transfer Conference, Jerusalem, Israel, vol 1 pp 395-418
4. Brumfield LK, Houze KN, and Theofanous TG (1975) Turbulent mass transfer at free, Gas-Liquid Interfaces, with Applications to Film Flows. Int. J. Heat Mass Transfer, vol 18 pp 1077-1081
5. Bunker RS and Carey VP (1986) Modeling of turbulent condensation heat transfer in the boiling water reactor primary containment, Nucl. Eng. Des., vol 91 pp 297-304
6. Churchill SW, Brier JC (1955) Convective heat transfer from a gas stream at high temperature to a cyrcular cylinder normal to the flow, Chem. Engng. Progr. Simp. Ser. 51, vol 17 pp 57-65
7. Fortescue GE and Pearson JRA (1967), On gas absorption into a turbulent liquid, Chem. Engng. Sci., vol 22 pp 1163-1176
8. Fourier J (1822) Theory analytique de la chaleur
9. Gnielinski V (1975) Berechnung mittlerer Waerme- und Stoffuebertragungskoeffizienten an laminar und turbulent ueberstroemenden Einzelkoerpern mit Hilfe einer einheitlichen Gleichung, Forsch. Ing. Wes, vol 41 no 5 pp 145-153
10. Hausen H (1958) Darstellung des Waermeueberganges in Roehren durch verallgemeinerte Potenzgleichungen, Verfahrenstechnik, vol 9 no 4/5 pp 75-79
11. Hobbhahn WK (Oct. 10-13, 1989) Modeling of condensation in light water reactor safety, Proc. of the Fourth International Topical Meeting on Nuclear Reactor Thermal-Hydraulics, Mueller U, Rehme K, Rust K, Braun G (eds) Karlsruhe, vol 2 pp 1047-1053
12. Jakob M, Linke W (1933) Der Waermeübergang von einer waagerechten Platte an sidendes Wasser, Forsch. Ing. Wes., vol 4 pp 75-81
13. Jensen RJ and Yuen MC (1982) Interphase transport in horizontal stratified concurrent flow, U.S. Nuclear Regulatory Commission Report NUREG/CR-2334
14. Kim HJ and Bankoff SG (Nov. 1983) Local heat transfer coefficients for condensation in stratified countercurrent steam - water flows, Trans. ASME, vol 105 pp 706-712

15. Kim MH, Corradini ML (1990) Modeling of condensation heat transfer in a reactor containment, Nucl. Eng. and Design, vol 118 pp 193-212
16. Lamont JC and Yuen MC (1982) Interfase transport in horizontal stratified concurrent flow, U. S. Nuclear Regulatory Commission Report NUREG/CR-2334
17. Pohlhausen E (1921) Der Waermeaustausch zwischen festen Koerpern und Fluessigkeiten mit kkleiner Reibung und kleiner Waermeleitung, Z. angew. Math. Mech., vol 1 no 2 pp 115 - 121
18. Petukhov BS, Popov VN (1963) Theoretical calculation of heat exchange and friction resistance in turbulent flow in tubes of an incompressible fluid with variable physical properties. High Temperature, vol 1 pp 69-83
19. Rohsenow WM, Choi H (1961) Heat, mass and momentum transfer, Prentice - Hall Publishers, New Jersey
20. Siddique M, Golay MW (May 1994) Theoretical modeling of forced convection condensation of steam in a vertical tube in presence of non condensable gas, Nuclear Technology, vol 106
21. Slattery JC 1990 Interfacial transport phenomena, Springer Verlag
22. Theofanous TG, Houze RN and Brumfield LK (1976) Turbulent mass transfer at free, gas-liquid interfaces, with application to open- channel, bubble and jet flows, Int. J. Heat Mass Transfer vol 19 pp 613-624
23. VDI-Waermeatlas (1984) 4. Auflage, VDI-Verlag.
24. Uchida U, Oyama A, Togo Y (1964) Evolution of post - incident cooling system of light water reactors, Proc. 3th Int. Conf. Peaceful Uses of Atomic Energy, International Atomic Energy Agancy, Vienna, Austria, vol 13 p 93

24. Condensation at cooled walls

24.1 Pure steam condensation

24.1.1 Onset of the condensation

Consider gas flow in a pipe. The heat transfer coefficient at the wall for turbulent flow without correction of the properties in the wall boundary layer is

$$h_{convective} = 0.023 Re_1^{0.8} Pr_1^{0.4} \frac{\lambda_1}{D_{hy}}, \qquad (24.1)$$

where the *Reynolds* and the *Prandtl* numbers are defined as a functions of the gas mixture properties

$$Re_1 = D_{hy} \rho_1 |w_1| / \eta_1, \qquad (24.2)$$

$$Pr_1 = \eta_1 c_{p1} / \lambda_1. \qquad (24.3)$$

For prescribed heat flux the wall temperature is

$$T_w = T_1 + \dot{q}_w'' / h_{convective}. \qquad (24.4)$$

If the wall temperature is less then the saturation temperature at the partial pressure of the steam

$$T_w \leq T_{cond}, \qquad (24.5)$$

where

$$T_{cond} = T'(p_{M1}), \qquad (24.6)$$

condensation at the wall occurs.

24.1.2 Condensation from stagnant steam (*Nusselt* 1916) at laminar liquid film

For laminar condensation on vertical wall with high H from stagnant steam *Nusselt* obtained in 1916 the following analytical solution

$$h_{satgnant} := \frac{\dot{q}_w''}{T_{cond} - T_w} = \frac{2}{3}\sqrt{2}\left[\frac{\rho'(\rho' - \rho'')\lambda'^3 g\cos\varphi\Delta h}{H\eta'(T_{cond} - T_w)}\right]^{1/4}, \qquad (24.7)$$

or

$$T_w = T_{cond} - \left(-\dot{q}_w''/C\right)^{4/3}, \qquad (24.8)$$

where

$$C = \frac{2}{3}\sqrt{2}\left[\frac{\rho'(\rho' - \rho'')\lambda'^3 g\cos\varphi\Delta h}{H\eta'}\right]^{1/4}, \qquad (24.9)$$

$\Delta h = h'' - h'$. The coefficient is set to 0.728 by *Isachenko* et al. [13] for external condensation on horizontal pipes by using as a characteristic length the external wetted diameter of the pipe $H = D_h$. *Bird* et al. [3] recommended the constant 0.725 for *external condensation on a horizontal single pipe* reporting $\pm$ 10% error compared with experiments. The correlation was also recommended for *bundles of horizontal pipes* [3] p 417. The *Nusselt* equation was modified for condensation *inside inclined pipe* by using a constant 0.296 taking into account the partial occupation of the pipe cross section by the flowing condensate and $H = D_h$ by *Collier* [8] in 1972.

Corrector for the subcooling of the film is introduced in the *Nusselt* theory by [18] p. 115 as follows:

$$\Delta h = h'' - h' + \frac{3}{8}c_p'(T_k' - T_w). \qquad (24.10)$$

Chato [9] come to 0.68 instead to 3/8, for $H = D_h$, and recommend for small inclinations $\Delta z/\Delta x < 0.002$ the constant in the *Nusselt* equation to be

$$0.468 f\left[\Pr_2, \frac{c_p'(T_k' - T_w)}{h'' - h'}\right]$$

a function of the liquid *Prandtl* number given graphically in his work. For larger inclinations $0.002 < \Delta z/\Delta x < 0.6$, 0.3 is used instead 0.468.

24.1.3 Condensation from stagnant steam at turbulent liquid film (*Grigul* 1942)

For large vertical pipes the heat transfer coefficient may exceed the computed one by 70% due to ripples that attain greatest amplitude on the long vertical tubes. It was found by *Kapitza* [14] that waves begin to form at $Re_{\delta 2} > 33$, and the film is turbulent at $Re_{\delta 2} > 1600$. Here the *Reynolds* number is computed using the hydraulic diameter of the film

$$Re_{2\delta} = \frac{\rho_2 w_2 4\delta_2}{\eta_2}. \tag{24.11}$$

Some authors modify the *Nusselt* theory for high film *Reynolds* numbers. The Eq. (24.3) was rewritten in the form

$$\frac{h_{satgnant}}{\lambda'} \left[\frac{\eta'^2}{\rho'(\rho'-\rho'')} \right]^{1/3} = 1.51 Re_{2\delta}^{-1/3}, \tag{24.12}$$

having in mind that

$$Re_{2\delta} = \frac{4}{\eta_2} \rho_2 w_2 \delta_2 = \frac{4}{\eta_2} \frac{\dot{q}_w'' H}{\Delta h}, \tag{24.13}$$

and then modified. So *Kutateladze* [16] proposed for the region $33 < Re_{\delta 2} < 1600$ to multiply the right hand site of Eq. (24.12) by $0.687 Re_{2\delta}^{0.11}$ to account for the effect of the waves. The result is

$$\frac{h_{satgnant}}{\lambda'} \left[\frac{\eta'^2}{\rho'(\rho'-\rho'')} \right]^{1/3} = 1.04 Re_{2\delta}^{-0.223}. \tag{24.14}$$

For the turbulent region, $Re_{\delta 2} > 1600$, *Labuntzov* [17] tackled in addition the dependence on the film *Prandtl* number

$$\frac{h_{satgnant}}{\lambda'} \left[\frac{\eta'^2}{\rho'(\rho'-\rho'')} \right]^{1/3} = 1.51 Re_{2\delta}^{1/4} Pr'^{1/2}. \tag{24.15}$$

Butterworth proposed in [6] the following set of correlations for the three regions:

1. Laminar film: *Nusselt* theory.

2. Wavy film: $33 < Re_{\delta 2} < 1600$,

$$\frac{h_{satgnant}}{\lambda'}\left[\frac{\eta'^2}{\rho'(\rho'-\rho'')}\right]^{1/3} = \frac{Re_{2\delta}}{1.08 Re_{2\delta}^{1.22} - 5.2}. \qquad (24.16)$$

3. Turbulent film: $Re_{\delta 2} > 1600$,

$$\frac{h_{satgnant}}{\lambda'}\left[\frac{\eta'^2}{\rho'(\rho'-\rho'')}\right]^{1/3} = \frac{Re_{2\delta}}{8750 + 58 Pr'^{-1/2}\left(Re_{2\delta}^{3/4} - 253\right)}. \qquad (24.17)$$

For turbulent condensation on vertical surfaces and

$$Re_{2\delta} > 1400, \qquad (24.18)$$

Grigul [11] recommend the following correlation

$$h_{stagnat\_steam/turbulent\_film} = 0.003\left[\frac{\rho'(\rho'-\rho'')\lambda'^3 g \cos\varphi H}{\eta'^3}\frac{T'-T_w}{h''-h'}\right]^{1/2}. \qquad (24.19)$$

24.2. Condensation from forced convection two-phase flow at liquid film

24.2.1 Down flow of vapor across horizontal tubes

As already mentioned, the *Nusselt* theory is valid also for condensation of stagnant vapor on a horizontal cylinder with the modification of the numerical coefficient to 0.728 by *Isachenko*. Several authors correlated their data by using as a measure the *Nusselt* heat transfer coefficient as follows:

Fijii et al. [10] : $\dfrac{h}{h_{Nu}} = 1.24\left(\dfrac{w_1 D_{hy}}{v_2}\right)\left(\dfrac{h_{Nu} D_{hy}}{\lambda'}\right)^{-0.2}$.

Berman and Tumanov [2]: $\dfrac{h}{h_{Nu}} = 1 + 0.0095\left(\dfrac{w_1 D_{hy}}{v_1}\right)^{11.8/\sqrt{\frac{h_{Nu} D_{hy}}{\lambda'}}}$.

Shekriladze and Geomelauri [20]: $\dfrac{h}{h_{Nu}} = \left\{\dfrac{1}{2}\left(\dfrac{h_{Sh}}{h_{Nu}}\right)^2 + \dfrac{1}{4}\left[\left(\dfrac{h_{Sh}}{h_{Nu}}\right)^4 + 1\right]^{1/2}\right\}^{1/2}$,

where

$$h_{Sh} = 0.9 \frac{\lambda_2}{D_{pipe}} \left(\frac{w_1 D_{hy}}{v_2} \right)^{1/2}.$$

Kutateladze [15]: $\dfrac{h}{h_{Nu}} = 1 + 0.004 \left(\dfrac{w_1^2 \rho_1 h_{Nu}}{g \rho_2 \lambda_2} \right)^{0.8}.$

24.2.2 Collier correlation

For forced convection in inclined pipe in the case of a *laminar film* inside Collier [8] (1972) proposed

$$h_{lam} = 0.825 \left[\frac{\rho'(\rho' - \rho'')\lambda'^3 g \cos\varphi}{GD_h} \right]^{1/3}, \qquad (24.12)$$

and in the case of *turbulent film*

$$h_{turb} = 0.065 \frac{\lambda' \rho''^{1/2}}{\eta'} \frac{0.023 \rho' w_1^2}{Re_1^{1/4}} Pr'^{1/2}, \qquad (24.13)$$

where the liquid properties are those of saturated liquid at the local pressure,

$$Pr' = \eta' c'_p / \lambda'. \qquad (24.14)$$

In practical computations the maximum of all three expressions is taken,

$$h = \max\left(h_{stagnant}, h_{lam}, h_{turb} \right), \qquad (24.15)$$

which is in fact using the *Nusselt* theory in cases where the forced convection correlations gives lower values than the values for stagnant steam condensation.

24.2.3 Boyko and Krujilin approach

Boyko and *Krujilin* [5] recommended in 1966 for forced convection condensation the following correlation

$$h = h_{20} \left[1 + X_1 \left(\frac{\rho'}{\rho''} - 1 \right) \right]^{1/2}, \qquad (24.16)$$

where

$$h_{20} = 0.024 Re_{20}^{0.8} Pr'^{0.4} \frac{\lambda_2}{D_{hy}}, \qquad (24.17)$$

$$Re_{20} = D_{hy} G / \eta_2. \qquad (24.18)$$

The correlation was verified with experimental data for $12.3 < p < 88$ bar, $1 < X_1 < 0.26$, $162\,000 < \dot{q}_w^{"\sigma 2} < 1\,570\,000$ W/m². The constant 0.024 was recommended for steel pipes and 0.032 for cupper pipes.

24.2.4 The *Shah* modification of the *Boyko* and *Krujilin* approach

Shah [19] reported a modification of this correlation

$$h = h_{20}\left[(1-X_1)^{0.8} + \frac{3.8 X_1^{0.76}(1-X_1)^{0.04}}{(p/p_{cr})^{0.38}}\right]. \qquad (24.19)$$

The correlation is reported to reproduce 474 data points within $0.002 < p/p_{cr} < 0.44$, $21 < T'(p) < 310°C$, $3 < w_1 < 300 m/s$, $0 < X_1 < 1$, $17.5 < G < 210.6 kg/s$, and $158 < \dot{q}_w^{"\sigma 2} < 1\,893\,000$ W/m² with mean deviation 15.4%.

The condensing mass per unit flow volume is then

$$\mu_{12} = \frac{4}{D_{hy}} \frac{\dot{q}_w^{"}}{\Delta h}. \qquad (24.20)$$

24.3 Steam condensation from mixture containing non-condensing gases

Consider forced convection in a pipe. Given is the gas temperature, the wall temperature and the local parameters. We ask for the mass transferred per unit time and mixture volume and for the power density removed from the wall. The overall heat transfer coefficient can be estimate also in this case by using the *Boyko* and *Krujilin* [5] heat transfer correlation. What changes in this case is the driving temperature difference across the water film. There is convective heat transfer between the gas and the interface

$$h_{convective}\left(T_1 - T_1^{\sigma 2}\right), \qquad (24.21)$$

24.3 Steam condensation from mixture containing non-condensing gases

where

$$h_{convective} = 0.023 Re_1^{0.8} Pr_1^{0.4} \frac{\lambda_1}{D_h}.$$ (24.22)

A diffusion controlled condensation mass flux is defined as follows

$$(\rho w)_{12} = \beta_{12}\rho_1 \left(C_{1M} - C_{1M}^{\sigma 2} \right),$$ (24.23)

where the condensation coefficient β_{12} is defined depending of geometry and local conditions. The procedure for its computation will be discussed later. Using the ideal gas approximation we may rewrite the definition equation in terms of the corresponding partial pressures, or alternatively using the local partial pressures and the gas temperature

$$(\rho w)_{12} = \beta_{12}\rho_1 \left(C_{1M} - C_{1M}^{\sigma 2} \right) = \beta_{12}\rho_1 \frac{p_{1M} - p'(T_1^{\sigma 2})}{p} = \frac{\beta_{12}}{R_1 T_1} \left[p_{1M} - p'(T_1^{\sigma 2}) \right].$$ (24.24)

The energy jump condition at the gas-liquid interface requires

$$h\left(T_1^{\sigma 2} - T_w\right) = (\rho w)_{12} \Delta h + h_{convective} \left(T_1 - T_1^{\sigma 2}\right).$$ (24.25)

This is a transcendental equation with respect to the interface temperature. The equation has to be solved by iterations. To construct a *Newton* iteration we rewrite Eq. (24.25) in the form

$$f\left(T_1^{\sigma 2}\right) = h\left(T_1^{\sigma 2} - T_w\right) - (\rho w)_{12} \Delta h - h_{convective} \left(T_1 - T_1^{\sigma 2}\right) = 0,$$ (24.26)

and search for the zeros. For this purpose we need the derivative

$$\frac{df\left(T_1^{\sigma 2}\right)}{dT_1^{\sigma 2}} = h + \frac{d(\rho w)_{12} \Delta h}{dT} + h_{convective},$$ (24.27)

which is used in

$$T_1^{\sigma 2} = T_{1,old}^{\sigma 2} - f \bigg/ \frac{df\left(T_1^{\sigma 2}\right)}{dT_1^{\sigma 2}}.$$ (24.28)

For small deviation of the interface temperature from the saturation temperature at the bulk vapor partial pressure the use of the *Clausius-Clapyron* equation is possible to transform pressure differences into temperature differences in both: the definition equation for the diffusion controlled mass transfer

$$(\rho w)_{12} \approx \beta_{12} \frac{p_1}{p} \frac{dp'}{dT} \left[T'(p_{1M}) - T_1^{\sigma 2} \right], \qquad (24.29)$$

and the energy jump condition

$$h\left(T_1^{\sigma 2} - T_w\right) = \beta_{12} \frac{p_1}{p} \Delta h \frac{dp'}{dT} \left[T'(p_{1M}) - T_1^{\sigma 2} \right] + h_{convective} \left(T_1 - T_1^{\sigma 2}\right). \qquad (24.30)$$

This allows direct computation of the interface temperature as follows

$$T_1^{\sigma 2} = \frac{hT_w + \beta_{12} \dfrac{p_1}{p} \Delta h \dfrac{dp'}{dT} T'(p_{1M}) + h_{convective} T_1}{h + \beta_{12} \dfrac{p_1}{p} \Delta h \dfrac{dp'}{dT} + h_{convective}}. \qquad (24.31)$$

Also in this case the dependence of the mass transfer coefficient on the interface temperature still remains and has to be resolved by iterations. Note that the iterative solution is very sensitive to the chosen initial values, and it has to be carefully controlled.

24.3.1 Computation of the mass transfer coefficient

24.3.1.1 Gas flow in a pipe

Bobe and *Solouhin* [4], see also in [12] p. 136, correlated experimental data for steam condensation from gas-steam mixtures inside pipes as follows:

$$\beta_{12} = \beta_{12,0} \left(\frac{p_1 \Sigma_n}{p_{1M} - p_{1M}^{\sigma 2}} \right)^m,$$

where

$$\beta_{12,0} = 0.0191 \left(\frac{w_1 D_h}{v_1} \right)^{0.8} \left(\frac{v_1}{D_{\Sigma n \to M}} \right)^{0.43} \frac{D_{1\Sigma n \to M}}{D_h} \left(\frac{p}{p_1 \Sigma_n} \right)^n$$

or

$$(\rho w)_{12} = \rho_1 \beta_{12,0} \left[\frac{p_{1M} - p'(T_1^{\sigma 2})}{p} \right]^{1-m},$$

where for $0.1 < \dfrac{p_1 \sum n}{p_{1M} - p'(T_1^{\sigma 2})} < 1$, $n = 0.6$, $m = 0.4$, and for $\dfrac{p_1 \sum n}{p_{1M} - p_{1M}^{\sigma 2}} \geq 1$, $n = 0.9$, $m = 0.1$. The derivative required for the iterative solution is then

$$\frac{d(\rho w)_{12}}{dT_1^{\sigma 2}} = -\beta_{12,0}(1-m)\frac{\rho_1}{p}\left[\frac{p}{p_{1M} - p'(T_1^{\sigma 2})}\right]^m \frac{dp'(T_1^{\sigma 2})}{dT_1^{\sigma 2}}.$$

24.3.1.2 Gas flow across a bundle of pipes

Berman and *Fuks* [1] correlated experimental data for steam condensation from gas-steam mixtures flowing perpendicular to horizontal internally cooled tubes:

$$\beta_{12} = const \left(\frac{w_1 D_h}{v_1}\right)^{0.8} \left(\frac{D_{1 \sum n \to M}}{D_h}\right)^{0.6} \left(\frac{p}{p_1 \sum n}\right)^{0.6} \left(\frac{p}{p_{1M} - p_{1M}^{\sigma 2}}\right)^{1/3},$$

or

$$(\rho w)_{12} = const \, \rho_1 \left(\frac{w_1 D_h}{v_1}\right)^{0.8} \left(\frac{p}{p_1 \sum n}\right)^{0.6} \left(\frac{p_{1M} - p'(T_1^{\sigma 2})}{p}\right)^{2/3} \frac{D_{1 \sum n \to M}}{D_h}$$

where $const = 0.94$ for single pipe, 1.06 for the first row of pipes, 1.64 for the third and the subsequent row of pipes.

24.3.1.3 The Vierow and Shrock explicit correlation

The simplest method for taking into account the resistance of the non-condensable gases was proposed by *Vierow* and *Shrock* [21]. It consists of three steps: 1) Compute the condensation from stagnant pure steam; 2) Correct the expression for the influence of the gas mixture forced convection; 3) Modify the resulting heat transfer coefficient by a multiplier being a function of the inert mass concentration. In particular, the *Nusselt* theory is used in the first step

$$h_{satgnant} = \dot{q}_w'' / (T_{cond} - T_w),$$

$$T_w = T_k' - (-\dot{q}_w'' / C)^{4/3},$$

$$C = \frac{2}{3}\sqrt{2}\left[\frac{\rho'(\rho' - \rho'')\lambda'^3 g \cos\varphi (h'' - h')}{H\eta'}\right]^{1/4}.$$

In the second step the multiplier

24. Condensation at cooled walls

$$f_1 = \min\left(2,\ 1 + 2.88^{-5} Re_1^{1.18}\right)$$

is used, where $Re_1 = D_h \rho_1 |w_1| / \eta_1$. In the third step an explicit correlation of experimental data is generated in the simple form by the second multiplier

$$f_2 = 1 - 10 C_{1n} \quad \text{for} \quad C_{1n} < 0.063,$$

$$f_2 = 1 - C_{1n}^{0.22} \quad \text{for} \quad C_{1n} > 0.6,$$

$$f_2 = 1 - 0.938 C_{1n}^{0.13} \quad \text{for} \quad 0.063 \leq C_{1n} \leq 0.6.$$

Recently *Choi* et al. [7] modified the multiplier

$$f_2 = 1 - 1.00962 C_{1n}^{0.13354} \quad \text{for} \quad 0.1 \leq C_{1n} \leq 0.65$$

in order to obtain better approximation of their own data base for $h_{cond} D_{hy,film} / \lambda_{film} < 30$ for stratified flow in inclined tubes (2.1° and 5°) – averaged percentage error 20 to 35%.

Nomenclature

Latin

a	$= \lambda / (\rho c_p)$, temperature conductivity, m²/s
a_{12}	$= 1/\Delta z$ gas-liquid interfacial area density in Cartesian coordinates, 1/m
C_{1M}	vapor mass concentration inside the gas mixture, dimensionless
C_{1n}	mass concentration of the non-condensing gas n inside the gas mixture, dimensionless
$C_{1M}^{\sigma 2}$	vapor mass concentration inside the gas mixture at the liquid interface, dimensionless
$D_{1\sum_{n \to M}}$	diffusion gas coefficient for the vapor inside the mixture consisting of condensable and non-condensable gases, m²/s
D_{hy}	pipe hydraulic diameter, m
g	acceleration due to gravity, m/s²
R_1	mixture gas constant, J/(kgK)

$Re_{2\delta}$	$= \dfrac{\rho_2 w_2 4\delta_2}{\eta_2}$, Reynolds number based on the hydraulic diameter of the film, *dimensionless*
H	height of the plate, *m*
h	specific enthalpy, *J/kh*
$h_{convective}$	convective heat transfer coefficient, $W/(m^2K)$
$h_{satgnant}$	heat transfer coefficient for stagnant steam, $W/(m^2K)$
$h_{stagnat\_steam/turbulent\_film}$	heat transfer coefficient for stagnant steam and turbulent condensate film, $W/(m^2K)$
h_{Nu}	heat transfer coefficient for stagnant steam in accordance with the *Nusselt* water skin theory, $W/(m^2K)$
h_{lam}	heat transfer coefficient for forced convection inclined pipe - laminar film, $W/(m^2K)$
h_{turb}	heat transfer coefficient for forced convection inclined pipe - turbulent film, $W/(m^2K)$
h_{20}	heat transfer coefficient for forced convection computed for flow having the mixture mass flow but consisting of liquid only, $W/(m^2K)$
$\dot{q}''_w$	wall heat flux from the interface into the gas, W/m^3
p	pressure, *Pa*
p_{cr}	critical pressure, *Pa*
p_{1M}	vapor partial pressure inside the gas mixture, *Pa*
$p_{1M}^{\sigma 2}$	vapor partial pressure inside the gas mixture at the liquid interface, *Pa*
$p_{1\sum n}$	partial pressure of the non-condensing gases inside the gas mixture, *Pa*
$p'\left(T_1^{\sigma 2}\right)$	vapor partial pressure inside the gas mixture at the liquid surface, *Pa*
Re_1	$= D_{hy}\rho_1|w_1|/\eta_1$, gas *Reynolds* number, *dimensionless*
Re_{20}	$= D_{hy}G/\eta_2$ total flow *Reynolds* number based on the liquid viscosity, *dimensionless*
Pr	$= \eta c_p/\lambda$, *Prandtl* number, *dimensionless*
T	temperature, *K*
$T_2^{1\sigma}$	liquid side interface temperature, *K*
$T_1^{\sigma 2}$	gas side interface temperature, *K*
T_{cond}	condensation initiation temperature, *K*
V_1	gas velocity, *m/s*
V_2	liquid velocity, *m/s*
w	axial velocity, *m/s*
X_l	mass flow concentration of the field *l* inside the mixture, *dimensionless*

Greek

β_{12}	condensation coefficient, m/s
δ	film thickness, m
φ	angle between the positive flow direction and the upwards directed vertical, rad
Δh	$= h'' - h'$, latent heat of vaporization, J/kg
η	dynamic viscosity of liquid, $kg/(ms)$
λ	thermal conductivity, $W/(mK)$
μ_{21}	steam mass generation per unit mixture volume and unit time due to evaporation, $kg/(sm^3)$
μ_{12}	liquid mass generation per unit mixture volume and unit time due to condensation, $kg/(sm^3)$
ρ	density, kg/m^3
$(\rho w)_{21}$	steam mass generation per unit surface and unit time due to evaporation, $kg/(m^2 s)$
$(\rho w)_{12}$	liquid mass generation per unit surface and unit time due to condensation, $kg/(m^2 s)$

Subscripts

1	gas
2	liquid film
2F	liquid film
M1	vapor inside the gas mixture
n1	inert components inside the gas

Superscripts

′	time fluctuation
'	saturated steam
"	saturated liquid

References

1. Berman LD and Fuks SN (1958) Teploenergetika, vol 8 pp 66-74; (1959) Teploenergetika, vol 7 pp 74-83
2. Berman LD and Tumanov VA (1962), Heat transfer during condensation of flowing steam in a tube bundle, Teploenergetika, vol 9 no 10 pp 77-83
3. Bird RB, Steward WE, Lightfoot EN (1960), Transport phenomena, John Wiley & Sons, New York

4. Bobe LS and Solouhin VA (1972) Heat and mass transfer during steam condensation from steam-gas mixture by turbulent pipe flow, Teploenergetika, vol 9 pp 27-30.
5. Boyko LD and Krujilin GN (1966), Izvestia akademii nauk SSSR Energetika I transport, no 2 pp 113-128. See also Boyko LD (1966) Teploobmen w elementach energeticheskih ustanovak, Nauka, Moskwa, pp 197-212
6. Butterworth D (1981) Condensers: Heat transfer and fluid flow, in heat exchanger, Kakac S, Bergeles AE and Mayinger F (eds) Hemisphere, Washington, pp 89-314
7. Choi KY, Chung HJ and No HC (2002) Nuclear Engineering and design, vol 211 pp 139-151
8. Collier JG (1972) Convection boiling and condensation, London: McGraw-Hill Book Company Inc.
9. Chato JC (1962) Laminar condensation inside horizontal and inclined tubes, ASHREE Journal, vol 4 pp 52-60
10. Fijii T, Honda H and Oda K (1979), Condensation of steam on horizontal tubes – influence of the oncoming velocity, Condensation heat transfer, 18th Nat. Heat Transfer Conf., San Diego August, pp 35-43
11. Grigul U (1942) Forsch Ing. Wes, vol 13 pp 49-47; (1942) Z. Ver. Dtsch. Ing., vol 86 pp 444
12. Isachenko VP (1977) Teploobmen pri kondensazij, Moskva, Energia
13. Isachenko VP, Osipova VA and Sukomel AS (1981) Teploperedacha, 4th Ed., Moskva, Energoisdat
14. Kapitza PL (1948) Zhur. Exper. Teoret. Fiz., vol 18 no 1
15. Kitateladze SS (1962), Heat transfer in condensation and boiling, AEC-tr-3778 chaps 5-7
16. Kutateladze SS (1963), Fundamentals of heat transfer, Academic, New York
17. Labuntzov DA (1957), Heat transfer in film condensation of pure steam, Teploenergetica, vol 4 no 7 pp 72-80
18. Rohsenow WM, Hartnett JP and Ganic EN (1985) Handbook of heat transfer fundamentals, second edition, McGraw-Hill Book Company
19. Shah MM (1979) A general correlation for heat transfer during film condensation inside pipes, Int. J. Heat Mass Transfer, vol 22 pp 547-556
20. Shekriladze IG and Gomelauri VI (1966), Laminar condensation on flowing vapor, Int. J. Heat Mass Transfer, vol 9 pp 581-591
21. Vierow KM and Shrock KE (1991) Condensation in a natural circulation loop with non condensable gases, Part 1 – Heat transfer, International Conference on Multi-Phase Flows, Tsukuba, Japan

25. Discrete ordinate method for radiation transport in multi-phase computer codes

25.1 Introduction

Several applications in technology require radiation transport analysis (RTA). Examples are processes with combustion and interaction of high temperature molten material with water in the safety analysis in the nuclear industry and in the metalurgy. The RTA is always a part of the general flow simulation and of the analysis of the flow interaction with the structure. Radiation transport in high temperature media may be the governing process of energy transfer. This Chapter is not intended to replace the famous book of *Siegel* and *Howell* [9] on this subject. The interested reader intending to reach a high level skill in this field has to go through this book. The Chapter rather gives some of the main ideas required to understand how the radiation transport theory can be implemented in multi-phase flow analysis. The Chapter is based on the experience gained by creating a radiation transport model for the computer code IVA [6]. This work started in 1998. In this year *Lanzenberger* [7] created as far as I know the first computer code for modeling of 3D radiation transport in three-phase flow as a part of the IVA4 code development. The code uses the first order *step* scheme setting the positive wall coefficient equal to zero. Strictly speaking the code was developed for simulation of Cartesian coordinates without any obstacles inside the computational region. The further development of this technique in IVA6 went through the following direction: (a) rewriting the code from Fortran 77 to Fortran 90, improving macroscopic data and coupling with the IVA6 computer code; (b) changing the algorithm so that it allows one to handle internal obstacles.

25.1.1 Dimensions of the problem

For technical application on the ground we consider each beam as being linear. It can travel in an arbitrary direction through a single point. Therefore the first three dimensions are coming from the spatial description. The volume of interest is also three dimensional. It introduces another three dimensions in the problem. The electromagnetic radiation energy flow transported per unit time across unit normal surface is depending on the wavelength. The continuous spectrum is usually split in n groups with prescribed boundaries. This gives the other n-dimensions of the problem. In the discrete ordinate radiation transport method the governing energy conservation equation is replaced by a set of equations for a finite number of ordinate directions, for a finite number of control volumes, and for finite number of spectral regions. It is the large number of dimensions that makes the RTA complicated.

25.1.2 Micro- versus macro-interactions

Any interactions of electromagnetic radiation with matter are described in physics by theoretical and experimental observation on a micro-scale. The engineer integrates the interactions over a control volume and prepares the information necessary for microscopic radiation transport on a macro-scale. If the characteristic interaction length is much smaller than the scale of the control volume, RTA on a macro-scale is not necessary because all the interaction happens inside the cell and no energy leaves the volume. An example is a relatively dense suspension of hot particles, which are shielding effectively the radiation inside large cells. If the characteristic interaction length is larger than the scale of the control volume then RTA is necessary because energy is leaving and entering the control volume. Examples are small concentrations of very hot spheres in water. The radiation energy may be transported up to 10 m in water up to full absorption. If the volume of interest is few cubic meters and the control volume of the discretization is of the order of liters the RTA is necessary.

25.1.3 The radiation transport equation (RTE)

The basic equation for the description of the radiation transport is the steady state energy conservation equation called radiation transport equation (RTE)

$$\frac{di'_\lambda(\mathbf{r},\mathbf{s})}{ds} = -a_\lambda i'_\lambda(\mathbf{r},\mathbf{s}) - \sigma_{s\lambda} i'_\lambda(\mathbf{r},\mathbf{s}) + a_\lambda i'_{\lambda b}(\mathbf{r},\mathbf{s})$$

$$+ \frac{\sigma_{s\lambda}}{4\pi} \int_{4\pi} i_\lambda(\mathbf{r},\omega_i) \Phi(\lambda, \omega, \omega_i) d\omega_i \,. \tag{25.1}$$

Here $i'_\lambda(\mathbf{r},\mathbf{s})$ is the spectral radiation intensity at the spatial location $\mathbf{r}$ along the direction $\mathbf{s}$ that is within an infinitesimal solid angle $d\omega$. Spectral – because it is associated with each particular wavelength λ. The dimensions of i'_λ are those of power flux through a square meter. It is in fact a heat flux. $a_\lambda, \sigma_{s\lambda}$ are the spectral absorption and scattering coefficients, Φ is the spectral scattering phase function and ω_i is the scattering angle. The scattering phase function describes which amount of incident radiation is scattered in all other directions, Fig. 25.1.

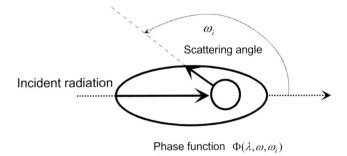

Fig. 25.1. Scattering of incident radiation

The intensity of a ray passing through an infinitesimal volume is attenuated by absorption. This is reflected by the first term on the right hand side of Eq. (25.1). The intensity of a ray passing through an infinitesimal volume is attenuated also by scattering. It is taken into account in the second term on the RHS. The intensity is increased by the emitted intensity of the medium reflected by the third term on the RHS, and by the scattering from any other direction to the considered direction of the vector **s**. For numerical integration of Eq. (25.1) it is convenient to introduce the so called spectral extinction coefficient

$$\beta_\lambda = a_\lambda + \sigma_{s\lambda} \tag{25.2}$$

and the source term

$$S' = a_\lambda i'_{\lambda b}(\mathbf{r},\mathbf{s}) + \frac{\sigma_{s\lambda}}{4\pi} \int_{4\pi} i_\lambda(\mathbf{r},\omega_i)\Phi(\lambda,\omega,\omega_i)d\omega_i \tag{25.3}$$

which simplify the notation

$$\frac{di'_\lambda(\mathbf{r},\mathbf{s})}{ds} = -\beta_\lambda i'_\lambda(\mathbf{r},\mathbf{s}) + S'. \tag{25.4}$$

In the discrete ordinate radiation transport method Eq. (25.4) will be replaced by a set of equations for a finite number of ordinate directions, for a finite number of control volumes, and for a finite number of spectral regions.

25.2 Discrete ordinate method

The spatial location in the space is defined by the position vector **r**. In a polar coordinate system as presented in Fig. 25.2 with radius $r = 1$, azimuthal angle φ, and polar angle θ the components of the unit vectors of **s** in a Cartesian coordinate system with coordinates x, y and z are defined as the direction cosines

$$\mathbf{s} = (\sin\theta\cos\varphi,\ \sin\theta\sin\varphi,\ \cos\theta) \equiv (s_x,\ s_y,\ s_z) \tag{25.5}$$

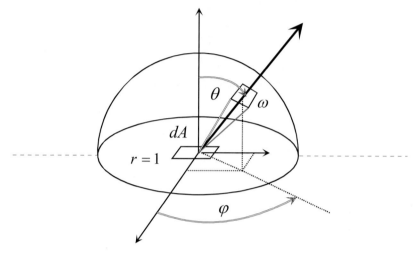

Fig. 25.2. Unit hemisphere, solid angle ω, angle φ and θ

The infinitesimal solid angle $d\omega$ is equal to the infinitesimal area on the surface of the unit hemisphere

$$d\omega = \sin\theta\, d\theta\, d\varphi\ . \tag{25.6}$$

Integrating over all possible solid angle gives

$$\int_{\varphi=0}^{2\pi}\int_{\theta=-\pi/2}^{\pi/2} \sin\theta\, d\theta\, d\varphi = 4\pi\ . \tag{25.7}$$

The total surface area of the unit hemisphere is 2π, and is denoted as the *total solid angle*. Let us divide the space into a discrete number of solid angles $\Delta\omega_{a,b}$, where

$$a = 1, a_{max}\ , \tag{25.8}$$

$$b = 1, b_{max}\ , \tag{25.9}$$

with a_{max} and b_{max} being even numbers. For equidistant discretization we have

$$\Delta\varphi = 2\pi / a_{\max}, \quad \varphi_a = \left(a - \frac{1}{2}\right)\Delta\varphi, \quad a = 1, a_{\max} \tag{25.10}$$

$$\Delta\theta = 2\pi / b_{\max}, \quad \theta_b = \left(b - \frac{1}{2}\right)\Delta\theta, \quad b = 1, b_{\max}, \tag{25.11}$$

and

$$\Delta\omega_{a,b} = \int_{\varphi_a - \Delta\varphi/2}^{\varphi_a + \Delta\varphi/2} \int_{\theta_b - \Delta\theta_b/2}^{\theta_b + \Delta\theta_b/2} \sin\theta \, d\theta \, d\varphi = 2\Delta\varphi_a \sin\theta_b \sin\frac{\Delta\theta_b}{2}$$

$$= \Delta\varphi_a \left[\cos\left(\theta_b - \frac{\Delta\theta_b}{2}\right) - \cos\left(\theta_b + \frac{\Delta\theta_b}{2}\right)\right]. \tag{25.12}$$

Here φ_a and θ_b define the specific ordinate direction denoted with subscripts a,b. Other notation is

$$\mathbf{s}_{a,b} = \left(s_{x,a,b}, \ s_{y,a,b}, \ s_{z,a,b}\right). \tag{25.13}$$

There is an alternative notation used frequently in the literature

$$\mathbf{s}_{a,b} = \left(\mu_{a,b}, \ \eta_{a,b}, \ \xi_{a,b}\right). \tag{25.15}$$

25.2.1 Discretization of the computational domain for the description of the flow

The discretization of the computational domain for the description of the flow is performed by using a curvilinear coordinate system. In many practical cases either a Cartesian or a cylindrical coordinate system is used. We give here examples with the Cartesian systems for simplicity, but this is not a limitation of the approach. The discretization is structured and non-equidistant in each direction. Each control volume has 6 faces with surface A_m. The surfaces are numbered 1 for high i, 2 for low i, 3 for high j, 4 for low j, 5 for high k and 6 for low k. The orientation of each surface is defined by the outwards directed unit vector

$$\mathbf{e} = \left(e_x, \ e_y, \ e_z\right). \tag{25.16}$$

For Cartesian coordinate system the 6 surfaces have the surface vectors

$$\mathbf{e}_1 = (1,\ 0,\ 0),\ \mathbf{e}_2 = (-1,\ 0,\ 0),$$
$$\mathbf{e}_3 = (0,\ 1,\ 0),\ \mathbf{e}_4 = (0,\ -1,\ 0),$$
$$\mathbf{e}_5 = (0,\ 0,\ 1),\ \mathbf{e}_6 = (0,\ 0,\ -1). \tag{25.17-22}$$

For discterization of the cylindrical computational domain the 6 surfaces have the surface vectors given below

$$\mathbf{e}_m = (\cos\theta_m,\ \sin\theta_m,\ 0),\ m = 1, 4,\ \mathbf{e}_5 = (0,\ 0,\ 1),\ \mathbf{e}_6 = (0,\ 0,\ -1).$$
$$\tag{25.23-28}$$

The angles θ_m are defined as follows $\theta_1 = \theta$, $\theta_1 = \pi + \theta$, $\theta_1 = \frac{3}{4}\pi + \theta + \frac{1}{2}\Delta\theta$, $\theta_1 = \frac{3}{4}\pi + \theta - \frac{1}{2}\Delta\theta$. Here θ is the azimuthal angle of the center of the control volume in the fluid domain. $\Delta\theta$ is the size of the control volume in the azimuthal direction. Note the difference between these two angles and the angles used in the polar coordinate system for definition of the discrete ordinates.

25.2.2 Finite volume representation of the radiation transport equation

Integrating the RTE over the control volume $dVol$ and over the angle $\Delta\omega_{a,b}$ we obtain

$$\iint_{\Delta\omega_{a,b}} \int_{\Delta Vol} \frac{di'_\lambda(\mathbf{r},\mathbf{s})}{ds} dVol\, d\omega = - \iint_{\Delta\omega_{a,b}} \int_{\Delta Vol} \beta_\lambda i'_\lambda(\mathbf{r},\mathbf{s}) dVol\, d\omega + \iint_{\Delta\omega_{a,b}} \int_{\Delta Vol} S' dVol\, d\omega$$
$$\tag{25.29}$$

or

$$\sum_{m=1}^{6} i'_{\lambda,a,b,m} A_m \iint_{\Delta\omega_{a,b}} (\mathbf{s}\cdot\mathbf{e})_m d\omega + \beta_\lambda i'_{\lambda,a,b} \Delta Vol\, \Delta\omega_{a,b} = S'_{a,b} \Delta Vol\, \Delta\omega_{a,b}. \tag{25.30}$$

Scalar product of the discrete directions vector and given surface vector: Now we are interested in the estimation of the integral

25.2 Discrete ordinate method

$$D_{a,b} = \iint_{\Delta\omega_{a,b}} \mathbf{s}\cdot\mathbf{e}\, d\omega = \int_{\varphi_a-\Delta\varphi/2}^{\varphi_a+\Delta\varphi_a/2} \int_{\theta_b-\Delta\theta_b/2}^{\theta_b+\Delta\theta_b/2} (\mathbf{s}\cdot\mathbf{e}) \sin\theta\, d\theta\, d\varphi =$$

$$\int_{\varphi_a-\Delta\varphi/2}^{\varphi_a+\Delta\varphi_a/2} \int_{\theta_b-\Delta\theta_b/2}^{\theta_b+\Delta\theta_b/2} \left(e_x \sin\theta\cos\varphi + e_y \sin\theta\sin\varphi + e_z \cos\theta \right) \sin\theta\, d\theta\, d\varphi$$

$$= \left(e_x \int_{\varphi_a-\Delta\varphi/2}^{\varphi_a+\Delta\varphi_a/2} \cos\varphi\, d\varphi + e_y \int_{\varphi_a-\Delta\varphi/2}^{\varphi_a+\Delta\varphi_a/2} \sin\varphi\, d\varphi \right) \int_{\theta_b-\Delta\theta_b/2}^{\theta_b+\Delta\theta_b/2} \sin^2\theta\, d\theta$$

$$+ e_z \Delta\varphi_a \int_{\theta_b-\Delta\theta_b/2}^{\theta_b+\Delta\theta_b/2} \cos\theta \sin\theta\, d\theta\ . \qquad (25.31)$$

Integrating we obtain

$$D_{a,b} = \left(e_x \cos\varphi_a + e_y \sin\varphi_a \right) 2 \sin\frac{\Delta\varphi_a}{2}\left[\left(1 - 2\cos^2\theta_b\right)\sin\frac{\Delta\theta_b}{2}\cos\frac{\Delta\theta_b}{2} + \frac{\Delta\theta_b}{2} \right]$$

$$+ e_z \Delta\varphi_a 2 \cos\theta_b \sin\theta_b \cos\frac{\Delta\theta_b}{2} \sin\frac{\Delta\theta_b}{2}\ . \qquad (25.32)$$

Working form of the finite differential form of the RTE: Using the above results the finite differential form of the RTE becomes

$$\sum_{m=1}^{6} i'_{\lambda,a,b,m} A_m D_{a,b,m} + \beta_\lambda i'_{\lambda,a,b} \Delta Vol\, \Delta\omega_{a,b} = S'_{a,b} \Delta Vol\, \Delta\omega_{a,b}\ . \qquad (25.33)$$

It is convenient to divide both sides by $\Delta Vol\, \Delta\omega_{a,b}$. The result is

$$\sum_{m=1}^{6} a_{a,b,m} i'_{\lambda,a,b,m} + \beta_\lambda i'_{\lambda,a,b} = S'_{a,b}\ , \qquad (25.34)$$

where

$$a_{a,b,m} = \frac{A_m\, D_{a,b,m}}{\Delta Vol\, \Delta\omega_{a,b}}\ , \qquad (25.35)$$

is geometry dependent only and is computed once at the beginning of the computation. Note that the ratio $D_{a,b,m}/\Delta\omega_{a,b}$ is dimensionless and the coefficients $a_{a,b,m}$ have dimension $1/m$ which is the same as that of the extinction coefficient. For Cartesian coordinates using Eqs. (25.17-22) the above equation simplifies as follows

$$\left(i'_{\lambda,a,b,1} - i'_{\lambda,a,b,2}\right)\frac{A_1}{\Delta Vol}\frac{s_{x,a,b}}{\Delta\omega_{a,b}} + \left(i'_{\lambda,a,b,3} - i'_{\lambda,a,b,4}\right)\frac{A_3}{\Delta Vol}\frac{s_{y,a,b}}{\Delta\omega_{a,b}}$$

$$+ \left(i'_{\lambda,a,b,5} - i'_{\lambda,a,b,6}\right)\frac{A_5}{\Delta Vol}\frac{s_{z,a,b}}{\Delta\omega_{a,b}} + \beta_\lambda\, i'_{\lambda,a,b} = S'_{a,b}. \qquad (25.36)$$

The general form is preferred here.

Note that the spectral intensities $i'_{\lambda,a,b,m}$ are defined at the cell boundaries and the spectral intensity $i'_{\lambda,a,b}$ is defined in the cell center. There are numerous methods giving different approximations for the relationship between $i'_{\lambda,a,b,m}$ and $i'_{\lambda,a,b}$. Three of them are discussed in the next section.

Weighting factors: Step (donor cell), diamond and positive weighting: The general form of the weighting factors for the first octant $s_x > 0$, $s_y > 0$, $s_z > 0$

$$i'_{\lambda,a,b} = f_x i'_{\lambda,a,b,1} + (1-f_x) i'_{\lambda,a,b,2} = f_y i'_{\lambda,a,b,3} + (1-f_y) i'_{\lambda,a,b,4}$$

$$= f_z i'_{\lambda,a,b,5} + (1-f_z) i'_{\lambda,a,b,6}, \qquad (25.37)$$

or

$$i'_{\lambda,a,b,1} = \frac{1}{f_x} i'_{\lambda,a,b} + \left(1 - \frac{1}{f_x}\right) i'_{\lambda,a,b,2}, \qquad (25.38)$$

$$i'_{\lambda,a,b,3} = \frac{1}{f_y} i'_{\lambda,a,b} + \left(1 - \frac{1}{f_y}\right) i'_{\lambda,a,b,4}, \qquad (25.39)$$

$$i'_{\lambda,a,b,5} = \frac{1}{f_z} i'_{\lambda,a,b} + \left(1 - \frac{1}{f_z}\right) i'_{\lambda,a,b,6}. \qquad (25.40)$$

Substituting in the discretized RTE and solving with respect to the intensity in the center of the cell $i'_{\lambda,a,b}$ we obtain finally

$$i'_{\lambda,a,b} = \left\{ \begin{array}{l} \left[S'_{a,b} - \left[a_{a,b,2} + a_{a,b,1}\left(1 - \dfrac{1}{f_x}\right)\right] i'_{\lambda,a,b,2} \right. \\ \left. - \left[a_{a,b,4} + a_{a,b,3}\left(1 - \dfrac{1}{f_y}\right)\right] i'_{\lambda,a,b,4} - \left[a_{a,b,6} + a_{a,b,5}\left(1 - \dfrac{1}{f_z}\right)\right] i'_{\lambda,a,b,6} \right] \end{array} \right\}$$

$$\times \left(a_{a,b,1}\dfrac{1}{f_x} + a_{a,b,3}\dfrac{1}{f_y} + a_{a,b,5}\dfrac{1}{f_z} + \beta_\lambda \right)^{-1}. \qquad (25.41)$$

Step difference scheme: Setting

$$f_x = f_y = f_z = 1 \qquad (25.42)$$

results in a first order *step* difference scheme

$$i'_{\lambda,a,b,1} = i'_{\lambda,a,b}, \quad i'_{\lambda,a,b,3} = i'_{\lambda,a,b}, \quad i'_{\lambda,a,b,5} = i'_{\lambda,a,b}, \qquad (25.43)$$

which resembles the *donor cell* scheme in the CFD. Consequently we have

$$i'_{\lambda,a,b} = \dfrac{S'_{a,b} - a_{a,b,2}i'_{\lambda,a,b,2} - a_{a,b,4}i'_{\lambda,a,b,4} - a_{a,b,6}i'_{\lambda,a,b,6}}{\beta_\lambda + a_{a,b,1} + a_{a,b,3} + a_{a,b,5}}$$

$$\equiv \dfrac{S'_{a,b} - a_{a,b,2}i'_{\lambda,a,b,i-1} - a_{a,b,4}i'_{\lambda,a,b,j-1} - a_{a,b,6}i'_{\lambda,a,b,k-1}}{\beta_\lambda + a_{a,b,1} + a_{a,b,3} + a_{a,b,5}}. \qquad (25.44)$$

Similar strategy is used for the other octants.

The optimum marching strategy: If the point *Gauss-Seidel* method is used to solve the above equation there is an optimum strategy of visiting the cells, and the ordinate directions – called the optimum marching strategy. The optimum marching strategy is direction dependent. First the octants are defined in which the discrete ordinate is, as given in the Table 25.1, and then the incident and the outlet walls of the cell. Then the marching strategy is defined in Tables 25.2 and 25.3.

Table 25.1. Definition of the octants. Incident and outlet walls in the cell

Angles	Octant	Incident wall cells i1,i2,i3	Outlet wall cells o1,o2,o3
$0 \le \theta_b < \pi/2$			
$0 \le \varphi_a < \pi/2$	1	2,4,6	1,3,5
$\pi/2 \le \varphi_a < \pi$	2	1,4,6	2,3,5
$\pi \le \varphi_a < 3\pi/2$	3	1,3,6	2,4,5
$3\pi/2 \le \varphi_a < 2\pi$	4	2,3,6	1,4,5
$\pi/2 \le \theta_b < \pi$			
$0 \le \varphi_a < \pi/2$	5	2,4,5	1,3,6
$\pi/2 \le \varphi_a < \pi$	6	1,4,5	2,3,6
$\pi \le \varphi_a < 3\pi/2$	7	1,3,5	2,4,6
$3\pi/2 \le \varphi_a < 2\pi$	8	2,3,5	1,4,6

Table 25.2. Marching strategy depending on the discrete ordinate direction

Octant	Marching strategy x-direction	y-direction	z-direction
1	$i = 2, i_{max} - 1$	$j = 2, j_{max} - 1$	$k = 2, k_{max} - 1$
2	$i = i_{max} - 1, 2, -1$	$j = 2, j_{max} - 1$	$k = 2, k_{max} - 1$
3	$i = i_{max} - 1, 2, -1$	$j = j_{max} - 1, 2, -1$	$k = 2, k_{max} - 1$
4	$i = 2, i_{max} - 1$	$j = j_{max} - 1, 2, -1$	$k = 2, k_{max} - 1$
5	$i = 2, i_{max} - 1$	$j = 2, j_{max} - 1$	$k = k_{max} - 1, 2, -1$
6	$i = i_{max} - 1, 2, -1$	$j = 2, j_{max} - 1$	$k = k_{max} - 1, 2, -1$
7	$i = i_{max} - 1, 2, -1$	$j = j_{max} - 1, 2, -1$	$k = k_{max} - 1, 2, -1$
8	$i = 2, i_{max} - 1$	$j = j_{max} - 1, 2, -1$	$k = k_{max} - 1, 2, -1$

Table 25.3. Marching strategy for the discrete ordinate

Octant	Marching strategy a-direction	b-direction
1	$a = 1, a_{max}/4$	$b = 1, b_{max}/2$
2	$a = a_{max}/4 + 1, a_{max}/2$	$b = 1, b_{max}/2$
3	$a = a_{max}/2 + 1, 3a_{max}/4$	$b = 1, b_{max}/2$

4	$a = 3a_{max}/4+1, a_{max}$	$b = 1, b_{max}/2$
5	$a = 1, a_{max}/4$	$b = b_{max}/2+1, b_{max}$
6	$a = a_{max}/4+1, a_{max}/2$	$b = b_{max}/2+1, b_{max}$
7	$a = a_{max}/2+1, 3a_{max}/4$	$b = b_{max}/2+1, b_{max}$
8	$a = 3a_{max}/4+1, a_{max}$	$b = b_{max}/2+1, b_{max}$

Therefore for each octant we have

$$i'_{\lambda,a,b} = \frac{S'_{a,b} - a_{a,b,i1}i'_{\lambda,a,b,i1} - a_{a,b,i2}i'_{\lambda,a,b,i2} - a_{a,b,i3}i'_{\lambda,a,b,i3}}{\beta_\lambda + a_{a,b,o1} + a_{a,b,o2} + a_{a,b,o3}}, \qquad (25.45)$$

where $i'_{\lambda,a,b,o1} = i'_{\lambda,a,b}$, $i'_{\lambda,a,b,o2} = i'_{\lambda,a,b}$, $i'_{\lambda,a,b,o3} = i'_{\lambda,a,b}$.

This method has been in use for a long time. It is very simple in design and stable in performance regardless of the ordinate set chosen or the local radiative properties, but it is reported to be inaccurate [8].

Lathrop's diamond difference scheme: Setting

$$f_x = f_y = f_z = \frac{1}{2} \qquad (25.47)$$

results in the second order *Lathrop* diamond difference scheme. The diamond schemes are known to be unstable without additional improvement of the weighting coefficients [8].

Fiveland scheme: Fiveland [2] proposed to use

$$f_x = f_y = f_z = f \qquad (25.48)$$

satisfying the following conditions

$$\Delta x < \frac{|s_{x,a,b}|}{\beta(1-f)}\phi, \quad \Delta y < \frac{|s_{y,a,b}|}{\beta(1-f)}\phi, \quad \Delta z < \frac{|s_{z,a,b}|}{\beta(1-f)}\phi, \qquad (25.49\text{-}51)$$

where

$$\phi = \frac{f^3 + (1-f)^2(2-5f)}{f}. \qquad (25.53)$$

It is thinkable to use a set of non-equal weighting factors like

$$f_x = \max\left[\left(1 - \frac{|s_{x,a,b}|}{\beta \Delta x}\right), \frac{1}{2}\right], \qquad (25.54)$$

$$f_y = \max\left[\left(1 - \frac{|s_{y,a,b}|}{\beta \Delta y}\right), \frac{1}{2}\right], \qquad (25.55)$$

$$f_z = \max\left[\left(1 - \frac{|s_{x,a,b}|}{\beta \Delta z}\right), \frac{1}{2}\right], \qquad (25.56)$$

to assure the positive scheme.

25.2.3 Boundary conditions

The radiation heat flux across the surface m is that computed by integrating over the discrete ordinates

$$\dot{q}''_{\lambda,m} = \Delta Vol \sum_{b=1}^{b_{max}} \sum_{a=1}^{a_{max}} a_{a,b,m} \Delta \omega_{a,b} i'_{\lambda,a,b,m}. \qquad (25.57)$$

The incident radiation heat flux from the inside of the computational cell is

$$\dot{q}''_{\lambda,m-incident} = \Delta Vol \sum_{b=1}^{b_{max}} \sum_{a=1}^{a_{max}} a_{a_i,b_i,m} \Delta \omega_{a_i,b_i} i'_{\lambda,a_i,b_i,m}. \qquad (25.58)$$

For Cartesian coordinate system the incident ordinate directions a_i, b_i are summarized in Table 25.4. At *diffusely emitting and reflecting* walls with temperature $T_{w,m}$ and emissivity coefficient ε_w the emitting intensity is

$$i'_{\lambda,a_o,b_o,m} = \varepsilon_w \frac{\sigma T_{w,m}}{\pi} + \left(1 - \varepsilon_w\right) \frac{\dot{q}''_{\lambda,m-incident}}{\pi}. \qquad (25.59)$$

The outlet ordinate directions are specified for each wall in Table 25.5 for a Cartesian coordinate system.

Setting

$$i'_{\lambda,a_o,b_o,m} = 0 \tag{25.60}$$

means a *transparent* wall. For *mirroring* walls we have

$$i'_{\lambda,a_{max}-a+1,b_{max}-b+1,m} = \varepsilon_w \frac{\sigma T_{w,m}}{\pi} + i'_{\lambda,a,b} \tag{25.61}$$

where $a,b \in m - incident$.

Table 25.4. Marching strategy for computation of the incident radiation at the cell walls required for definition of boundary conditions

Wall Nr.	Marching strategy, incident a-direction	b-direction
1	$a = 1, a_{max}/4$	$b = 1, b_{max}$
	$a = 3a_{max}/4+1, a_{max}$	
2	$a = a_{max}/4+1, 3a_{max}/4$	$b = 1, b_{max}$
3	$a = 1, a_{max}/2$	$b = 1, b_{max}$
4	$a = a_{max}/2+1, a_{max}$	$b = 1, b_{max}$
5	$a = 1, a_{max}$	$b = 1, b_{max}/2$
6	$a = 1, a_{max}$	$b = b_{max}/2+1, b_{max}$

Table 25.5. Marching strategy for computation of the outlet radiation at the cell walls required for definition of boundary conditions

Wall Nr.	Marching strategy, outlet a-direction	b-direction
1 (e)	$a = a_{max}/4+1, 3a_{max}/4$	$b = 1, b_{max}$
2 (w)	$a = 1, a_{max}/4$	$b = 1, b_{max}$
	$a = 3a_{max}/4+1, a_{max}$	
3 (n)	$a = a_{max}/2+1, a_{max}$	$b = 1, b_{max}$
4 (s)	$a = 1, a_{max}/2$	$b = 1, b_{max}$
5 (u)	$a = 1, a_{max}$	$b = b_{max}/2+1, b_{max}$
6 (d)	$a = 1, a_{max}$	$b = 1, b_{max}/2$

For mirroring the incident directions required are $a = a_{max} - a + 1$, $b = b_{max} - b + 1$

25.3 Material properties

25.3.1 Source terms – emission from hot surfaces with known temperature

The black body intensity in any direction is computed by integrating *Planck*' s spectral distribution emissive power within the prescribed boundaries as follows

$$i'_{1b} = \frac{1}{\pi} \int_{\lambda_1}^{\lambda_2} \frac{2\pi C_1}{\lambda^5 \left(e^{\frac{C_2}{\lambda T}} - 1 \right)} d\lambda , \qquad (25.62)$$

where

$$C_1 = 0.59552197 \times 10^{-16} \, m^2 / sr , \qquad (25.63)$$

$$C_2 = 0.014338769 \times 10^{-16} \, mK . \qquad (25.64)$$

Here index b stands for black body. We use three bands for the source terms. The boundaries are selected as follows

$$\lambda_1 = 7 \times 10^{-7} \, m, \; \lambda_2 = 1.2 \times 10^{-6} \, m , \qquad (25.65,66)$$

for the first band,

$$\lambda_1 = 1.2 \times 10^{-6} \, m, \; \lambda_2 = 1.8 \times 10^{-6} \, m , \qquad (25.67,68)$$

for the second band, and

$$\lambda_1 = 1.8 \times 10^{-6} \, m, \; \lambda_2 = 20 \times 10^{-6} \, m , \qquad (25.69,70)$$

for the third band. Other choices are of course possible. Thus the source term due to radiating melt particles assuming

$$\Phi(\lambda, \omega, \omega_i) = 1 \qquad (25.71)$$

is

$$S' = f_{r,3} a_{3\lambda} i'_b(T_3) + \frac{\sigma_{3s\lambda}}{4\pi} \sum_{a=1}^{a_{\max}} \sum_{b=1}^{b_{\max}} i'_{\lambda,a,b} = f_{r,3} a_{3\lambda} i'_b(T_3) + \frac{\sigma_{3s\lambda}}{4\pi} i'_{\lambda,total} . \qquad (25.72)$$

25.3.2 Spectral absorption coefficient of water

The experimentally measured index of refraction, n, and the index of absorption, k, as a function of the incident wave length are taken from *Hale* and *Marvin* [5] (1973) and *Zolotarev* and *Dyomin* [12] (1977). The data of *Hale* and *Marvin* are used for wavelengths in the region of 0.7 to 7 μm and the data of *Zolotarev* and *Dyomin* for the region of 7 to 17 μm. The spectral absorption coefficient is then computed as follows

$$a_\lambda(\lambda) = \frac{4\pi k}{\lambda}. \tag{25.73}$$

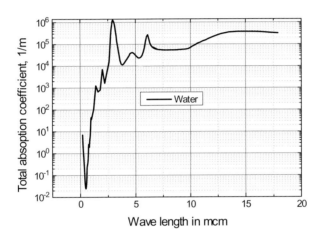

Fig. 25.3. Spectral absoption coefficient for water as a function of the wavelength

The so obtained absorption coefficient is presented in Fig. 25.3. The argumentation that the spectral properties of water are not strong functions of water temperature and pressure summarized by *Fletcher* [3] is acceptable. The total absorption coefficient is then computed as follows

$$a'(T_3) = \frac{\int_{\lambda_{\min}}^{\lambda_{\max}} i'_{1b}(T_3,\lambda)\left[1-e^{-a_\lambda(\lambda)s}\right]d\lambda}{\int_{\lambda_{\min}}^{\lambda_{\max}} i'_{1b}(T_3,\lambda)d\lambda}. \tag{25.74}$$

We use for the limits the following values $\lambda_{\min} = 0.7\mu m$ and $\lambda_{\max} = 33\mu m$ and perform the integration numerically using 1000 steps. This is a very time consuming procedure. To avoid computation cost *Fletcher* [4] tabulated the results obtained in similar manner for three different radiation surface temperatures, namely 1000, 2500 and 3500 K as a function of the water thickness s. *Vaeth* [11] approximated the absorption coefficients obtained by *Fletcher* [4] for three different radiation surface temperature, namely 1000, 2500 and 3500 K by an analytical function. We compared *Vaeth*'s approximation with the analytically obtained coefficients and found that the coefficients for 3500 K are with about 30% lower and for 2500 K about 10% lower then those computed analytically. The modified *Vaeth* approximation is then recommended here

$$a(T_3,s) = a(T_3 = 1000K, s)\frac{1}{15}\left[32 + 4\frac{T_3}{1000}\left(-6 + \frac{T_3}{1000}\right)\right]$$

$$+ a(T_3 = 2500K, s)\frac{1}{3}\left[-7 + \frac{T_3}{1000}\left(9 - 2\frac{T_3}{1000}\right)\right]$$

$$+ a(T_3 = 3500K, s)\frac{1}{5}\left[5 + \frac{T_3}{1000}\left(-7 + 2\frac{T_3}{1000}\right)\right], \qquad (25.75)$$

where

$$a(T_3 = 1000K, s) = 1.395086 \ln\left(s^{0.01374877} + 1\right), \qquad (25.76)$$

$$a(T_3 = 2500K, s) = 1.1 \times 0.8641743 \ln\left(s^{0.1427472} + 1\right), \qquad (25.77)$$

$$a(T_3 = 3500K, s) = 1.3 \times 0.4948444 \ln\left(s^{0.2694667} + 1\right). \qquad (25.78)$$

Note that this correlation cannot be extrapolated for water thickness less than 1 mm because of the unacceptable error increase - see Fig. 25.5. Values for this region are given in Fig. 25.4. We see that for water thickness between 1 and 30 μm the error of the modified *Vaeth* approximation is less than 8%.

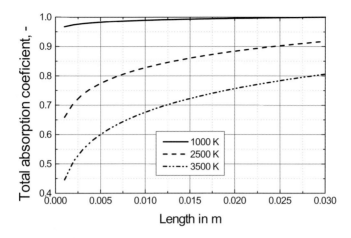

Fig. 25.4. Analytically computed total absorption coefficient as a function of the water layer thickness. Parameter – temperature of the radiation sources

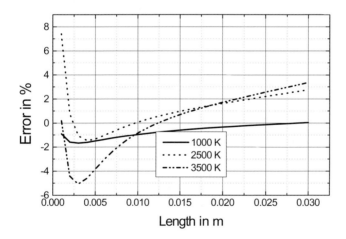

Fig. 25.5. Error of the modified Vaeth approximation compared to the analytically computed total absorption coefficient as a function of the water layer thickness. Parameter – temperature of the radiation sources

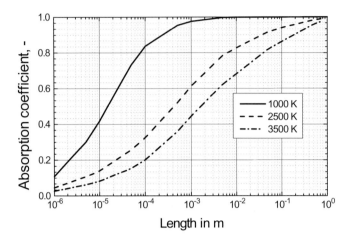

Fig. 25.6. Analytically computed total absorption coefficient as a function of the water layer thickness. Parameter – temperature of the radiation sources

Partitioning of the absorbed energy between the bulk and the surface: Now let us estimate the band absorption coefficients for three different bands as presented in Figs. 25.7 a) through c). The figures demonstrated the strong heterogeneity of the absorption depending on the wavelength. Thermal radiation in the first band can cover much longer distance without being absorbed than the radiation in the second and in the third band. In band no.2 the temperature dependence is not as strong as in the other two bands.

The problem of how much radiation energy is deposited at the surface and immediately transferred into evaporation mass is not solved up to now and has to be attacked in the future. Fig. 25.7 d) demonstrates that thermal radiation with wavelengths larger 3 μm are completely absorbed within a 0.5 *mm* water layer. *Vaeth* [11] proposed to consider all this energy deposited onto the surface which is obviously true only for lengths larger then 0.5 *mm*. The *Vaeth* approximation of the part of the energy deposited onto the surface is

$$f_{surface} = \min\left(1, \frac{\left(\dfrac{T_3}{1000}\right)}{\left(\dfrac{T_3}{1000}\right)^2 - 0.4611115\left(\dfrac{T_3}{1000}\right) + 0.8366274}\right). \quad (25.79)$$

If we arbitrary take all the deposit energy in 0.01 *mm* as surface energy the absorbed fractions of the incident energy analytically computed are 0.417, 0.141 and 0.0815 for the temperatures 1000, 2500 and 3500 *K*, respectively.

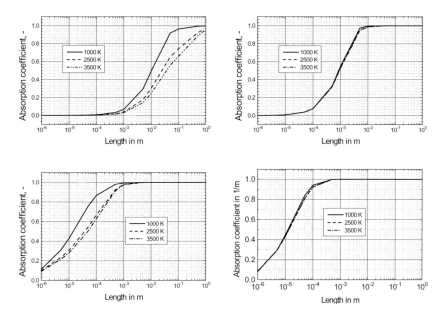

Fig. 25.7. Analytically computed partial absorption coefficient as a function of the water layer thickness. Parameter – temperature of the radiation sources. a) Band no.1 7×10^{-7} to 1.2×10^{-6} m; b) Band no.2 1.2×10^{-6} to 1.8×10^{-6} m; c) Band no.3 1.8×10^{-6} to 20×10^{-6} m; d) Infrared 3×10^{-6} to 200×10^{-6} m.

25.3.3 Spectral absorption coefficient of water vapor and other gases

Information about the total emissivity and absorptivity for carbon dioxide, water vapor and their mixtures is provided by *Steward* and *Kocaefe* [10]. Information about the gas emissivity for wide range of process conditions is provided by *Docherty* [1].

25.4 Averaged properties for some particular cases occurring in melt-water interaction

During interaction of molten material with water some specific geometrical arrangement called flow pattern may happen. For some of them we provide the averaged properties across a control volume with finite size. We will consider three different cases as examples starting with the simplest and ending with a more complicated one:

1) Gas sphere inside a molten material.
2) Concentric spheres of water droplets, surrounded by vapor, surrounded by molten material.
3) Spherical particles of hot radiating material surrounded by a layer of vapor surrounded by water.

The first two cases are typical examples for local interaction. Whatever happens with the radiation transport it happens within the cavity. The third one is a typical example for interactions across the faces of the control volumes. It is the idealized film boiling case.

25.4.1 Spherical cavity of gas inside a molten material

Consider the case presented in Fig. 25.8.

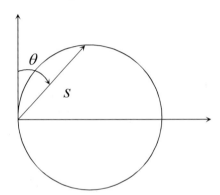

Fig. 25. 8. Spherical cavity of gas inside a molten material

The gas cavity is surrounded completely by melt. Given is the diameter of the cavity D_1, the temperatures of the melt T_3 and of the gas T_1 and their radiation properties. The task is to estimate how much energy per unit time Q_A is transferred from the melt into the gas.

We select an arbitrary point at the interface and draw one axis going through the point an the center of the sphere and one vertical axis perpendicular to it. With respect to the vertical axis we define the angle θ. All possible beam directions in the horizontal cross section are defined within the angle interval: $\theta_1 = 0$, $\theta_2 = \frac{\pi}{2}$. The beam length s is then depending only on the angle $s = D_1 \sin \theta$. The averaged beam length is then defined as follows

25.4 Averaged properties for some particular cases occurring in melt-water interaction

$$\overline{s} = \frac{\int_0^{2\pi} \int_{\theta_1}^{\theta_2} s_1 \cos\theta \sin\theta \, d\theta \, d\varphi}{\frac{1}{2}(\sin^2\theta_2 - \sin^2\theta_1)} = 2\int_0^{\frac{\pi}{2}} s\cos\theta \sin\theta \, d\theta = \frac{2}{3} D_1 . \qquad (25.80)$$

The integration over the angle φ is in the boundaries 0 and 2π because we have axis symmetry. Note in this case that $\sin^2\theta_2 - \sin^2\theta_1 = 1$. With this averaged beam length we compute the $\alpha'_1(\overline{s})$ spectral-averaged absorption coefficient and the spectral-averaged emissivity coefficient of the vapor $\varepsilon'_1(\overline{s})$. The transferred power through the vapor sphere is therefore

$$Q = \pi D_1^2 \sigma \left\{ [1 - \alpha'_1(\overline{s})] \varepsilon_3 T_3^4 + \varepsilon'_1(\overline{s}) T_1^4 \right\} . \qquad (25.81)$$

The absorbed power inside the sphere is

$$Q_A = \pi D_1^2 \sigma \left[\alpha'_1(\overline{s}) \varepsilon_3 T_3^4 - \varepsilon'_1(\overline{s}) T_1^4 \right] . \qquad (25.82)$$

25.4.2 Concentric spheres of water droplets, surrounded by vapor, surrounded by molten material

Consider a continuous melt with water and vapor enclosed inside the melt as presented in Fig.25.9.

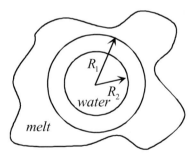

Fig. 25.9. Sphere of water droplet, surrounded by vapor, surrounded by molten material

The volume fraction of the vapor α_1, and of the water α_2 are known. The water diameter is known. We assume that the vapor is surrounding the water concentrically. The vapor diameter is then

$$D_1 = D_2 \sqrt[3]{1 + \frac{\alpha_1}{\alpha_2}}, \quad \alpha_2 > 0. \tag{25.83}$$

The melt radiates through the vapor and through the droplet. It is to be computed how much power radiated from the melt is absorbed by the vapor and how much is absorbed by the water droplet. Again we select an arbitrary point at the melt interface and draw an axis going through the point to the center of the sphere and one vertical axis perpendicular to it. With respect to the vertical axis we define the angle θ. All possible beam directions in the horizontal cross section are defined within the angle interval: $\theta_1 = 0$, $\theta_2 = \frac{\pi}{2}$ but in contrast with the previous case there are two characteristic zones. In the first the beam crosses gas only and in the second gas, water, gas successively. We perform the integration within these two zones separately.

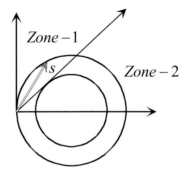

Fig. 25.10. Zone 1 - beam crosses vapor only, zone 2 - beam crosses vapor-liquid-vapor

First zone - vapor only: This zone is defined within the angles

$$\theta_1 = 0, \quad \theta_2 = \frac{\pi}{2} - \arcsin\left(\frac{D_2}{D_1}\right), \quad \sin^2\theta_2 - \sin^2\theta_1 = 1 - \left(\frac{D_2}{D_1}\right)^2.$$

The beam length inside the zone is

$$s_{v,1.zone} = D_1 \sin\theta. \tag{25.84}$$

The averaged beam length is

$$\overline{s}_{v,1.zone} = \frac{\int_0^{2\pi}\int_{\theta_1}^{\theta_2} s_1 \cos\theta \sin\theta \, d\theta \, d\varphi}{\frac{1}{2}\left(\sin^2\theta_2 - \sin^2\theta_1\right)}$$

25.4 Averaged properties for some particular cases occurring in melt-water interaction

$$= 2 \frac{\int_0^{\frac{\pi}{2}-\arcsin\left(\frac{D_2}{D_1}\right)} s_{2,1.zone} \cos\theta \sin\theta d\theta}{1-\left(\frac{D_2}{D_1}\right)^2} = \frac{2}{3} D_1 \left[1-\left(\frac{D_2}{D_1}\right)^2\right]^{1/2}. \quad (25.85)$$

The transferred power through the vapor is

$$Q = \pi D_1^2 \sigma \left\{[1-\alpha'_v(\overline{s}_{v,1.zone})]\varepsilon_3 T_3^4 + \varepsilon'_v(\overline{s}_{v,1.zone})T_1^4\right\}\left(1-\frac{D_2^2}{D_1^2}\right). \quad (25.86)$$

The absorbed power inside the vapor is

$$Q_A = \pi D_1^2 \sigma \left\{\alpha'_v(\overline{s}_{v,1.zone})\varepsilon_3 T_3^4 - \varepsilon'_v(\overline{s}_{v,1.zone})T_1^4\right\}\left(1-\frac{D_2^2}{D_1^2}\right). \quad (25.87)$$

Second zone – vapor/droplet: The second zone is defined within the interval

$$\theta_1 = \frac{\pi}{2}-\arcsin\left(\frac{D_2}{D_1}\right) \text{ and } \theta_2 = \frac{\pi}{2}, \; \sin^2\theta_2 - \sin^2\theta_1 = \left(\frac{D_2}{D_1}\right)^2.$$

Vapor: The beam length inside the vapor is

$$s_{v,2.zone} = \frac{1}{2} D_1 \left[\sin\theta - \left(\left(\frac{D_2}{D_1}\right)^2 - \cos^2\theta\right)^{1/2}\right]. \quad (25.88)$$

The averaged beam length results then in

$$\overline{s}_{v,2.zone} = \frac{\int_{\theta_1}^{\theta_2} s_{v,2.zone} \cos\theta \sin\theta d\theta}{\frac{1}{2}(\sin^2\theta_2 - \sin^2\theta_1)} = \frac{2\int_{\theta_1}^{\theta_2} s_{v,2.zone} \cos\theta \sin\theta d\theta}{\left(\frac{D_2}{D_1}\right)^2}.$$

$$= \frac{1}{3} D_1 \frac{1 - \left(\frac{D_2}{D_1}\right)^2 - \left[1 - \left(\frac{D_2}{D_1}\right)^2\right]^{3/2}}{\left(\frac{D_2}{D_1}\right)^2}. \qquad (25.89)$$

The transferred power through the vapor is

$$Q = \pi D_1^2 \sigma \left\{ [1 - \alpha_v'(\overline{s}_{v,2.zone})] \varepsilon_3 T_3^4 + \varepsilon_v'(\overline{s}_{v,2.zone}) T_1^4 \right\} \left(\frac{D_2}{D_1}\right)^2. \qquad (25.90)$$

The absorbed power inside the vapor is

$$Q_A = \pi D_1^2 \sigma \left\{ \alpha_v'(\overline{s}_{v,2.zone}) \varepsilon_3 T_3^4 - \varepsilon_v'(\overline{s}_{v,2.zone}) T_1^4 \right\} \left(\frac{D_2}{D_1}\right)^2. \qquad (25.91)$$

Water: The beam length inside the water is

$$\Delta s_{w,2.zone} = D_1 \left(\left(\frac{D_2}{D_1}\right)^2 - \cos^2 \theta \right)^{1/2}. \qquad (25.92)$$

The averaged beam length is then

$$\Delta \overline{s}_{w,2.zone} = 2 \frac{\int_{\theta_1}^{\theta_2} s_{w,2.zone} \cos\theta \sin\theta d\theta}{\left(\frac{D_2}{D_1}\right)^2} = \frac{2}{3} D_2. \qquad (25.93)$$

The transferred power through the water droplet is

$$Q = \pi D_1^2 \sigma \left\{ \begin{array}{l} [1 - \alpha_w'(\overline{s}_{w,2.zone})][1 - \alpha_v'(\overline{s}_{v,2.zone})] \varepsilon_3 T_3^4 \\ + [1 - \alpha_w'(\overline{s}_{w,2.zone})] \varepsilon_v'(\overline{s}_{v,2.zone}) T_1^4 \\ + \varepsilon_w'(\overline{s}_{w,2.zone}) T_2^4 \end{array} \right\} \left(\frac{D_2}{D_1}\right)^2. \qquad (25.94)$$

25.4 Averaged properties for some particular cases occurring in melt-water interaction

The absorbed power by the water droplet is

$$Q_{Aw} = \pi D_1^2 \sigma \begin{Bmatrix} \alpha_w'(\overline{s}_{w,2.zone})[1-\alpha_v'(\overline{s}_{v,2.zone})]\varepsilon_3 T_3^4 \\ +\alpha_w'(\overline{s}_{w,2.zone})\varepsilon_v'(\overline{s}_{v,2.zone})T_1^4 - \varepsilon_w'(\overline{s}_{w,2.zone})T_2^4 \end{Bmatrix} \left(\frac{D_2}{D_1}\right)^2. \quad (25.95)$$

The absorbed total power by the vapor is the sum of the absorbed power in the both vapor regions

$$Q_{Av} = \pi D_1^2 \sigma \begin{bmatrix} \{\alpha_v'(\overline{s}_{v,1.zone})\varepsilon_3 T_3^4 - \varepsilon_v'(\overline{s}_{v,1.zone})T_1^4\}\left(1-\frac{D_2^2}{D_1^2}\right) \\ +\{\alpha_v'(\overline{s}_{v,2.zone})\varepsilon_3 T_3^4 - \varepsilon_v'(\overline{s}_{v,2.zone})T_1^4\}\left(\frac{D_2}{D_1}\right)^2 \end{bmatrix}. \quad (25.96)$$

25.4.3 Clouds of spherical particles of radiating material surrounded by a layer of vapor surrounded by water –Lanzenberger's solution

Consider a cloud of equidistantly arranged hot spherical particles. The cloud is flying in a water and gas mixture. Compute the power radiated by a hot spherical particle and its environment. Compute what part of this energy is absorbed by water and what part is absorbed in the gas as a function of the characteristic scales inside the three-phase mixture. Compute the length required for almost complete length absorption.

The task can formally be described in the following way: Consider an artificially defined sphere with diameter D_2 containing concentrically two other spheres as presented in Fig. 25.11. This resembles film boiling of hot particles in water. We use this structure as an abstract model for computing the radiation properties of the mixture consisting of particles, vapor and water. This elementary cell is called sometimes a *Wigner* cell. It contains as much vapor and water as the total vapor and water in the computational cell divided by the number of the particles inside the cell, respectively. We will first compute what happens inside the *Wigner* cell and how much radiation energy is leaving the cell. Then we will ask the question what happens with the radiation energy in consequently interacting *Wigner* cells.

The inner most sphere, having diameter D_3, is radiating. We call this sphere a "particle". The space between the first and the second spherical surfaces is filled with vapor. The second sphere has diameter D_1. The space between the second and the external sphere (*Wigner* cell) is filled with water. Now consider a space filled equidistantly with particles having volume fraction $\alpha_3 > 0$ surrounded by vapor with volume fraction $\alpha_1 > 0$ and water having volume fraction $\alpha_2 > 0$.

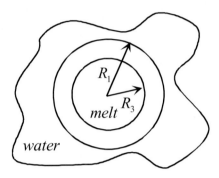

Fig. 25.11. Radiating sphere concentrically surrounded by vapor and water

For homogeneously distributed particles we have

$$D_1 = D_2 \sqrt[3]{\alpha_1 + \alpha_3} = D_3 \sqrt[3]{1 + \frac{\alpha_1}{\alpha_3}}, \quad \alpha_3 > 0, \tag{25.97}$$

$$D_2 = \frac{D_3}{\sqrt[3]{\alpha_3}}, \quad \alpha_3 > 0. \tag{25.98}$$

Frequently used ratios in the following text are

$$\frac{D_1}{D_2} = (\alpha_1 + \alpha_3)^{1/3}, \tag{25.99}$$

$$\frac{D_3}{D_2} = \alpha_3^{1/3}. \tag{25.100}$$

In this section we will present *Lanzenberger's* results reported in [7] simplifying in some places the resulting integrals for convenient use.

25.4.3.1 Space-averaged beam lengths for the concentric spheres radiation problem

The averaging procedure for computing the averaged beam length results from the radiation theory is

$$\overline{s} = \frac{\int_0^{2\pi} \int_{\theta_1}^{\theta_2} s(\varphi,\theta)\cos\theta \sin\theta \, d\theta \, d\varphi}{\int_0^{2\pi} \int_{\theta_1}^{\theta_2} \cos\theta \sin\theta \, d\theta \, d\varphi} = \frac{\int_{\theta_1}^{\theta_2} s(\theta)\cos\theta \sin\theta \, d\theta}{\frac{1}{2}\left(\sin^2\theta_2 - \sin^2\theta_1\right)}. \qquad (25.101)$$

It gives the averaged beam length over the angles $0 < \varphi < 2\pi$ and $\theta_1 < \theta < \theta_2$. The formal procedure contains (a) specification of θ_1 and θ_2, (b) derivation of the relationship $s_1(\varphi,\theta)$ and (c) finally integration in order to obtain $\overline{s}$. Two specific geometries presented in Fig. 25.2 and Fig. 25.3 represent all cases. The results of the computations are presented below:

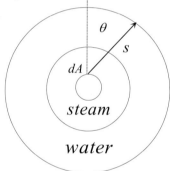

Fig. 25.12. Radiating hot sphere through steam and water. The first Wigner cell.

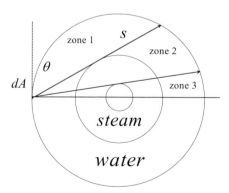

Fig. 25. 13. Radiation entering the second Wigner cell from outside sources

25.4.3.1.1 Beam emitted from the particle surface

Angle interval: $\theta_1 = 0$ and $\theta_2 = \pi/2$, $\sin^2\theta_2 - \sin^2\theta_1 = 1$.

Vapor: $0 < s < s_1$

Beam length: $s_1 = \dfrac{1}{2}D_3 \left[\left(\left(\dfrac{D_1}{D_3}\right)^2 - \sin^2\theta \right)^{1/2} - \cos\theta \right]$. (25.102)

Averaged beam length:

$$\overline{s_1} = 2\int_0^{\pi/2} s_1 \cos\theta \sin\theta\, d\theta = \dfrac{1}{3}D_3 \left[\left(\dfrac{D_1}{D_3}\right)^3 - \left(\left(\dfrac{D_1}{D_3}\right)^2 - 1\right)^{3/2} - 1 \right].$$ (25.103)

Water: $s_1 < s < s_2$

Beam length:

$$s_2 - s_1 = \dfrac{1}{2}D_3 \left[\left(\left(\dfrac{D_2}{D_3}\right)^2 - \sin^2\theta \right)^{1/2} - \left(\left(\dfrac{D_1}{D_3}\right)^2 - \sin^2\theta \right)^{1/2} \right].$$ (25.104)

Averaged beam length:

$$\Delta\overline{s_2} = 2\int_0^{\pi/2} (s_2 - s_1)\cos\theta \sin\theta\, d\theta$$

$$= \dfrac{1}{3}D_3 \left\{ \left(\dfrac{D_2}{D_3}\right)^3 - \left(\dfrac{D_1}{D_3}\right)^3 + \left(\left(\dfrac{D_1}{D_3}\right)^2 - 1\right)^{3/2} - \left(\left(\dfrac{D_2}{D_3}\right)^2 - 1\right)^{3/2} \right\}.$$ (25.105)

25.4.3.1.2 Beam entering the external sphere (*Wigner* cell) surface

First zone - water only:

Angle interval: $\theta_1 = 0$, $\theta_2 = \dfrac{\pi}{2} - \arcsin\left(\dfrac{D_1}{D_2}\right)$, $\sin^2\theta_2 - \sin^2\theta_1 = 1 - \left(\dfrac{D_1}{D_2}\right)^2$.

Beam length: $s_{2,1.zone} = D_2 \sin\theta$. (25.106)

25.4 Averaged properties for some particular cases occurring in melt-water interaction

Averaged beam length:

$$\overline{s}_{2,1.zone} = \frac{\int_0^{2\pi}\int_{\theta_1}^{\theta_2} s_1 \cos\theta \sin\theta \, d\theta \, d\varphi}{\frac{1}{2}\left(\sin^2\theta_2 - \sin^2\theta_1\right)} =$$

$$= 2\frac{\int_0^{\frac{\pi}{2} - \arcsin\left(\frac{D_1}{D_2}\right)} s_{2,1.zone} \cos\theta \sin\theta \, d\theta}{1 - \left(\frac{D_1}{D_2}\right)^2} = \frac{2}{3} D_2 \left[1 - \left(\frac{D_1}{D_2}\right)^2\right]^{1/2}. \qquad (25.107)$$

Second zone:

Angle interval: $\theta_1 = \frac{\pi}{2} - \arcsin\left(\frac{D_1}{D_2}\right)$ and $\theta_2 = \frac{\pi}{2} - \arcsin\left(\frac{D_3}{D_2}\right)$,

$$\sin^2\theta_2 - \sin^2\theta_1 = \left(\frac{D_1}{D_2}\right)^2 - \left(\frac{D_3}{D_2}\right)^2.$$

Region 1 – water:

Beam length: $s_{21,2.zone} = \frac{1}{2} D_2 \left[\sin\theta - \left(\left(\frac{D_1}{D_2}\right)^2 - \cos^2\theta\right)^{1/2}\right]. \qquad (25.108)$

Averaged beam length:

$$\overline{s}_{21,2.zone} = \frac{\int_{\theta_1}^{\theta_2} s_{21,2.zone} \cos\theta \sin\theta \, d\theta}{\frac{1}{2}\left(\sin^2\theta_2 - \sin^2\theta_1\right)} = \frac{2\int_{\theta_1}^{\theta_2} s_{21,2.zone} \cos\theta \sin\theta \, d\theta}{\left(\frac{D_1}{D_2}\right)^2 - \left(\frac{D_3}{D_2}\right)^2}$$

$$= \frac{1}{3} D_2 \frac{\left(1-\left(\frac{D_3}{D_2}\right)^2\right)^{3/2} - \left[\left(\frac{D_1}{D_2}\right)^2 - \left(\frac{D_3}{D_2}\right)^2\right]^{3/2} - \left(1-\left(\frac{D_1}{D_2}\right)^2\right)^{3/2}}{\left(\frac{D_1}{D_2}\right)^2 - \left(\frac{D_3}{D_2}\right)^2}. \quad (25.109)$$

Region 2 – vapor:

Beam length: $\Delta s_{1,2.zone} = D_2 \left(\left(\frac{D_1}{D_2}\right)^2 - \cos^2\theta\right)^{1/2}.$ (25.110)

Averaged beam length:

$$\Delta \overline{s}_{1,2.zone} = 2 \frac{\int_{\theta_1}^{\theta_2} s_{21,2.zone} \cos\theta \sin\theta d\theta}{\left(\frac{D_1}{D_2}\right)^2 - \left(\frac{D_3}{D_2}\right)^2} = \frac{2}{3} D_2 \left[\left(\frac{D_1}{D_2}\right)^2 - \left(\frac{D_3}{D_2}\right)^2\right]^{1/2}.$$

(25.111)

Region 3 – water:

Beam length: The same as in region 1.

Averaged beam length: $\Delta \overline{s}_{22,2.zone} = \overline{s}_{21,2.zone}.$ (25.112)

Third zone:

Angle interval: $\theta_1 = \frac{\pi}{2} - \arcsin\left(\frac{D_3}{D_2}\right)$ and $\theta_2 = \frac{\pi}{2}$, $\sin^2\theta_2 - \sin^2\theta_1 = \left(\frac{D_3}{D_2}\right)^2.$

Region 1 –water only:

Beam length: $s_{2,3.zone} = \frac{1}{2} D_2 \left[\sin\theta - \left(\left(\frac{D_1}{D_2}\right)^2 - \cos^2\theta\right)^{1/2}\right].$ (25.113)

Averaged beam length:

25.4 Averaged properties for some particular cases occurring in melt-water interaction

$$\overline{s}_{2,3.zone} = \frac{\int_{\theta_1}^{\theta_2} s_{2,3.zone} \cos\theta \sin\theta d\theta d\varphi}{\frac{1}{2}\left(\sin^2\theta_2 - \sin^2\theta_1\right)} = \frac{2\int_{\theta_1}^{\theta_2} s_{2,3.zone} \cos\theta \sin\theta d\theta d\varphi}{\left(\frac{D_3}{D_2}\right)^2}$$

$$= \frac{1}{3}D_2 \frac{1-\left(\frac{D_1}{D_2}\right)^3 + \left[\left(\frac{D_1}{D_2}\right)^2 - \left(\frac{D_3}{D_2}\right)^2\right]^{3/2} - \left(1-\left(\frac{D_3}{D_2}\right)^2\right)^{3/2}}{\left(\frac{D_3}{D_2}\right)^2}. \qquad (25.114)$$

Region 2 –vapor only:

Beam length: $s_{1,3.zone} = \frac{1}{2}D_2 \left\{ \left[\left(\frac{D_1}{D_2}\right)^2 - \cos^2\theta\right]^{1/2} - \left[\left(\frac{D_3}{D_2}\right)^2 - \cos^2\theta\right]^{1/2} \right\}.$

$$(25.115)$$

Averaged beam length:

$$\overline{s}_{1,3.zone} = \frac{\int_{\theta_1}^{\theta_2} s_{1,3.zone} \cos\theta \sin\theta d\theta}{\frac{1}{2}\left(\sin^2\theta_2 - \sin^2\theta_1\right)} = \frac{2\int_{\theta_1}^{\theta_2} s_{1,3.zone} \cos\theta \sin\theta d\theta d\varphi}{\left(\frac{D_3}{D_2}\right)^2}$$

$$= \frac{1}{3}D_2 \frac{\left(\frac{D_1}{D_2}\right)^3 - \left(\frac{D_3}{D_2}\right)^3 - \left[\left(\frac{D_1}{D_2}\right)^2 - \left(\frac{D_3}{D_2}\right)^2\right]^{3/2}}{\left(\frac{D_3}{D_2}\right)^2}. \qquad (25.116)$$

25.4.3.2 Solution of the RTE

25.4.3.2.1 Directional spectral intensity inside the vapor

Along the beam $0 < s < s_1$ the analytical solution of the RTE equation without scattering gives the heat flux per unit solid angle, per unit wavelength and per unit area as a function of the beam length

$$i'_\lambda(\lambda, s, T_1, T_3) = i'_{3,\lambda}(\lambda, T_3) e^{-a_{1,\lambda}s} + i'_{1,\lambda}(\lambda, s, T_1)\left(1 - e^{-a_{1,\lambda}s}\right). \tag{25.117}$$

Across the infinitesimal cross section $dA\cos\theta$, for the infinitesimal solid angle $d\omega$, and within the wavelength region $d\lambda$ we have

$$d^3Q = i'_\lambda(\lambda, s, T_1, T_3) dA \cos\theta \, d\omega \, d\lambda$$

$$= \left[i'_\lambda(\lambda, 0, T_3) e^{-a_{1,\lambda}s} + i'_{\lambda b}(\lambda, T_1)\left(1 - e^{-a_{1,\lambda}s}\right)\right] dA \cos\theta \, d\omega \, d\lambda. \tag{25.118}$$

Integrated over all wavelengths of the emitting media we have

$$d^2Q = \left[\int_0^\infty i'_\lambda(\lambda, 0, T_3) e^{-a_{1,\lambda}s} d\lambda + \int_0^\infty i'_{\lambda b}(\lambda, T_1)\left(1 - e^{-a_{1,\lambda}s}\right) d\lambda\right] dA \cos\theta \, d\omega. \tag{25.119}$$

Estimation of the first integral: Having in mind the definition of the emissivity coefficient of the particle surface

$$\varepsilon_3(T_3) = \frac{\int_0^\infty i'_\lambda(\lambda, T_3) d\lambda}{\int_0^\infty i'_{\lambda b}(\lambda, T_3) d\lambda}, \tag{25.120}$$

and the definition of the vapor absorption coefficient of black body radiation with temperature T_3

$$a'_1(T_3, s) = \frac{\int_0^\infty i'_{\lambda b}(T_3, \lambda)\left[1 - e^{-a_{1,\lambda}s}\right] d\lambda}{\int_0^\infty i'_{\lambda b}(T_3, \lambda) d\lambda}$$

25.4 Averaged properties for some particular cases occurring in melt-water interaction 677

$$= 1 - \frac{\int_0^\infty i'_{\lambda b}(T_3,\lambda)e^{-a_{1,\lambda}s}d\lambda}{\int_0^\infty i'_{\lambda b}(T_3,\lambda)d\lambda} = 1 - \frac{\int_0^\infty i'_{\lambda b}(T_3,\lambda)e^{-a_{1,\lambda}s}d\lambda}{\frac{\sigma T_3^4}{\pi}} \quad (25.122)$$

we obtain

$$\int_0^\infty i'_\lambda(\lambda,0,T_3)e^{-a_{1,\lambda}s}d\lambda = \varepsilon_3(T_3)\int_0^\infty i'_{\lambda b}(\lambda,T_3)e^{-a_{1,\lambda}s}d\lambda$$

$$= \varepsilon_3(T_3)\left[1 - a'_1(T_3,s)\right]\frac{\sigma T_3^4}{\pi}. \quad (25.123)$$

Estimation of the second integral: For the second integral we use again the definition of the total emittance for vapor radiating with temperature T_1

$$\int_0^\infty i'_{\lambda b}(\lambda,T_1)\left(1 - e^{-a_{1,\lambda}s}\right)d\lambda = \varepsilon_1(T_1,s)\frac{\sigma T_1^4}{\pi}. \quad (25.124)$$

Thus, after the integration over the total wavelength region we obtain

$$d^2Q = \left[\varepsilon_3(T_3)\left[1 - a'_1(T_3,s)\right]\frac{\sigma T_3^4}{\pi} + \varepsilon_1(T_1,s)\frac{\sigma T_1^4}{\pi}\right]dA\cos\theta d\omega. \quad (25.125)$$

The first term gives the intensity at the point s coming from the particle with temperature T_3. The second term is the net contribution of the gas radiating with temperature T_1 at point s.

25.4.3.2.2 Directional spectral intensity inside the water

Along the beam $s_1 < s < s_2$ the analytical solution of the RTE equation without scattering gives the heat flux per unit solid angle, per unit wavelength and per unit area as a function of the beam length

$$i'_\lambda(\lambda,s,T_1,T_2,T_3) = i'_\lambda(\lambda,s_1)e^{-a_{2,\lambda}s} + i'_{2,\lambda}(\lambda,s,T_2)\left(1 - e^{-a_{2,\lambda}s}\right). \quad (25.126)$$

Across the infinitesimal cross section $dA\cos\theta$, for the infinitesimal solid angle $d\omega$, and within the wavelength region $d\lambda$ results in

$$d^3Q = i'_\lambda(\lambda, s, T_1, T_2, T_3) dA \cos\theta d\omega d\lambda$$

$$= \left[i'_\lambda(\lambda, s_1) e^{-a_{2,\lambda}s} + i'_{2,\lambda}(\lambda, s, T_2)\left(1 - e^{-a_{2,\lambda}s}\right) \right] dA \cos\theta d\omega d\lambda. \tag{25.127}$$

Integrated over all wavelengths of the emitting media results in

$$d^2Q = \left[\int_0^\infty i'_\lambda(\lambda, s_1) e^{-a_{2,\lambda}s} d\lambda + \int_0^\infty i'_{\lambda b}(\lambda, T_2)\left(1 - e^{-a_{2,\lambda}s}\right) d\lambda \right] dA \cos\theta d\omega.$$

Estimation of the first integral: The directional spectral intensity entering the water at point s_1 is

$$i'_\lambda(\lambda, s_1, T_1, T_3) = i'_{3,\lambda}(\lambda, T_3) e^{-a_{1,\lambda}s_1} + i'_{b\lambda}(\lambda, s_1, T_1)\left(1 - e^{-a_{1,\lambda}s_1}\right). \tag{25.128}$$

The total intensity at the same point is

$$i'(s_1, T_1, T_3) = \int_0^\infty i'_\lambda(\lambda, s_1) d\lambda = \varepsilon_3(T_3)\left[1 - a'_1(T_3, s_1)\right] \frac{\sigma T_3^4}{\pi} + \varepsilon_1(T_1, s_1) \frac{\sigma T_1^4}{\pi}. \tag{25.129}$$

Now note the special definition of the absorption coefficient in water required here

$$\bar{a}'_2(s, T_1, T_3) = \frac{\int_0^\infty i'_\lambda(\lambda, s_1)\left(1 - e^{-a_{2,\lambda}s}\right) d\lambda}{\int_0^\infty i'_\lambda(\lambda, s_1) d\lambda} \approx a'_2(s, T_3), \tag{25.130}$$

which can be replaced by the usual definition of the water absorption coefficient only as an *approximation*. Thus we obtain finally

$$\int_0^\infty i'_\lambda(\lambda, s_1) e^{-a_{2,\lambda}s} d\lambda$$

$$= \left[1 - a'_2(T_3, s)\right]\left\{ \varepsilon_3(T_3)\left[1 - a'_1(T_3, s_1)\right] \frac{\sigma T_3^4}{\pi} + \varepsilon_1(T_1, s_1) \frac{\sigma T_1^4}{\pi} \right\}. \tag{25.131}$$

25.4 Averaged properties for some particular cases occurring in melt-water interaction

Estimation of the second integral: For the second integral we use again the definition of the total emittance of the water radiating with temperature T_2

$$\int_0^\infty i'_{\lambda b}(\lambda, T_2)\left(1 - e^{-a_{2,\lambda}s}\right)d\lambda = \varepsilon_2(T_2, s)\frac{\sigma T_2^4}{\pi}. \tag{25.132}$$

Thus, after the integration over the total wavelength region we obtain

$$d^2Q = \left[1 - a'_2(T_3, s)\right]\left\{\begin{array}{l}\varepsilon_3(T_3)\left[1 - a'_1(T_3, s_1)\right]\dfrac{\sigma T_3^4}{\pi} \\ \\ +\varepsilon_1(T_1, s_1)\dfrac{\sigma T_1^4}{\pi}\end{array}\right\} + \varepsilon_2(T_2, s)\dfrac{\sigma T_2^4}{\pi}\right]dA\cos\theta d\omega. \tag{25.133}$$

The first term gives the intensity at the point s coming from the particle with temperature T_3 and crossing the vapor with temperature T_1. The second term is the net contribution of the water radiating with temperature T_2 at point s.

25.4.3.2.3 Total absorbed intensity

Vapor: The total absorbed directional intensity in the vapor is therefore that emitted by the particle minus that leaving the vapor sphere at s_1

$$d^2Q_{1a} = \left[\varepsilon_3(T_3)a'_1(T_3, s_1)\frac{\sigma T_3^4}{\pi} - \varepsilon_1(T_1, s_1)\frac{\sigma T_1^4}{\pi}\right]dA\cos\theta d\omega. \tag{25.134}$$

Water: The total absorbed directional intensity in the water is therefore that entering from the internal sphere minus that leaving at s_2.

$$d^2Q_{2a} = \left\{\begin{array}{l}\varepsilon_3(T_3)\left[1 - a'_1(T_3, s_1)\right]a'_2(T_3, s_2)\dfrac{\sigma T_3^4}{\pi} \\ \\ +a'_2(T_3, s_2)\varepsilon_1(T_1, s_1)\dfrac{\sigma T_1^4}{\pi} - \varepsilon_2(T_2, s_2)\dfrac{\sigma T_2^4}{\pi}\end{array}\right\}dA\cos\theta d\omega. \tag{25.135}$$

Integration over the solid angle

The integration over the solid angle gives an expression in which optical properties have to be averaged over the solid angle. For the case of the absorption coefficient it will result in

$$\int_0^{2\pi} \int_{\theta_1}^{\theta_2} a'(T,s)\cos\theta \sin\theta \, d\theta \, d\varphi \approx \bar{a}'(T) \int_0^{2\pi} \int_{\theta_1}^{\theta_2} \cos\theta \sin\theta \, d\theta \, d\varphi, \quad (25.136)$$

where

$$\bar{a}'(T) = \frac{\int_0^{2\pi} \int_{\theta_1}^{\theta_2} a'(T,s)\cos\theta \sin\theta \, d\theta \, d\varphi}{\int_0^{2\pi} \int_{\theta_1}^{\theta_2} \cos\theta \sin\theta \, d\theta \, d\varphi}. \quad (25.137)$$

The computation of the averaged absorption coefficient is very expensive. That is why the following approximation will be used. The averaged optical property is set as the optical property being a function of the averaged beam distance. We illustrate this in the following example.

$$\bar{a}'(T) \approx a'(T, \bar{s}), \quad (25.138)$$

where

$$\bar{s} = \frac{\int_0^{2\pi} \int_{\theta_1}^{\theta_2} s(\theta,\varphi)\cos\theta \sin\theta \, d\theta \, d\varphi}{\int_0^{2\pi} \int_{\theta_1}^{\theta_2} \cos\theta \sin\theta \, d\theta \, d\varphi}. \quad (25.139)$$

The appropriate averaged beam lengths are already given in Section 25.4.3.1.

Vapor:

Performing the averaging of the optical properties over the solid angle results in

$$\int_0^{2\pi} \int_{\theta_1}^{\theta_2} a_1'(T_3,s_1)\cos\theta \sin\theta \, d\theta \, d\varphi \approx a_1'(T_3,\bar{s}_1) \int_0^{2\pi} \int_{\theta_1}^{\theta_2} \cos\theta \sin\theta \, d\theta \, d\varphi$$

$$= a_1'(T_3,\bar{s}_1)\pi\left(\sin^2\theta_2 - \sin^2\theta_1\right) \quad (25.140)$$

25.4 Averaged properties for some particular cases occurring in melt-water interaction 681

$$\int_0^{2\pi}\int_{\theta_1}^{\theta_2} \varepsilon_1(T_1,s_1)\cos\theta\sin\theta d\theta d\varphi \approx \varepsilon_1(T_1,\overline{s}_1)\pi\left(\sin^2\theta_2 - \sin^2\theta_1\right). \quad (25.141)$$

In particular, integrating for $\theta_1 = 0$ and $\theta_1 = \pi/2$ we obtain

$$\int_0^{2\pi}\int_{\theta_1}^{\theta_2} a_1'(T_3,s_1)\cos\theta\sin\theta d\theta d\varphi \approx a_1'(T_3,\overline{s}_1)\pi, \quad (25.142)$$

$$\int_0^{2\pi}\int_{\theta_1}^{\theta_2} \varepsilon_1(T_3,s_1)\cos\theta\sin\theta d\theta d\varphi \approx \varepsilon_1(T_3,\overline{s}_1)\pi, \quad (25.143)$$

and consequently

$$d^2Q_{1a} = \left[\varepsilon_3(T_3)a_1'(T_3,\overline{s}_1)\sigma T_3^4 - \varepsilon_1(T_1,\overline{s}_1)\sigma T_1^4\right]dA. \quad (25.144)$$

Integrating over the surface of the particle results in

$$Q_{1a} = \pi D_3^2 \left[\varepsilon_3(T_3)a_1'(T_3,\overline{s}_1)\sigma T_3^4 - \varepsilon_1(T_1,\overline{s}_1)\sigma T_1^4\right]. \quad (25.145)$$

The emitted total power from the particle is

$$Q_{3\to\infty} = \pi D_3^2 \varepsilon_3(T_3)\sigma T_3^4. \quad (25.146)$$

The emitted power from the vapor interface into the water is

$$Q_{3\to\infty} - Q_{1a} = \pi D_3^2 \sigma\left\{\left[1 - a_1'(T_3,\overline{s}_1)\right]\varepsilon_3(T_3)T_3^4 + \varepsilon_1(T_1,\overline{s}_1)T_1^4\right\}. \quad (25.147)$$

Water:

Similarly integrating for $\theta_1 = 0$ and $\theta_1 = \pi/2$ we obtain for the averages over the solid angle of the water optical properties

$$\int_0^{2\pi}\int_{\theta_1}^{\theta_2} a_2'(T_3,s_2)\cos\theta\sin\theta d\theta d\varphi \approx a_2'(T_3,\Delta\overline{s}_2)\pi, \quad (25.148)$$

$$\int_0^{2\pi}\int_{\theta_1}^{\theta_2} \varepsilon_2(T_3,s_2)\cos\theta\sin\theta d\theta d\varphi \approx \varepsilon_2(T_3,\Delta\overline{s}_2)\pi, \quad (25.149)$$

and consequently

$$d^2Q_{2a} = \sigma \begin{Bmatrix} \varepsilon_3(T_3)\left[1-a_1'(T_3,\bar{s}_1)\right]a_2'(T_3,\Delta\bar{s}_2)T_3^4 \\ +a_2'(T_3,\Delta\bar{s}_2)\varepsilon_1(T_1,\bar{s}_1)T_1^4 - \varepsilon_2(T_2,\Delta\bar{s}_2)T_2^4 \end{Bmatrix} dA. \qquad (25.150)$$

Integrating over the surface of the particle results in

$$Q_{2a} = \pi D_3^2 \sigma \begin{Bmatrix} \varepsilon_3(T_3)\left[1-a_1'(T_3,\bar{s}_1)\right]a_2'(T_3,\Delta\bar{s}_2)T_3^4 \\ +a_2'(T_3,\Delta\bar{s}_2)\varepsilon_1(T_1,\bar{s}_1)T_1^4 - \varepsilon_2(T_2,\Delta\bar{s}_2)T_2^4 \end{Bmatrix}. \qquad (25.151)$$

The emitted power from the vapor sphere is

$$Q_{3\to\infty} - Q_{1a} = \pi D_3^2 \sigma \left\{\left[1-a_1'(T_3,\bar{s}_1)\right]\varepsilon_3(T_3)T_3^4 + \varepsilon_1(T_1,\bar{s}_1)T_1^4\right\}. \qquad (25.152)$$

The difference between this power and that absorbed by the water is the power radiated from the water surface into the environment of the first cell

$$Q_{3\to\infty} - Q_{1a} - Q_{2a}$$

$$= \pi D_3^2 \sigma \begin{Bmatrix} \left[1-a_2'(T_3,\Delta\bar{s}_2)\right]\left[1-a_1'(T_3,\bar{s}_1)\right]\varepsilon_3(T_3)T_3^4 \\ +\left[1-a_2'(T_3,\Delta\bar{s}_2)\right]\varepsilon_1(T_1,\bar{s}_1)T_1^4 + \varepsilon_2(T_2,\Delta\bar{s}_2)T_2^4 \end{Bmatrix}. \qquad (25.153)$$

Second cell

The power emitted by the first *Wigner* cell is entering the second one – Fig. 25.13 There are three characteristic space angle regions for computing the absorbed and the transmitted power in the second *Wigner* cell. We consider the three regions separately. The total directional intensity of the entering radiation is

$$\int_0^\infty i_\lambda'(\lambda,T_1)d\lambda = \frac{Q_{3\to\infty} - Q_{1a} - Q_{2a}}{\pi D_2^2 \pi}. \qquad (25.154)$$

First zone: Water only

$$i'_\lambda(\lambda,s) = i'_\lambda(\lambda)e^{-a_\lambda s} + i'_\lambda(\lambda,s)\left(1-e^{-a_\lambda s}\right) \qquad (25.155)$$

$$Q = \pi D_2^2 ([1-\alpha'_w(\overline{s}_{11})]\frac{Q_1}{\pi D_2^2} + \varepsilon'_w(\overline{s}_{11})\sigma T_2^4)\left(1-\frac{D_1^2}{D_2^2}\right), \qquad (25.156)$$

$$Q_A = \pi D_2^2 (\alpha'_w(\overline{s}_{11})\frac{Q_1}{\pi D_2^2} - \varepsilon'_w(\overline{s}_{11})\sigma T_2^4)\left(1-\frac{D_1^2}{D_2^2}\right). \qquad (25.157)$$

Second zone:

Region 1: water

$$Q = \pi D_2^2 ([1-\alpha'_w(\overline{s}_{21})]\frac{Q_1}{\pi D_2^2} + \varepsilon'_w(\overline{s}_{21})\sigma T_2^4)\left(\frac{D_1^2}{D_2^2} - \frac{D_3^2}{D_2^2}\right) \qquad (25.158)$$

$$Q_A = \pi D_2^2 (\alpha'_w(\overline{s}_{21})\frac{Q_1}{\pi D_2^2} - \varepsilon'_w(\overline{s}_{21})\sigma T_2^4)\left(\frac{D_1^2}{D_2^2} - \frac{D_3^2}{D_2^2}\right). \qquad (25.159)$$

Region 2: vapor

$$Q = \pi D_2^2 ([1-\alpha'_v(\overline{s}_{22})] [1-\alpha'_w(\overline{s}_{21})]\frac{Q_1}{\pi D_2^2}$$

$$+[1-\alpha'_v(\overline{s}_{22})]\,\varepsilon'_w(\overline{s}_{21})\,\sigma T_2^4 + \varepsilon'_v(\overline{s}_{22})\,\sigma T_1^4\,)\left(\frac{D_1^2}{D_2^2} - \frac{D_3^2}{D_2^2}\right) \qquad (25.160)$$

$$Q_A = \pi D_2^2 (\alpha'_v(\overline{s}_{22})[1-\alpha'_w(\overline{s}_{21})]\frac{Q_1}{\pi D_2^2}$$

$$+\alpha'_v(\overline{s}_{22})\,\varepsilon'_w(\overline{s}_{21})\,\sigma T_2^4 - \varepsilon'_v(\overline{s}_{22})\,\sigma T_1^4\,)\left(\frac{D_1^2}{D_2^2} - \frac{D_3^2}{D_2^2}\right) \qquad (25.161)$$

Region 3: water

$$Q = \pi D_2^2 ([1-\alpha_w'(\bar{s}_{23})]Q^* + \varepsilon_w'(\bar{s}_{23})\sigma T_2^4) \left(\frac{D_1^2}{D_2^2} - \frac{D_3^2}{D_2^2} \right) \quad (25.162)$$

$$Q_A = \pi D_2^2 (\alpha_w'(\bar{s}_{23})Q^* - \varepsilon_w'(\bar{s}_{23})\sigma T_2^4) \left(\frac{D_1^2}{D_2^2} - \frac{D_3^2}{D_2^2} \right) \quad (25.163)$$

with

$$Q^* = [1-\alpha_v'(\bar{s}_{22})][1-\alpha_w'(\bar{s}_{21})]\frac{Q_1}{\pi D_2^2} + [1-\alpha_v'(\bar{s}_{22})]\varepsilon_w'(\bar{s}_{21})\sigma T_2^4 + \varepsilon_v'(\bar{s}_{22})\sigma T_1^4$$

$$(25.163)$$

Third zone:

Region 1: water

$$Q = \pi D_2^2 ([1-\alpha_w'(\bar{s}_{31})]\frac{Q_1}{\pi D_2^2} + \varepsilon_w'(\bar{s}_{31})\sigma T_2^4) \frac{D_3^2}{D_2^2}, \quad (25.164)$$

$$Q_A = \pi D_2^2 (\alpha_w'(\bar{s}_{31})\frac{Q_1}{\pi D_2^2} - \varepsilon_w'(\bar{s}_{31})\sigma T_2^4) \frac{D_3^2}{D_2^2}. \quad (25.165)$$

Region 2: vapor

$$Q = \pi D_2^2 ([1-\alpha_v'(\bar{s}_{32})][1-\alpha_w'(\bar{s}_{31})]\frac{Q_1}{\pi D_2^2}$$

$$+ [1-\alpha_v'(\bar{s}_{32})]\varepsilon_w'(\bar{s}_{31})\sigma T_2^4 + \varepsilon_v'(\bar{s}_{32})\sigma T_1^4) \frac{D_3^2}{D_2^2}, \quad (25.166)$$

$$Q_A = \pi D_2^2 (\alpha_v'(\bar{s}_{32})[1-\alpha_w'(\bar{s}_{31})]\frac{Q_1}{\pi D_2^2} + \alpha_v'(\bar{s}_{32})\varepsilon_w'(\bar{s}_{31})\sigma T_2^4 - \varepsilon_v'(\bar{s}_{32})\sigma T_1^4) \frac{D_3^2}{D_2^2}.$$

$$(25.167)$$

25.4.4 Chain of infinite number of *Wigner* cells

Knowing what happens inside a single cell we can now compute the energy that is transported $Q_{T,1}$ and that is absorbed by the vapor $Q_{av,1}$ and by the water $Q_{aw,1}$ if such cells are virtually interacting in a subsequent chain. The *Lanzenberger* solution is given below

$$Q_{T,i} = a_T Q_{T,i-1} + b_T, \quad i = 2, n, \tag{25.168}$$

$$Q_{av,i} = a_v Q_{T,i-1} + b_v, \quad i = 2, n, \tag{25.169}$$

$$Q_{aw,i} = a_w Q_{T,i-1} + b_w, \quad i = 2, n, \tag{25.170}$$

$$Q_{T,2} = a_T Q_{T,1} + b_T, \tag{25.171}$$

$$Q_{T,3} = a_T a_T Q_{T,1} + a_T b_T + b_T, \tag{25.172}$$

$$Q_{T,4} = a_T a_T a_T Q_{T,1} + a_T a_T b_T + a_T b_T + b_T, \tag{25.173}$$

$$Q_{T,i} = a_T^{i-1} Q_{T,1} + b_T \sum_{m=2}^{i} a_T^{m-2}, \tag{25.174}$$

$$Q_{av,i} = a_v Q_{T,i-1} + b_v, \quad i \geq 2, \tag{25.175}$$

$$Q_{av,2} = a_v Q_{T,1} + b_v, \tag{25.176}$$

$$Q_{av,i} = a_v a_T^{i-1} Q_{T,1} + a_v b_T \sum_{m=2}^{i} a_T^{i-2} + b_v, \quad i \geq 3, \tag{25.177}$$

$$Q_{av} = Q_{av,1} + Q_{av,2} + \sum_{i=3}^{n} Q_{av,i},$$

$$= Q_{av,1} + a_v Q_{T,1} + b_v + \sum_{i=3}^{n} a_v a_T^{i-1} Q_{T,1} + \sum_{i=3}^{n} a_v b_T \sum_{m=2}^{i} a_T^{i-2} + \sum_{i=3}^{n} b_v$$

$$= Q_{av,1} + (n-1)b_v + a_v \left[Q_{T,1}\left(1 + \sum_{i=3}^{n} a_T^{i-1}\right) + b_T \sum_{i=3}^{n}\sum_{m=2}^{i} a_T^{i-2} \right]. \qquad (25.178)$$

25.4.5 Application of *Lanzenbergers*'s solution

Consider a cloud of equidistantly arranged hot spherical particles. The cloud is flying in a water and gas mixture. Compute the radiated power by a hot spherical particle and its environment. Compute what part of this energy is absorbed by water and what in the gas as a function of the characteristic scales inside the three-phase mixture. Compute the length required for almost complete light absorption. Figure 25.15 presents the total absobtion depth in three-phase particle-vapor-water flow as a function of the particle volume fraction. The gas volume fraction is 1%. The hot particle size is $0.005m$. We see from Figs. 25.15 and 25.16 an important behaviour: increasing the particle concentration leads to reduction of the length neccesary for complete absorbtion of the radiation. That is why for cases with large particle concentrations the discrete ordinate method is not necessary and a local in-cell description of the radiated energy is sufficient. This is not the case for small particle concentrations where the absorbtion length can be considerably larger than the cell size. In this case doing so as every thing happens in the cell leads to an overestimation of the local energy transport and therefore to larger local vapor production with all the consequences to the final macroscopic results.

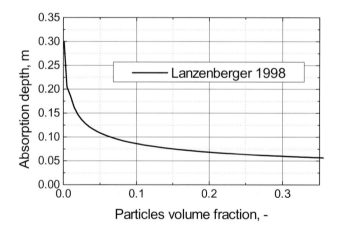

Fig. 25.14. Total absobtion depth in three-phase particle-vapor-water flow as a function of the particle volume fraction. Gas volume fraction 1%. Particle size $0.005m$

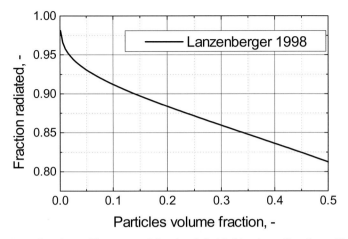

Fig. 25.15. Radiated specific power of the cloud divided by the radiated specific power of a single particle as a function of the partcles volume fraction

Nomenclature

Latin

A_m surface area at side m of the cell, m^2

a_{max} number of the discrete angle in the azimuthal direction

$a_\lambda(\lambda) = \dfrac{4\pi k}{\lambda}$, spectral absorption coefficient, $1/m$

$a_{a,b,m} = \dfrac{A_m}{\Delta Vol} \dfrac{D_{a,b,m}}{\Delta\omega_{a,b}}$, m-surface geometry coefficient, $1/m$

b_{max} number of the discrete angles in the polar direction

$c_0 = 2.998 \cdot 10^8$, speed of light in vacuum, m/s

c speed of light, m/s

$D_{a,b,m}$ $\iint\limits_{\Delta\omega_{a,b}} (\mathbf{s} \cdot \mathbf{e})_m \, d\omega$, scalar product of the (a,b)-directed unit vector and the outward directed m-surface unit vector integrated over the angle $\Delta\omega_{a,b}$, sr.

D_3 diameter, m

e	$=(e_x, e_y, e_z)$, outward directed unit normal to a surface
$i'_\lambda(\mathbf{r},\mathbf{s})$	spectral radiation intensity at the spatial location **r** along the direction **s** that is within an infinitesimal solid angle $d\omega$, in other words heat flux per unit solid angle, per unit wavelength, and per unit area, $W/(m^3 sr)$
k	index of absorption
n	index of refraction - ratio of propagation velocity c_0 in vacuum related to the propagation velocity c inside the medium
$\dot{q}''_{\lambda,m-incident}$	incident radiation heat flux from the inside of the computational cell, W/m^2
r	position vector, m
r	radius, m
s	direction vector
$T_{w,m}$	wall temperature at the m-face of the control volume
x	x-coordinate, m
y	y-coordinate, m
z	z-coordinate, m

Greek

Δ	finite difference
ΔVol	cell volume, m^3
β_λ	$=a_\lambda + \sigma_{s\lambda}$, spectral extinction coefficient, $1/m$
ε_w	wall emissivity coefficient
Φ	spectral scattering phase function
φ	azimuthal angle, rad
$\sigma_{s\lambda}$	spectral scattering coefficients, $1/m$
σ	$=\dfrac{2C_1\pi^5}{15C_2^4} = 5.67051\times 10^{-8}$, *Stefan-Boltzmann* constant, $W/(m^2 K^4)$
θ	polar angle, rad
ω	solid angle, rad
$d\omega$	infinitesimal solid angle, rad

Indices

a	integer index for the discrete angle in the azimuthal direction
b	integer index for the discrete angle in the polar direction
b	black body
w	wall

x	in x-coordinate
y	in y-coordinate
z	in z-coordinate
1	vapor, gas
2	water
3	melt, solid particle

References

1. Docherty P (6-10.9.1982) Prediction of gas emissivity for wide range of process conditions, Proc. of the 7th Heat Transfer Conference, Munich, Hemisphere Publ. Corp., Wash., vol 2 p 481-485
2. Fiveland WA (1988) Three-dimensional radiative heat transfer solutions by discrete-ordinates method, Journal of Thermophysics and Heat Transfer, vol 2 no 4 pp 309-316
3. Fletcher DF (1999) Radiation absorption during premixing, Nuclear Engineering and Design, vol 189 pp 435-440
4. Fletcher DF (1985) Assesment and development of the Bankoff and Han coarse mixing model, Culham Laboratory Report, CLM-R252
5. Hale GM and Marvin RQ (1973) Optical constants in water in the 200-nm to 200-µm wavelength region, Appl. Opt., vol 12 no 3 pp 555-563
6. Kolev NI (October 3-8, 1999) Verification of IVA5 computer code for melt-water interaction analysis, Part 1: Single phase flow, Part 2: Two-phase flow, three-phase flow with cold and hot solid spheres, Part 3: Three-phase flow with dynamic fragmentation and coalescence, Part 4: Three-phase flow with dynamic fragmentation and coalescence – alumna experiments, Proc of the Ninth International Topical Meeting on Nuclear Reactor Thermal Hydraulics (NURETH-9), San Francisco, California, , Log. Nr. 315
7. Lanzenberger K (1997) Thermal radiation in multiphase flow - Application to the severe accident scenario of molten fuel coolant interaction (MFCI), Siemens AG, Power Generation (KWU), 1997 Karl Wirtz Prize for yang scientists
8. Mathews KA, (1999) On the propagation of rays in discrete ordinates, Nuclear Science and Engineering, vol 132 pp 155-180
9. Siegel R and Howell JR (1996) Thermal radiation heat transfer, Third edition, Taylor& Francis, Washington
10. Steward FR and Kocaefe YS (17.-22.8.1986) Total emisivity and absoptivity for carbon dioxide, water vapor and their mixtures, Proc. 8th Heat Transfer Conf., San Francisco, Springer Berlin, vol 2 p 735-740
11. Vaeth L (Maerz 1995) Radiative heat transfer for the transient three-phase, three-component flow model IVA-KA, Forschungszentrum Karlsruhe, Internal report, INR-1914, PSF-3212
12. Zolotarev VM and Dyomin AV (1977) Optical constants of water on wide wavelength range 0.1 A to 1 m, Optics ans Spectroscopics, vol 43 no 2

Index

acceleration induced bubble fragmentation 226
accommodation coefficient 588, 590
activated nucleation site density 362
active nucleation site density 484
active nucleation site density as a function of superheat 441, 442, 443
active nucleation sites density 362, 366
Adiabatic flows 6
Altshul formula 72
aluminum 327
amount of melt surrounded by continuous water 293
analogy between heat and mass transfer 153
annular channel 510
annular channels 511
annular dispersed flow 124
annular film flow 9
annular film flow with entrainment 9
annular flow 13, 20, 49
annular two-phase flow 133
aspect ratio of a bubble 28
Avdeev 412
average center-to-center spacing 426
average mass conservation 168
averaged beam length 664
averaged film boiling heat transfer coefficient 536
averaged film thickness 557
averaged mass source term 407
averaged vapor film thickness 501
Avogadro number 574

bag - and - stamen breakup 203
bag and stamen breakup 200, 216
bag breakup 200, 202, 214
bag mode 201
Baroczy correlation 78
black body 658
black body intensity 658

blow down 474
boiling flow 80
boiling of subcooled liquid 489, 518
boiling with mutually bubble interaction 432
Boltzmann constant 348
Bowring 519
breakup period for droplets in gas 209
breakup time 207
bubble coalescence mechanisms 179
bubble collapse in subcooled liquid 304
bubble condensing in a subcooled liquid 400
bubble departure diameter 421, 445, 484
bubble departure diameter as a function of pressure 432
bubble departure diameter as a function of superheat 430
bubble departure diameter as function of mass flow rate 430
bubble departure diameter during boiling or flashing 421
bubble departure frequency 447
bubble disappearance 471
bubble dynamics 306
bubble entrainment from the vapor layer 472
bubble entrainment size 472
bubble flow 6, 14, 19, 21, 49, 102
bubble flow in a annular channel 106
bubble fragmentation 470
bubble growth 375
bubble growth in superheated liquid 375
bubble growth in the bulk 469
bubble inclination angle 428
bubble rise velocity in pool 5
bubbles generated due to nucleation at the wall 468
bubbles in asymmetric flow field 304

bubbles in turbulent liquid 247
Bubble-to-slug flow transition 18
bubbly flow 3, 119, 518
bundles of horizontal pipes 632

cap bubbles 31, 34
catastrophic breakup 200, 216
chain of infinite number of Wigner cells 685
change of the bubble number density due to condensation 409
channel flow 2, 133
Channel flow – inclined pipes 10
Channel flow – vertical pipes 6
churn turbulent flow 6, 8, 35, 119
churn-to-annular, dispersed flow transition 18
churn-turbulent flow 102
Clausius - Clapeyron equation 345
cloud of equidistantly arranged hot spherical particles 669, 686
coalescence 171, 177, 470
coalescence escaping regime 254
coalescence frequency 173, 256, 471
coalescence probability 173, 254, 256, 257, 471
coalescence probability of small droplets 257
coalescence rate 173
coalescence time 252
Coddington and Macian 103
Colebrook and Witte formula 73
collision frequency 173, 175, 471
collision time 252
condensation at cooled walls 631
condensation from forced convection two-phase flow at liquid film 634
condensation from stagnant steam at turbulent liquid film 633
condensation inside a vertical pipe 632
condensation of a pure steam bubble in a subcooled liquid 399
condensation of pure steam on subcooled film 623
condensing moving bubble in a subcooled liquid 402
conservation equation of the droplets 181
contact angle 350, 351
contact heat transfer 289, 310
contact time 252, 312

contaminated systems 31
convection inside the droplet 582, 585
convective boiling of saturated liquid 505
convective heat transfer 607
converging disperse field 171
converging-diverging nozzle 473
coolant 289
coolant fragmentation 317
coolant interface classification 320
coolant viscosity increase 329
corrector for the subcooling of the film 632
creeping flow 96
critical bubble size 423
critical heat flux 513, 516
critical mass flow rate 360
critical mass flow rate in short pipes, orifices and nozzles 360
critical nucleation size 570
critical size 361
critical Weber number 189
cross flow in vertical rod bundle 79
cross-section averaged vapor velocity profile 532
crust straight 316

Darcy's law 45
dense packed regime 44
departure from nucleate boiling 514
deposition correlatins 157
deposition in annular two-phase flow 153
deposition mass flow rate 154
deposition rate 138
depressurization 356
diffusely emitting and reflecting walls 656
diffusion controlled collapse 414
diffusion controlled evaporation 386
diffusion mass flow rate at macroscopic surface 598
diffusion velocities for algebraic slip models 89
dimensional analysis for small scale motion 238
direct contact condensation of pure steam on subcooled droplet 585
direction cosines 648
discrete ordinate method 647

discrete ordinate radiation transport method 647
disintegration of the continuum 170
dissolved inert gases 360
distorted bubble regime 30, 34
distorted particles 30, 34
distribution parameter 100
distribution parameter for annular flow 109
Dittus-Boelter equation 505
donor cell scheme 653
down flow of vapor across horizontal tubes 634
drag 90
drag coefficient 27, 36, 92, 95
Drag coefficient for single bubble 28
drag forces 27
drift flux correlation 95
drift flux models 100, 119
droplet evaporation in gas mixture 594
droplet growth 575
droplet mass conservation equation 182
droplet pulsation velocity 155
droplet size after the entrainment 144
droplets in turbulent liquid 247
droplets-gas system 36
dry out 514
duration of fragmentation 170
dynamic friction velocity 618

effective kinematic viscosity of turbulence 240
efficiency of coalescence of bubbles 252
emisivity and absoptivity for carbon dioxide 663
empirical flow map 15
entrainment 133
entrainment from liquid film for very high gas velocity 277
entrainment velocity 134
entrapment 305
entrapment triggers 302
equation of Klevin and Laplace 345
equilibrium droplet concentration 138
equilibrium steam mass flow rate concentration 489
Ergun's equation 45
evaporation of saturated droplets in superheated gas 591

evaporation of saturated film in superheated gas 622
evaporation of the entire droplet 592
external condensation on horizontal single pipe 632
external triggers 295

family of particles in continuum 32
Fauske 311
film boiling 20, 517, 527
film boiling heat flux 540
film boiling heat transfer coefficient 499, 555
film boiling in horizontal upwards-oriented plates 500
film boiling model and data base 546
film boiling on a horizontal cylinder 502
film collapse dynamics 306
film flashing 472
film flashing bubble generation 471
film thickness 9, 146, 536, 605
film to fine dispersed flow transition 518
film velocity 125
film-gas force 50
film-gas interface in the multi-phase flows 605
film-wall force 49
final fragment velocity 203
fine dispersed flow 518
finite differential form of the RTE 651
finite volume representation of the radiation transport equation 650
Fiveland scheme 655
flashing in adiabatic pipes 467
flashing inception pressure 360
flow boiling 431
flow pattern transition criteria for non adiabatic flow 18
flow patterns 1, 2, 167
flow regime transition 4
Flow regime transition criteria 1
forced convection 489, 535
forced convection boiling 505
forced convection condensation 635
forced convection DNB and DO correlations 517
forced convection film boiling 507, 510, 512

forced convection film boiling at vertical plate 567
forced convection film boiling in channels with very low mass flux 501
forced-convection film boiling data 553
Fourier equation 581
Fourier series 582
fragmentation 170, 189
fragmentation and coalescence 167
fragmentation frequency 171
fragmentation mechanism 264
fragmentation modes 200
fragmentation of melt in coolant 289
free falling droplets in gas 9
free falling sphere 92
free particles regime 38
free path of a steam molecule 576
free rising bubble velocity 472
free settling velocity 92, 121
frequency of bubble departure 446
frequency of coalescence of single bubble 179
friction coefficient of turbulent flow 411
friction pressure drop 71, 119
friction pressure loss 79
friction pressure loss coefficient 71
frictional pressure loss 81
full-range drift flux correlations 110
fully developed steady state flow 183

gas jet disintegration in pools 280
gas jet in a turbulent liquid stream 246
gas side averaged shear stress 618
gas-to-liquid velocity ratio 122
Gauss-Seidel method 653
generated nuclei 573
geometrical film-gas characteristics 605
geometry definition to film boiling at sphere 548
Gibbs number 348, 572
gradually applied relative velocities 194
Groeneveld 511
group of particles 99

Hagen and Poiseuille law 71
heat and mass transfer at the film-gas interface 605
heat conduction due to turbulence 613

heat flux as a function of superheat 433, 449
heat flux as a function of superheat for saturated water 450
heat flux as a function of the wall superheating 439
heat released during the condensation 578
heat transfer across droplet interface 578
heat transfer between interface and gas 608
heat transfer coefficient 493
heat transfer coefficient on the surface of moving solid sphere and water droplets 579
heat transfer coefficient on the surface of solid sphere moving in a liquid 401
heat transfer from the interface into the bulk liquid 611
heated channels 18, 80
Helium II 382
Henry and Fauske 313
heterogeneous nucleation 349, 350, 467
highly energetic collisions 224
homogeneous local vapor volume fraction 501
homogeneous nucleation 348, 349, 573
homogeneous nucleation temperature 500
homogeneous turbulence characteristics 237
horizontal or inclined pipes 60
hydraulic diameters 56
hydrodynamic stability theory 514
hydrodynamic theory of boiling 515
hydrodynamic theory of critical heat flux 451

inclined pipes 118
inert gases 328
inertially controlled bubble growth 345
initiation of the visible nucleate boiling 489
initiation of visible boiling 489
inner scale or micro scale of turbulence 238
instability of jets 263
interaction of molten material with water 663

Index   695

interface solidification 301
interfacial area density 50, 133
interfacial disturbances 542
interfacial instability due to bubble collapse 309
inverted annular flow 55, 517
isolated bubble boiling 432
isotropic turbulence 614

Jakob number 577
jet atomization 264
jet disintegration due to asymmetric waves 264
jet disintegration due to symmetric interface oscillations 264
jet fragmentation in pipes 279
jet of molten metal 273
jets causing film boiling 275

Kelvin - Helmholtz instability 190
Kelvin-Helmholtz gravity long wave theory 12
Kelvin-Helmholtz instability 542
Kolmogoroff 238
Kutateladze 5
Kutateladze terminal velocity 9, 97

laminar condensation on vertical wall 632
laminar film 633, 635
Lanzenberger 's solution 669
Lanzenberger solution 685
Laplace and Kelvin equation 343, 572
Laplace constant 81
Laplace number 203
large scale eddies 237
large scale motion 237
larger bubble sizes 5
Lathrop's diamond difference scheme 655
lava-water interaction 527
lean systems 294
Lellouche 104
length scale 1
Levich equation 244
liquid - liquid system 206
liquid and gas jet disintegration 263
liquid boundary layer 537
liquid droplets 569
liquid jet disintegration in pools 263
liquid-liquid systems 246

Liu and Theofanous 561
local drift velocity 101
local vapor film thickness 535
Lockhart and Martinelli 74
look-up table for critical heat flux 521

Mamaev 10, 118
Marangoni effect 319
Martinelli - Nelson method 81
Martinelli and Nelson 76
Martinelli factor 505
mass concentration of droplets 154
mass flow rate 489
mass transfer coefficient 638
maximal achievable superheating 356
maximum duration of the bubble collapse 413
maximum packing 32
maximum packing density 3
maximum superheat 356
mean free path 155
mean site-to-site distance 484
mechanical fragmentation 289
mechanism of the thermal fragmentation 306
melt 289
melt with water and vapor enclosed inside 665
mercury droplets in air 192
micro-convection velocity 427, 447
Mikic 379
minimum film boiling temperature 294, 499, 501
minimum heat flux 499
mirroring walls 657
mixed convection film boiling 554
mixed-convection film boiling at spheres 547
mixed-convection film boiling on vertical walls 527
mixture of gas, film and droplets 124
mole weight 576
mono-disperse 168
moving bubble 401
multi-group approach 169
mutual interaction of the growing and departing bubbles 432

natural convection 549
natural convection film boiling 499

natural convection film boiling at vertical plate 567
natural convection film boiling on a sphere 502
natural-convection film boiling 551
nearest-neighbor distance 367
net evaporation 492
net mass flow rate of the spontaneous evaporation 588
Newton's regime 37, 40
Nigmatulin 135
Nikuradze diagram 10, 71
non solidifying droplet 311
non-averaged source terms 406
non-condensable gases 307, 624
non-condensing gas 413
Non-oscillating particles 3
non-stable liquid 342
nucleate boiling 439
nucleate boiling heat flux 505
nucleate boiling heat transfer coefficient 506
nucleate pool boiling 444
nucleation 569
nucleation energy 342
nucleation frequency 574
nucleation in liquids 341
nucleation in presence of non-condensable gasses 360
nucleation kinetic 348
nucleation kinetics 573
nucleation site density 443
nucleation sites density 441
nucleation theories 352
nucleus capable to grow 344
Nukiama-Tanasava law 409
Nukuradze diagram 59
number density after the fragmentation 170
Nusselt 632

onset of the condensation 631
optimum marching strategy 653
oscillating particles 4
Oseen equation 29
over-entrained regime 137
oversaturated 315
oxidation 327

particle production rate 214

particle production rate during the thermal fragmentation 322
particle production rate in case of entrainment 180
particle sink velocity in pipes 119
particle size after thermal fragmentation 321
particle size formation in pipes 180
particle size in the steady state flow 182
particles in film boiling 20
partitioning of the absorbed energy between the bulk and the surface 662
penetration length 265
permeability coefficients 46
Pilch 201
pipe forces 475
pitting of the surface 305
Planck's spectral distribution 658
Poiseuille flow 45
Poisson distribution law 445
pool boiling of saturated liquid 446
pool flow 2
pool flows 3
porous media 19
positive scheme 656
potentially explosive mixtures 291
Prandtl - Nikuradze 74
pressure drop 71
primary breakup 217, 219
probability of capture 589
probability of departure from the averaged state 573
probability of escape 589
probability of origination of a single bubble 349
probability of origination of single nucleus 573
production rate 170, 220
propagation velocity 312
pulsation velocity 613
pure liquid 31
pure natural convection 535
pure steam bubble drifting in turbulent continuous liquid 410
pure steam condensation 631

radiated specific power of the cloud 687
radiating sphere concentrically surrounded by vapor and water 670
radiation from the sphere 550
radiation transport analysis 645

radiation transport equation 646
Raleigh-Taylor wavelength 4
Rankine and Hugoniot equation 500
Rayleigh equation 307
rectangular channels 56
reduction of the driving forces of the evaporation 598
relative permeabilities 47
relative permeability multipliers 49
relative velocity 101
resistance of the non-condensable gases 639
resisting force between film and gas 63
Riccati 204
Richardson and Zaki 94
Richardson-Zaki 100
rippled film-to-micro film transition 19
rod bundles 21
roll wave regime 59

Sakagushi 121
saturated liquid boiling 517
scattering phase function 646
self condensation stop 577
shear stress at the wall 617
shear stress in the film 618
sheet stripping 200, 203, 216
single bubble terminal velocity 95
single particle terminal velocity 99
single-phase flow 71
single-phase liquid convection heat flux 505
sinuous jet breakup 264
size initialization by flow pattern transition 199
size of the entrained droplets 134, 144
size of the ligaments 268
slightly contaminated liquid 31
slip 122
slip models 122
slug bubble diameter 93
slug flow 7, 20, 22, 35, 49, 98, 119
slug flow in a tube 107
slug-to-churn flow transition 18
small bubble sizes 5
small scale motion 238
Smolin 517
Smoluchowski 175, 241
solid particles 38, 44, 125
solid particles in a gas 37
solid particles in bubbly flow 40

solidifying droplet 313
Souter mean diameter 215
space averaged beam lengths for the concentric spheres radiation problem 671
Spalding mass transfer number 598
spectral absoption coefficient for water as a function of the wavelength 659
spectral absorption coefficient 659
spectral absorption coefficient of water 659
spectral absorption coefficient of water vapor 663
spectral averaged absorption coefficient 665
spectral averaged emissivity coefficient 665
spectral extinction coefficient 647
spectral radiation intensity 646
sphere interface temperature during film boiling 556
spherical cap 97
spherical cavity of gas inside a molten material 664
spinoidal line 572
spontaneous condensation of pure subcooled steam 569
spontaneous evaporation of the superheated film 621
spontaneous flashing of superheat liquid 375
spontaneous flashing of superheated droplet 587
stability criterion for bubbles in continuum 244
stable particle diameter after the fragmentation 170
stagnant bubble 399
static contact angles 442
statistical theory of turbulence 410
steam condensation from gas-steam mixtures flowing perpendicular to horizontal internally cooled tubes 639
steam condensation from gas-steam mixtures inside pipes 638
steam condensation from mixture containing non-condensing gases 636
step difference scheme 653
stochiometric 315

stochiometric liquid volume fraction 316
Stokes regime 29, 33, 36, 39
stratified flow 10, 56, 60, 118
stratified flows 10
stratified horizontal flow 118
stratified wavy flow 10, 13, 15
stratified-intermittent transition 15
stripping of droplets 202
stripping of the ligaments 202
strong disturbance on the film-to-small ripples transition 18
sub cooled boiling 80
subcooled nucleate boiling 18
suddenly applied relative velocity 198
superheated steam 385
surface entrainment effect 309, 322
surface tension 571
surface tension of water 342
surface vectors 649
surfactants 329
suspension volumetric flux 121

Taylor bubble 7, 97
temperature correction factor 574
temperature profile in the vapor layer 532
terminal speed of a spherical particle 92
the Taylor bubble velocity 12
thermal controlled bubble growth 376, 576
thermal controlled collapse 413
thermal fragmentation 324, 327
thermal resistance inside the drop 587
thermo-capillar velocity 320
thermo-capillary flow in the drop 319
thermo-mechanical fragmentation 289, 294, 320
thick and thin vapor film 290
thin thermal boundary layer 378
three dimensional flow in porous structure 79
three phase flow 125
three velocity fields 124
three-dimensional flow 33
three-phase disperse flow 43
three-phase flows 82
threshold pressure 300
time-averaged coalescence rate 173
total absobtion depth in three phase particle-vapor-water flow 686

total solid angle 648
transition boiling 303
transition boiling in tubes 512
transition conditions 17
transition criteria 10
transition to annular flow 15
transition to bubble flow 16
transition-boiling correlation 513
transparent wall 657
transport enhancement factor 589
tubes 511
turbulence dissipation rate 239
turbulence induced droplet fragmentation in channels 248
turbulence induced particle fragmentation and coalescence 237
turbulence Reynolds number 237
turbulent condensation on vertical surfaces 634
turbulent diffusion constant 154
turbulent film 634, 635
turbulent fluctuation velocity 470
turbulent gas pulsation 153
turbulent length scale 613
turbulent viscosity 240
two parallel plates 11
two-phase friction pressure drop 74

umbrella mode 202
under-entrained regime 137
undersaturated 315
undisturbed particles 30, 34

vapor collapse 296
vapor thickness in film boiling 291
vapor-coolant instability 289
varicose mode of water jet breakup 269
vertical flow around rod bundles 511
vertical rod bundle 21
vibration breakup 200, 202, 214
Vierow and Shrock 639
violent explosion 303
virtual mass coefficient 48
viscous limit 238, 239
viscous regime 29, 33, 36, 39

waiting time 447
wakes behind particles 250
wall friction force 95
wall-gas 63
wall-liquid forces 63

Wang and Dhir 441
wave crest striping 216
wave crest stripping 203
wave crests stripping 200
wavy film 634
weighted mean velocity 100
weighting factors 652

Wigner cell 669
work reduction factor 352

Zaichik 156
Zuber and Findlay model 101

Printing: Mercedes-Druck, Berlin
Binding: Stein+Lehmann, Berlin